A primate radiation:
evolutionary biology of the African guenons

A primate radiation: evolutionary biology of the African guenons

Edited by
ANNIE GAUTIER-HION,
FRANÇOIS BOURLIÈRE,
JEAN-PIERRE GAUTIER

*Université de Rennes, Station Biologique de Paimpont,
Plélan-le-Grand, France*

JONATHAN KINGDON

*Animal Ecology Research Group, Department of Zoology,
University of Oxford, UK*

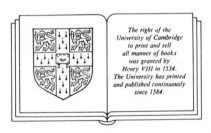

The right of the
University of Cambridge
to print and sell
all manner of books
was granted by
Henry VIII in 1534.
The University has printed
and published continuously
since 1584.

CAMBRIDGE UNIVERSITY PRESS

Cambridge

New York   New Rochelle

Melbourne   Sydney

Published by the Press Syndicate of the University of Cambridge
The Pitt Building, Trumpington Street, Cambridge CB2 1RP
32 East 57th Street, New York, NY 10022, USA
10 Stamford Road, Oakleigh, Melbourne 3166, Australia
© Cambridge University Press 1988

First published 1988

Printed in Great Britain at
the University Press, Cambridge

British Library cataloguing in publication data

A primate radiation: evolutionary biology of the
African guenons.
1. Cercopithecus—Evolution
I. Gautier-Hion, Annie
599.8'2   QL737.P93

Library of Congress cataloguing in publication data

A Primate radiation: evolutionary biology of the African guenons
edited by Annie Gautier-Hion . . . [et al.].
    p.   cm.
Includes index.
ISBN 0–521–33523–X
1. Cercopithecus—Africa—Evolution. 2. Mammals—Evolution.
3. Mammals—Africa—Evolution. I. Gautier-Hion, Annie.
QL737.P93P75 1988
599.8'2—dc 19   87–35503CIP

ISBN 0521 33523 X

CONTENTS

CONTRIBUTORS

Francois Bourlière, *Station Biologique de Paimpont, 35380 Plélan-le-Grand, France*

Thomas M. Butynski, *Impenetrable Forest Conservation Project, Zoology Department, Makerere University, PO Box 7062, Kampala, Uganda*

Janice Chism, *Department of Zoology, University of California, Berkeley, California 94720, USA*

Marc C. Colyn, *Laboratorium voor Algemene Dierkunde, Rijkuniversitair Centrum Antwerpen, Groenenborgerlaan 171, B-2020 Antwerpen, Belgium*

Marina Cords, *Department of Zoology, University of California, Berkeley, California 94720, USA*

Jérome Couturier, *UA 620 CNRS, Institut Curie, 26 rue d'Ulm, 75231 Paris Cedex 05, France*

Bernard Dutrillaux, *UA 620 CNRS, Institut Curie, 26 rue d'Ulm, 75231 Paris Cedex 05, France*

Laurence Fedigan, *Faculté Saint-Jean, University of Alberta, Edmonton, Alberta, T6G 2E1, Canada*

Linda M. Fedigan, *Department of Anthropology, University of Alberta, Edmonton, Alberta, T6G 2E1, Canada*

Jean-Pierre Gautier, *UA 373 CNRS, Station Biologique de Paimpont, 35380 Plélan-le-Grand, France*

Annie Gautier-Hion, *UA 373 CNRS, Station Biologique de Paimpont, 35380 Plélan-le-Grand, France*

Alan C. Hamilton, *Department of Environmental Studies, University of Ulster, Coleraine, Co Londonderry BT52 1SA, UK*

Jonathan Kingdon, *Animal Ecology Research Group, Department of Zoology, University of Oxford, South Parks Road, Oxford OX1 3PS, UK*

Meave Leakey, *National Museums of Kenya, PO Box 40658, Nairobi, Kenya*

Lysa Leland, *New York Zoological Society, Kibale Forest Project, Bronx Park, New York, New York 10460, USA*

Jean-Marc Lernould, *Parc Zoologique, 51 rue du Jardin Zoologique, 68062 Mulhouse, France*

Jean-Noël Loireau, *Station Biologique de Paimpont, 35380 Plélan-le-Grand, France*

Jeremiah S. Lwanga, *Kibale Forest Project, Department of Zoology, Makerere University, PO Box 7062, Kampala, Uganda*

Ann M. MacLarnon, *Department of Anthropology, University College London, Gower Street, London WC1E 6BT, UK*

Jerald E. Maiers, *Department of Anthropology, University of Wisconsin, PO Box 413, Milwaukee, Wisconsin 53201, USA*

Robert D. Martin, *Anthropologisches Institut und Museum, Universität Zürich-Irchel, Winterthurerstrasse 190, CH-8057 Zürich, Switzerland*

Carol S. Mott, *Department of Anthropology, University of Wisconsin, PO Box 413, Milwaukee, Wisconsin 53201, USA*

Martine Muleris, *UA 620 CNRS, Institut Curie, 26 rue d'Ulm, 75231 Paris Cedex 05, France*

John F. Oates, *Department of Anthropology, Hunter College of CUNY, 695 Park Avenue, New York, New York 10021, USA*

Martin Pickford, *Institut de Paléontologie, 8 rue de Buffon, 75005 Paris, France, and Johannes Gutenberg Universität, Institut für Geowissenschaften, Saarstrasse 21, D-6500 Mainz 1, Federal Republic of Germany*

Thelma E. Rowell, *Department of Zoology, University of California, Berkeley, California 94720, USA*

Maryellen Ruvolo, *Department of Biological Chemistry, Harvard Medical School, Cambridge, Massachusetts 02138, USA*

Brigitte Senut, *UA 49 CNRS, Laboratoire d'Anthropologie, Museum National d'Histoire Naturelle, 8 rue de Buffon, 75005 Paris*

Thomas T. Struhsaker, *New York Zoological Society, Kibale Forest Project, Bronx Park, New York, New York 10460, USA*

Trudy T. Turner, *Department of Anthropology, University of Wisconsin, PO Box 413, Milwaukee, Wisconsin 53201, USA*

INTRODUCTION

Why study African guenons?

For a long time, African guenons were, together with the New World monkeys, the left-overs of the primate world. Even their most common vernacular name, guenons, is not particularly appealing, not being considered as a very gratifying nickname on either side of the English Channel. This is indeed an undeserved judgment, as most members of supergenus *Cercopithecus* are graceful animals; far from looking ugly or dull, some of them even rank among the most elegant and colourful monkeys.

The disregard of primatologists had more serious causes. Those mostly concerned with primate mental abilities and cognitive perform-ances felt that guenons were not the right kind of animals to start working with, as none of them was apparently as clever as apes, or even capuchin monkeys. On the other hand, the ecologists and sociobiologists interested by the influence of various demographic, social, and environmental parameters on the social organization of non-human primates were, in the fifties, led to conclude that intensive field work in tropical rain forests was a next to impossible task; consequently, they quite naturally, started observing 'savanna mon-keys' such as macaques, baboons, or common Indian langurs.

It was not until the 1960s that the first intensive field studies of guenons were initiated on forest species, though mention must be made of the pioneer work done in the late forties by A. J. Haddow and A. P. Buxton on the Uganda redtail, in relation to the epidemiology of yellow fever. During the past two decades it has been possible to make up the lost time, and the number of field studies has steadily increased, though many interesting species remain to be studied.

To take stock of the actual state of our knowledge was therefore becoming imperative, all the more since it quickly became apparent that the supergenus *Cercopithecus* was actually far less homogeneous than it looked at first sight. Its numerous species and subspecies have successfully colonized the widest possible spectrum of African habitats, from the lowland rain forests to the mountain forests (up to 3300 m), and from the swamps of Zaïre and Rwanda to the edge of the Sahara and Kalahari deserts. One species is almost completely terrestrial and a few semi-terrestrial, while most of the 'true' guenons are arboreal, some of them almost never setting foot on the ground. Some can swim and even exploit mangroves, and others cannot. Some have a very restricted distribution, whereas others range over wide areas, the green monkeys even thriving through savannas from the Senegal to the Cape of Good Hope. The cercopithecine monkeys definitely appear now as a paradigm of diversity among African Primates.

Moreover, whatever their particular research interest, most of the scientists working on guenons became intuitively convinced that the whole group had recently undergone a rapid adaptive radiation, and might even still be in an active stage of speciation. If this was the case, the group would provide a unique opportunity to identify and rank the major determinants of this process, and to evaluate their respective importance.

Finally, while savanna guenons, and perhaps some semi-terrestrial forest guenons, live in monospecific troops, some forest species frequently form semi-stable 'mixed troops', an uncommon situation in terrestrial vertebrates. Not only the behavioural mechanisms that allow closely related species to live together while remaining reproductively isolated needed to be scrutinized, but the adaptive value, if any, of their mutualistic interactions had to be clearly established.

In order to proceed to a comprehensive and balanced evaluation of our present knowledge, and eventually to propose new lines of research, Professor R. D. Martin and the editors of the present book took advantage of the convening of the XIXth International Ethological Conference, to organize a 'satellite' symposium on the 'Biology, phylogeny and speciation of forest cercopithecines'.

This meeting, which was restricted to a limited number of participants representing most of the research groups working at that time on guenons, took place at the Station Biologique de Paimpont, France, from 20 to 22 August 1985. The amount of information presented at this meeting was so large, the exchange of views so stimulating, some opinions voiced so new, if not iconoclastic, that it was unanimously decided to produce a book on the subject, and an editorial committee was appointed.

The subject matter of this volume was subsequently broadened to embrace both forest and savanna guenons, and to serve as a comprehensive introduction to the evolutionary biology of the whole group. A number of new authors were asked to join the initial participants, and during the year that followed everyone worked hard to prepare successive versions of his own chapter(s), and to comment on and criticize the contributions of the others. At this refereeing stage, however, we did not attempt always to reach unanimous agreement on controversial issues, preferring to leave the door open to alternative interpretations. This cross-referring process unfortunately took longer than expected, as many contributors were back in the wilds of Africa, and the circulation of manuscripts was sometimes hampered by political turmoils. Nevertheless, the book is now finished and we hope that it will stimulate the thoughts of ecologists as well as of behavioural scientists.

It is now the most pleasant duty of the editors to thank all their co-authors for their contributions, sometimes completed under difficult circumstances. The success of this book will be theirs. We are also particularly grateful to Anh and Gérard Galat for the loan of four of their pictures and for making available to us many of their unpublished observations. Our warmest thanks too to Elena Kingdon and Mike Harrison for their help in improving the English versions of some manuscripts, and to the Wellcome Trust for providing money for the colour plates.

We would also like to extend our thanks to the Syndics of Cambridge University Press for accepting this volume for publication, to Dr Robin Pellew, then Senior Editor for Biological Sciences, and to Mr Martin Walters, for their unfailing help.

Part I

Guenons and the African environment: past and present

I.1

Fossil evidence for the evolution of the guenons

MEAVE LEAKEY

Fossil guenons are rare and none has been reported older than 3 Myr. The diversity seen today in the genus *Cercopithecus*, with at least 25 species and more than 70 subspecies, probably appeared relatively recently, perhaps during the last million years. A brief summary of the fossil evidence for the evolution of the Cercopithecidae provides a framework for a discussion of the evolution of the guenons.

Cercopithecid monkeys first make their appearance in the fossil record during the early Miocene at a time when the diversity of the 'apes' was at a peak. Specimens are known from sites in North Africa (Wadi Moghara, Egypt and Gebel Zelten, Libya) and East Africa (Napak, Buluk, Maboko, Nyakach and Loperot) which are between 15 and 20 Myr old. These early monkeys possess a number of features characteristic of the subfamily Victoriapithecinae (Von Koenigswald, 1969; Leakey, 1985) which distinguishes them from the modern subfamilies, the Colobinae and Cercopithecinae. Although the North African taxa, *Prohylobates tandyi*, represented by three damaged mandibles from Wadi Moghara, Egypt (Simons, 1969) and *P. simonsi*, represented by one mandibular fragment from Gebel Zelten, Libya (Delson, 1979), were considered generically distinct from the East African *Victoriapithecus* (Von Koenigswald, 1969), recent finds from Buluk, Kenya (Leakey, 1985), have shown that the distinctions are not clear and these taxa may well be congeneric. The remains of these early monkeys are generally very rare or absent at Miocene sites in East Africa where hominoids are relatively common. Almost all the *Victoriapithecus* specimens have come from a single site on Lake Victoria, Maboko, which has yielded an exceptionally large number of primates. Higher primates make up 40% of the mammalian specimens at Maboko and monkeys constitute 30%. The number of *Victoriapithecus* crano-dental specimens recovered (438) (Harrison, 1986) far exceeds the total number of hominoid specimens recovered from any Miocene locality in western Kenya (the maximum number of hominoid specimens from a single locality is 299 from the Hiwegi Formation, Rusinga) (Pickford, 1986).

The Maboko monkeys were originally interpreted as representing two distinct species (Von Koenigswald, 1969) and later it was suggested that the differences were evidence of an early divergence of the modern subfamilies with one species being ancestral to the colobines and the other to the cercopithecines (Delson, 1975a; Szalay & Delson, 1979). Recent detailed analysis of the original Maboko collection and of more recent collections indicates that all the *Victoriapithecus* dental specimens belong either to one species, or, if two species are present they cannot be distinguished on the basis of teeth alone (Benefit & Pickford, 1986). The divergence of the modern subfamilies of Cercopithecidae probably took place later in the middle Miocene.

The earliest unequivocal colobines were reported recently from two sites in Kenya dated between 10.5 and 8.5 Myr (Benefit & Pickford, 1986). These specimens, a complete mandibular body of *Microcolobus tugenensis* from Ngeringerowa and an isolated lower molar from Nakali, are the main evidence for African fossil monkeys between 15 and 7 Myr although Pickford (1986) reports *Victoriapithecus* from Nyakach (13.4±1.3 Myr) and Majiwa. (Examination of the specimens from Majiwa shows they are not cercopithecid.) The vertebrate fossil record in Africa between 14 and 4 Myr is relatively sparse; few sites are known and the fossil remains are generally poor. It is not until 4 Myr that there is again a rich fossil record and by this time both colobine and cercopithecine monkeys are relatively common and diverse and the non-hominid apes are absent.

The evolution of the family Cercopithecidae is linked to significant changes in the primate fauna which took place in the late middle Miocene and late Miocene. During this time the majority of species of Miocene 'apes', which are considered to have occupied many of the niches which the cercopithecines inhabit today, became extinct and the monkeys diversified. Unfortunately, because of the paucity of the fossil record between 14 and 4 Myr there is little evidence for this but when we again have a good fossil record in the Plio–Pleistocene there is considerable diversity among the Cercopithecidae. Contrary to the popular belief that early monkeys were arboreal folivores, evidence from the dentition (B. R. Benefit, personal communication) and the postcranial remains of *Victoriapithecus* suggests that the earliest monkeys were semi-terrestrial frugivores and it is possible that their early evolution was largely in open woodland and savanna habitats.

It is in the Pliocene that we find the earliest evidence of the guenons, in sedimentary deposits in the Lake Turkana basin. At the Omo Valley, White Sands locality, in deposits of the Usno Formation (de Heinzelin, Haesaerts & Howell, 1976) aged approximately 2.9 Myr, two small upper molars which show similarities to *Cercopithecus aethiops* and *C. ascanius schmidti* have been recovered (Eck & Howell, 1972). Also from the Omo Valley but slightly higher in the section in the Shungura Formation (de Heinzelin *et al.*, 1976) four isolated teeth were found in deposits 2.69 Myr old in Member B and a very small mandibular fragment from deposits about 2.5 Myr old in Member C (Eck, 1987). At East Turkana in the Koobi Fora Formation (Brown & Feibel, 1986) a small *Cercopithecus* mandible with M_{2-3} about the size of that of a talapoin was recovered from deposits approximately 2.6 Myr old. These are the oldest known guenons but the remains are so fragmentary that they provide little information other than to confirm that the guenons had evolved as a separate group by this time. It should be noted that the identification of fossil *Cercopithecus* specimens is necessarily tentative. Most extant guenon species are remarkably similar in their skeletal morphology so that the identification of species from fossils which are almost always fragmentary is in most cases impossible. Fragments are often difficult to distinguish even from other genera. When the specimens consist of isolated teeth or fragments of upper or lower jaw it may be impossible to distinguish *Cercopithecus* from similar sized *Colobus* species especially if the teeth are worn. Guenon lower third molars are distinctive, however, because they lack a hypoconulid and the talapoin sized 2.6 Myr old mandible from East Turkana which preserves M_{2-3} is attributed to *Cercopithecus* with some confidence.

Younger deposits at both the Omo and East Turkana have yielded additional *Cercopithecus* remains. At Omo, from the Shungura Formation, these include the left and right sides of a mandible (Eck & Howell, 1972), an isolated upper molar and a male upper canine from Member G (1.98 Myr) and fragments of mandible and maxilla which are probably associated from Member J (*c.* 1.34 Myr) (Eck, 1987). At East Turkana, from the Koobi Fora Formation, an isolated upper molar, a complete femur and a proximal ulna have come from deposits dated between 1.9 and 1.6 Myr. These specimens are generally larger than the earlier ones but as with the previous material they cannot be identified to species. Even more recent late Pleistocene deposits at Olduvai Gorge, Tanzania, and Loboi, Kenya, have yielded more complete maxillary and mandibular material.

Pilgrim (1915) assigned material from the Dhok Pathan zone of the Middle Siwaliks to *Cercopithecus*, but this attribution has been shown to be incorrect by subsequent workers (Eck & Howell, 1972; Simons, 1972). Known fossil and extant *Cercopithecus* are restricted to Africa and there is no evidence to support Hill's suggestion (1966) that either *Cercopithecus* or a common ancestral cercopithecoid evolved in Asia and later migrated to Africa.

It is unlikely that the relative rarity of *Cercopithecus* in the Pliocene fossil record is related to a bias in the collections of fossils recovered. It is more likely to reflect a true rarity of these monkeys at this time. If early guenons were inhabiting the savanna at Laetoli 3.5 Myr ago alongside the relatively common *Parapapio* and *Paracolobus* (Leakey & Delson, 1987) some evidence would have been recovered in the fossil record. Rodents and insectivores have been recovered in large numbers from this site, so the absence of *Cercopithecus* cannot be related to small size. Similarly, the absence of guenons from Pliocene sites such as Lothogam and Kanapoi which have yielded other monkeys appears to reflect a true rarity or absence of guenons in the fauna. It is possible that during the Pliocene guenons were confined to forests and that their remains are thus not recorded in sites preserving faunas of savanna or open woodland habitats, but a more likely explanation is that at that time the guenons were rare or absent and the diversity and distribution that typifies the group today evolved later.

The earliest subsaharan fossil evidence for most genera of extant cercopithecids appears later than that for *Cercopithecus*. At East Turkana the earliest *Colobus* appears about 1.6 Myr ago while at the Omo the earliest evidence is from roughly contemporary deposits 1.54 Myr old (Eck, 1976). Similarly, *Cercocebus* first appears contemporary with the earliest *Colobus* at East Turkana, but it has not been reported from the Omo. *Cercocebus* was previously reported from Laetoli (Hopwood, 1933), but additional cercopithecine material recovered recently has shown that this specimen is actually *Parapapio* (Leakey & Delson, 1987). The earliest recorded occurrence of the genus *Papio* (Leakey, 1969) has recently been questioned by Eck & Jablonski (1984), who consider that many of the early specimens attributed to *Papio* are in fact *Theropithecus*. They suggest that if other specimens are referred to *Dinopithecus* the genus *Papio* may disappear from the fossil record of eastern Africa. The identification of *Papio* is difficult without good cranial material because the teeth cannot be distinguished from those of other papionines such as *Parapapio*, *Dinopithecus* and *Gorgopithecus*. Although the evidence is limited, *Cercopithecus* may appear earlier than any other modern sub-saharan genera, with the exception of *Theropithecus* which appeared before 3 Myr ago.

Although monkeys are frequently recovered in the Plio–Pleistocene fossil faunas the genera and species represented were not those common today. In the Pliocene of East Africa, at sites such as Laetoli, and in the Plio–Pleistocene of South Africa, *Parapapio* was the dominant genus in the fossil faunas. Later, *Theropithecus* became the dominant cercopithecid genus in East Africa. Initially, *T. brumpti* was dominant in the Turkana Basin (the Omo Valley, East Turkana and West Turkana) inhabiting the woodland and riverine forest prevalent in this area between about 2.6 and 2.0 Myr. After a climatic change this species was replaced by *T. oswaldi*, a savanna adapted species, which was dominant from about 1.9 to less than 1.0 Myr. Today *T. oswaldi* has been replaced by *Papio* while only one species of the genus *Theropithecus*, *T. gelada*, remains and is restricted to the highlands of Ethiopia. Colobines were common in the Plio–Pleistocene of East Africa but they were also represented by genera now extinct. Many were large (Leakey, 1982) and some show significant terrestrial adaptations of the postcranial skeleton (Birchette, 1981) in contrast to the modern small arboreal *Colobus*.

Today African monkeys are considered to be primarily forest living arboreal animals but this is largely due to the great diversity of arboreal guenons inhabiting the forests across central Africa. The majority of species of extant African monkeys are guenons and very few of these, such as the patas monkey (*C. patas*) and the vervet monkey (*C. aethiops*), are semi-terrestrial. It is likely that for most of their evolutionary history monkeys were predominantly semi-terrestrial, open country animals with only a few species inhabiting the forests. This is why both colobines and cercopithecines are common and diverse in the Plio–Pleistocene faunal assemblages which are generally sampling savanna or woodland habitats. The majority of the extant forest adapted species are guenons and their evolution is probably a recent phenomenon; the fossil record is in accord with the hypothesis that this may have occurred within the last million years.

In summary, although fossil evidence for the evolution of the guenons is sparse, the following points are pertinent:

1. The extinction of many species of Miocene 'apes' which are considered to have occupied the forest habitats filled by cercopithecids today took place some time during the late middle Miocene and late Miocene.

2. An increase in the diversity of the Cercopithecidae occurred during the same time period.
3. These events are poorly documented because the primate fossil record during the period 14–4 Myr is sparse, but in the rich fossil record of the Plio–Pleistocene there is considerable diversity among the Cercopithecidae.
4. The early cercopithecids were probably semi-terrestrial frugivores inhabiting woodland and open country. Many of the species known from the Plio–Pleistocene are similarly open woodland or savanna living forms. It is probable that for most of their evolutionary history cercopithecids were primarily semi-terrestrial animals.
5. The genus *Cercopithecus* first appears in the fossil record 2.9 Myr ago, but fossil specimens are rare and fragmentary until the late Pleistocene when more complete specimens have been recovered.
6. The *Cercopithecus* monkeys evolved as a diverse but primarily forest adapted genus. Although the fossil record is sparse it supports the hypothesis that this diversity of the guenons appeared some time during the last million years.

Acknowledgements

I thank the Government of Kenya and the Governors of the National Museums of Kenya. I am grateful to my husband Richard for stimulating discussions and critical comments on the manuscript.

I.2

Guenon evolution and forest history

A. C. HAMILTON

Introduction

It is generally assumed that multiplication of species in guenons involves the isolation and divergence of populations. Such isolation could occur when a period of adverse climate reduces a formerly continuous forest cover and splits it into remnants. Isolated populations of guenons will then start to diverge. If contact between remnant forests is later re-established, populations of guenons may move out from their separate refugia and meet; speciation can be said to have taken place if successful interbreeding does not occur.

To match forest history with the guenon phylogenetic tree depends on knowledge of the timing of past changes in vegetation and quantification of a number of biological parameters. The latter include the minimum size of a forest needed for guenon survival, the forest characteristics required by individual species, the rates at which species can move out of refugia when their potential ranges expand and the times required for pairs of isolated populations to attain non-breeding thresholds. There are so many uncertainties concerning these biological parameters that it is unlikely that the geography of earlier guenon speciation will ever be established in any detail. Only for the most recent past does it seem possible to relate some features of the evolution and modern distribution of forest guenons to particular events in forest history.

There is a complication in that the differentiation of a guenon population may not always require vegetational change. Forests can be semi-isolated rather than completely isolated from one another and guenons may occasionally be able to move across intervals of some unfavourable habitat, for instance along strips of riverine forest. This is analogous to the occasional transportation of plants and animals to distant oceanic islands and their subsequent establishment and independent evolution.

Most and possibly all guenons do not require the presence of a particular forest type in a floristic sense, but rather are thought to respond more to forest structure and a sufficient supply of fleshy fruits (Bourlière, 1985). The lowland/montane divide is not significant to guenons, although few species attain very high altitudes. The present restriction of some species to swamp forest or mangrove forest may not indicate their total potential habitat; such restriction could be due to historical factors or competition from other species.

The period of time considered here is the last 8 million years, focusing on more recent times. Some detailed matching of forest history with features of guenon distribution seems possible for the last 20 000 years. The geological periods covered include the Holocene (0–10 000 BP), the Pleistocene (10 000–1.9 Myr), the Pliocene (1.9–5.4 Myr) and the late Miocene.

Several lines of evidence aid in reconstructing the past vegetation of Africa and these are considered in five sections. The first section looks at the modern patterns of distribution of forest organisms in Africa. These patterns can suggest where forest was able to survive during episodes of arid climate. The second section examines climatic and vegetational history from the time of the last world ice maximum at 18 000 BP to the middle of the Holocene (5000 BP); this relatively well understood period includes phases of both major forest retraction and major forest expansion. the latter half of the Holocene is then considered; this has witnessed reduction in the extent of forest in Africa, in some cases leading to fragmentation of the ranges of guenons. The fourth section examines the longer-term history of vegetation and climate, based largely on evidence from deep-sea cores and climatic modelling. Finally, plant fossils dating to between 1 and 8 Myr are discussed; these provide direct insight into older vegetation, acting as a test for climatic models established largely from other sources of information.

A few words are necessary about the limitations of pollen analysis. There are often problems with dating, sometimes because of inadequate stratigraphic study or a shortage of material suitable for isotopic analysis. Even when dating control is adequate, the reconstruction of past vegetation from pollen diagrams can present many problems. Examination of modern pollen samples and comparisons of fossil pollen spectra with plant macrofossil assemblages shows that many taxa are unrepresented in pollen diagrams even when common in the vegetation (Hamilton, 1972; Hamilton & Perrott, 1980; Bonnefille, 1984). Some pollen types are well dispersed and can be abundant in the pollen rain at a considerable distance from their source. Except in the most favourable circumstances, pollen analysis provides only a coarse-grained and sometimes disputable picture of the past.

Patterns of biotic diversity in African forests

The modern distribution of animals and plants has arisen through events in the past and an analysis of modern distributions should sometimes provide evidence of environmental history. The distributions of various groups of African forest organisms have been examined on a range of geographical scales and these have revealed a number of pattern features believed by many workers to have historical significance. Inequalities in the numbers of species and endemics in different areas may be due to longer establishment of forest in some places than in others. Disjunct distributions may be due to the presence of species in intervening areas in the past.

A major concept is that of a refuge or core area (latter terminology after Haffer, 1977), which is an area of forest relatively rich in numbers of species and endemics and which tends to include the ranges of isolated populations of disjunct taxa. There are often gradients of decreasing biotic diversity extending away from core areas. Core areas are usually situated where modern rainfall is particularly high and their characteristics must to some extent be a response to modern environments. However, this does not account for all their features and it is generally held that core areas represent places of relative climatic stability which continued to carry forest during arid periods in the past. There is evidence that the main features of atmospheric circulation over tropical Africa did not differ much in their geographical distribution between wet and dry periods in the Quaternary, though there were major contrasts in the moisture contents of air masses and in their rates of flow (see below). Topographic changes have been relatively minor during the last million or so years and thus it is not surprising that areas which are exceptionally wet now have also been exceptionally wet in the past.

Details of the locations of refuge areas can differ between authors, depending partly on how well the taxonomy and distributions of species are known and on how much weight is given to species with unusual occurrence. Figure 2.1, based on a wide survey of literature, shows the main forest refugia and associated gradients of decreasing

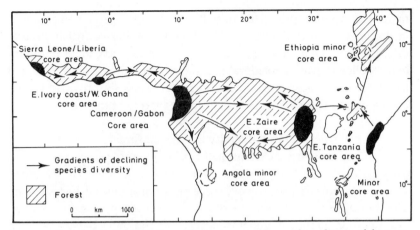

Figure 2.1. Distribution of forest, core areas and gradients of decreasing biotic diversity in tropical Africa. Adapted from Hamilton (1976, 1982).

biotic diversity. Two major refugia are located on either side of the Zaïre River basin (in Cameroon/Gabon and eastern Zaïre) and there are two biologically somewhat more impoverished refugia in West Africa (in Sierra Leone/Liberia and eastern Ivory Coast/western Ghana). Forests near the Indian Ocean, especially in eastern Tanzania, constitute another important refugium, or perhaps more correctly refugia, since here some groups of organisms show a pattern of small-scale local endemics. The places designated as minor core areas on Figure 2.1 are of much less importance than those already mentioned. Some gradients of decreasing biotic diversity are extremely well defined, including for example those extending into the centre of the Zaïre Basin from its flanks and that which reaches eastwards from Zaïre across Uganda and into western Kenya.

Some features of the distribution of passerine birds in African forests (Figure 2.2; Diamond & Hamilton, 1980) show many similarities to those of Figure 2.1, which is indeed partly derived from the ornithological evidence. It is obvious in passerines that evolutionary divergence is associated overwhelmingly with isolation of populations in the refugia. For example, of the many taxa disjunctly distributed across the Zaïre Basin some of the two populations are apparently identical, some are recognized as distinct varieties or subspecies, and some are taken to be different species.

Figure 2.3 covers central Africa from Nigeria to Lake Victoria and shows the disposition of refugia based on the distribution of guenons

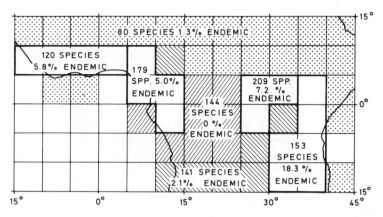

Figure 2.2. Avifaunal divisions of African forests, based on passerine birds (Diamond & Hamilton, 1980).

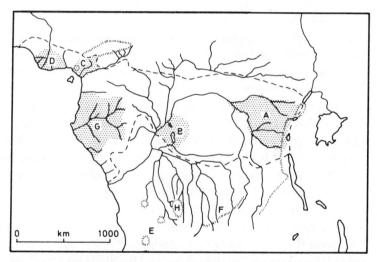

Figure 2.3. Forest refuges during arid periods (after Kingdon, 1980). A: Central refuge; D. C. G: Cameroon/Gabon refuge; D: Niger section; C: Cameroon section; G: Gabon or Ogoowé basin section; B: Southern Zaïre basin refuge; E: North Angola refuge; F: Southern scarps of Zaïre basin; H: Lunda Plateau.

and other forest mammals (Kingdon, 1980). Two additional refugia are recognized off the map in West Africa (in Liberia and Ghana). The Cameroon/Gabon refuge differs from that shown on Figure 2.1 mainly in being broken up into three sections, one of which, the Niger Section, is much poorer in species of most taxonomic groups than the others, the Cameroon and Gabon Sections. The refugia in North Angola and in the southern Zaïre Basin are impoverished and according to Kingdon unlikely to be very old. Forest may have been sustained in the southern Zaïre Basin by groundwater during climatic aridity. Kingdon also suggests that there may have been small forests along rivers draining off the elevated southern margin of the Zaïre Basin during periods of dry climate.

An issue raised by comparison of Figure 2.3 with Figure 2.1 concerns the weight given to the odd endemic or slightly enhanced biotic richness in the recognition of refugia. Such features could be related to the survival of relatively impoverished forest patches during past aridity but, because of their poor definition, other factors such as under-recording or failure to recognize features of the modern environment which might encourage diversity might also be responsible. In some cases it is conceivable that the presence of endemics does not necessarily indicate the existence of refugia; species may have become recently extinct within refugia but survive outside.

A detailed analysis of the distribution of forest mammals in East Africa has confirmed the biological richness of the Central Refuge centred in Eastern Zaïre (Figure 2.4; Rodgers, Owen & Homewood 1982). Elsewhere, the forests fall into three groups, aligned along two geographical axes interpreted as routes of migration out of the Central Refuge (see also Kingdon 1971). A northern route extends eastwards as far as Mt Kenya and is characterized by decreasing numbers of species from west to east and very low endemism. A southern route loops around the south of Tanzania to reach Usambara and the Tana River; again, there is increasing poverty away from the west, but endemism is higher. Rodgers *et al.* (1982) believe that the northern route has seen recent large scale extinctions under a dry climate, followed by a wave of colonization from the west. Forests along the southern route have been at least locally more persistent, but barriers to the movement of species have been more formidable.

Not all groups of organisms show the same patterns of distribution in East Africa. This can be illustrated with reference to Usambara in eastern Tanzania, which is exceptionally rich in species and endemics of all taxonomic groups except mammals (Rodgers & Homewood,

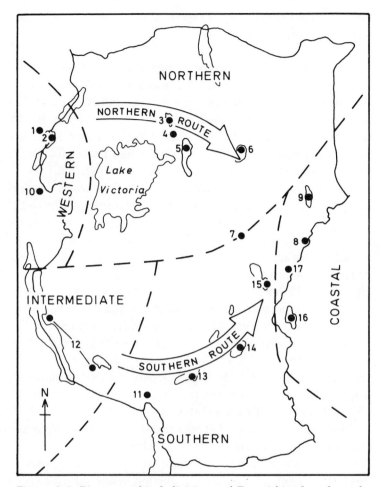

Figure 2.4. Biogeographical divisions of East Africa based on the distribution of forest mammals (Rodgers *et al.*, 1982). The number of species decreases along both northern and southern routes. Forests: 1. Congo; 2. Ruwenzori; 3. Elgon; 4. Kakamega; 5. Mau; 6. Mt Kenya; 7. Kilimanjaro; 9. Sokoke; 9. Tana River; 10. Kivu; 11. Southern Highlands; 12. Makari Mts and Ufipa; 13. Uzungwa; 14. Uluguru; 15. Usambara; 16. Zanzibar; 17. SE Kenya.

1982). The degree of taxonomic differentiation in some other groups suggests that well developed contact with the main Guineo-Congolian forests to the west is ancient, certainly pre-Quaternary. Factors which might be responsible for the exceptional status of the mammals include a lower ability to survive in reduced forest during arid phases, a higher ability to cross intervals of unfavourable habitat and a greater ability of newly arrived species to penetrate the established ecosystems and replace species already present.

Forest expansion since ice-age aridity at 18 000 BP

Patterns of distribution of modern forest organisms indicate that there have been periods of great forest reduction in tropical Africa and there is no doubt from a diversity of information that the most recent of these occurred at the time of the last major world glaciation with its maximum at 18 000 BP. Deglaciation proceeded rapidly after 16 000–13 000 BP (Ruddiman & Duplessy, 1985), with the modern world ice volume being reached at about 7000 BPbp. The period of ice-age aridity in tropical Africa ended abruptly at 12 500–12 000 BP, leading to the expansion of forest. Many cases of disjunction shown by African forest organisms must date back at least to the time of the last glacial maximum; many gradients of decreasing biotic diversity must be largely the result of differential expansion of species out of refugia after 12 500–12 000 BP.

Surface temperatures of the equatorial Atlantic were decreased by 4–5 °C during the last glacial maximum, with even greater reduction off northwest and southwest Africa where upwelling of cold bottom water was intensified and pushed further towards the equator (CLIMAP, 1976; Van Zinderen Bakker, 1982). Decrease in sea temperatures of this magnitude would have greatly lowered evaporation (Flohn, 1973) which, together with a weakened southwesterly monsoon (Flohn & Nicholson 1980), is regarded as a major factor responsible for ice-age aridity in West and Central Africa. The Indian Ocean differed from the Atlantic in that surface temperatures were lowered by only 1 or 2 °C (Prell *et al.*, 1980) and the great reduction in rainfall which is recorded from East Africa (see below) may have been due predominantly to a much weaker southern monsoon. The latter is evident both from studies of marine sediments (Prell *et al.*, 1980; Van Campo, Duplessy & Rossignol-Strick, 1982), and from the direction of ice dip on formerly glaciated East African mountains (Hamilton & Perrott, 1979; Hurni, 1981). It is the southerly monsoon which today brings most moisture to this part of Africa.

Temperature depression on the continent at 18 000 BP has been calculated from the altitudes of past glaciers and analysis of plant fossils (Figure 2.5). Estimates from past glaciers are 6.7–9.5 °C for Uganda, Kenya and northern Tanzania and 7 °C for northern Ethiopia, in all cases making allowance for likely changes in precipitation (Livingstone, 1980; Hurni, 1981). Pollen diagrams from altitudes

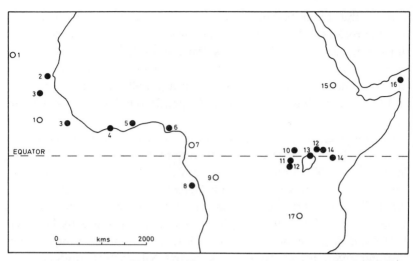

Figure 2.5. Sites with pollen or plant macrofossil evidence of more arid vegetation during the last glacial maximum (closed circles); some other localities mentioned in the text (open circles). 1. Stein & Sarnthein, 1984; 2. Rossignol-Strick & Duzer, 1979; 3. Agwu & Beug, 1984; 4. Assémien *et al.*, 1970; 5. Talbot *et al.*, 1984; 6. Sowunmi, 1981a; 7. Kadomura, 1982; 8. Giresse & Lanfranchi, 1984; 9. De Ploey, 1968; 10. Livingstone, 1967; 11. Morrison, 1968; 12. Hamilton, 1982; 13. Kendall, 1969; 14. Coetzee, 1967; 15. Hurni 1981; 16. Van Campo *et al.*, 1982; 17. Livingstone, 1971.

between 2000 and 4000 m in montane East Africa show a lowering of vegetation zones by an altitude of about 1000 m, equivalent to a fall in temperature by about 6°C (Morrison, 1968; Hamilton, 1972; Van Zinderen Bakker & Coetzee, 1972). The lowest altitude locality for which there is fossil evidence of temperature depression lies below 1000 m in Ghana; the presence of *Olea capensis* pollen and pooid grass cuticles points to temperature depression by several degrees Centigrade (Talbot *et al.*, 1984). Temperature depression at all altitudes is actually predicted by meteorological theory; indeed, the environmental lapse rate would have been even greater than now in most areas because of the drier air.

It can be predicted that temperature depression associated with the Ice Ages would have caused major extinctions of African forest organisms, not so much during the last glaciation, by which time the damage had already been done, as at earlier times. A eustatic drop in sea-level by 100 m subtracted from an altitudinal lowering of biological zones by 1000 m means that 900 m of lowland forest was eliminated.

Today, the basal 900 m of forest in Africa might be expected to constitute a single biological zone, the result of ecosystem relaxation during the present and previous warm periods. The actual altitude of this basal zone is presumably not the same in all places. For instance, the Central Refuge has a minimum altitude of about 500 m and the basal biological zone should extend up to an altitude of about 1400 m.

Aridity over tropical Africa during the last glacial maximum is demonstrated by a widespread lowering of lake levels (Kendall, 1969; Butzer *et al.*, 1972; Livingstone, 1975; Gasse, Rognon & Street, 1980; Servant & Servant-Valdary, 1980; Tiercelin *et al.*, 1981; Talbot *et al.*, 1984). Since temperatures were lower, there is no doubt that there was a major reduction in precipitation. Lake levels rose greatly between 12 500 and 10 000 BP, sometimes overflowing and contributing to much stronger river flows. An example is the Nile, which underwent a sudden and massive increase in flow at 12 500–12 000 BP (Rossignol-Strick *et al.*, 1982). The establishment of overflow down the Nile at 12 500–12 000 BP (Kendall, 1969; Livingstone, 1980) occurred just as forest started to develop in southern Uganda (Kendall, 1969) and presumably the river would have proved an obstacle to the eastward movement of some forest species. However, this potential barrier could have been easier to cross during a brief period at about 10 000 BP when river-flow out of Lake Victoria briefly ceased but some forest remained (Kendall, 1969).

Aridity in tropical Africa during the last glacial maximum is further shown by the extensive development of dunes in both hemispheres (Sarnthein, 1978). In the Northern Hemisphere fixed dunes dating to 20 000–12 000 BP lie 400–600 km south of the modern Saharan dune-limit (Sarnthein, 1978; Mainguet, Canon & Chemin, 1980; Talbot, 1980). If the forest zone had experienced southerly movement of a similar magnitude, forest would have been all but eliminated in West Africa, which in fact did not happen, as shown by the biological evidence discussed in the last section. The alignment of fixed dunes and the distribution of Saharan dust in Atlantic sediments suggest that the pattern of atmospheric circulation over west tropical Africa at glacial maximum was the same as now, that is with about the same position for the subtropical high pressure gyre and the same degree of northward penetration by the intertropical convergence zone in summer (Talbot, 1980; Sarnthein *et al.*, 1981; Stein & Sarnthein, 1984). Aridity can be attributed partly to a drier and less vigorous southwesterly monsoon.

Pollen diagrams from East Africa show that forest was absent from the northern Lake Victoria hinterland and in the lowlands around Ruwenzori for at least several thousands of years prior to 12 500–12 000 BP (Livingstone, 1967; Kendall, 1969; Hamilton, 1972). There was virtually no forest on Mt Elgon between >23 000 and <14 000 BP and on the Cherangani Hills between 28 000 and 12 500 BP (Coetzee, 1967; Hamilton, 1982), though a fresh look at the pollen evidence for Elgon does suggest that there could have been very small pockets of upper montane or dry montane forest on the lower slopes of the mountain (Hamilton unpublished data).

The date of forest expansion in western and southern Uganda is established at 12 500–12 000 BP (Livingstone, 1967; Kendall, 1969). On Mt Elgon forest spread after 14 000 BP (probably after 12 500 BP) and was extensive by 11 000 BP (Hamilton, 1982 and unpublished data). Thus the date of the most recent forest expansion along at least the first part of the northern route of forest mammal movement across East Africa is well established.

Mt Elgon and the Cherangani Hills were exceptional among those montane areas which have been examined palynologically in East Africa, in the severity of montane forest reduction during the last glacial maximum. These other areas are Ruwenzori (vegetation history known back to 15 000 BP), southwestern Uganda, Rwanda and Mt Kenya. In all cases there was, however, some forest reduction (Coetzee, 1967; Livingstone, 1967; Morrison, 1968; Hamilton, 1982). It is no surprise that the western sites near the Central Refuge were more forested than Mt Elgon and of greater interest in the present context is the clear palynological evidence that some forest, albeit of a relatively dry type, was present on Mt Kenya (Coetzee, 1967; Van Zinderen Bakker & Coetzee, 1972; Hamilton, 1982). Meteorologically, a wetter Mt Kenya compared with Mt Elgon could be related to its greater proximity to the Indian Ocean and the absence of Lake Victoria, evaporation from which contributes to a high rainfall on Elgon today. Mt Kenya lies at the eastern end of the northern route of forest mammal movement in East Africa and its paucity in mammalian endemics, in spite of evidence of some forest survival during the last Ice Age, could have a number of causes. Forest reduction may have been more drastic during earlier arid periods, the types of forest present during phases of contraction may have been unsuitable for many species, or mammals surviving on the mountain during arid periods may have proved uncompetitive in the face of immigrants moving in from the Central Refuge.

The upper Quaternary history of vegetation and climate along the southern route of forest mammal movement in East Africa has hardly been investigated by pollen analysis. In northern Zambia a pollen diagram from a small lake within *Brachystegia* woodland shows little vegetation change during the last >21 000 years (Livingstone, 1971). Forest and very arid communities never occurred, but the poor pollen representation of many savanna trees means that considerable change could have occurred within the savanna vegetation, which has not been detected.

Central Africa, from Gabon and Cameroon to the Western Rift, has received relatively little attention from Quaternary workers, partly because there are thought to be few good sites for pollen analysis. There are, however, indications of forest contraction during the last glacial maximum. Marine sediments off the mouth of the River Zaïre record decreased amounts of forest pollen before 12 500 BP (Giresse & Lanfranchi, 1984). Geomorphological work in southwest Cameroon and near Stanleypool show that forest was once absent; dating is difficult but such evidence as exists points to the last unforested phase as occurring during the last glacial maximum (De Ploey, 1968; Kadomura, 1982). The Cameroon study was carried out between the Cameroon and Gabon Sections of the Cameroon/Gabon Refuge (Figure 2.3), giving support to Kingdon's view that this refuge was fragmented. The exceptional richness of the fish fauna of the River Zaïre makes it likely that some river flow has been maintained for a long time (Livingstone, Rowland & Bailey, 1982). The river serves as a barrier to some terrestrial mammals today, for instance to some species of the *Cercopithecus (mona)* group (Kingdon, 1980). The river would have constituted a less formidable barrier during periods of lower flow.

Plant fossils from several sites show reduction of mesic vegetation in West Africa during the last glacial maximum. Lake Bosumtwi lies in Ghana outside the area in East Ivory Coast/West Ghana which is believed to have constituted a forest refuge during arid periods. Before 9000 BP and extending back to at least 19 000 BP the lake sediments contain abundant pollen and cuticles of grasses, suggesting a basically non-forested environment, though small quantities of forest tree pollen point to small patches of montane-type forest, possibly in valleys (Maley & Livingstone, 1983). The presence of lowland forest after 9000 BP is attested by abundant pollen and leaves of forest trees (Talbot & Hall, 1981). A core from the Niger Delta is interpreted as showing retreat of forest and other more mesic vegetation types at some time between 35 000 and 8000 BP; forest was established by 7600 BP

(Sowunmi, 1981a). Mangrove pollen is recorded throughout the sequence and perhaps persistent mangrove forest during the last glacial maximum might have provided a refuge for guenons.

Pollen diagrams from on- and offshore southern Ivory Coast show rather open forest during the latest Pleistocene and earliest Holocene and replacement by more continuous forest at 9000–8000 BP (Assémien *et al.*, 1970; Fredoux & Tastet, 1976 – both quoted in Talbot & Hall, 1981). Pollen analysis of a series of cores of marine sediment collected between 8°N and 33.5°N off the west coast of Africa has indicated that the southern boundary of the Sahara was 150–550 km south of its present position during the last glacial maximum, when forests and woodlands were less extensive than during either the postglacial or the preceding interglacial (Agwu & Beug, 1984). Marine cores off Senegal also register enhanced aridity during the last glacial maximum, with a strong decrease in the quantity of lowland forest pollen before 12 500 BP (Rossignol-Strick & Duzer, 1979).

To summarize, palynological and other evidence of vegetational and climatic history have demonstrated that throughout much of tropical Africa the last period of great forest reduction occurred during the time of the last world glacial maximum. Rainfall increased suddenly at about 12 500 BP, leading to forest expansion. The latter was delayed in some places for reasons which have yet to be fully analysed.

Forest reduction during the upper Holocene

The latter half of the Holocene has witnessed a reduction in the extent of mesic vegetation in Africa, partly due to a drier climate and partly due to increasing destruction by agricultural man. One consequence has been the fragmentation of the forest cover, especially in peripheral areas. Today populations of guenons in forest remnants may be as effectively isolated from one another as were populations of their ancestors in Ice-Age refugia. There are opportunities for investigating rates of evolutionary divergence, although rarely will it prove possible to establish the exact times of separation of populations.

Hastenrath & Kutzbach (1983) have carried out a sensitivity analysis of the factors determining lake-level change in Kenya during the Holocene and have concluded that increased precipitation was the main cause of high stands during the early Holocene; the probably slightly lower temperatures of the time are calculated to have rather little impact on the water budget. Lower lake-levels during the later Holocene, especially after 4000 BP, are widely reported from tropical Africa; Nile levels were also reduced (Hecky & Degens, 1973;

Adamson *et al.*, 1980; Gasse *et al.*, 1980; Servant & Servant-Valdary, 1980; Talbot & Delibrias, 1980; Talbot *et al.*, 1984). Following Hastenrath & Kutzbach's analysis, reduction in rainfall can be taken to be the main causative factor.

Considerable areas of Africa, especially in the relatively dry zone around the margins of the main forest blocks, are believed to be climatically suitable to forest but carry non-forest vegetation (Aubréville, 1949; Keay, 1959; Hopkins, 1962; Langdale-Brown, Osmaton & Wilson, 1964). The main cause is believed to be destruction of forest by man. Today the forest/savanna boundary is usually sharp, but types of vegetation intermediate between forest and savanna woodland occur locally and at one time are thought to have been extensive (Lebrun & Gilbert, 1954; Devred, 1958; Keay, 1959; Werger & Coetzee, 1978; White & Werger, 1978). Primates which today are regarded as strictly forest species may once have been able to range more widely; clarification of the forest/savanna boundary could have forced some guenons to adopt a more strictly arboreal role. Burning of savanna is a major factor keeping the forest boundary distinct today and, since most savanna fires are started by people, the acquisition and development of burning technology by man could have influenced the evolution of forest organisms. Man has possessed the use of fire for hundreds of thousands of years.

Little is known about the date of the origin of agriculture in tropical Africa. The oldest evidence of forest destruction by agricultural man is from southwestern Uganda where fire was used to help clear lower moist montane forest soon before 5000 BP (Hamilton, Taylor & Vogel, 1986). Increases in the pollen of oil palm at 3000 BP in Nigeria (Sowunmi, 1981a) and at 3500 BP in Ghana (Talbot, 1983) are probably due to agriculture. This is a useful reminder that agriculture or other types of human disturbance within the forest zone may not always be to the disadvantage of forest species. Hall & Swaine (1981) suggest that there is no forest in Africa which has not been subject to some human influence.

Pollen diagrams from a number of areas show replacement of more mesic by less mesic vegetation during the upper Holocene. These include diagrams from marine sediments off the coasts of Senegal and Zaïre (Rossignol-Strick & Duzer, 1979; Giresse & Lanfranchi, 1984). The northern boundary of savanna woodland in northwest Sudan has retreated southwards by 500 km since the early Holocene (Ritchie, Eyles & Haynes, 1985). Drier types of montane forest increased at the expense of more humid types on many of the East African mountains

(Hamilton, 1982). In some cases, especially where the catchment areas of sites used for pollen analysis are small or where pollen accumulation rates are known in terms of deposition per unit time, forest retreat is clearly seen to be due to man (Kendall, 1969; Morrison & Hamilton, 1974; Hamilton *et al.*, 1986), but in other cases the relative contributions of human activities and increased climatic aridity are difficult to assess. A clear distinction between these potential causes of forest retreat may in any case not be always valid. Man himself has certainly influenced the climate of the upper Holocene through his effects on the hydrological and energy characteristics of the earth's surface (Charney, Stone & Quirk, 1975).

To summarize this section, the forest boundary has receded in many parts of Africa during the last 5000 years and there have been changes in floristic composition due to a drier climate and human influence. Some populations of guenons have become isolated in forest remnants, providing opportunities for studying divergent evolution.

Ocean cores, climatic modelling and long-term forest history

Studies based mainly on sites in the African interior have provided a reasonably clear picture of the vegetation and climate of tropical Africa between the time of the last world glacial maximum and the present day. The continental record has proved less useful for earlier times, being less studied, fragmentary and often difficult to date. For a longer perspective we must rely at present mainly on the study of offshore sediments and on climatic modelling. The records of environmental history contained within deep-sea sediments are remarkable and their investigation during the last twenty years has revolutionized knowledge of the Quaternary. Actually, very long sedimentary records are probably available in several lake basins in Africa, but have not yet been studied (Livingstone, 1981).

A major result of research on deep-sea sediments has been to establish that there have been many more glacial periods during the Quaternary than was once believed (Shackleton & Opdyke, 1973). Van Donk (1976) has estimated 21 glacials or near-glacials during the last 2.3 Myr. The Quaternary has undoubtedly been a period of exceptional climatic instability.

A second major advance has been the widespread acceptance that the fundamental cause of the glacial/interglacial cycle has been variation in the amount of solar radiation reaching the world related to changes in the earth's orbit and axial orientation (the Croll–Milankovitch Theory). The periodicities of the parameters concerned are known and the timing of Ice Ages and other past climatic events can be predicted, providing opportunities to test the theory. There is such good agreement between such climatic models and reality, as provided by the detailed records found in some deep-sea sediments and cores of polar ice, that there is no doubt of the basic correctness of the astronomical theory (Hays, Imbrie & Shackleton, 1976; Imbrie & Imbrie, 1980).

The strong association of aridity in tropical Africa with the last glacial maximum and of a moister climate with the postglacial might suggest that all glacial periods during the Quaternary have been dry in Africa and all interglacials wet. Actually, there are so many factors which can influence climate that such an hypothesis should only be accepted with caution, pending development of detailed models relating to tropical climatic history, backed up by good field evidence. Rainfall in tropical Africa is monsoonal and in recent years has tended to be high when solar insolation has also been high (Rossignol-Strick, 1983); high solar insolation produces lower than average pressures in the 'heat trough' of the intertropical convergence zone and stronger than average pressure gradients between the heat trough and subtropical high pressure areas. Rossignol-Strick has computed variations in tropical insolation for the last 464 000 years and, on the basis of her understanding of the controls over rainfall, has calculated expected variations in rainfall for this period. All interglacials are predicted to be wet and, interestingly and perhaps unexpectedly, so too do two periods during glacial times (at 176 000 and 220 000 BP). This model has been tested through the examination of the ages of sapropels in marine sediments in the East Mediterranean Sea (Rossignol-Strick et al., 1982). Sapropels are black organic-rich layers and those in the East Mediterranean are believed to have formed during times of heavy Nile discharge, itself determined by high rainfall in Africa. Sapropels occur in the sediments every time high rainfall is predicted by Rossignol-Strick's model and at no other times. Furthermore, pollen and other fossils show that the climate at 176 000 and 220 000 BP was colder than at other times of sapropel formation, also as predicted. These empirical findings are strongly supportive of the model.

The oceans of the last interglacial, which are quite well known from analysis of deep sea cores (CLIMAP, 1984), were similar to modern oceans, contrasting greatly with those of the intervening glacial (CLIMAP, 1976). Pollen analysis of deep-sea sediments from the Arabian Sea has confirmed that northeast Africa was drier during the last glaciation than during the preceding interglacial or following postglacial (Van Campo et al., 1982). Analysis of aeolian material in cores from the Atlantic off the northwest and equatorial coasts of Africa indicates that wettest conditions were attained during times of deglaciation (e.g. the very late Pleistocene/early Holocene) and driest during times of ice-growth (Pokras & Mix, 1985).

The implications of this work for forest history and guenon evolution in Africa are considerable. The correlations established between interglacials and a wet tropical climate and between most glacials and aridity mean that there must have been many (roughly 20) cycles of major forest spread and retreat in Africa during the last 2.3 Myr. Opportunities for the isolation and divergence of populations have been numerous. Indications that a few of the wet periods were also cold will be of interest to many biogeographers, who have long been interested in the dispersal of montane species between isolated highland areas.

It would be especially useful to have more precise information about forest distribution in Africa between the time of the end of the last interglacial at 70 000 BP and the time of the last major glacial period (Wurm II), centred on 18 000 BP. The interglacial before 70 000 BP was wet and forest was probably about as extensive as it has been during the Holocene. Deep-sea cores off the African coast show that the Wurm as a whole (70 000–12 500 BP) has been relatively dry, but the oceanic record can lack fine resolution and there could have been relatively brief, but biologically significant, periods of moister climate when forest was extensive. This is relevant when considering the times of divergence of closely related forest species, such as members of the superspecies *Cercopithecus (mona)* and *C. (cephus)*. These species have certainly been geographically separate for 20 000 years, but did isolation of their ancestral populations occur soon before that date, or at 70 000 BP, or even earlier? The continental African record demonstrates that the period 20 000–40 000 BP (the latter being the normal lower limit for radiocarbon dating) was not climatically uniform, but rather contained episodes of varying degrees of wetness (Gasse *et al.*, 1980; Hamilton, 1982; Perrott & Street-Perrott, 1982). An assessment of likely changes in forest extent during this period and further back to the end of the preceding interglacial awaits more evidence and analysis.

In addition to climatic changes associated with earth orbital variations there are longer-term climatic trends which have greatly influenced the distribution and evolution of terrestrial organisms. An important determinant of these longer trends appears to have been the movement of continents, drifting them slowly into new climatic zones and altering patterns of atmospheric and oceanic circulation. The movement of Antarctica to its present polar position led to the gradual build-up of ice during the Cenozoic, producing cooling all over the globe (Flohn, 1973). One effect of an ice-covered Antarctic has been to destabilize the global climate, so that climatic oscillations associated with orbital variations became more dramatic (Woodruff, Savin & Douglas, 1981).

Mineralogical and other analysis of cores collected from between 5° N and 35° N off the west coast of Africa have helped to reveal the following long-term picture of the West and central African climate (largely after Stein & Sarnthein, 1984):

1. From 8.8 to 6.4 Myr the climate was relatively warm and stable and the amplitudes of short-term (Croll–Milankovitch) variations generally low. West Africa was characteristically humid, as during Quaternary interglacials.

2. 6.4–4.6 Myr. The Mediterranean Sea became isolated from other oceans and evaporated leaving behind vast deposits of salt. The remaining oceans became less saline, contributing to a major increase in the volume of Antarctic ice. This 'Mediterranean salinity crisis' produced enhanced aridity around the Mediterranean and indeed, by less direct climatic connections, in Africa as a whole. There was probably a major expansion of African savanna (Hsu *et al.*, 1977). Climatic fluctuations of the Croll–Milankovitch type became more marked, with oscillations between warmer, more humid and colder, drier periods in Africa.

3. 4.6–2.43 Myr. The climate was generally similar to that of 8.8–6.4 Myr, being relatively warm and wet, though with a shift towards being drier and cooler after 3.5 Myr.

4. 2.43–1.0 Myr. 2.43 Myr marks the beginning of major glaciation in the Northern Hemisphere, probably a consequence of northward drift of the semi-closed Arctic Ocean to a point where its southern margin lay north of the permanent snowline (Rea & Schrader, 1985; Shackleton *et al.*, 1984). The world became generally colder and drier and there were pronounced oscillations between a warm wet and a cold dry climate in tropical Africa.

5. 1.0 Myr to the present. This was similar to the preceding period, but with climatic oscillations at twice their former amplitude. Stein & Sarnthein (1984) stress that both before and after 1 Myr, it is not so much the climate of the

interglacials which has changed as the depth of aridity reached during intervening dry phases.

In summary, deep-sea cores and climatic modelling show that climatic variability has itself been variable, becoming more pronounced at 6.4 and 2.43 Myr (with reversion to greater stability at 4.6 Myr) and especially at 1 Myr. There has been a trend over this time towards greater aridity. As a result, the fragmentation of forest guenons into isolated populations has become increasingly probable, as has the chance of refugia disappearing completely during arid periods. There have been numerous opportunities for the isolation and later expansion of populations. Our knowledge of the rates of processes which can influence guenon evolution is so limited that for earlier times it is not yet possible to match particular episodes of habitat fragmentation with particular phylogenetic splits. Geographically, the determination of the localities which are associated with particular genealogical events becomes hazardous for all but the uppermost branches of the evolutionary tree.

Plant fossils, 8–1 Myr BP

A number of sites from Ethiopia to Tanzania has yielded plant fossils, producing an invaluable record which can be compared with expectations based on climatic modelling and analysis of deep-sea cores. The ages of some of the fossil beds are well known from potassium–argon dating of interbedded volcanic rocks.

Yemane, Bonnefille & Faure (1985) have published a pollen diagram from a Late Miocene (about 8 Myr) lacustrine deposit in the middle of the north-western plateau of Ethiopia. The diagram is rich in pollen types and shows the presence of dense lowland forest (or, according to me, perhaps moist montane forest). The forest plants include some which no longer grow in Ethiopia, but are still present in the main Guineo-Congolian forests. The climate was certainly wetter than now and temperatures at the sample site were higher, the latter being partly due to major post-depositional uplift of the sediments. This site fully supports the climatic picture painted by Stein & Sarnthein (1984) based on analysis of deep-sea cores.

There is macrofossil evidence that forest and moist woodland in the Turkana areas of S Ethiopia/N Kenya retained some exotic elements even after the time of the Mediterranean salinity crisis. These 'exotic elements' are found today in the main Guineo-Congolian forest blocks or in moist woodlands of the Southern Hemisphere, but are not

present in Ethiopia. Fossil fruits of the forest tree *Antrocaryon* (near
A. micraster) and fossil shells of the prosobranch gastropod *Potadoma*,
as well as associated forest rodents and prosimians, provide evidence
of a more diverse forest flora and fauna and a wetter climate near Lake
Turkana at 3.4–3.3 Myr (Bonnefille & Letouzey, 1976; Williamson,
1985). Seventy-four species of tree are known from fossil wood dating
to between 4 and 1.5 Myr (mostly 2–1.8 Myr) from around Lake
Turkana (Bonnefille, 1984; Dechamps & Maes, 1985). Much of this
wood is believed to have originated from riverine vegetation, which is
shown to have been richer in woody species than now; some of the taxa
have become extinct in Ethiopia and Kenya but survive today in the
Guineo-Congolian forests. Floristic impoverishment of riverine forest
in S Ethiopia/N Kenya continued after 1.5 Myr, perhaps only reaching
its present species-poor state after aridity became intensified at 1 Myr.

Pollen analysis of a range of Pliocene and lower Pleistocene sites
from Ethiopia to northern Tanzania shows that the broad pattern of
modern vegetation has been established from 3.7 Myr onwards, with
savanna in the lowlands, denser woody vegetation along rivers and
montane forest on mountains (Bonnefille, 1976, 1979, 1984). In contrast
to the macrofossil studies, pollen analysis has failed to reveal any
'exotic elements', undoubtedly partly because of the lower precisions
usually possible in identification and partly for taphonomic reasons.
Within the overall vegetational context mentioned above, the pollen
evidence does show that there were major variations in vegetation and
climate. Pollen samples dating to 3.6–3.2 Myr from Hadar in the
sub-desert of northeast Ethiopia show the presence of much more
mesic vegetation than occurs there today; forest grew close to the
sample site. At Gadep in the highlands of southern Ethiopia a pollen
diagram shows climatic cooling and major depression of the vegetation
belts during a period dated to sometime between 2.51 and 2.35 Myr
(Bonnefille, 1983). This is a highly significant discovery in that here we
have the earliest evidence for such major cooling in tropical Africa and
the cooling could have been contemporaneous with the first major
glaciation of the northern hemisphere (which began at 2.43 Myr).

The overall picture presented by these fossil data supports conclusions reached from deep-sea cores and climatic models. There has been
a trend towards greater aridity since Miocene times, causing reduction
in the area of forest and biotic impoverishment in marginal areas such
as Ethiopia. It is uncertain when Ethiopian forests were last connected
to the Guineo-Congolian forest block, but even contact by tenuous
riparian forest seems unlikely during the last 1 Myr.

Apart from changes in the moistness and vigour of air masses, vegetational and climatic patterns on the African continent will have been influenced by changes in topography. Andrews and Van Couvering (1975) have summarized evidence that topography during the early Miocene was subdued, with a low-lying watershed between westward and eastward flowing rivers lying in Kenya and Tanzania roughly along the line of the modern Eastern Rift Valley. These authors argue that moist winds from the Atlantic would then have been able to penetrate further to the east, allowing forest to be more extensive and possibly continuous from the Atlantic to the Indian Ocean. Perhaps some of the central African/East Coast disjunctions date back to the Miocene. By the Pliocene, uplift and rifting in East Africa were well advanced, but exactly how and when the modern topography became established in essentially its modern form is uncertain. There has been major uplift along the Western Rift Valley during the Pleistocene (Andrews & Van Couvering, 1975); this in itself would have decreased rainfall to the east, though this effect would have been counterbalanced in southern Uganda by increased rainfall following the development of Lake Victoria.

Why are there so few species of forest guenons?

If every phase of forest contraction had resulted in the formation of a new species of guenon in each refugium and if subsequent to forest expansion each new guenon had spread out to occupy the whole forest area, then there would be hundreds of species of guenons in Africa today. This has obviously not happened. Factors which may have helped to limit the numbers of species include slow rates of differentiation of populations, slow rates of species movements out of refugia, hybridization between isolates and restriction of ranges and extinction caused by competition from other species.

It may be possible to quantify aspects of the evolution of taxa associated with cycles of habitat fragmentation and coalescence. In guenons there seems to be an upper limit to the number of species (4–6) which may coexist in any particular locality, possibly related to the carrying capacity of the environment. In passerine birds there is a similarity between the representation of endemics in the three major refuge areas of West Africa (5.8%), Cameroon/Gabon (5.0%) and eastern Zaïre (7.2%), despite considerable variations in the numbers of species (Diamond & Hamilton, 1980). The high endemism in East Africa is not comparable, there being many local endemics in southern and eastern Tanzania associated with small isolated forests each of which is presumed to be a refuge area. In passerines there are more species in Cameroon/Gabon (179) and eastern Zaïre (209) than in West Africa (120), probably related to variations in the degree of habitat diversity in each refugium during times of forest contraction. Comparative studies between groups of organisms should constitute a fruitful field for future research.

Acknowledgements

I am very grateful to Naomi Hamilton and David Taylor for reading the manuscript and to Kilian McDaid and Nigel McDowell for assistance with the artwork.

I.3

Habitat and locomotion in Miocene cercopithecoids

MARTIN PICKFORD and BRIGITTE SENUT

Introduction

The early evolution of the Cercopithecoidea is still extremely poorly documented, the earliest known fully derived members of the order being *Victoriapithecus* from Maboko, Nachola, and Loperot in Kenya, and *Prohylobates* from Wadi Moghara and Gebel Zelten in northern Africa and Buluk in Kenya (Fourtau, 1920; Von Koenigswald, 1969; Delson, 1975b; Szalay & Delson, 1979; Leakey, 1985).

Although there has been considerable debate concerning the question of whether cercopithecoid roots can be traced as far back as the late Eocene Fayum (Egypt) Primates, it seems that any phylogenetic links, even if demonstrable (Simons, 1967, 1970, 1972, 1974, 1985) are very remote (Delson, 1974, 1975b). It is usually assumed that the earliest members of the order Cercopithecoidea evolved some time during the Oligocene period, i.e. some time later than the period of accumulation of the Fayum sequence and before the period represented by Napak (Uganda), Maboko (Kenya), Wadi Moghara (Egypt), and other early middle Miocene sites in Africa. It is unfortunate that the Oligocene period is exceptionally poorly represented by sediments in Africa and at present the chances seem remote that the original members of the order will ever be found.

There is also debate about the origins of the two extant subfamilies of Old World monkeys, the Cercopithecinae and the Colobinae. There are two main hypotheses to be found in the literature; the first is that two taxa represented in the middle Miocene collections from Maboko (Kenya) provide evidence for the subdivision of the family Cercopithecidae into the two extant subfamilies (Delson, 1975b; Napier & Napier, 1985); the opposite view is that the Maboko and other early

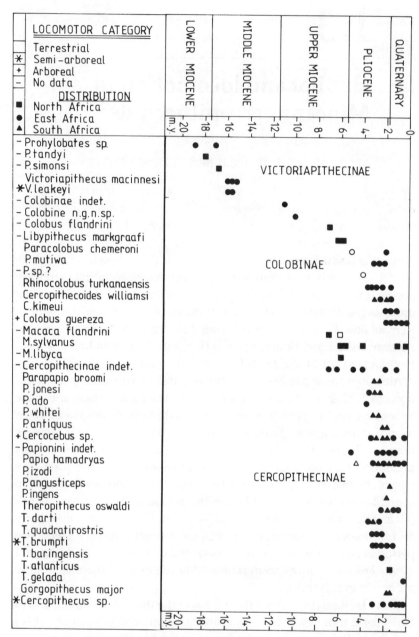

Figure 3.1. Chronological distribution of fossil monkeys in the Miocene and Plio–Pleistocene of Africa. Solid symbols = confident identification; open symbols = less sure identification.

middle Miocene monkeys belong in a subfamily of their own, the Victoriapithecinae (Von Koenigswald, 1969; Leakey, 1985), with no special connections to extant subfamilies. Evidence from upper Miocene deposits (8–10 million years ago) indicates that both extant subfamilies had evolved by then (Figure 3.1).

Previous hypotheses

From the palaeoenvironmental perspective there have been two main hypotheses about the origins of the Old World monkeys. Napier (1970a) suggested that the cercopithecoids, especially the colobines, were primitively forest adapted animals and that it was inconceivable that they should have evolved from a ground-living stock. For the cercopithecines, Napier (1970a) held an opposite view, suggesting that they stemmed from a distant ground-living species of the Miocene. Delson (1975b) and Andrews (1981a), in contrast, considered that both subfamilies of cercopithecoids had a more terrestrial ancestry, and that extant forest-living colobines and cercopithecines represent a secondary invasion of the forests by these groups.

From the dietary perspective there seems to be little argument in the literature that the basic adaptation of the Old World monkeys concerns the development of a bilophodont cheek dentition, but there is a certain amount of debate as to what precisely bilophodonty means in terms of diet. For Jolly (1966) the original dietary adaptive niche of the monkeys was leaf-eating, from which he proposed that the colobine molar pattern should be considered the primitive or plesiomorphic state. Delson (1975b) however, considered that the earliest cercopithecoids were probably not fully committed folivores, but were more eclectic in their diet, which he visualised as being essentially omnivorous. Napier (1970b) held a similar view about the importance of dietary specializations in the early monkeys, and he suggested that in seasonal forests, the ability to subsist on leaves rather than fruits when necessary would have been a great selective advantage. Thus, the ancestors of the cercopithecoids may not have been completely dependent on leaves, but were able to supplement their diets of fruits and other items (animal proteins, for example) with foliage during seasonal fluctuations in food supply or in those marginal environments where 'preferred' food items might be in short supply most of the year.

It is important when considering the early adaptations of cercopithecoids to distinguish between locomotor adaptations and dietary adaptations. Part of the debate in the literature stems from a failure to keep the two separate. Thus, although colobines seem to be plesiomorphic (at least to some extent) in their post-cranium, they appear to be highly derived in their dentition and possession of sacculated stomach and other modifications of the digestive tract (Kuhn, 1964). The opposite seems to be the case with cercopithecines, the post-cranium being more derived when compared to Fayum primates for example, while the dentition and digestive tract is plesiomorphic (Andrews, 1981b) except perhaps for the development of cheek pouches. To speak of colobines or cercopithecines as being more or less derived compared with one another, without first of all specifying which part of the body one is discussing, surely leads to much confusion.

Nevertheless, despite the debates, there seems to be a consensus among primatologists who have written about the early evolution of monkeys, that both dietary and locomotor modifications took place from a more primitive catarrhine or pre-catarrhine condition represented by a fully arboreal, possibly frugivorous, or at least non-folivorous, primate. It is further generally agreed that the initial adaptive selections took place during the Oligocene period some time between 32 and 20 millions years ago. Arguments in favour of *Parapithecus*, from late Eocene deposits in Egypt, being a cercopithecoid (Simons, 1971, 1985) are not entirely convincing (Delson, 1975b), whereas there seems to be universal agreement that *Victoriapithecus* and *Prohylobates* from Miocene deposits are fully derived cercopithecoids.

Fossil evidence

The post-Eocene African fossil record improves dramatically in the lower Miocene (20 Myr ago) after a tremendously long Oligocene fossil void (some 14 Myr long) (Figure 3.2). In the early Miocene sites (Drake *et al.*, 1987) there is a dearth of monkey fossils in the deposits, in sharp contrast to the wealth of hominoid primates (Pickford, 1981). It is not until the early parts of the middle Miocene that fossil monkeys become abundant at certain sites such as Maboko (Kenya), where they occur side by side with hominoids.

The earliest known cercopithecoids come from North African sites at Wadi Moghara (Egypt) (Fourtau, 1920; Simons, 1969) in which a primitive species of the anthracothere genus *Brachyodus* is found, suggesting a lower Miocene age. Perhaps a little later in time is Napak (Uganda) which has yielded a few monkey fragments including a molar, a canine and proximal parts of associated radius and ulna (Pickford *et al.*, 1986). Somewhat later in time is the locality of Buluk, northern Kenya, dated at about 17.2 Myr or a little older (MacDougall & Watkins, 1985) from which Leakey (1985) reported the presence of

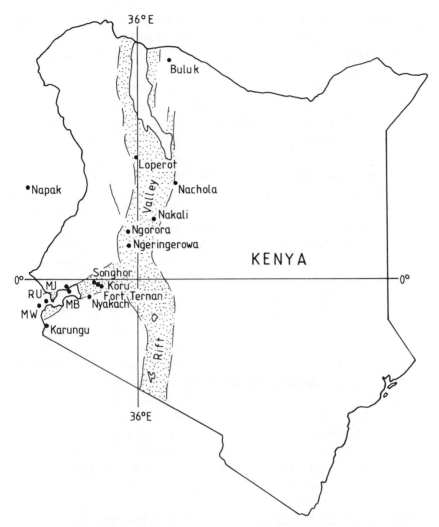

Figure 3.2. Distribution of fossil localities of Miocene age in East Africa, which have yielded fossil monkeys or which are discussed in the text. RU = Rusinga; MB = Maboko; MJ = Majiwa; MW = Mfwangano.

Prohylobates. Probably somewhat younger than Buluk is the site of Maboko, western Kenya, and allied sites at Majiwa and Nyakach which have yielded several hundred monkey specimens (MacInnes, 1943; Von Koenigswald, 1969, Pickford, 1986). New localities at Nachola have also yielded *Victoriapithecus*, while a large species of *Prohylobates* has been recorded from Gebel Zelten, Libya (Delson, 1977), which is possibly about 17–16 million years old.

There is rather a significant gap in the fossil record after the Maboko–Nyakach period during which no monkeys have been found (Figure 3.1) and the next youngest monkey fossils have been found at Ngorora, Ngeringerowa and Nakali in upper Miocene strata of the Gregory Rift Valley, by which time fully derived colobines have evolved. At approximately 7–8 million years ago in North Africa, some sites such as Menacer (Marceau) have yielded both colobines and cercopithecines (Arambourg, 1959; Delson, 1975b, Thomas & Petter, 1986).

The first major radiation of the cercopithecoids seems, however, to have occurred appreciably later in the Miocene and the early parts of the Pliocene period (Figure 3.1), when abundant genera and species appear in North, East and South Africa. It is unlikely that our perception of a radiation at the Miocene/Pliocene boundary is purely an artefact of the fossil record. Upper Miocene sites in Kenya and North Africa have been well studied from the palaeontological point of view and have yielded abundant mammalian fossils, but relatively few monkeys. In Pliocene deposits of East and South Africa, in contrast, monkeys are common and diverse.

The Lower Miocene dearth of monkeys

It has often been remarked that the exceptionally fossiliferous strata of lower Miocene age in western Kenya have yielded no monkeys: reports of the presence of monkeys in these deposits are based on faulty identifications (Leakey, 1958) or incorrectly provenienced fossils (MacInnes, 1943; Von Koenigswald, 1969) and can confidently be discounted. Simons (1972) wondered why this should be so, and listed six possible explanations which might account for their rarity or absence from sites which are known to yield hominoid primates in abundance. Among these, explanations such as collecting bias can immediately be discounted, since if anything the deposits in question have been more thoroughly examined than many other deposits in Africa.

From the palaeoenvironmental perspective, two of Simons' six explanations can now be investigated. The first was his suggestion that the rarity of monkeys in the Miocene deposits might be due to the fact that the earliest Old World monkeys were adapted to live principally in rain forests wetter than East Africa in Miocene times. Secondly, he thought it might be possible that Miocene African monkeys may have been generally adapted to a drier climate and more terrestrial way of life than was available in the East African Miocene. Of these two possibilities, Simons (1972) preferred the first.

1. *Cercopithecus erythrotis camerunensis*
Red eared guenon – Cameroon.

2. Signal patterns and postures of African guenons showing geometry of design and the role of colour and tonal contrast to polarise attention, especially on the head and genital regions.

3. *Cercopithecus pogonias pogonias*
Crowned guenon – Cameroon.

4. Schematic profile views of *Cercopithecus pogonias* (left) and *Cercopithecus wolfi* (right) with sonograms of the contact call. Relatively small alterations in the position and angle of facial markings or changes in pitch in vocal signals make species look and sound distinct.

With the newly available palaeoenvironmental reconstructions for many lower and middle Miocene sites of East Africa, we can now investigate the questions raised by Simons and others, while with recently augmented collections of post-cranial skeletal elements, we can examine the locomotor hypotheses.

Several major problems have traditionally hampered the solution of these quandaries. Apart from imperfections of the fossil record, including the Oligocene fossil void, there have been few palaeoenvironmental reconstructions until recently (Verdcourt, 1963; Pickford, 1982, 1983, 1984, 1985) and the available post-cranial evidence has been scanty (Szalay & Delson, 1979). Recently two things have happened which have helped to throw light on the subject. The collection and study of terrestrial gastropods (Verdcourt, 1963, 1984) has yielded abundant evidence concerning the palaeoenvironments of many of the lower and middle Miocene sites of Kenya (Pickford, 1984, 1985) and improved collections of fossils from Maboko, Nyakach and Napak have yielded many additional monkey fossils, including post-cranial remains. Although a great deal remains to be done with the new collections, we feel that we can reach some preliminary conclusions about early and middle Miocene monkey adaptations.

Palaeoenvironments

Palaeoenvironmental reconstructions have now been made for a number of fossil sites in western Kenya (Verdcourt, 1963; Pickford, 1982, 1983, 1984, 1985), principally using fossil terrestrial gastropod assemblages as guides to past environmental conditions. Fossil gastropod assemblages have now been made at 85 different localities, at 35 of which the assemblages are based on abundant specimens belonging to numerous taxa. Fossil primates have been recovered at most of these localities.

Despite the large number of fossil sites in western Kenya, there is only a limited variety of snail assemblages. Pickford (1982), recognized three main gastropod assemblages in the Miocene strata of western Kenya, to which must now be added a fourth. In addition, one of the previously recognized assemblages can be subdivided into two sub-categories as discussed by Tassy & Pickford (1983).

Each of these gastropod assemblages can be closely matched to extant East African assemblages (Table 3.1) each of which appears to be habitat specific today. The assumption is that the fossil assemblages signify the same or similar habitats for past times as they do today.

Snail assemblage I is characterised by the abundant presence of the

Table 3.1. *Distribution of gastropods in Western Kenya*

Taxon	Snail assemblage			
	I	II	III	IV
Primigulella	XXXXX			
Achatina leakeyi	XXXXX			
Gonospira	XXXXX			
Opeas	XXXXX			
Krapfiella	XXXXX			
Trochozonites	XXXXX	?		
Tayloria	XXXXX	X		
Marconia	XXXXX	X		
Gulella small spp.	XXXXX	X	X	X
Thapsia	XXXXX	X	X	X
Gonaxis protocavalii	XXXXX	X	X	
Ptychotrema	XXXXX	X	X	
Subulinidae	XXXXX	XXXXX	X	XXXXX
Burtoa nilotica	XXXXX	XXXXX	XXXXX	XXXXX
Trochonanina	XXXXX	XXXXX	X	X
Gonaxis aff. *craveni*	X	XXXXX	X	
Halolimnohelix	X	XXXXX		
Maizania	X	XXXXX		
Tropidophora	X	XXXXX	X	
Edouardia	X	XXXXX	X	
Edentulina		XXXXX	X	
Chlamydarion		X		XXX
Urocyclidae	X	XXXXX	X	
Cerastua majus		XXXXX		
Cerastua miocenica		XXXXX		
Limicolaria aff *martensiana*		X	XXX	?
?Zingis		X	XXXXX	
Rhachistia cf *rhodotaenia*			XXXXX	
Pupoides			XXXXX	
Nothapalinus			XXXXX	
Curvella				XXXXX
Homorus	?	?	?	XXXXX

XXXXX = abundant; XXX = common; X = present but rare; ? = possible record; blank = no record

genera *Primigulella, Gonospira, Trochozonites* and *Krapfiella*. Sub-assemblage IA lacks the genera *Tropidophora (Ligatella)* and *Maizania*, while sub-assemblage IB contains both these taxa, although *Maizania* is never abundant therein. Snail assemblage I is today found typically in wet to dry forest and humid woodland, IA being characteristic of higher altitudes (above 1200 m) while IB is more typically found at lower altitudes (below 1200 m).

Table 3.2. *Distribution of hominoids and cercopithecoids in Western Kenya*

Region	Age (Myr)	Number of sites known	Snail Ass.	Number of sites with homs	Number of sites with cercs	Vegetation inferred from snail evidence
Koru-Songhor	19–20	50	I	36	0	Humid forest
Rusinga	18	40	II	15	0	Dry forest to woodland
Other Rangwa sites	18	32	II	7	0	Dry forest to woodland
Maboko–Nyakach	16–17	27	III	6	8	Semi-arid woodland with gallery forest
Fort Ternan	14	6	IV	3	0	Upland humid woodland
Totals		155		67	8	

Snail assemblage II is characterised by the genera *Cerastua, Edentulina, Edouardia* and abundant *Maizania* and *Tropidophora*. Today this sort of snail assemblage is characteristic of lowland woodlands to coastal forests.

Snail assemblage III is characterised by an abundance of the genera *Pupoides, Nothapalinus* and *Rhachistia* which today are typically confined to semi-arid habitats below 1200 m, such as the modern *Acacia/Commiphora* woodlands of lowland eastern Kenya.

Finally, snail assemblage IV is characterised by the presence of the genera *Chlamydarion, Curvella* and *Homorus*, similar assemblages occurring today in upland humid woodland and forest.

In western Kenya, out of 155 known Miocene sites (Pickford, 1984) only eight have yielded monkeys (Table 3.2) while 67 have yielded hominoids. The restricted distribution of monkeys in comparison with hominoids prompts some questions. Were monkeys very rare in early Miocene times, or were they restricted in their habitat preferences? Are there any patterns in their distributions in these West Kenya sites? The answers to these questions are as follows. Although monkeys have been found at only a few sites in West Kenya, they are very common at some of them. For example, at Maboko, more than 700 specimens have already been collected, and the site shows little sign of exhaustion. At this site and neighbouring occurrences at Majiwa and Nyakach, monkeys are more common as fossils than hominoids. Conversely, at numerous lower Miocene sites at Koru, Songhor, Rusinga, Karungu and Mfwangano, where hominoids occur, sometimes in abundance, no fossil monkeys have been reliably recorded. The only differences between the sites which yield monkeys and those that do not appear to be chronological and palaeoenvironmental (Figure 3.5). The Maboko and Nyakach Formations are younger than the other sites in the region, with the exception of Fort Ternan, which is younger than Maboko. No monkeys have been reported from the latter site, although hominoids are common.

The only exception to this observation concerns the record of monkeys at Napak (Uganda) dated about 19 Myr (Bishop, Miller & Fitch, 1969) and therefore contemporary with the Songhor and Koru assemblages. At Napak, monkey fossils (three or four specimens) are outnumbered by hominoids (40+ specimens).

Apart from age, the major significant difference between the sites that yield monkeys and those that do not, is that the monkey sites all yield snail assemblage III, while none of the others do.

All the sites in western Kenya which have yielded snail assemblages IA, IB, II and IV, totalling 128 fossiliferous occurrences, have so far failed to yield a single monkey fossil, although they have yielded an abundance of fossil hominoids (more than 1000 specimens from 67 sites). Conversely, of all the West Kenya sites which have yielded snail assemblage III, totalling 27 sites, eight have yielded monkeys (more than 700 specimens) and fewer hominoids (207 specimens). It is difficult to evade the conclusion that the major factor responsible for this distribution pattern of lower and middle Miocene primates was environmental in nature. We have examined depositional environments, chronology, associated faunas and other possible correlates in order to account for the common occurrence of monkeys with snail assemblage III and their virtually complete absence from assemblages I, II and IV, and we found no reason other than palaeoenvironment to account for their distribution. We wondered, as did Simons (1972) whether competition from hominoids might not have led to the dearth of fossil monkeys in so many sites in western Kenya. Although at first glance this seems an attractive hypothesis, we consider it to be unimportant because where we find fossil monkeys we also find numerous hominoids. The distribution of hominoids and monkeys is not mutually exclusive in the West Kenya Miocene sites. Undoubtedly competition from hominoids may have been a factor in limiting the distribution of monkeys, but it was not the primary one.

It is our conclusion that middle and early Miocene monkeys were adapted to semi-arid conditions, and that they entered the West Kenya fossil record only when local conditions became semi-arid. When local environments subsequently reverted to more humid conditions, as at Fort Ternan (14 Myr old) they once more disappeared from the fossil record of western Kenya. In contrast, throughout the lower and middle Miocene periods, hominoids were abundant in western Kenya; not only were they able to live in forested areas from which monkeys seem effectively to have been excluded, but some of them were adapted to more arid conditions where they occurred side by side with monkeys.

Present day environmental analogy

What sort of environment is indicated by the presence of snail assemblage III at Maboko, Majiwa and Kaloma? The closest modern analogue that we could identify is the regional *Acacia/Commiphora* woodland of lowland eastern Kenya, locally known as the 'Nyika', especially where it is traversed by gallery forest. The Nyika is a monotonous spread of low trees and bushes interspersed with grass patches, the monotony in vegetation type being broken principally by gallery forests which fringe the drainage lines and permanent rivers. Where these galleries are very narrow (one to two trees wide) such as in the middle reaches of the Tana River in Kenya (Figure 3.3), we find

Figure 3.3. Photograph of middle reaches of the Tana river (Kenya) showing regional *Acacia–Commiphora* woodland (steppe) with gallery forest fringing the river. Vervet monkeys are predominantly restricted to the gallery forest whereas baboons are also found in the semi-arid woodland.

today two species of monkey, the vervet (*Cercopithecus aethiops*) and the baboon (*Papio anubis*). The vervets are pretty well confined to these gallery fringes, but the baboons range further afield into the true 'nyika'.

Lower down its course, where it enters its floodplain, the Tana River is fringed by a gallery forest which is up to two kilometres wide (Andrews, Groves & Horne, 1975) containing many tree species characteristic of lowland forests elsewhere in Africa. In the wide Tana gallery forest occur five species of monkey; the vervet, the baboon, the Sykes monkey (*Cercopithecus mitis*), the Tana River mangabey (*Cercocebus galeritus*) and the red colobus *Colobus badius rufomitratus*). The former two species spend a great deal of the daytime in terrestrial activity, moving into the trees at night or for safety, or for foraging during the day. The latter three species are more fully committed tree dwellers, coming down to the ground much less frequently.

Since the depositional environment of the Maboko and Nyakach strata was floodplain in its affinities (Pickford, 1985) and the regional environment seems on the basis of the snail evidence to have been semi-arid, we postulate that the local and regional palaeoenvironments might have been rather similar to the modern conditions which prevail in the lower reaches of the Tana River in eastern Kenya. We envisage that the two or perhaps more monkey species at Maboko and Nyakach may well have interacted with the environment in much the same way as the vervet and the Tana River mangabey do today.

What seems clear from the fossil evidence is that during the lower and middle Miocene, cercopithecoids were not adapted to the cooler more humid woodlands and forests of East Africa, which were occupied by an abundance of hominoid taxa of various body sizes (Pickford, 1983). Instead, they were adapted to the more marginal primate habitats found in relatively hot, semi-arid conditions. We postulate that the monkeys possibly owe their origin to the development of semi-arid woodlands in Africa during the Oligocene, and that they moved into western Kenya as already fully developed monkeys only when local conditions became semi-arid in the early part of the middle Miocene. We doubt that they evolved *in situ* in West Kenya, either from an undiscovered ancestor, or from one of the hominoid taxa.

It is unfortunate that most of the late Miocene and Plio–Pleistocene fossil sites of East Africa which have yielded fossil monkeys are not as amenable to comparable palaeoenvironmental analyses as the Miocene sites. The later sites have seldom yielded fossilised terrestrial gastropods, the only exception being Laetoli in Tanzania, which is about 3.7 Myr old. The latter site has yielded a characteristic savanna snail fauna with gallery forest species in abundance (Verdcourt, personal communication). Ngeringerowa, (about 10 Myr old) has yielded a single terrestrial gastropod in close association with the fossil jaw of a diminutive colobine, *Microcolobus tugenensis* which is the size of *Presbytis obscurus* (Benefit & Pickford, 1986). The gastropod concerned, *Bloyetia*, is characteristic of drier parts of Kenya, although some related taxa occur in forested areas. Clearly, this is not a great deal upon which to base palaeoenvironmental reconstructions.

The post-cranial evidence

Previous researches on the Maboko monkey post-cranial sample (Von Koenigswald, 1969; Delson, 1975b, 1979; Senut, 1986a, b) have indicated the likely presence of two morphological entities in the collections: one more terrestrially adapted form, the other less terrestrially adapted. One of the drawbacks of the Maboko–Nyakach monkey sample is that the post-cranial bones cannot be satisfactorily matched with the dentitions, even on the basis of size or commonality. Indeed, there is a debate about the number of species or genera represented in the abundant dental collections (Benefit & Pickford, 1986). But for the purposes of this study, it is not necessary to postulate associations between the dental remains and the post-cranial bones, and we prefer to leave that question aside for the moment until associated or articulated skeletal material has been found.

Delson (1975b) suggested that the sample of larger post-cranial bones from Maboko (a humerus, two ulnae and some robust phalanges) possibly represented a relatively terrestrial cercopithecine-like monkey of moderate size, while the sample of smaller bones (a gracile humerus and a phalanx) suggested the presence of a more arboreal, possibly colobine-like monkey (at least in its humeral morphology). Delson further postulated the possible association of the larger more cercopithecine-like teeth with the more terrestrially adapted limb bones, while for the more derived (colobine-like) teeth he suggested a possible association with the more gracile primitive post-cranial bones. While these postulated associations are possible and allow interesting scenarios to be put forward, they are by no means proven. Indeed, the possibility of arguing in circles is obvious. Nevertheless, Szalay & Delson (1979) have suggested that the existence of two morphometric groups of post-cranial bones at Maboko indicate the presence of two taxa at the site, from which they concluded that the

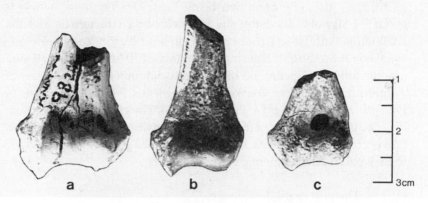

Figure 3.4. Middle Miocene cercopithecoid left distal humeri. *a*: KNM NC 9824 from Nyakach (Kenya), *b*: KNM MB 3 and *c*: KNM MB 19 from Maboko (Kenya). *a* is more terrestrial; *b* and *c* are semi-terrestrial.

dichotomy between colobines and cercopithecines could be traced back to the middle Miocene.

We agree that there are at least two varieties of post-cranial bones represented at Maboko, and that both of them represent monkeys (Figure 3.4). But we prefer to be more cautious about the question of subfamilial assignment, at least until the fossil evidence improves. In the recently augmented collections, which are still very poor by any standards, we have access to three distal humeri and two proximal ulnae in good condition, plus five fragmentary humeri and ulnae.

The larger of the post-cranial bones suggest that the animals to which they belonged were slightly smaller than males of *Presbytis rubiconda*. Morpho-functional analysis of these bones (Senut, 1986a, b) suggests that locomotion in this taxon was predominantly terrestrial, although in one of the specimens some indications of arboreality can be discerned. The distal humerus from Nyakach in particular is close to those of cercopithecines and is suggestive of a relatively long prior history of cercopithecid evolution before Nyakach times, at least for the elbow joint.

The group of smaller post-cranial bones from Maboko represents an animal about the size of females of *Cercopithecus petaurista* whilst the morpho-functional analysis indicates a mainly arboreal species in which arboreal features are overprinted over the terrestrial characters.

Both sets of post-cranial bones differ markedly from early Oligocene primate post-cranial from the Fayum deposits of Egypt, identified as

Propliopithecus and *Aegyptopithecus* (Fleagle, 1983). They are also readily distinguished from the 'so-called' monkey-like post-cranial bones of *Proconsul* and other lower Miocene hominoids of western Kenya (McHenry & Corruccini, 1975), which are in our opinion not particularly like those of Old World monkeys, but are perhaps a bit more like those of the New World monkeys (Rose, 1983). The Maboko monkey post-cranial bones are already derived towards typical Old World monkey morphology, a feature which suggests a substantial (though unknown) prior period of separation from other primate groups. It is because of their distinctiveness that we feel confident that the bones do not belong, for example, to the small-bodied apes which occur at Maboko and Majiwa (Pickford, 1982; Harrison, 1982, 1986).

It is our opinion that both types of monkey post-cranial bones at Maboko each possess characters which represent adaptations towards terrestrial locomotion; but that the group of larger specimens indicates a taxon which is more fully derived in this direction, while the smaller specimens represent a species less committed to terrestrial life, and therefore retaining morphological features indicative of arboreal locomotion. Both taxa however, possess more morpho-functional features suggestive of terrestriality than do the post-cranial bones assigned to *Aegyptopithecus* from Egypt.

Although there are two types of cercopithecoid post-cranial skeleton at Maboko, we feel that the evidence they afford is not complete enough, nor compelling enough for us to postulate, as others have done, that the dichotomy between colobines and cercopithecines can be demonstrated to have taken place prior to Maboko times. However, we might add that the Nyakach distal humerus of equivalent age appears to be fully derived in the direction of Cercopithecinae and it might ultimately afford evidence for the existence of this subfamily in the strata.

Upper Miocene to Plio–Pleistocene monkeys

The earliest evidence we have for the existence of fully derived colobine dentitions is provided by a small colobine *Microcolobus tugenensis* Benefit & Pickford, 1986 from Ngeringerowa, Kenya (Figures 3.1, 3.2) aged about 10 Myr old, and probably significantly older than the colobine *Mesopithecus pentilici* from upper Miocene deposits of Eurasia (Szalay & Delson, 1979) although it is probably a contemporary of the Wissberg colobine from early upper Miocene strata in Germany (Tobien, 1986).

The earliest evidence from East Africa for fully derived cerco-pithecines (discounting the Maboko sample, about which there is debate) is from upper Miocene deposits in Kenya (Mpesida and Lukeino (Pickford, 1975), and Lothagam (Szalay & Delson, 1979)) dated about 7 to 5 Myr old (Figure 3.1). Early cercopithecines have also been recorded from North Africa (Arambourg, 1959) at Menacer, Algeria, possibly about 7 Myr old. However, the significance of these dates is not great, since there are many gaps in the fossil record and the absence of monkey fossils during the period 15 to 7 Myr ago is undoubtedly due in great part to imperfections of the fossil record, rather than to real absence.

By latest Miocene times and during the Plio–Pleistocene both col-obines and cercopithecines underwent radiations an order of mag-nitude greater than anything before (Figure 3.1). Many of the Plio–Pleistocene species are not represented by post-cranial elements, so it is difficult to say anything of value concerning their locomotor adapta-tions. But of the species for which the locomotor patterns can reason-ably be deduced, the majority appear to have been terrestrial. Out of 25 species for which the locomotor repertoire can be estimated, two were arboreal, two were semi-arboreal and 21 were probably terrestrial. These figures suggest to us that the terrestrial adaptation is the basic cercopithecoid one, and that arboreality is a relatively recent departure. Conversely, the fossil record could be playing one of its tricks on us, especially as the bulk of our samples come from eastern and southern Africa, which were probably more openly vegetated than West-Central Africa where no fossils of significance to our investigation have been found (Hooijer, 1963, records a single monkey tooth from Ongoliba, (Zaïre)).

Conclusions

There has been a long-standing debate concerning the initial radiation of the cercopithecoid monkeys. Most of the debaters have argued from the basis of adaptations observed in extant Old World monkeys, and have projected their scenarios backwards into the fossil record, some authors favouring initial radiation as far back as the late Eocene deposits of the Fayum (Egypt), although we see no apomorphic cercopithecoid characters in any of the Fayum primates. There is little question that the Miocene monkeys found at Maboko, Nyakach, Buluk and Loperot in Kenya, and Wadi Moghara and Gebel Zelten in North Africa, are derived cercopithecoids, both dentally, and where known, post-cranially. There do remain, however, questions about their status within the superfamily, with some authors placing these early and middle Miocene monkeys in extant subfamilies, while others, with whom we tend to agree, place them in a subfamily of their own, the Victoriapithecinae. Accordingly, we suggest that the initial radiation of the cercopithecoids took place prior to Napak and Maboko–Buluk times (about 17–20 Myr ago) and probably later than Fayum time (about 31–32 Myr ago). Unfortunately, Oligocene strata are poorly represented in Africa, and the chances seem remote that we will locate representatives of these very early monkeys. In this respect, recognition that the site of Malembe, Angola is early Oligocene in age is of some interest, especially since it has yielded primates among other terrestrial mammals (Pickford, 1986).

An observation which surely needs examination concerns the virtually complete absence of monkey fossils from lower Miocene strata in East Africa. Up to now a single upper molar, a canine and a radio-ulna (Pickford, *et al.*, 1986) have been reported from a single site at Napak (Uganda). This paucity of monkeys seems not to be due to sampling error: all the lower Miocene strata of East Africa have yielded abundant fossils including numerous higher primates (Pickford, 1986). Middle Miocene strata of East Africa, in strong contrast, have yielded abundant monkey fossils of at least two species if not genera.

Can we explain this distribution in terms of autochthonous evolution of monkeys from one or other of the lower Miocene higher primates from the same region? We consider that the answer to this question must be 'no', considering the morphological distance between *Victoriapithecus* and all lower Miocene primates from West Kenya. We consider it to be more likely that cercopithecoids evolved elsewhere in Africa and subsequently migrated to western Kenya as fully derived monkeys, when environmental conditions in West Kenya became suitable for them.

Palaeoenvironmental reconstructions on the basis of terrestrial gastropods indicate that all the lower Miocene sites in western Kenya, which are devoid of monkey fossils, were forested or heavily wooded. In sharp contrast, the middle Miocene sites which yield abundant monkey fossils appear, on the basis of the gastropods, to have been semi-arid, hot and seasonal.

We consider it likely, therefore, that the primary cercopithecoid adaptation was for life in hot, semi-arid, seasonally variable habitats (Figure 3.5), and that during the lower Miocene they were effectively excluded from the more humid, better vegetated regions of Africa for a variety of reasons, perhaps including competition from the abundant

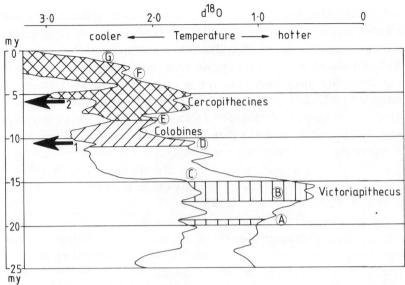

Figure 3.5. Concordance between cercopithecoid history and a δO^{18} curve for benthic foraminifera of the Atlantic and Pacific oceans. The earliest known cercopithecoids are about 19–20 million years old (A) but they do not become common as fossils in East Africa until the Maboko Event (B) a marked period of aridity in Western Kenya and elsewhere. Cercopithecoids disappear from the East African record during the mid-Miocene cooling plunge (C). Several million years later (D) the earliest colobines appear in East Africa. Even later in time (E) the earliest known cercopithecines make their appearance. Colobines and cercopithecines undergo a major radiation at the end of the Messinian Crisis (F) and the *Cercopithecus* group begin their major radiation, especially in forested environments, during the Pleistocene (G). Colobines (1) and cercopithecines (2) emigrated to Eurasia during two late Miocene cooling events, the first about 10–11 Myr ago and the second about 6–5 Myr ago.

and diverse hominoid fauna. Aspects of the elbow joint of two known varieties of middle Miocene monkeys indicate that terrestriality was important in one taxon, and less important in the other. Such diversity in locomotor adaptations might be expected in semi-arid regions with gallery forests, where the ability to cross open ground would be a common if not an essential aspect of daily activity.

In the large sample of Plio–Pleistocene monkeys from eastern and southern Africa, it seems that the majority of species was terrestrial, but a few fully arboreal and semi-arboreal species are known from deposits as old as 3.5 Myr.

If our arguments are correct, or nearly so, then we would argue that the present-day occupation of forests and humid woodlands by a great diversity of cercopithecines and colobines results from a radiation into the forests subsequent to the middle Miocene, and probably as late as the early Pliocene period five million years ago. We feel that evolutionary scenarios concerning forest adaptations in cercopithecines should probably be formulated within a time-frame of less than six million years. For colobines, we feel that the forest adaptations are perhaps more ancient, possibly as old as 10 or 11 million years.

Acknowledgements

We are anxious to thank the editors for inviting us to participate in the symposium entitled 'Biologie, Phylogénie et Spéciation chez les Cercopithecinae forestiers africains'.

We thank Prof. P. Taquet (Institut de Paléontologie, Paris), Prof. J. Biegert and Dr P. Schmid (Anthropologisches Institut der Universität Zürich), and R. E. F. Leakey (Head of the National Museums of Kenya, Nairobi) for permission to study materials under their care.

Funds were kindly provided by the Chaire d'Anthropologie et de Préhistoire du Collège de France (Prof. Y. Coppens), and the L.S.B. Leakey Foundation.

I.4

Classification and geographical distribution of guenons: a review

JEAN-MARC LERNOULD

Introduction

The guenons are difficult to classify, as testified by the number of classifications so far proposed and by the number of synonyms for many taxa (Hill, 1966; Napier & Napier, 1967; Dandelot, 1971; Kingdon, 1980; Napier, 1981). In this chapter, I will not so much make another attempt to propose a new classification, as try to improve the existing ones and to clarify the geographical distribution of many species, on the basis of new published and unpublished information and by taking into consideration the proposals of phylogenetic relationships between guenons based on genetic, anatomical and behavioural criteria (see Dutrillaux, Muleris & Couturier, Chapter 9; Ruvolo, Chapter 7; Martin & MacLarnon, Chapter 10; Kingdon, Chapters 11, 13; Gautier, Chapter 12).

In the following pages, I endeavour to use the hierarchy of taxonomic categories defined by Mayr (1969, 1974): genus, subgenus, super-species, species, subspecies and local populations. However, species and subspecies are living entities still subject to natural selection, and the taxonomist may be confronted with such intermediate categories as 'incipient' species or 'semi-species' or 'clusters' of related forms, all of which suggest that a group is still in an active stage of speciation. In addition, classifying some specimens within any scheme remains difficult for many reasons: old faded study skins, specimens in poor condition, subadult specimens wrongly labelled as adults, cases of melanism, erythrism or flavism, and even hybrids. This last phenomenon is not uncommon among guenons, as shown by the great number of hybrid guenons recorded in the literature (Appendix 4.1);

hence the possibility that some unique type specimens of very scarce 'species' are merely hybrids.

The geographical distribution of the various forms is often not easy to establish either, as many parts of Africa have not been well explored. Museum specimens may be sources of errors: many old skins mention only the country of origin, place names have often changed and more than one collecting locality in a given country can bear the same name. Furthermore, many early travellers brought back specimens with a single mention of the names of villages where they were obtained, although this does not mean that the animals were living in their neighbourhood.

The habitat preferences of a species are even more difficult to establish than its range, without careful field study; accurate information was seldom recorded on old museum labels; furthermore, forest destruction has been proceeding at a great rate during the last 50 years or so in Africa, and it is often difficult to infer, from present day conditions, the environmental conditions prevailing in a given locality some decades ago.

Classification and geographical distribution

Modern taxonomists agree that guenons are a very distinctive group of species among Old World monkeys, and Hill (1966) restricts the subfamily Cercopithecinae to the guenons only, the mangabeys, baboons and macaques being grouped together into the subfamily Papioninae. Conversely, Napier & Napier (1967), Dandelot (1971) and Napier (1981) group all these monkeys together in the subfamily Cercopithecinae, made of two tribes: Cercopithecini and Papionini. However, Dutrillaux *et al.*'s finding (1982) that Cercopithecinae and Papioninae must share a common ancestor means that one cannot consider any living species of these two subfamilies as an ancestor to the others, although the frequency of hybridization between captive guenons, macaques and mangabeys (Appendix 4.1) indicates a close relationship between them.

Hill (1966) recognizes four genera of guenons: *Cercopithecus, Miopithecus, Allenopithecus* and *Erythrocebus*. Napier & Napier (1967) retain only two genera: *Cercopithecus* with *Cercopithecus, Miopithecus, Allenopithecus* as subgenera, and *Erythrocebus*; Napier (1981) then reverts to the opinion of Hill. Dandelot (1971) considers that all guenons belong to genus *Cercopithecus*. If one considers the potential for interbreeding in guenons and their karyological phylogeny, it

Table 4.1. *Classification of guenons adopted in this chapter*

Genus	Subgenus	Super-species	Species	Subspecies
Cercopithecus	Allenopithecus	—	nigroviridis	
	Miopithecus	talapoin	talapoin	
			(?)	
	Erythrocebus	—	patas	baumstarki, patas, pyrrhonotus, villiersi
	Cercopithecus	diana?	diana	diana, roloway
			dryas	
			salongo	
		aethiops	aethiops	aethiops, djamdjamensis, hilgerti, matschiei
			pygerythrus	arenarius, callidus, centralis, cynosuros, excubitor, helvescens, johnstoni, marjoriae, nesiotes, ngamiensis, pygerythrus, rubella, rufoviridis, zavattarii
			sabaeus	
			tantalus	budgetti, marrensis, tantalus
		lhoesti	lhoesti	
			preussi	insularis, preussi
			solatus	
		—	neglectus	
		—	hamlyni	hamlyni, kahuziensis[1]
		cephus	ascanius	ascanius, atrinasus, katangae, schmidti, whitesidei
			cephus	cephus, cephodes
			erythrogaster	
			erythrotis	erythrotis, camerunensis, sclateri?
			petaurista	buettikoferi, petaurista
			sclateri?	
		mona	campbelli	campbelli, lowei
			mona	
			pogonias	grayi, nigripes, pogonias
			wolfi	denti, elegans, pyrogaster, wolfi
		nictitans	nictitans	nictitans, martini, stampflii
			mitis	boutourlinii, doggetti, heymansi, kandti, mitis, opisthotictus, schoutedeni, stuhlmanni
			albogularis	albogularis, albotorquatus, erythrarchus, francescae, kinobotensis, kolbi, labiatus, moloneyi, monoïdes, nyasae, phylax, zammaranoi

becomes reasonable to retain only a single genus, *Cercopithecus*, including *Allenopithecus*, *Miopithecus* and *Erythrocebus* as subgenera in order to emphasize the morphological and/or physiological features of these monkeys which are now considered as closer to the ancestral guenons than the 'true' *Cercopithecus*.

Previous classifications have also tried to group the most closely related species by using the concepts of superspecies or species-group, which bring together all borderline cases such as semi-species or incipient species. Following Hill (1966), the sequence of species in this chapter is based on karyological criteria. Data refer to the work of

Dutrillaux and his co-workers and species are ranked on the basis of their increasing number of chromosomes, those with the smallest numbers being assumed to be the most primitive (Dutrillaux, Couturier & Chauvier, 1980; Dutrillaux *et al.*, 1982; Muleris *et al.*, 1985; Dutrillaux *et al.*, Chapter 9). When several species of a superspecies share a common karyotype, the sequence follows the alphabetical order of the specific name. The name retained for the superspecies is the name of the species first described. The rank order for subspecies is alphabetical. Table 4.1 summarizes the classification retained. Localities' and rivers' names used in the text are mapped on Appendix 4.2.

Cercopithecus nigroviridis (Figure 4.1)

The genus *Allenopithecus* was established by Lang (in Hill, 1966), but Verheyen (1962) and Dandelot (1971) kept it only as a subgenus of *Cercopithecus*. Whereas the presence of *C. nigroviridis* is certain along the Zaïre River and many of its tributaries, the limits of its range remain poorly known. Colyn's observations (Chapter 6) clarified its distribution in the Zaïre basin, while Gautier (1985) extended its southern limit to 6°30′ S, from the report of a local hunter.

Cercopithecus talapoin, Northern talapoin (undescribed form Figure 4.1)

I. Geoffroy St Hilaire established the genus *Miopithecus* for the talapoin on the basis of a dental character which was later proven to be incorrect (Hill, 1966). Its small size, proportionally larger head with an infantile type of mandible and a shorter muzzle, relatively large scrotum, and cyclical sexual swellings in females, are the external characters which set this monkey apart from other guenons (Hill, 1966).

Four subspecies were previously recognized: *talapoin*, north of the Zaïre River, in Congo, Gabon, and Cameroon; *ansorgei* in Angola; *vleeschouwersi* in southwestern Zaïre, and *pilettei* from Ruwenzori. However, Machado (1969) followed by Dandelot (1971) concluded that there were only two forms: the first, which includes *ansorgei, pilettei* and *vleeschouwersi*, must be called *C. talapoin* because the original description was that of an Angolan specimen; the other corresponds to the monkeys living north of the Zaïre River which Machado considers specifically distinct: this northern talapoin, so called by Napier (1981), has not yet been given a scientific name, although it is present throughout the forested part of southern Cameroon, even north of the Sanaga River (Gartlan, in Wolfheim, 1983), in Equatorial Guinea, in Gabon, and in the west, north and northeast parts of the Congo

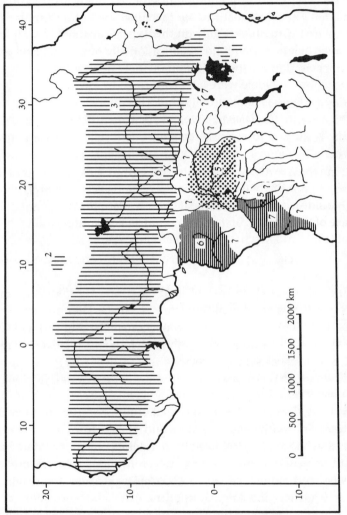

Figure 4.1. Distribution of *Cercopithecus patas patas* (1), *C. p. villiersi* (2), *C. p. pyrrhonotus* (3), *C. p. baumstarcki* (4), *C. nigroviridis* (5), northern talapoin (6), *C. talapoin*, southern talapoin (7).

(Malbrant & Maclatchy, 1949). Whether the northern talapoin and *C. nigroviridis* are sympatric or not between the Sangha and Ubangui Rivers (Central African Republic, CAR) is still undecided. Blancou (1958) saw one talapoin in the gallery forest of the upper Uaka River; however, neither Carroll (1986) nor Galat (1977) found the species in the southwest of this country, although pygmies knew it from further south (G. Galat & A. Galat-Luong, personal communication). The presence of the talapoin, not only on the upper Uaka, but also farther north, nevertheless remains possible as Blancou was a very reliable observer, whose unpublished information (field notes) on the presence of other guenons at Ndélé were later confirmed by Fay (1985).

Machado (1969) has shown that the northernmost localities for the southern talapoin (*C. talapoin*) were very close to those where *C. nigroviridis* is known to exist. Moreover, one locality (Panzi) is near those where *C. nigroviridis* could also be present (Kasongo-Lunda: Gautier, 1985). Was there any confusion between *C. talapoin* and *C. nigroviridis*, or are the two species actually sympatric in some areas, as already considered possible for the northern talapoin?

The case of the Ruwenzori talapoin, known from a single skin which Dandelot (1971) and Machado consider identical to *C. talapoin*, remains in doubt.

Cercopithecus patas (Figure 4.1)

This monkey ranges from Senegal to western Ethiopia (but not to Somalia: Funaioli & Simonetta, 1966), and extends southwards to northern Tanzania, in open wooded steppes and savannas. It is also able to live in rocky habitats, as in the Aïr mountains of southern Sahara (Dekeyser, in Hill, 1966).

The genus *Erythrocebus* was established on the basis of a number of morphological characters by Trouessart (in Hill, 1966), but Verheyen (1962) considered it as congeneric with *Cercopithecus* on the basis of their skull similarities. Among recent classifications, only Dandelot (1971) considered *Erythrocebus* as a subgenus of *Cercopithecus*.

Up to 14 forms of patas were described until 1912. Since Schwarz's revision, taxonomists have agreed to consider only one species, with three or four subspecies (Hill, 1966). Taking into account the great variability of the coat of captive animals, Dandelot (1971, see also Loy in Napier, 1981) agreed with Hill, but stressed the need for a taxonomic revision of this species.

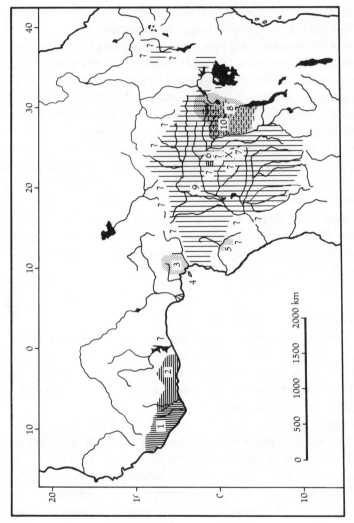

Figure 4.2. Distributions of *C. diana diana* (1), *C. d. roloway* (2), *C. preussi preussi* (3), *C. p. insularis* (4), *C. solatus* (5), *C. salongo* (6), *C. dryas* (7), *C. lhoesti* (8), *C. neglectus* (9), *C. hamlyni* (10).

Cercopithecus diana, C. dryas, C. salongo (Figure 4.2)

These species are here treated together, since Hill (1966) considered *dryas* as a subspecies of *C. diana*, and Thys van den Audenaerde (1977) treated *salongo* as a Central African representative of *C. diana*. In fact so little is known of *C. dryas* and *C. salongo* that it is presently impossible to have a clear idea of their true taxonomic status.

The distribution of *C. diana* is discussed by Oates (Chapter 5). Known from Sierra Leone to Ghana, its presence in Guinea is documented in the southeast of the country (Roche, 1971), and it is also found up to an altitude of 1400 m on the northern side of Mt Nimba, following gallery forests upstream with *C. petaurista* and *C. campbelli* (G. Galat & A. Galat-Luong, personal communication). Its occurrence in southern Cameroon (Jeannin, 1936) is doubtful while it should be present in Togo (personal information).

C. dryas is still known by the single skin of a young individual collected in central Zaïre (Hill, 1966). Its exact geographic origin is, however, doubtful (Colyn, Chapter 6). *C. salongo* is known only from the region of Djolu (Thys van den Audenaerde, 1977; Kuroda, Kano & Muhindo, 1985; Colyn, Chapter 6) about 300 km north of the type locality of *C. dryas*.

The patterns of the holotypes of *C. dryas* and *C. salongo* are different but we cannot entirely reject the possibility of *dryas* being the juvenile pelage of *salongo*, as Thys van den Audenaerde did, especially if one compares the faces of the two guenons (Schouteden, 1947; Kuroda *et al.*, 1985). The opinion of Kuroda *et al.* remains ambiguous since they found it 'almost impossible to include them (*salongo* and *diana*) in the same superspecies as Hill did', although only *dryas* and *diana* were known to Hill at that time. Does it mean these authors consider *dryas* and *salongo* as belonging to the same species?

Cercopithecus aethiops, C. pygerythrus, C. sabaeus, C. tantalus (Figure 4.3)

The classification of the savanna guenons of the *aethiops* group is very complex and the habit of field primatologists often speaking of '*aethiops*' without more precision does not help to clarify the situation. My personal observations in Ethiopia (1967–9) exemplify the complexity of the situation. Both grivets (*C. aethiops, sensu stricto*) and vervets (*C. pygerythrus*) were found there and were easily distinguished (see Dandelot, 1959; Hill, 1966). Among grivets, differences in coat colour were obvious: some were pale grey and pale olive dorsally (*aethiops*), whereas others were fawn, more or less russet (*matschiei*), or dark olive

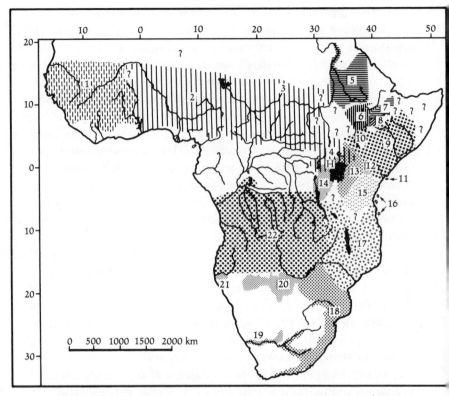

Figure 4.3. Distributions of *C. sabaeus* (1), *C. tantalus tantalus* (2), *C. t. marrensis* (3), *C. t. budgetti* (4), *C. aethiops aethiops* (5), *C. a. matschiei* (6), *C. a. hilgerti* (7), *C. a. djamdjamensis* (8), *C. pygerythrus arenarius* (9), *C. p. zavattarii* (10), *C. p. excubitor* (11), *C. p. rubella* (12), *C. p. callidus* (13), *C. p. centralis* (14), *C. p. johnstoni* (15), *C. p. nesiotes* (16), *C. p. rufoviridis* (17), *C. p. pygerythrus* (18), *C. p. marjoriae* (19), *C. p. ngamiensis* (20), *C. p. helvescens* (21), *C. p. cynosuros* (22), hybridization area (H).

green (*hilgerti*). A high altitude form (*djamdjamensis*), close to *matschiei* by its coat colour (but with reduced and not falciform whiskers and a shorter, black-tipped tail), was also 'rediscovered' (see Dandelot & Prevost, 1972). This latter form has nothing in common with *C. pygerythrus*, as supposed by Napier (1981).

From this personal field experience I consider that the classification proposed by Dandelot (1959), who studied museum specimens, observed captive animals and collected many photographs of living monkeys, is still the most satisfactory. Dandelot concluded that the various forms of savanna guenons could not be considered just as subspecies of *C. aethiops* as Schwarz did (1926). He used the concept of superspecies to indicate the close relationships between the three species *sabaeus, aethiops* and *pygerythrus*, the two latter including a number of subspecies. Hill (1966), Napier & Napier (1967) followed him, but later Dandelot (1971) subsequently raised to specific status the *tantalus* monkey, previously considered by him as a subspecies of *aethiops*. More recently, Napier (1981) reconsidered her position and regrouped all 'green' monkeys into a single species, with four subspecies corresponding to Dandelot's species.

The great affinities among the *aethiops* forms which are able to interbreed (this is the case for *tantalus* and *pygerythrus* in Uganda; Dandelot, 1959, 1965, and possibly for *matschiei* and *pygerythrus* in Ethiopia: Dandelot & Prevost, 1972) could lead to considering them as semi-species, all the more so since all have the same karyotype. However, the four forms are really closer to species than to subspecies on the basis of coat patterns, and also because three of them have their own subspecies, *sensu* Mayr (1969). Within such subspecies, there are also variations in coat colour (as in *matschiei*, personal observation) in relation to altitude or location.

In Ghana, *C. sabaeus* is well separated from *C. tantalus* by the Volta River (Booth, 1956b) but, further north, we do not know whether the two species hybridize or not. *C. tantalus* has a broader distribution from Ghana to Sudan, Uganda, and possibly Kenya. It is not known whether *C. t. marrensis* from Jebel Marra is clearly isolated from *C. t. tantalus*; whether there is a clear limit, and where, between the latter and *C. t. budgetti*; nor whether the Nile River is a barrier to dispersal between *tantalus* and *aethiops* in central Sudan. In the south, *C. t. budgetti* exists east of the Nile (Dandelot, 1959); it is present north and northeast of the Zaïre rainforest, and even enters it (Schouteden, 1947; Dandelot, 1959). Schouteden also indicates some localities along the Lualaba River, but M. Colyn (personal communication) was unable to confirm the existence of the species in this area.

Along the Nile River, in northern Sudan, *C. aethiops* reaches the extreme northern extension of all guenons, but it has now disappeared from many localities following deforestation (Kock, 1969). We do not know how far north *C. a. aethiops* ranges in central Sudan and whether *C. a. matschiei* enters into southeastern Sudan. Similarly, the subspecies distribution in Ethiopia remains poorly known. In the southeast, *C. pygerythrus* (probably *arenarius*, personal observation) follows the Webi Shebele, the Ganale River and their tributaries upstream. Two specimens were obtained South of Goba (Napier, 1981) and monkeys were observed 40 km east of this town, at 2000 m a.s.l. (personal observation). On the southern slopes of Harrar Mountains, it is very likely that *C. pygerythrus* meets *C. a. hilgerti*. In the Rift Valley, I saw vervets along the western shore of Lake Abaya, and specimens collected there were considered as possible hybrids with *C. a. matschiei* (Dandelot & Prevost, 1972). Conversely, Napier (1981) attributes specimens collected in the same region to *C. p. zavattarii*, a subspecies localised in the southwest of Ethiopia.

From Uganda to Cape Province and to Angola, a number of other subspecies of *C. pygerythrus* have been described whose distribution is not always known and would require a thorough study. The list retained in this chapter is based on Dandelot (1971) and Napier (1981). The malbrouck monkey (*C. p. cynosuros*) is easily distinguished from the vervet, and Dandelot (1971) feels that these two forms are probably sympatric at Victoria Falls. However, A. de B. Machado (personal communication) considers the *cynosuros* he saw there were not typical.

Cercopithecus lhoesti, C. preussi, C. solatus (Figure 4.2)

Dandelot (1971) considers *C. preussi* as to be a subspecies of *C. lhoesti*, while Eisentraut (1973) feels that *preussi* is probably more closely related to the *C. albogularis* group than to *C. lhoesti*. The recent study of the karyotype of a female *C. preussi* which was found identical to that of *C. lhoesti* (Dutrillaux *et al.*, Chapter 9) confirms the relationship between these two taxa. In 1984, a new species *solatus*, apparently related to *lhoesti* and *preussi*, was discovered in Gabon (Forêt des Abeilles), by Harrison (1984b, 1988). Its karyotype differs from those of *lhoesti* and *preussi* although it shares the same number of chromosomes.

C. p. preussi lives in a restricted area on the mountains of western Cameroon and in the surrounding lowland forests, and it even reaches Nigeria further north (Oates, Chapter 5). *C. p. insularis*, from Bioko, is very similar to the nominate form. This is probably due to its recent isolation, the island being only *c.* 11 000 years old (Moreau, 1966).

C. lhoesti is a mountain monkey reaching the bamboo zone in Uganda, Rwanda, Burundi, and eastern Zaïre, but its range extends to lowland rain forests in Zaïre (Colyn, Chapter 6). Its extension further south is probably limited to large forest blocks.

C. solatus is mainly found on the left bank of the Ogoué River (Gautier, Loireau & Moysan, 1986); the southern limits of its range remain to be clarified.

Cercopithecus neglectus (Figure 4.2)

The de Brazza monkey differs from other guenons by its general aspect, its age-related pelage pattern (only *C. hamlyni* and to a lesser extent *C. lhoesti* share this peculiarity) and its ecological and behavioural characteristics (Gautier-Hion & Gautier, 1978). The distribution of this forest monkey, which is mostly found along watercourses, is very large and no subspecies have been described. In CAR, it exists in the south (Blancou, 1958; Galat, 1977, 1978) and in the extreme north (Fay, 1985) where the gallery forests reach their northernmost limits. Thus it can be expected almost everywhere in the riverine forest of the Zaïre and Chad basins. It could even reach the Sudan border.

In Gabon, *C. neglectus* is presently restricted to the northeast (Malbrant & Maclatchy, 1949; A. Gautier-Hion, personal communication), although an old skin has been reported from the coast (Napier, 1981).

Cercopithecus hamlyni (Figure 4.2)

C. hamlyni has no obvious affinity with any other guenon species; however, coat colour modification with aging and many of its vocalizations (Gautier, Chapter 12) show relationships with *C. neglectus*. Its distribution is very close to that of *C. lhoesti*.[1]

Cercopithecus ascanius, C. cephus, C. erythrogaster, C. erythrotis, C. petaurista (Figure 4.4)

These species are grouped together here, as by Napier (1981), because they have a broadly equivalent ecological niche and replace each other geographically. They share the same karyotype.

C. petaurista

This species is mostly restricted to the Upper Guinea forest block. To the north, its range within the savanna–forest mosaic is not well documented while to the east, it is present in the forest patches of Benin within the Dahomey Gap (see Oates, Chapter 5). Two subspecies are recognized: *petaurista* in the east and *buettikoferi* in the west. Individuals of both forms as well as intermediates have been collected in Ivory Coast, between the Nzo, Sassandra and Cavally rivers (Booth, 1956a).

C. erythrogaster

Until recently (Oates, 1982, 1985) little information was available on this form, which was supposed to vary greatly in some of its coat features. Oates (1985) has shown that *C. nictitans signatus* must not be considered as a synonym of *C. erythrogaster* (Hill, 1966) and concluded that the latter was a valid species within the *cephus* group. Its known range is limited to southern Nigeria, west of the Niger Delta,

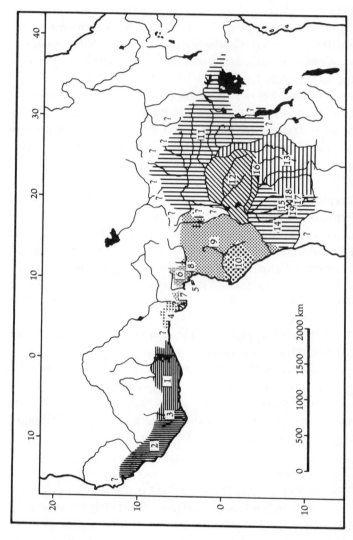

Figure 4.4. Distributions of *C. petaurista petaurista* (1) *C. p. buettikoferi* (2), *C. petaurista*, transition area (3), *C. erythrogaster* (4), *C. erythrotis erythrotis* (5), *C. e. camerunensis* (6), *C. e. sclateri* (7), hybridization area: *C. e. camerunensis × C. cephus cephus* (8), *C. c. cephus* (9), *C. c. cephodes* (10), *C. ascanius schmidti* (11), *C. a. whitesidei* (12), *C. a. katangae* (13), *C. a. ascanius* (14), *C. a. atrinasus* (15), transition area between *C. a. whitesidei-katangae* (16), *C. a. katangae-ascanius* (17), *C. a. katangae-atrinasus* (18), *C. a. ascanius-atrinasus* (19).

but it may extend east of the delta and reach Benin in the west (Sayer & Green, 1984; Oates, Chapter 5) and eventually Togo (personal information).

C. erythrotis

C. e. erythrotis is restricted to Bioko Island and is very close to *C. e. camerunensis* which lives on the continent between the Cross and the Sanaga Rivers. However, one skin of the latter form was collected south of the Sanaga (Napier, 1981), and not far from this area, Struhsaker (1970) recorded intermediates between *C. erythrotis* and *C. cephus* that led him to consider these two forms as conspecific. *C. e. sclateri* is found between the Niger Delta and the Cross River. It is very different from the two other subspecies and has been considered as a full species by Kingdon (1980). Oates (1985) has started to investigate the case of *sclateri* but more data are needed to clarify its taxonomic status. If *sclateri* is not a hybrid between *erythrogaster* and *erythrotis*, then it must be one of the most endangered species in Africa.

C. cephus

Two subspecies are known; *C. c. cephodes* is found in southwest Gabon and adjacent Congo, and is completely encircled by *C. c. cephus* from which it differs only by a grey (not russet) tail (Malbrant & Maclatchy, 1949). In Cameroon, *C. cephus* is separated from *C. erythrotis* by the lower course of the Sanaga River which is crossed by the latter species but not by *cephus* except in its upper course (Oates, Chapter 5). The latter probably does not stray very far from the forest block, only entering the gallery forests in central Cameroon and southwest CAR where it meets with *C. ascanius schmidti*. In the extreme southwest of the CAR, the presence of *C. cephus* west of the Sangha River is confirmed by recent sightings (Galat, 1977; R. W. Carroll, personal communication) and it is also proven east of this river (Galat, 1977; Napier, 1981, in Bambio; J. M. Fay, personal communication). However, Fay found that *C. ascanius* was more common than *C. cephus* which, apparently, drops out further east. Fay proves that there is an area of sympatry between *C. c. cephus* and *C. a. schmidti*. While Malbrant & Maclatchy (1949) stated that *cephus* was present everywhere in the Congo, and did not mention the presence of *ascanius*, the observations of Galat (1977), Carroll (1986) and Fay (personal communication) suggest that *C. a. schmidti* is present in the northeastern part of this country and that the area of sympatry with *C. cephus* extends in Congo.

South of its junction with the Sangha, the Zaïre River limits the extension of *C. cephus* to the east, but the lower course of this river is no longer a barrier as this guenon is found in northern Angola (Machado, 1969). Contact or even sympatry between *C. c. cephus* and *C. a. ascanius* would be possible in this area.

C. ascanius

Following the discovery of *C. a. atrinasus* (Machado, 1965), taxonomists consider five subspecies: *ascanius, katangae, whitesidei, schmidti* and *atrinasus*. The subspecies *montanus* from the mountains between Lake Tanganyika and the Lualaba River (Hill, 1966) is now considered as a synonym of *schmidti* (Dandelot, 1971; Napier & Napier, 1967).

The northernmost subspecies *schmidti* is the only one without a temporal whorl; it is separated from the others by the Zaïre and Lualaba rivers. We do not know how far it extends to southeast Zaïre. To the east, it almost reaches the Rift Valley, but the northeastern, northern and western limits are poorly known. Kock (1969) recorded it in the extreme southwest of Sudan. In CAR, *C. a. schmidti* is found almost everywhere in riverine forest east of the Sangha (Blancou, 1958; Galat, 1977, 1978; Fay, 1985; Carroll, 1986). Carroll (personal communication) observed it on both sides of this river. However, M. Harrison (personal communication) never saw or heard of *C. ascanius* in Cameroon about 100 km west of the Sangha. *C. a. schmidti* is probably also present in northeast Congo, where it meets *C. c. cephus*. The *C. ascanius* reported from Cameroon by Jeannin (1936) is clearly a *C. erythrotis*.

C. a. whitesidei is included inside the bend of the Lualaba and Zaïre rivers, up to the Kasai and Sankuru rivers; south of the Zaïre and Kasai lives *C. a. ascanius* while *C. a. katangae* is found south of the Sankuru and west of the Lualaba (Schouteden, 1947). In Angola, *C. a. katangae* reaches the Cuilo River, and to the west, intermediates with *C. a. ascanius* are known (Machado, 1965, 1969). In that region, the localized and very distinct *atrinasus* form is found along the Wamba River and its tributaries. Intermediates between the latter form and the neighbouring *ascanius* and *katangae* exist. It is possible that either *atrinasus* or intermediate populations enter Zaïre along the Wamba, as I have seen a young male bought in Kinshasa very similar to the intermediate individual figured by Machado (1969, p. 129).

Machado considered the existence of a transitional population between *katangae* and *whitesidei*, localised between the Lukenie and

Sankuru Rivers, while Colyn (Chapter 6) found that specimens collected between the Lualaba and Lomani Rivers were somewhat intermediate between *whitesidei* and *katangae*. Indeed, a specimen photographed by J.-P. Gautier near Mbandaka in the recognized range of *whitesidei*, as well as a good number of captive individuals I have seen in Europe, possess a deep orange nose-patch which differs from the buff or yellowish nose-patch of *whitesidei*, and a head profile and face which differ from those of *katangae* in the importance of the temporal whorl. It seems necessary, therefore, to reconsider the taxonomic position of the *whitesidei* and *katangae* populations.

Cercopithecus campbelli, C. mona, C. pogonias, C. wolfi (Figure 4.5)

All taxonomists group these monkeys in the superspecies *mona*. I follow Dandelot (1971) who considered *denti* a subspecies of *wolfi*. These species replace each other from Senegal to Uganda following a different pattern from that observed for the *cephus* group. Only *C. campbelli* parallels *C. petaurista* with its two subspecies (*campbelli* and *lowei*) replacing each other in the area between the Cavally and Sassandra rivers where intermediate individuals are known (see Booth, 1956a; Oates, Chapter 5).

C. mona

Without any subspecies, it ranges from Ghana to Cameroon, unrestricted by the Dahomey Gap or the Sanaga River, and follows gallery forest far from the rain forest itself. There is a small zone of sympatry with *C. c. lowei* in Ghana and a wide area of sympatry with *C. pogonias* in Cameroon.

C. pogonias

C. p. pogonias is found from southwest Nigeria (Wolfheim, 1983) down to Equatorial Guinea, including Bioko where it is supposedly shorter-tailed (Eisentraut, 1973). *C. p. nigripes* occurs from northern Equatorial Guinea to southern Congo. According to Malbrant & Maclatchy (1949), it occurs throughout Gabon and western Congo, although Dandelot (1971) considers it restricted to western Gabon, and Haltenorth & Diller (1985) give its range as south of the Ogoué River. However, monkeys from northeastern Gabon as well as another individual from Franceville are referable to *nigripes* (J.-P. Gautier, personal communication documented by photographs). Further east, in Congo, the limit between *nigripes* and *grayi* is also imprecisely known. Dandelot (1971) notes the possible existence of an unnamed

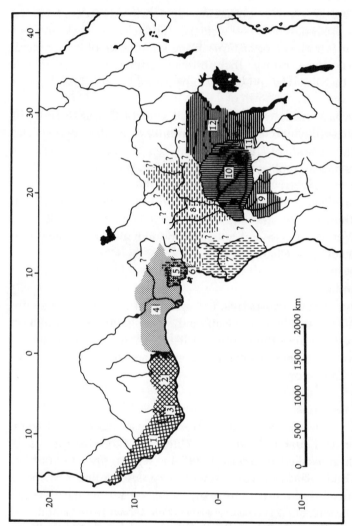

Figure 4.5. Distributions of *C. campbelli campbelli* (1), *C. c. lowei* (2), *C. campbelli*, transition area (3), *C. mona* (4), *C. pogonias pogonias* (5), *C. pogonias* ssp.? (6), *C. p. nigripes* (7), *C. p. grayi* (8), *C. wolfi pyrogaster* (9), *C. w. wolfi* (10), *C. w. elegans* (11), *C. w. denti* (12).

form of *C. pogonias* south of the Sanaga along the coast in southern Cameroon. South of *nigripes* ranges *schwarzianus*, known only from the type that Dandelot considers synonymous with *nigripes*. *C. p. grayi* is found from southern Cameroon to the Zaïre River and further east to a tributary, the Itimbiri, where it is replaced eastward by *C. w. denti* (Schouteden, 1947; Colyn, Chapter 6). Blancou (in Malbrant, 1952) has corrected a former identification of *C. denti* from CAR: this was a specimen of *grayi*, clearly shown by the coloured drawing in his unpublished field notes. Unfortunately this correction passed unnoticed to Hill (1966) and others. Fay (1985) also found a '*mona*' monkey in the extreme north of CAR which, according to his description is a *C. p. grayi*. The exact extent of *grayi* in this country is still unknown, as is the limit between its distribution and that of *denti*.

C. wolfi

The distribution of *C. w. wolfi* is similar to that of *C. a. whitesidei*. *C. w. elegans* lives between the Lualaba and Lomani Rivers but only south of the main forest block, while *wolfi* ranges north of *elegans* (Colyn & Verheyen, 1987a). I found no additional information on the distribution of *pyrogaster* to that given by Schouteden (1947). This form seems to be limited to the area between the Kasai and its tributaries, the Kwango and the Lulua rivers in Zaïre. *C. w. denti* occupies the eastern part of Zaïre, east of the Lualaba and Zaïre rivers, and part of Uganda. The lower course of the Itimbiri separates this guenon from *C. p. grayi* but *denti* crosses its upper course northwards (Colyn, Chapter 6). North of the Uele River, its range remains unknown.

Cercopithecus nictitans, C. mitis, C. albogularis (Figure 4.6)

Napier & Napier (1967) grouped *C. nictitans, C. petaurista, C. ascanius, C. erythrotis* and *C. erythrogaster*. Hill (1966), Dandelot (1971) and Napier (1981) include *nictitans, mitis* and *albogularis* within the superspecies *mitis* and Kingdon (1980) within the superspecies *nictitans*. Both Kingdon and Napier consider *albogularis* to be a synonym of *mitis*, because there are cases of hybridization between *C. mitis stuhlmanni* and *C. a. kolbi*, some *kolbi* crossing the Rift Valley (Kingdon, 1971), and because Booth (1968) considered the population around the Lake Manyara as being hybrid between *stuhlmanni* and *albogularis kinobotensis*. Karyological studies by B. Dutrillaux (unpublished data) show that the chromosome count of *albogularis* differs from those of the other *mitis* forms studied, thus it is fair to consider *albogularis* as a distinct species.

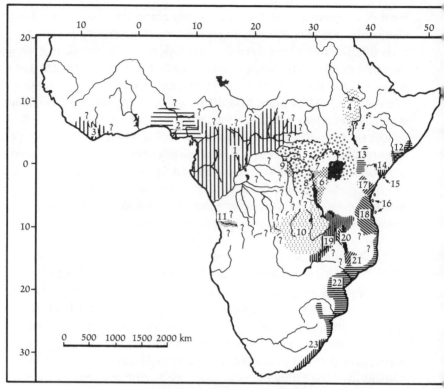

Figure 4.6. Distributions of C. *nictitans nictitans* (1), C. *n. martini* (2), C. *n. stampflii* (3), C. *mitis boutourlini* (4), C. *m. stuhlmanni* (5), C. *m. doggetti* (6), C. *m. kandti* (7), C. *m. schoutedeni* (8), C. *m. heymansi* (9), C. *m. opisthostictus* (10), C. *m. mitis* (11), C. *albogularis zammaranoi* (12), C. *a. kolbi* (13), C. *a. albotorquatus* (14), C. *a. phylax* (15), C. *a. albogularis* (17), C. *a. kibonotensis* (17), C. *a. monoides* (18), C. *a. moloneyi* (19), C. *a. francescae* (20), C. *a. nyasae* (21), C. *a. erythrarchus* (22), C. *a. labiatus* (23).

C. nictitans

Three subspecies are recognized (*nictitans, martini* and *stampflii*), but Dandelot (1971) and Kingdon (1980) considered *martini* and *stampflii* as synonyms. C. *n. signatus* known from three zoo specimens of unknown origin (Oates, 1985) was seen by Dandelot (1971) as a possible mutant of *martini*. Schwarz (in Hill, 1966) considers it as a synonym of C. *erythrogaster* while Oates (1985) feels that it might be a hybrid between *nictitans* and one form of *cephus*.

The C. *nictitans* from Bioko is said to differ slightly from C. *n. martini* and to be closer to C. *n. nictitans* (Eisentraut, 1973). However, the small number of specimens examined prevented Eisentraut from proposing subspecific status for this isolated population.

The distribution of *C. nictitans* in West Africa is discussed by Oates (Chapter 5). Curiously, Dorst & Dandelot (1970) extend its range to Senegal, although *stampflii* is only known from Liberia and Ivory Coast; this statement may be based on the presumed observation of *nictitans* in Senegal by Dupuy (1972), who probably meant *petaurista*, judging from an older publication (Dupuy 1971). Wolfheim (1983) states that *C. nictitans* is present in Sierra Leone, but this may also be a confusion with *C. petaurista*, although *nictitans* could exist in the east of the country (A. C. Davies, personal communication) since it has been collected in Liberia near Kailahun (Kuhn, 1965). Its possible presence in southeastern Guinea can be expected on an environmental basis but this remains unproven since Galat & Galat-Luong (personal communication) did not find it on the northern slope of Mt Nimba and Roche (1971) did not collect it at Seredou.

The limit between *nictitans* and *martini* in Cameroon and Nigeria is not easy to define precisely since the Sanaga seems to act as a barrier only along its lower course. With the possible exception of *stampflii* (Booth, 1956a), *C. nictitans* is able to live in gallery forests very far from the major forest blocks. Blancou (1958) as well as Fay (1985) found it in the gallery forests east of Ndélé in northern CAR. Therefore, the eastern and northern limits of *C. n. nictitans*, as well as the northern and western limits of *C. n. martini*, may extend further than previously supposed. For example, it is not certain that all the observations of heart-nosed monkeys made in Sudan (Kock, 1969) refer to *C. ascanius*, as Blancou (1958) found *nictitans* in the southeast of CAR at Zemio.

To the south, the range of *C. n. nictitans* is limited by the Zaïre River; north of this, it seems to be separated from *C. m. stuhlmanni* by the lower course of the Itimbiri (Schouteden, 1947); however, north of the upper course of this river, the limit between these two guenons is not precisely known. *C. m. maesi* (now *heymansi*) was reported to live north of the Zaïre River, west of the Itimbiri (Schouteden, 1947) and to be sympatric with *C. nictitans*. This is highly questionable as no specimen of *C. mitis* exists from this region (Colyn, Chapter 6). Kingdon (1980) mentions the possibility of rare relict populations of *C. mitis* being absorbed genetically by *C. nictitans* along the little known northwestern distributional margins of *C. mitis*, while individuals that are possible intermediates between these two species have reached the pet market in Belgium in the last few years (B. van Puijenbrook, personal communication).

C. mitis

Eight subspecies have been described. *C. m. boutourlini* in the southwest of Ethiopia, as *C. m. mitis* in Angola, are isolated populations whose distributions are not well known.[2] *C. m. heymansi* is known with certainty only from the north of the forest block between the Lualaba and Lomani rivers (Colyn & Verheyen, 1987b). We do not know whether *C. m. stuhlmanni* and *C. m. opisthostictus* actually enter into contact. *C. m. kandti* has a very restricted distribution around the Virunga Volcanoes, being surrounded by *stuhlmanni*, *dogetti* and *schoutedeni*, this last subspecies being limited to an island in Lake Kivu and to a small area north of this lake (Rahm, 1970). Kingdon (1971) considers that *schoutedeni* might be an intermediate between *kandti* and *dogetti*.

Further study is needed to understand better the geographic distribution and evolution of the *mitis* group, within which different karyotypes have been found in supposed subspecies such as *stuhlmanni* and *opisthostictus*.

C. albogularis

From Somalia to South Africa, up to fourteen subspecies have been described, from which Dandelot (1971) retains eleven and to which I add *C. a. phylax* from Patta Island. Booth (1968) considered that there was a clinal variation between the Kenya highlands and Zanzibar along a range of altitudes from 3000 m to sea-level; individuals from the two extreme localities (*kolbi* and *albogularis*) look different, while the intermediate populations (*kinobotensis* and *monoïdes*) vary progressively in pelage, body weight and tail length. All four were then grouped by Booth into a single subspecies, *albogularis*. Kingdon (1971) also found a clinal variation from *moloneyi*, *monoïdes*, *kinobotensis* to *kolbi*. All these populations may actually be in a more or less advanced stage of subspeciation. Thus while the island subspecies, *phylax* and *albogularis*, are well isolated, they are not yet well differentiated morphologically from mainland subspecies, probably as a consequence of the relatively recent isolation of these islands (*c.* 10 000 years for Zanzibar, *c.* 8000 for Mafia: Moreau, 1966). There is little information on *C. albogularis* in Mozambique. We still do not know the boundary between *monoïdes* and *erythrarchus*, and whether there is a clinal variation from *erythrarchus* to *labiatus*.

Conclusion

The goal of any classification is to provide an inventory of all living forms within a genus (or between related genera), to express their relationships and to understand their phylogeny. This applies to guenons as to other animals, but guenons beautifully demonstrate that animals are not always easy to classify according to the stated rules of taxonomy. It is not often easy to discriminate between a species and a subspecies. This is further complicated by the possibilities for interspecific matings among sympatric guenons (Struhsaker, Chapter 18). Thus, I do not consider that two species of guenons, which clearly differ in their morphology and are well separated by geographical and ecological barriers, have to be seen as subspecies on the basis of occasional interbreeding, especially when such species are split into a number of subspecies.

Let us take as an example the case of the hybrids between *C. erythrotis* and *C. cephus* found south of the Sanaga River (Struhsaker, 1970). It is easy to imagine that an individual *erythrotis* crossing the river will try to join a group of *cephus* in the absence of a sexual partner of its own species. Would such an event take place, however, if the two species came naturally into contact? A careful study of the monkey troops in the known area of sympatry between *C. cephus* and *C. ascanius*, would be very informative. Similar problems are also raised by the *aethiops* and *nictitans* forms.

For the great naturalists of the 18th and 19th centuries who unravelled the wonders of life in the tropics, the classification of species was an objective in itself, which led to the elaboration of the concept of evolution. More recently, however, taxonomy has come to be considered as a rather subsidiary activity by most biologists. Modern taxonomy, though, has nothing to do with 'stamp collecting'; it allows us to understand better 'evolution in action'. The taxonomy of guenons provides a good example of this.

The study of museum specimens is no longer sufficient to improve our understanding of guenon classification. Modern genetic, anatomical, ecological and behavioural investigations have already contributed to our knowledge of the phylogenetic affinities within this group, as evidenced in this book. Much more can be expected in the future and it is the role of the modern taxonomist to integrate this new information.

In trying to do so, I have often been surprised at the paucity of basic information on the natural history of many species of the African tropical rain forest, as shown by the discovery of *C. salongo* in Zaïre in 1977 and *C. solatus* in Gabon in 1984. It is no longer possible (nor desirable) to collect large series of museum specimens but a great deal can be learned from a careful study of the 'bag' of local hunters – monkeys unfortunately being their 'choice game'. Showing skins or photographs to villagers can also give good indications about geographical distribution, while the observation of captive animals remains a precious source of information.

In conclusion, it is difficult not to express our growing concern for the future of forest primates. The major environmental catastrophe of our times is not so much due to people hunting monkeys and other 'game' for food, but rather to the massive deforestation resulting from increasing human populations, from the action of big timber companies and, more generally, from our selfish behaviour. It is only when scientists will be able to convince politicians that environmental destruction is damaging not only for wild animals but for ourselves that there will be some hope of reversing the trend. Therefore, as expressed by Thorington & Groves (1970): 'in its function as a focus for biological data and in its emphasis on adaptation and the ecological setting of species, taxonomy has found a new role for itself, in what may become one of the most fundamental questions of our time'.

Acknowledgements

My thanks go first to Annie and Jean-Pierre Gautier for inviting me to take part to the Paimpont Symposium in 1985 and to contribute to this book. Very special thanks to Pierre Dandelot, who lent me maps of Africa with vegetation zones on which he figured the location of all museum specimens of guenons he studied years ago mainly in the British Museum, Musée Royal de l'Afrique Centrale and Museum National d'Histoire Naturelle; to J. Blancou who kindly put at my disposal the original field notes of his father; to A. G. Davies for the distribution maps of the guenons in Sierra Leone. I would also like to acknowledge the valuable information provided by R. W. Carroll, M. Colyn, P. Dandelot, A. G. Davies, B. Dutrillaux, J. M. Fay, G. Galat & A. Galat-Luong, J.-P. Gautier, A. Gautier-Hion, M. Harrison and A. de Machado. Last, but not least I wish to express my thanks to Prof. F. Bourlière and Dr A. Gautier-Hion for their help and advice for this publication and H. Grillou for the artwork.

Appendix 4.1.

Interspecific and intraspecific wild hybrids

C. aethiops matschiei × C. pygerythrus ssp.? (Dandelot & Prévost, 1972)
C. aethiops budggetti × C. pygerythrus centralis (Dandelot, 1959, 1965)
C. campbelli campbelli × C. campbelli lowei (Booth, 1968)
C. petaurista petaurista × C. p. buettikoferi (Booth, 1956a)
C. mitis stuhlmanni × C. m. albogularis (Booth, 1968)
C. mitis stuhlmanni × C. m. kolbi (Kingdon, 1971)
C. mitis stuhlmanni × C. ascanius schmidti (Aldrich-Blake, 1968; Struhsaker, Chapter 24)
C. cephus cephus × C. erythrotis camerunensis (Struhsaker, 1970)
C. mona × C. pogonias grayi (Struhsaker, 1970)
C. mona × C. pogonias pogonias (Struhsaker, 1970)
C. ascanius ascanius × C. ascanius atrinasus (Machado, 1969)
C. ascanius katangae × C. a. atrinasus (Machado, 1969)
C. ascanius katangae × C. a. ascanius (Machado, 1969)
C. ascanius katangae × C. a. whitesidei (Machado, 1969)

Appendix 4.1 *continued* .

Interspecific and intraspecific captive hybrids
A. *nigroviridis* × C. *aethiops* (Chiarelli in Hill, 1966)
E. *patas* × C. *aethiops* (IZY, Vol. 14, 1974)
E. *patas* × C. *cephus* (IZY, Vol. 2, 1961)
E. *patas patas* × E. *p. pyrrhonotus* (IZY, Vol. 22, 1982)
C. *diana diana* × C. *d. roloway* (IZY, Vol. 10, 1970)
C. *aethiops* × E. *patas* (IZY, Vol. 19, 1979; Vol. 20, 1980)
C. *aethiops* × C. *mona* (IZY, Vol. 8, 1968)
C. *aethiops aethiops* × C. *a. pygerythrus* (IZY, Vol. 18, 1978)
C. *aethiops pygerythrus* × C. *a. sabaeus* (IZY, Vol. 21, 1981; Vol. 22, 1982; Vol. 23, 1983)
C. *sabaeus* × C. *aethiops* (IZY, Vol. 9, 1969)
C. *sabaeus* × C. *albogularis* (IZY, Vol. 12, 1972)
C. *aethiops tantalus* × C. *a. sabaeus* (IZY, Vol. 17, 1977)
C. *hamlyni* × C. *lhoesti* (Crandall, 1964)
C. *cephus* × C. *petaurista* (IZY, Vol. 18, 1978)
C. *erythrogaster* × C. *mona* (IZY, Vol. 10, 1970)
C. *petaurista buettikoferi* × C. *nictitans martini* (IZY, Vol. 10, 1970)
C. *mona* × C. *cephus* (Hill, 1966)
C. *mona* × C. *nictitans* (IZY, Vol. 7, 1967)
C. *mona* × C. *mitis doggetti* (IZY, Vol. 5, 1965)
C. *mona* × C. *mitis albogularis* (IZY, Vol. 4, 1963)
C. *mona* × C. *leucampyx* (IZY, Vol. 4, 1963)
C. *mona wolfi* × C. *neglectus* (Crandall, 1964)
C. *pogonias* × C. *ascanius* (IZY, Vol. 20, 1980)
C. *mitis* × C. *albogularis* (IZY, Vol. 14, 1974)
C. *mitis stuhlmanni* × C. *m. albogularis* (Booth, 1968)
C. *albogularis* × C. *neglectus* (IZY, Vol. 17, 1977)
C. *albogularis albogularis* × C. *a. moloneyi* (IZY, Vol. 13, 1973; Vol. 15, 1975, Vol. 23, 1983)

Hybrids between captive guenons and papionines
C. *aethiops* × *Macaca mulatta* (Chiarelli in Gray, 1972)
C. *sabaeus* × *M. mulatta* (Zuckerman in Hill, 1966 and in Gray, 1972)
C. *mitis* × *Cercocebus torquatus torquatus* (Montagu in Hill, 1966)
M. *irus* × C. *pygerythrus* (Gunning in Hill, 1966)
M. *radiata* × C. *aethiops* (Chiarelli in Gray, 1972)
M. *radiata* × C. *pygerythrus* (Gunning in Gray, 1972)
M. *sinica* × C. *aethiops* (Chiarelli in Gray, 1972)
M. *sinica* × C. *pygerythrus* (Gunning in Hill, 1966 and in Gray, 1972)

The names of the taxa have been given as published; C. *leucampyx* = C. *mitis*; IZY = International Zoo Yearbook.

    Notes (added in proof)
    1. A new form (C. *hamlyni kahuziensis*) has recently been described by Colyn and Rahm (in preparation). It lacks the white nose stripe and is limited to the bamboo forest of the Kahuzi–Biega National Park in Zaïre.
    2. C. *m. mitis* extends in the coastal part of Angola between 5 and 15° S and C. *m. opisthostictus* enters the far East of the country (J. Crawford-Cabrol and A. de B. Machado, personal communication).

Appendix 4.2.

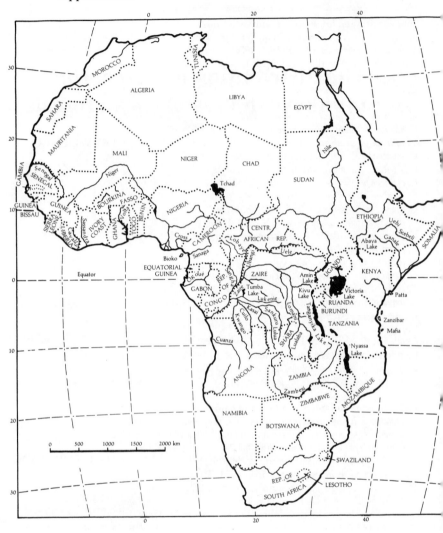

I.5

The distribution of *Cercopithecus* monkeys in West African forests

JOHN F. OATES

Introduction

Between 1954 and 1958, Angus Booth published a classic series of papers on the primates of West Africa, focusing on their patterns of distribution and attempting to explain these patterns (Booth, 1954a,b, 1955, 1956a,b, 1957, 1958a,b). These papers formed an important source of comparative evidence and ideas for studies of other West African vertebrates, notably the treefrogs (Schiøtz, 1967) and birds (Moreau, 1969), and they have influenced analyses of African-wide distribution and speciation patterns (e.g. Hamilton, 1976; Grubb, 1978b).

Booth (1958b) treated the West African primates as essentially either 'High Forest' or 'Savannah' species and paid most attention to those in the high forest (equivalent to the 'Lowland Rain Forest' of White, 1983). Booth's main conclusion was that the present distribution patterns of forest primates in West Africa are best explained as resulting from fluctuations of climate during the Pleistocene 'leading to the alternate formation and resolution of discontinuities in the High Forest zone,' with monkey populations in particular differentiating from one another in forest refuges during dry periods. Booth also concluded that the River Niger had acted as a barrier preventing forest areas west of the river from being recolonized by mammals spreading from the east after the last presumed dry period.

I have undertaken a new analysis of some primate distribution patterns in West Africa, incorporating data not considered by Booth and more careful mapping techniques. This analysis has produced a picture more complex than that painted by Booth, one that cannot be explained entirely by a simple forest refuge model. In particular, this analysis suggests that a habitat-classification of West African primates into either forest or savanna forms obscures some significant variation. Each 'forest' form inevitably has a unique range of environmental tolerances as a result of a unique evolutionary history. These different tolerances produce different geographical ranges, only some of which show close congruence with one another. Here, I present the results of my analyses and discuss them in relation to some of the most recent evidence on past changes in the climate of West Africa.

Area of analysis – its vegetation and zoogeography

This analysis covers almost exactly the same area as that examined by Booth (1958b), that is, Africa south of the Sahara, north of the Sanaga River, and west of latitude 12° E (see Figure 5.1). The only major difference is that I include the island of Bioko (Fernando Poo), which Booth omitted. The Sanaga coincides approximately with a transition between the Cameroon and West Equatorial faunal regions, which contain several different species and subspecies of primates (Oates, 1986).

Within this area of analysis are several distinct faunal and floral communities. Fringing some, but not all, of the West African coast are areas of mangrove and freshwater swamp. Inland, the vegetation

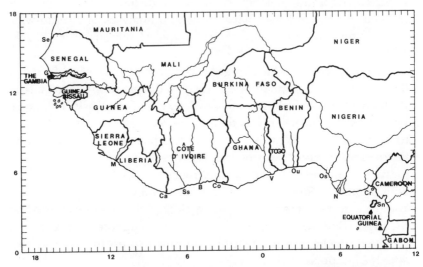

Figure 5.1. West Africa, showing national boundaries and major rivers. Rivers are coded as follows: B: Bandama; Ca: Cavally; Co: Comoé; Cr: Cross: G: Gambia; M: Moa; N: Niger; Ou: Ouémé; Os: Osse; Sn: Sanaga; Ss: Sassandra; V: Volta. Marginal calibrations are half-degrees of latitude and longitude.

changes from south to north as a consequence of decreasing rainfall. As climate becomes drier, the number of deciduous species in the lowland forest increases. The most northerly part of the original forest cover has been largely replaced (probably in the main by human activity) by a savanna woodland where semi-evergreen forest survives only along rivers and in scattered patches; this is known as the forest–savanna mosaic or forest–woodland transition zone; its southern portions include the 'derived savanna' of Keay (1953). North of these forest zones is Sudanian woodland, and north of this the Sahel. Embedded within these latitudinal zones are areas of montane and inland swamp vegetation, with the largest area of montane vegetation occurring in the Cameroon Highlands. Figure 5.2 displays the distribution of these major vegetation types, replotted from the map of White (1983) on a grid of half-degree latitude/longitude cells to allow direct quantitative comparisons with primate distribution data. In the area of West Africa south of 14°N covered in this analysis, the frequency of different vegetation types as plotted in Figure 5.2 is as follows: Sudanian

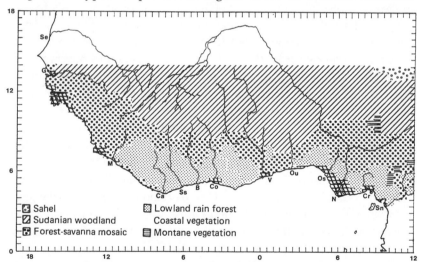

Figure 5.2. Vegetation of West Africa south of 14° N latitude, plotted by half-degree cells of latitude and longitude. Cells have been allocated to one of six vegetation types based on the predominant vegetation shown for that cell in the map of White (1983). One exception has been made. The vegetation of Bioko island (Fernando Poo) has been plotted as rain forest rather than as the forest–grassland mosaic shown by White. Although much of the original lowland forest vegetation on Bioko has been heavily modified by human activity, this modification has predominantly involved conversion to cocoa plantations, in which many large trees are left standing. It would be misleading to classify this land with savanna.

Woodland 467 (48.3%), Forest–savanna Mosaic 267 (27.6%), Rain Forest 167 (17.3%), Coastal vegetation 47 (4.9%), Montane vegetation 8 (0.8%), Sahel 10 (1.0%).'Coastal vegetation' includes mangrove, swamp forest and coastal grassland.

As this chapter will show, different forest monkeys are affected in different ways by these vegetation patterns. Some species (such as *Cercopithecus diana*) are largely restricted to the lowland rain-forest zone, while others thrive in forests within the forest–savanna mosaic. One species (*C. preussi*) does not extend far beyond the montane forests of Cameroon and Bioko.

Within the lowland forest zone a major gap occurs on the coastal plains from near Accra in Ghana to the border between Nigeria and the Republic of Benin (formerly Dahomey). Here, there is an annual rainfall of only 700–800 mm, possibly due to the combined effects of cool offshore currents and the 'shadowing' effect of Cape Three Points on rain-bearing westerly winds (Church, 1980). Although dry savanna vegetation predominates in this 'Dahomey Gap', the gap does contain an area of moist hill forest on the Ghana–Togo border and this forest has tenuous links to relic patches in Benin via river galleries. This gallery forest is particularly well developed on the Mono River and its tributaries, where *Cercopithecus mona* occurs (personal observation, 1985). The Dahomey Gap has long been recognized as forming a significant zoogeographic barrier to forest animals (e.g. Booth, 1954b) and will be discussed in greater detail below.

The forest area to the west of the Dahomey Gap has traditionally been called the Upper Guinea forest district, or region, while the forests extending from east of the gap, and across western equatorial Africa and the Congo Basin have been called the Lower Guinea forest (Chapin, 1923). Within the Upper Guinea region there is a partial gap in the Baoulé area of central Côte d'Ivoire, where the forest–savanna mosaic reaches to within 100 km of the coast. This is the Baoulé-V (Booth, 1958b). Several rivers in this area, notably the Sassandra and Bandama, have been suggested to be significant faunal barriers, but it is hard to distinguish their effects (and they are not of great size) from those of general ecosystem changes. For instance, at a latitude of 7°N the annual rainfall at the Sassandra is about 1600 mm; 100 km to the west it is 1800 mm and 100 km to the east it is 1300 mm (Aubréville, 1949).

At the eastern edge of the Upper Guinea forest region, the Volta River has been considered as a potential faunal boundary, while further east, in the Lower Guinea region, the Niger and Cross rivers have been cited as having zoogeographical significance (Rosevear,

1953; Booth, 1958b; Moreau, 1969). The effects of these potential barriers on forest monkeys will be discussed below.

Species considered

In this contribution, I will only consider in detail the distribution patterns of *Cercopithecus* species restricted to forest habitats, but comparisons will be made with the distributions of other forest monkeys. The species classification used is the same as that employed by Booth, except that the *preussi* form in the *Cercopithecus lhoesti* group is regarded as specifically distinct (following Napier, 1981). The species considered in detail are:

Diana monkey, *Cercopithecus diana* Linnaeus, 1758
Preuss's monkey, *Cercopithecus preussi* Matschie, 1898
Putty-nosed guenon, *Cercopithecus nictitans* Linnaeus, 1766
Spot-nosed guenon, *Cercopithecus petaurista* Schreber, 1774
White-throated guenon, *Cercopithecus erythrogaster* Gray, 1866
Red-eared guenon, *Cercopithecus erythrotis* Waterhouse, 1838
Moustached guenon, *Cercopithecus cephus* Linnaeus, 1758
Campbell's monkey, *Cercopithecus campbelli* Waterhouse, 1838
Mona monkey, *Cercopithecus mona* Schreber, 1774
Crowned monkey, *Cercopithecus pogonias* Bennett, 1833

Methods

The sources of data used in this analysis are of three major kinds: (1) locality information accompanying specimens in museum collections in the United States and Europe, (2) published records, and (3) personal observations in the field. Data from these different sources have been collated, and then plotted on a grid matrix of half-degree cells of latitude and longitude.

Data were gathered in the following collections: American Museum of Natural History, New York; Anatomisches Institut, University of Göttingen; British Museum (Natural History), London; Musée Royal de l'Afrique Centrale, Tervuren; Muséum National d'Histoire Naturelle, Paris; Museum für Naturkunde, E. Berlin; The Powell-Cotton Museum, Birchington; Rijkmuseum van Natuurlijke Historie, Leiden; and the United States National Museum, Washington, DC. Locality data were transcribed from skin labels and/or catalogues in these collections, and the species identities of the skins checked.

Many publications were consulted. Among the most important were Booth's *The distribution of primates in the Gold Coast* (1956), Kuhn's *A provisional checklist of the mammals of Liberia* (1965), Rosevear's *Checklist and atlas of Nigerian mammals* (1953), Dupuy's *Statut actuel des primates au*

Sénégal (1971), Sayer & Green's *The distribution and status of large mammals in Benin* (1984), Roche's *Recherches mammalogiques en Guinée forestière* (1971), Monard's *Résultats de la mission scientifique du Dr Monard en Guinée Portugaise 1937–38* (1938), and several works by Galat and Galat-Luong on the primates of Côte d'Ivoire (e.g. Galat-Luong & Galat, 1978; Galat-Luong, 1983; Galat & Galat-Luong, 1985).

I collected additional locality data by direct observation during surveys in Nigeria (1966–7, 1981, 1982), Sierra Leone (1979, 1980, 1982), The Gambia (1979), and Togo (1985).

Where any serious doubt exists as to the validity of a distribution record (for instance, due to uncertain identification, an anecdotal account, or an ambiguous place-name), the record has been rejected from the analysis.

Apart from the ease with which it allows data to be recorded and analysed, a half-degree grid has been used because it is judged to produce a plot whose accuracy is consistent with the accuracy of the data. Many 'precise' localities on specimens or in the literature are towns or villages that were collectors' bases, or the residences of hunters from whom specimens were obtained. Specimens are likely to be culled at considerable distances from such a place. In the area covered by this analysis, a half-degree cell has a side of 55–60 km. Despite the probable inaccuracy of much of the data, it is very unlikely that any specimen whose locality I have mapped originated outside either the cell in which it has been plotted or an immediately adjacent cell.

Many parts of West Africa have still not been closely surveyed for their primates, and in many areas where forest monkeys once occurred they are no longer present as a result of the destruction of their habitats and/or hunting. The maps reproduced here should therefore be regarded as showing the *patterns* of distribution of the West African forest monkeys in the first half of the twentieth century, rather than as an exact record of past or present distribution.

Results

Individual Cercopithecus *distributions*

Table 5.1 summarizes the distribution of each species in relation to vegetation. Each species, with the exception of *Cercopithecus mona*, has over 50% of its distribution within the lowland rain-forest zone. The remainder of this section will examine individual distributions in more detail.

Table 5.1. *Distribution of West African forest* Cercopithecus *monkeys across vegetation zones. The table shows the frequency of occurrence (Number of cells, and percentage of total cells occupied) of each species in the zones mapped in Figure 5.2.*

Species	Habitat					
	Sudanian woodland	Forest–savanna mosaic	Lowland rain forest	Coastal vegetation	Montane vegetation	Total cells occupied
C. diana		7 (20.0)	27 (77.1)	1 (2.9)		35
C. preussi		1 (12.5)	6 (75.0)	1 (12.5)		8
C. nictitans		11 (29.7)	22 (59.5)	3 (8.1)	1 (2.7)	37
C. petaurista		26 (31.7)	52 (63.4)	4 (4.9)		82
C. erythrogaster				4 (80.0)	1 (20.0)	5
C. erythrotis			10 (83.3)	2 (16.7)		12
C. cephus		1 (100.0)				1
C. campbelli		27 (35.5)	45 (59.2)	4 (5.3)		76
C. mona	6 (11.8)	21 (41.2)	17 (33.3)	6 (11.8)	1 (2.0)	51
C. pogonias			8 (88.9)	1 (11.1)		9

Cercopithecus diana

Taxonomy. Two subspecies of *C. diana* are generally recognized, *C. d. diana* east of the Sassandra River and *C. d. roloway* to the west. *C. d. roloway* has yellow rather than red inner thighs and a longer beard than *diana*. However, there are some specimens from the vicinity of the Sassandra that show features intermediate between *diana* and *roloway*. For instance, some specimens from Gagnoa, east of the Sassandra, have quite reddish thighs, while one from Duékoué between the Sassandra and Nzo has orange thighs.

Distribution. The distribution of *C. diana* is shown in Figure 5.3. It is closely congruent with the area of the Upper Guinea forest region, and does not occur within or to the east of the Dahomey Gap.

Habitat. The majority of cells in which I have recorded this species are classified as rain forest (Table 5.1). The seven forest–savanna mosaic cells in which *C. diana* has been recorded are in Sierra Leone. Patches of rain forest and evergreen riverine forest occur throughout most of Sierra Leone and in the southern and eastern parts of the country these forests often contain diana monkeys. Since 1815, extensive deforestation has occurred in Sierra Leone through the combined effects of timber exploitation and farming (Dorward & Payne, 1975), leaving behind patchy relics in a large area of farmbush and derived savanna woodland.

Comment. Competitive exclusion may occur between this species and *C. nictitans* (see below).

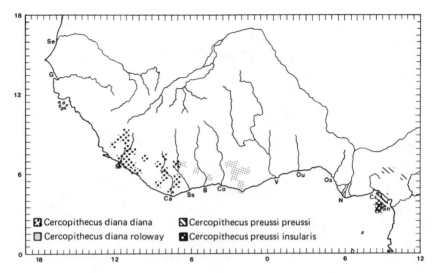

Figure 5.3. Distributions of *Cercopithecus diana* and *C. preussi.*

Cercopithecus preussi

Taxonomy. This species is a close relative of *C. lhoesti* of E. Zaïre, Rwanda and southwestern Uganda. A monkey recently discovered in the Forêt des Abeilles in Gabon (Harrison, 1984b) is evidently a third member of this group (see Dutrillaux *et al.*, Chapter 9; Gautier, Chapter 12).

Distribution. This species has a very restricted distribution. I have found records from only eight cells, in Cameroon, Nigeria and on Bioko island (Figure 5.3).

Habitat. In East Africa, *C. lhoesti* occurs in medium-altitude and montane rain forests. Although none of the cells in which I have recorded *C. preussi* classifies as montane forest in this analysis (due to a preponderance in the cell of other vegetation types), all the cells in which it occurs either contain, or are adjacent to, mountainous terrain on Mt Cameroon, the Cameroon Highlands, and the mountains of Bioko. Rain forest at medium and high elevation would therefore seem to be the preferred habitat of this species, although it has been recorded in some low-elevation sites (e.g. 100–300 m at Idenau by Gartlan & Struhsaker, 1972). The Gabonese form occurs at a lowland site.

Cercopithecus nictitans

Taxonomy and distribution. Many classifications recognize three subspecies of putty-nosed guenons: *nictitans*, *martini*, and *stampflii. C. n. stampflii* is usually considered to be restricted to western Côte

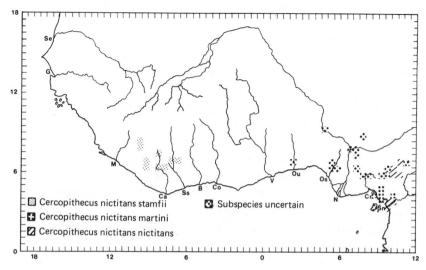

Figure 5.4. Distribution of *Cercopithecus nictitans*.

d'Ivoire and Liberia, with *C. n. martini* occurring from the Dahomey Gap to the Sanaga and *C. n. nictitans* to the east and south of the Sanaga (Booth, 1958b; Kuhn, 1965; Napier, 1981). All three subspecies have a white nose, and both *stampflii* and *martini* have a white or whitish chest, but in typical *stampflii* light coloration continues down the venter to the groin. The proximal dorsal half of the tail in *stampflii* is more grizzled than it is in *martini* or *nictitans*. In typical *martini*, only the chest is white and there is a distinct subcaudal patch of reddish hair which is lacking in *stampflii*; typical *C. n. nictitans* have a grey chest and no subcaudal red patch.

Some putty-nosed guenon populations do not fit well into one of these three subspecific groups, however, and the Sanaga and Dahomey Gap are not clear-cut barriers to their distribution. The species has been recorded from within the Dahomey Gap (Sayer & Green, 1984) and there is a *nictitans nictitans* population in the Cameroon Highlands between Bamenda and the Donga River valley, far to the north of the Sanaga (specimens in the British Museum) (see Figure 5.4). Although animals that classify unambiguously as *martini* occur around Mt Cameroon, populations between Mt Cameroon and Liberia display a mixture of *martini* and *stampflii* features. Specimens from central and western Nigeria, although resembling *martini* in many respects, lack red subcaudal hairs, while one Côte d'Ivoire specimen has a few reddish-brown hairs at the tail base. Some Nigerian specimens are like *stampflii* in showing a marked contrast between the creamy-white of their ventral surfaces and their much darker lateral pelage. The intermediacy of this central and western Nigerian population led Hayman (in Sanderson, 1940) to refer it 'temporarily' to *stampflii*. The affinities of the Bioko population are also ambiguous. Although the type specimen of *martini* in the British Museum is labelled 'Fernando Po', the skin (which is incomplete) may have come from western Cameroon (Eisentraut, 1973). Eisentraut describes several differences between four definite Bioko specimens and Cameroonian *martini*. One of these differences is the lack of red subcaudal hairs, as in central and western Nigerian specimens. It seems that the taxon *C. nictitans martini* encompasses populations which, in pelage at least, show considerable variation, and some of these populations are quite similar to *stampflii*. Pending a more substantial taxonomic review, the inclusion of *stampflii* within *martini* by Dandelot (1971) and Kingdon (1980) is not an unreasonable course.

Habitat. Of the 37 cells in which *nictitans* was recorded in this study (Table 5.1), a relatively large number (29.7%) are classified as forest–savanna mosaic. Many of these cells are in central Nigeria, where *C. nictitans* lives in relic patches of high forest and in forest along rivers (see, e.g. Oates, 1982).

Comment. Although in Cameroon and Nigeria *nictitans* seems to be a widespread and relatively successful species, it may suffer from competition with *diana* in the west of its range (see A. Gautier-Hion, Chapter 23). They are similarly-sized animals, with adult females averaging close to 4 kg body weight in both species (*C. nictitans nictitans nictitans* 4.2 kg, $N = 9$, Gautier-Hion, 1975; *C. diana diana* 3.9 kg, $N = 11$, unpublished data from museum specimens collated by this author). Both typically exploit the upper part of the forest canopy (Gartlan & Struhsaker, 1972; Galat & Galat-Luong, 1985), and both include a relatively high percentage of leaves in their diet (*C. diana* 7.7%: Galat & Galat-Luong, 1985; *C. nictitans* 16.5%: Gautier-Hion, 1983). Five of the nine cells occupied by *stampflii* in Côte d'Ivoire and Liberia are not occupied by *diana*. Most of the *stampflii* specimens have come from the northern part of the rain-forest zone (Kuhn, 1965), and it seems probable that *C. nictitans* is better adapted than is *C. diana* to relatively dry forests and to gallery forests in the savanna zone. One key to this could be an ability to obtain a significant protein component of the diet from foliage. This could explain the occurrence of *nictitans* not only on either side of the Dahomey Gap, but in forest patches within the gap (Sayer & Green, 1984). In moist forest, however, it may be that *nictitans* is competitively excluded by *diana*.

Cercopithecus petaurista

Taxonomy. This species was mistakenly classified with the very different *C. nictitans* by Schwarz (1928b). It is a member of a superspecies (the *cephus* group) that also includes *C. erythrogaster, C. erythrotis, C. cephus* and *C. ascanius* (Oates, 1985). Two subspecies are generally recognized: *C. p. petaurista* and *C. p. buettikoferi; buettikoferi* differs from *petaurista* in having a patch of yellow-speckled hairs on the cheek and in lacking a black band across the nape of the neck (Napier, 1981).

Distribution. As shown in Figure 5.5, this species occurs from northwestern Sierra Leone to the Republic of Benin. Its range probably extends further north along the coast, but there have been few surveys of this area. According to Monard (1938) there is a specimen from Bubaque in the Bijagos Islands, Guinea-Bissau, and Frade (1949) reports it from Rubane Island in the same archipelago. It is possible that this is an introduced population. According to P. Grubb (personal communication) there is a specimen in the British Museum (Natural History) that was probably collected (by G. Fenwick Owen) in the vicinity of what is now the Niokolo–Koba National Park on the Guinea–Senegal border. Otherwise there are no museum specimens from either Senegal or The Gambia. A record from Seleti, just south of the Senegal–Gambia border (Dupuy, 1972), is of questionable validity. Dupuy's record is based on a 1970 sighting and report from Mr Eddie Brewer of The Gambia Wildlife Conservation Department. When I

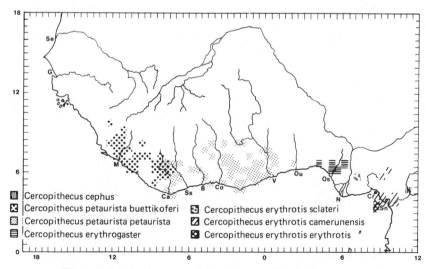

Figure 5.5. Distributions of *Cercopithecus petaurista, C. erythrogaster, C. erythrotis* and *C. cephus.*

visited Seleti in 1979 I observed a group of C. *campbelli*. Until more careful surveys are conducted in the area, the presence of *petaurista* should be regarded as uncertain.

The lower Cavally River on the Liberia–Côte d'Ivoire border seems to mark the boundary between the subspecies *buettikoferi* and *petaurista*, but the range of *buettikoferi* crosses the upper Cavally, and meets that of *petaurista* between the Cavally and Nzo rivers. In this region, Booth also collected an intermediate individual (BMNH 1956.309).

Habitat. Although predominantly a species of the rain-forest zone, it is associated particularly with thick second-growth vegetation and riverine forest and is relatively uncommon in mature high forest. During surveys in southeastern Sierra Leone in 1979 and 1980 I encountered this species more frequently than any other monkey, and 80% of 40 sightings were in young secondary and riverine forest. C. *petaurista* was also found in more grid cells (82) than any other species in this analysis. 63% of these cells are classified as rain forest and 32% as forest–savanna mosaic.

Comment. This species' ability to flourish in riverine forest and edge habitats probably accounts for its presence outside the main block of Upper Guinea rain forest, and for its presence in forest patches within the Dahomey Gap (Sayer & Green, 1984).

Cercopithecus erythrogaster

Taxonomy. C. *erythrogaster* is a distinct taxon within the *cephus* superspecies (Oates, 1985).

Distribution. This species is only known definitely from southwestern Nigeria (Figure 5.5), although there is a possibility that it extends into eastern Benin (Sayer & Green, 1984; Oates, 1985). The Dahomey Gap and the River Niger seem to form the western and eastern limits of its range, although some hunters' reports suggest that it may occur in the eastern part of the Niger Delta (Oates, 1985).

Habitat. C. *erythrogaster* is most often seen in thick secondary growth and riverine forest within the rain-forest zone (Oates, 1985). Of five cells plotted in this study (Table 5.1), four are classified as rain forest and one as coastal vegetation.

Comment. C. *erythrogaster* seems to be a more strictly rain-forest species than C. *petaurista*. Its presence in southwest Nigeria has been used as evidence for the hypothesis that a Pleistocene refuge was located in this area (Booth, 1958b).

Cercopithecus erythrotis

Taxonomy. This is another member of the *cephus* superspecies. Three subspecies are generally recognized: *erythrotis, camerunensis,* and *sclateri. C. e. erythrotis* of Bioko differs from *camerunensis* only in its smaller size and darker and longer coat, but *sclateri* is very different in many features of its colour pattern, leading Kingdon (1980) to regard it as a distinct species.

Distribution. C. erythrotis erythrotis occurs only on Bioko island, and *sclateri* occurs only in eastern Nigeria between the Niger and Cross rivers (in fact, only a single wild-collected individual, from the Okigwi area, is known). *C. e. camerunensis* has a wider, but still quite restricted distribution, from the Cross River down to the Sanaga; it crosses the lower Sanaga, where some hybridization with *C. cephus* occurs (Struhsaker, 1970).

Habitat. Of the grid cells in which *erythrotis* was recorded in this study, 83% are classified as rain forest. Gartlan & Struhsaker (1972) noted that in Cameroon *C. e. camerunensis* occupied the lower forest strata. It was not encountered in young secondary forest at Idenau. The only field record for *sclateri* (BMNH 1935.4.2.1) gives its habitat as 'orchard bush country', although the area has been classified by White (1983) as lowland rain forest. *C. erythrotis* occurs in both lowland and montane forests on Bioko (Eisentraut, 1973), and labels on some Bioko specimens in the British Museum refer to a 'thick bush' habitat.

Comment. The status of *sclateri* needs to be elucidated. Otherwise *erythrotis* seems to be a rain-forest specialist, with a distribution very similar to that of several other members of a 'Cameroon' faunal region: from the Cross to the Sanaga, with a population on Bioko.

Cercopithecus cephus

Taxonomy. This is the nominate form of the *cephus* superspecies. Napier (1981) recognizes two subspecies, *C. c. cephus* and *C. c. cephodes.*

Distribution. C. cephus is a member of the Western Equatorial primate community (Oates, 1986) and occurs only at the far eastern edge of the area considered here. One specimen in the American Museum (AMNH 120381), collected by F. McC. Grisset is labelled as coming from Bafia (4°40' N, 11°05' E), just north of the Upper Sanaga. Grubb (1973) has cast doubt on the veracity of the same locality for a specimen of *Mandrillus sphinx* also collected by Grisset. There seems no special reason to doubt the authenticity of the specimens, however, especially since Grisset provides additional information ('100–105 miles northwest of Yaoundé'), and vegetation maps (e.g. White, 1983, and see Figure 5.2) show Bafia as lying close to a block of rain-forest vegetation that is contiguous with the forest south of the River Sanaga but separated by savanna from the western Cameroon forests (inhabited by *C. erythrotis* and *M. leucophaeus*). The single cell plotted for *C. cephus* in Figure 5.5 is more than 50% made up of forest–savanna mosaic and therefore is classified as such in Figure 5.2, but in fact the cell does contain large areas of rain forest.

Cercopithecus campbelli

Taxonomy. With the species *mona* and *pogonias*, Campbell's monkey forms part of the *Cercopithecus mona* superspecies. Two subspecies are widely recognized, *campbelli* and *lowei*. The rump, lateral surfaces of the thighs and arms, and the dorsal surface of the tail are almost black in *lowei*, grey in *campbelli*; *lowei*'s browband is yellow, *campbelli*'s white.

Distribution. This is a widespread species (Figure 5.6), occurring from the Gambia–Senegal border east to the Volta River in Ghana. Both the overall distribution pattern and the pattern of subspecific change are very similar to those found in *petaurista*, with *C. c. campbelli* west of the Cavally River and *lowei* east of the Sassandra–Nzo. Between

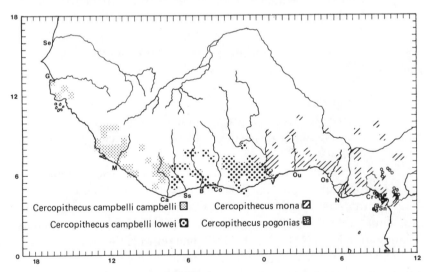

Figure 5.6. Distributions of *Cercopithecus campbelli*, *C. mona* and *C. pogonias*.

these rivers both forms occur, as well as some individuals showing features intermediate between the two.

Habitat. As noted by Booth (1956a), this species occurs in primary and secondary rain forest, forest outliers, and river-fringing forest in the savanna. In Sierra Leone it is found in coastal mangroves (Jones, 1950). This wide habitat tolerance is reflected in the large number of cells recorded for *C. campbelli* in this analysis (82) and the relatively high percentage of these cells (36%) classified as forest–savanna mosaic (Table 5.1).

Comment. East of the Volta, this species is replaced by *C. mona*, with which it is sympatric in a limited area west of the Volta. Booth (1955, 1956a) noted that in this area of overlap there is a slight habitat differentiation, with *mona* tending to favour the river banks, and *campbelli* the forest behind. This sympatry is a major reason for the classification of *mona* and *campbelli* as separate species (Booth, 1955). The narrowness of the overlap zone suggests another instance of competitive exclusion.

Cercopithecus mona

Taxonomy. This is a close relative of the preceding and subsequent species, and shows no subspecific variation.

Distribution. The range of this species is centred on Nigeria, but it occurs west to, and across, the Volta (Booth, 1955, 1956a) and south to, and across, the Sanaga (specimens in the British and Powell-Cotton Museums, and observations by Struhsaker, 1970).

Habitat. This species is most frequent in the forest–savanna mosaic zone (41% of cells in Table 5.1) and in Nigeria and the Republic of Benin it ranges further north than any other forest *Cercopithecus*, extending into the Sudanian woodland zone via gallery forests (Figure 5.6). In Cameroon, *C. mona* prefers secondary rain forest, and is particularly frequent in mangrove swamps, where it has been observed swimming across a creek (Gartlan & Struhsaker, 1972).

Comment. See *C. campbelli*.

Cercopithecus pogonias

Taxonomy and distribution. This is the third member of the *mona* superspecies in our area, where it is represented by the subspecies *C. p. pogonias*. *C. p. pogonias* occurs only on Bioko and in the vicinity of Mt Cameroon (Figure 5.6). South of the Sanaga *C. p. grayi* is found.

Habitat. In our area this is a rain-forest species; eight of the nine cells in which it has been recorded are classified as rain forest.

Comment. Struhsaker (1970) notes hybrids between this species and *C. mona* in the Idenau area of Cameroon. However, the wide sympatry of the two species with only occasional interbreeding, and their different vocal repertoires and niches (Struhsaker, 1970; see also Gautier, Chapter 12), argue for the validity of their separation at the species level.

Other forest monkeys

In addition to the *Cercopithecus* species, 10–13 other species of forest-living cercopithecids occur in the area of West Africa considered in this chapter. These are the talapoin (*Miopithecus talapoin*), three species of mangabey (*Cercocebus*), the drill (*Mandrillus leucophaeus*) and possibly the mandrill (*M. sphinx*), and five to seven species of colobus monkey.

M. talapoin is essentially a species of the Western Equatorial forest region (Oates, 1986), but it has been seen on the north bank of the Sanaga River (Gartlan & Struhsaker, 1972). Of the mangabeys, *C. atys* has a similar distribution to *Cercopithecus campbelli*, with one subspecies, *atys* occurring from Senegal to the Sassandra and another subspecies, *lunulatus*, occurring to the east. However, the distribution of the *lunulatus* subspecies seems to be more restricted than that of *C. campbelli lowei* (perhaps as a result of selective hunting). The closely related *Cercocebus torquatus* has an unusual distribution, occurring from western Nigeria eastward, crossing the Sanaga and extending down to Gabon.

The drill occurs from east of the Cross River to the Sanaga; south of the Sanaga the mandrill is found. However, there is at least one apparently valid specimen of a drill from Cameroon south of the lower Sanaga (Grubb, 1973), and there are unconfirmed hunters' reports of drills from Gabon (A. Gautier-Hion and C. Tutin, personal communications). There are also disputed records of mandrills from the Bafia region, north of the Sanaga (see above, and Grubb, 1973). The drill is a predominantly rain-forest species which may competitively exclude *C. torquatus* from dry-land forests in Cameroon. *C. torquatus* seems to be mainly associated with swamp forests in Cameroon and Rio Muni (Jones & Sabater Pi, 1968; Gartlan & Struhsaker, 1972), but seems equally at home in swamp and dry-land forests in southern Nigeria where *M. leucophaeus* is absent (personal observation). *M. leucophaeus* occurs on Bioko, where *C. torquatus* is absent.

Oates & Trocco (1983) recognize four species of black-and-white colobus in West Africa: *Colobus polykomos, C. vellerosus, C. guereza* and

C. satanas. Many authors regard the western forms (*polykomos* and *vellerosus*) as conspecific (e.g. Napier, 1985). Together, these western forms have a somewhat similar distribution pattern to *Cercopithecus petaurista*; a genetic transition occurs in the vicinity of the Sassandra River, and *vellerosus* enters the Dahomey Gap; however, the *vellerosus* form extends much further north (along gallery forests in the savanna) and east (into western Nigeria) than does *petaurista. Colobus guereza* only occurs in the far east of our area, where it is particularly associated with gallery forests around the Cameroon Highlands, while *C. satanas* occurs only on Bioko and south of the Sanaga.

The red colobus show a unique pattern. Two species may be recognized in West Africa, *Procolobus badius* and *P. pennantii* (Dorst & Dandelot, 1970; Oates, 1986). *P. badius* is the only forest monkey to have a distinct form, *temminckii*, in Senegal, Guinea-Bissau and The Gambia. Further east, *P. badius badius* is replaced by *P. b. waldroni* in the vicinity of the Bandama River. *P. b. waldroni* has a similar distribution to *Cercopithecus diana roloway.* There is then a huge gap in red colobus range to western Cameroon and Bioko, where *P. pennantii* occurs. *P. pennantii* has different forms in Cameroon (*preussi*) and on Bioko (*pennantii*). To the south and east there is another large gap in red colobus distribution, with the nearest confirmed population to that in western Cameroon occurring close to the River Congo in the Congo Republic.

Yet another distinctive pattern is shown by the olive colobus, *Procolobus verus.* It occurs only to the south of 8.5° N, but extends from Sierra Leone to eastern Nigeria without any obvious subspecific variation, spanning the Volta, Niger and the Dahomey Gap (Oates, 1981).

Emergent patterns
Although no two species examined have identical distributions, some patterns do emerge from this analysis:

1. *Cercopithecus campbelli, C. diana, C. petaurista, Cercocebus atys,* and *Procolobus badius* occur only to the west of the Dahomey Gap. Each of these species changes subspecifically between eastern and western Côte d'Ivoire. While most of these transitions occur between the Cavally and Sassandra rivers, *P. badius* subspeciates further east, between the Bandama and Comoé. *Colobus polykomos* is replaced by *C. vellerosus* in the vicinity of the Sassandra.
2. While *Cercopithecus campbelli* and *C. petaurista* extend into

the forest–savanna mosaic well outside the lowland rain-
forest zone, *C. diana* is mainly restricted to the rain forest.

3. *Cercopithecus mona*, *C. nictitans* and *Colobus vellerosus* all
   occur within the Dahomey Gap as well as east and west of
   it. Although each has a different overall distribution pattern, each also extends well to the north of the rain forest
   block in parts of its range.

4. *Cercopithecus erythrogaster* has a unique distribution,
   restricted to the forests of southwestern Nigeria.

5. Restricted to the area around Mt Cameroon and the island
   of Bioko are *Cercopithecus preussi*, *C. pogonias pogonias*,
   Cercopithecus erythrotis (excluding the *sclateri* form in eastern Nigeria), *Mandrillus leucophaeus*, and the *Procolobus*
   forms *pennantii* and *preussi*.

6. *Cercopithecus cephus*, *Cercocebus albigena* and *Colobus satanas*
   occur very patchily in the Cameroon area but have most of
   their distribution to the south and east.

7. *Procolobus verus* has a unique pattern, occurring without
   subspecific variation through the forest zone from southern Sierra Leone to Ghana, and also in southeastern
   Nigeria.

Explanation of patterns

Refuges and dispersals

To what extent are these patterns consistent with previous
hypotheses attempting to explain the distribution of West African
forest primates and other vertebrates?

The most salient hypothesis is that developed by Booth (1958a,b),
invoking climatic change during the Pleistocene. Booth argued that
during dry periods in the Pleistocene the West African lowland forest
('High Forest') became restricted to three major 'refuges', one in
Liberia, one in eastern Côte d'Ivoire and western Ghana, and one in the
region of Mount Cameroon. Booth postulated a secondary, minor
refuge in the Benin area of western Nigeria. Between the Liberian and
Côte d'Ivoire–Ghana refuges, Booth postulated that a tongue of
savanna, the Baoulé-V, reached the sea, while the present Dahomey
Gap occupied most of the area between western Ghana and eastern
Nigeria (with the minor Benin refuge intervening, possibly in riverine
forest). According to this scenario, populations that were isolated in
refuges during dry periods differentiated from one another. When

climate ameliorated, populations spread again, sometimes meeting and hybridizing in a narrow contact zone, sometimes achieving sympatry over a wide area without hybridizing, and sometimes being limited in their spread by major rivers (particularly the Niger). Some general aspects of refuge theory are discussed elsewhere in this book by Hamilton (Chapter 2).

Booth's hypothesis is supported by Schiøtz's (1967) analysis of treefrog distributions. Like Booth, Schiøtz found that the Baoulé-V produces a less significant faunal discontinuity than the Dahomey Gap, but he also points out that in treefrogs there is a distinct 'farmbush' fauna occupying young secondary forest within the rain-forest zone, as well as the northern fringes of the rain forest and gallery forests within the savanna. Schiøtz argues that human agricultural activity has expanded the available habitat for members of this fauna, many (but not all) elements of which extend without change from Sierra Leone to the region of the Cross River in Nigeria.

Moreau (1969), while finding the fish distribution data of Clausen (1964) to be consistent with Booth's hypothesis, showed that forest birds have a somewhat different pattern. The Baoulé-V has almost no effect on them, and many species and subspecies cross the Dahomey Gap unchanged. However, several taxa are limited by the Gap and a number subspeciate at the Niger. Moreau argued that the Niger has been an important faunal barrier, possibly due to the existence of a broad estuary in place of the present delta during times of raised sea-levels (i.e. interglacial episodes). Moreau suggested that this estuary may have reached inland to the savanna zone during a southward advance of the Sahara. This suggestion is based on Moreau's view that interglacials were the driest periods of the African Quaternary, and glacials (due to lowered temperatures) the wettest.

Grubb (1978b) has taken a rather different view to Booth and Moreau. He has argued that the speciation of African forest mammals is most likely to have been initiated not at the event of isolation in a refuge, but rather during the expansion of populations into unoccupied habitats when forest itself expanded at the end of dry climatic episodes. Since Grubb accepted the evidence that glacial episodes were dry in Africa, these periods of forest expansion would have coincided with deglaciations.

Barriers to gene flow
Whatever the events may have been surrounding speciation, both refuge and dispersal models require the presence of barriers that for a greater or lesser period impede gene flow. The major potential

geographic barriers to West African forest mammals would seem to be: (1) gaps in the forest occupied by savanna or montane vegetation, (2) rivers, and (3) the ocean (of particular significance in the case of the island of Bioko).

How significant are these barriers? There is little doubt that the interglacial sea channel (presently 32 km wide) between Bioko and Cameroon is an effective barrier to gene flow for non-human primates in the absence of 'rafting' (for which there is no evidence in this case). There is considerable uncertainty, however, about the significance of savanna vegetation and rivers as dispersal barriers. Addressing this issue, Robbins (1978) re-evaluated data on high forest mammals and concluded that the Volta and, especially, the Niger have had more influence on West African forest mammal distributions than has the Dahomey Gap.

The evidence presented in this chapter is not consistent with Robbins' conclusions. Curiously, while re-evaluating the significance of the Dahomey Gap, Robbins only partially assessed the effect on mammals of the vegetation transition which produces the gap. He noted that 28 mammalian taxa (50% of them bats) bridge the gap as well as the Volta and Niger Rivers, and he analyzed the effect of the Volta and Niger as barriers, but he failed to consider whether any mammals are limited by the gap itself. Robbins listed four forest monkeys among 10 mammalian taxa for which the Volta is stated to be an eastern range boundary. These are *Cercopithecus campbelli, C. diana roloway, Procolobus badius waldroni* and *Procolobus verus*. Booth's 1956 paper and my own analysis show that *C. d. roloway* and *P. b. waldroni* are not limited by the Volta; they do not extend east of the moist forests of southwestern Ghana – in other words, they *are* limited by the Dahomey Gap. On the other hand, *P. verus* is found on both sides of the Volta and east of the Niger. *C. campbelli* does have its eastern limit near the Volta, but its eastward spread may be affected by competition with *C. mona*, which it meets in this area. Robbins listed 10 primates out of 19 mammalian taxa for which the Niger is an eastern or western range boundary, but this analysis is also flawed. Robbins included *Galago crassicaudatus, Cercopithecus petaurista* and *Perodicticus potto potto* (none of which occur in Nigeria), and placed *C. erythrogaster* and *P. potto juju* on the wrong sides of the river. Despite these errors, the Niger does indeed seem to limit prosimian distributions (Oates & Jewell, 1967; Grubb, 1978a), probably because these animals are less inclined to swim than are monkeys. *C. erythrogaster*, while possibly separated from *C. erythrotis sclateri* by the Niger, does not enter the Dahomey Gap to the west of its range (Oates, 1985).

Primate evidence supports Booth and Moreau in their view that the Dahomey Gap forms a significant faunal discontinuity for forest animals and that other major discontinuities occur in the vicinity of the Baoulé-V and the River Niger. Another significant discontinuity occurs in the vicinity of the Cross River, probably not because the river itself is a significant barrier, but because of a drier climate and a high human population density (with intensive cultivation and no large forest blocks) to the west of the river (Schiøtz, 1967).

At the south of the area considered here, the Sanaga River has also been proposed as a major faunal boundary. However, several forms which have their southern limit in this area occur on both the north and south banks of the lower Sanaga, e.g. *Cercopithecus erythrotis camerunensis*, *C. mona*, *C. pogonias pogonias* and *Mandrillus leucophaeus* (Gartlan & Struhsaker, 1972; Grubb, 1973). Conversely, some forms for which the Sanaga has been considered a northern limit definitely or possibly occur north of the river, e.g. *C. cephus*, *Miopithecus talapoin*, *Cercocebus albigena* and *Mandrillus sphinx* (Gartlan & Struhsaker, 1972, and evidence presented in this chapter). On the other hand, although Gartlan & Struhsaker saw *C. nictitans martini* on the north bank of the Sanaga and *C. nictitans nictitans* on the south bank, *C. n. nictitans* does occur to the north of the upper Sanaga.

In general there is little evidence that any rivers are presently major barriers to the dispersal of forest monkeys in West Africa, although they obviously could serve to somewhat impede gene flow. All *Cercopithecus* monkeys are able to swim, but some species (such as *C. neglectus*) will take to the water much more readily than others (A. Gautier-Hion, personal communication). The importance of rivers as zoogeographical boundaries in West Africa has probably been over-rated because of an *assumption* that they are significant. When a species' range ends in the vicinity of a river it has been too readily assumed that the river itself is a dispersal barrier. In fact, other environmental discontinuities in the vicinity of the river may be having a more significant effect. While the Niger and its large delta apparently do form a significant barrier to some mammals (see also the recent paper of Happold, 1985), it is possible that other factors than the river itself play important roles in limiting distributions in this area, and that these factors are reinforced by the position of the river. Southern Nigeria, where the Niger enters the Atlantic has a complex history and geography. Its climate, the flow of the Niger and the shape of the delta have been subject to profound changes, but the exact pattern of change and their effects on the fauna are difficult to reconstruct on present evidence (Grove & Warren, 1968; Sowunmi, 1981b; Oates, 1985).

Added to the potential effects of vegetation and water barriers to the dispersal of these primates may be competition. *Cercopithecus mona* and *C. nictitans* seem to be limited in their spread in the Upper Guinea forests by competition with *C. campbelli* and *C. diana* respectively. Local extinction may also have played a role in producing differences between the Upper Guinea and Cameroon faunas. While several species show vicariance between the two areas (e.g. *Cercopithecus petaurista* and *C. erythrotis*, *Cercocebus atys* and *C. torquatus*, *Colobus vellerosus* and *C. guereza*, *Procolobus badius* and *P. pennantii*), others do not (e.g. *Cercopithecus diana* and *Mandrillus leucophaeus*). These latter species may have once had a wider distribution and have since become locally extinct, and/or they may have been limited in their dispersal from a point of origin by their environmental tolerances (including their ability to compete with other species).

Divergence times

Some features of West African forest monkey distributions are consistent with Booth's forest refuge model. Prominent among these features is the Baoulé-V, where six forest monkey species have different subspecies in eastern and western Côte d'Ivoire. Such a pattern is reasonably explained by the separation of forest-dependent populations by a non-forest barrier for a period or periods sufficiently long to allow for significant genetic divergence; it is less consistent with Grubb's model, which would require six events of population dispersal to be accompanied by a similar level of genetic divergence in more or less the same place.

If populations did diverge on either side of an expanded Baoulé-V during a period of dry climate, when might this have happened? The degree of difference shown by the monkey subspecies here is similar to that between the mammals of Bioko (Fernando Poo) and the mainland. The Bioko mammals have presumably diverged from mainland forms since the rise in sea level after the height of the last glaciation 18 000 years BP. Isolation on the order of 10 000 years is consistent with a period of increased aridity in northwestern Africa reflected in Atlantic sediments deposited 24 000–13 000 years ago (Pokras & Mix, 1985). On the other hand, it is possible that the last glaciation simply reinforced a process of divergence set in motion during one or more previous glaciations.

The faunal divergence of the Upper Guinea forests from those of Lower Guinea to the east around Mt Cameroon is more profound than that in the vicinity of the Baoulé-V and is almost certainly of longer standing. According to the classification followed in this chapter, only two forest monkey species, *Cercopithecus mona* and *C. nictitans*, are common to both forests and of these *C. mona* only touches the extreme eastern edge of the Upper Guinea forest zone. Both species occur in gallery forests and forest patches within the Dahomey Gap and north of the Niger–Benue Rivers in Nigeria. Apparently these species have features which allow them to flourish in an adaptive zone which other species have not been able to exploit so successfully.

Between the Upper Guinea and Cameroon forests there are, or have been, a series of potential barriers, not just one. The effects of the Dahomey Gap are compounded by the effects of the lower Niger and its delta, and by the climatic and vegetational change in the vicinity of the Cross River referred to above. Whatever the history of connection between the Upper Guinea and Cameroon regions may have been, it seems very likely that the present differences in monkey fauna are the result of processes that have been in operation since long before the last glacial maximum.

Discussion and conclusions
While a forest-refuge model may help to explain the differentiation and present distribution of some rain-forest dependent species, it does not provide a complete explanation for the present distribution of all African forest monkeys. Species such as *Cercopithecus diana*, apparently dependent on relatively large blocks of lowland rain forest, do seem susceptible to 'refuging', but other species (such as *C. mona* and *C. nictitans*, which do well in gallery forests and forest outliers) do not. These gallery forest species share some features with Schiøtz's farmbush fauna. They may be subject to isolation (with possible genetic differentiation) during extremely arid periods, but in other circumstances they seem well able to cross gaps between forest blocks. For instance, *C. campbelli*, *C. petaurista*, *Cercocebus atys* and *Colobus polykomos* all extend well to the west of the area usually mapped by botanists as lowland rain forest. This probably indicates the ability of these species to exploit gallery forests and forest patches in coastal Guinea, Guinea-Bissau and southern Senegal, as they do in northern Sierra Leone (Harding, 1984). A similar gallery forest monkey fauna from the northern Central African Republic was recently described by Fay (1985). Gallery forest and 'dry' forest are clearly an important habitat for many African primates generally considered to be 'rain forest' species, but their ecological and evolutionary significance has been relatively neglected. Species in these forests must be adapted to

lower diversity and more seasonal food resources than species restricted to moist lowland rain forest. See Colyn, Chapter 6.

So, at least three different major monkey faunas can be recognized in West Africa, each probably affected differently by changes in climate and vegetation. These are: (1) a strict lowland rain-forest fauna, (2) a forest fauna that extends well into the gallery forests of the savanna woodland zone, and (3) a savanna woodland fauna (members of the *C. aethiops* group, savanna baboons, and *Erythrocebus patas*). Two species (*Procolobus verus* and *Cercopithecus preussi*) do not fit well into any one of these groups. *P. verus* bridges the Dahomey Gap and the Niger, but does not occur as far north or west as elements of the gallery forest fauna; nor is it found in the Cameroon region (although it is possible that it once occurred there and is now extinct). Its distribution is similar in some respects to that of the pygmy hippopotamus (*Choeropsis liberiensis*), the distribution of which seems to be tied in part to swampy areas within the rain-forest zone. *C. preussi* is particularly associated with montane forests on the eastern edge of our area.

The presence of *C. preussi* in the Mt Cameroon area and on Bioko, highly disjunct from its relative *C. lhoesti* in the medium- and high-altitude forests on the eastern rim of the Congo Basin and from its recently discovered relative in Gabon is difficult to explain by the most widely accepted model of the history of African forest refuges, which argues (from considerable evidence) that forest was reduced during dry episodes that were contemporary with glacial maxima. This model, described by Livingstone (1975), Hamilton (1976, 1981, 1982) and others, is at variance with Moreau's earlier model (1966) proposing that montane forest was spread more widely at these times. However, there is now a good deal of evidence that the climate of Africa has not responded uniformly to glacial events (see Hamilton, Chapter 2). For instance, while northwestern Africa (including the area discussed in this chapter) seems to have been generally most arid during times of high ice volume (and therefore lowered sea-levels), central equatorial Africa seems to have been driest during ice-growth phases and to have been maximally humid during deglaciations (Pokras & Mix, 1985). Evidence from Lake Bosumtwi in Ghana of montane tree pollen in sediments dated between 15 000 and 8 000 BP (Maley & Livingstone, 1983) indicates that montane forest also spread more widely in north-western Africa during deglaciations.

Such evidence leads to the unsurprising conclusion that the history of African forest vegetation has been very complex. Among other possibilities, it seems very probable that montane-type forest was linked across or around the Congo Basin during some cool, wet phases, including the last deglaciation. Such a scenario is consistent with White's (1981) evidence of a relatively recent dispersal track for montane forest trees between the Cameroon Highlands and the East African mountains around the southern edge of the Congo Basin and through Gabon and Congo-Brazzaville. This evidence of complexity should warn us against accepting overly simple models to explain current primate distribution patterns. Not only does each present-day species have a different set of environmental tolerance limits, it also inevitably has a different history in space and time. Certain histories and present-day tolerances do show major overlap and these overlaps produce patterns of species clustering in a mapping process. While these clusters can help us to identify phenomena of evolutionary significance, we must obviously be careful not to force species into clusters in which they sit uncomfortably, thereby obscuring variation which is itself of evolutionary significance.

Acknowledgements

The research discussed here is part of a larger programme of study on the evolutionary ecology of West African rain-forest primates that has been funded by a series of grants from the Research Foundation of the City University of New York and the National Science Foundation. I am indebted not only to those organizations, but to the museum curators who have given me access to the specimens in their care. I thank Jörg Ganzhorn and Sigrid Müller-Friauf for translating part of Eisentraut's monograph, and I would like to thank Jean-Pierre and Annie Gautier for inviting me to participate both in the stimulating meeting at Paimpont and in the book that has evolved from that meeting.

I.6

Distribution of guenons in the Zaïre–Lualaba–Lomami river system

MARC M. COLYN

Introduction

Our knowledge of the taxonomic status and geographical distribution of mammals of the Zaïre River Basin is still far from satisfactory, despite the activities of a number of scientific expeditions. The most famous of them is unquestionably that carried out by H. Lang and J. P. Chapin (1909–15), who explored the areas north and northeast of the Zaïre River. Later on, emphasis was mostly laid on the eastern part of the country, the 'Central African Rift', where extensive collections and observations were made by the 'Swedish Zoological Expedition' (1921), G. F. de Witte (1933–5), S. Frechkop (1938), the 'Lund University Congo Expedition' (1951–2), F. Bourlière and J. Verschuren (1957–62) and U. Rahm (1960–9). Conversely, the lowland rain forests within the Zaïre Basin remained almost unexplored. This is readily apparent on the maps of the excellent work of H. Schouteden (1947), especially south of the Zaïre River.

Despite these shortcomings, the taxonomy of guenons in Zaïre was discussed by Matschie (1912, 1913), Lönnberg (1919), Schwarz (1928a,b,c), Schouteden (1947), Booth (1955), Verheyen (1962), Dandelot (1968), and Rahm (1970). More recently, a new species, *Cercopithecus salongo*, was described from Wamba by Thys van den Audenaerde (1977): and the very rare *C. wolfi elegans* was rediscovered by Colyn & Verheyen (1987a). More information on the distribution of other species of the *mona* group was provided by Schwarz (1928a) and Booth (1955) as well as on *C. hamlyni*, *C. lhoesti*, and *C. mitis* from the eastern part of the country (Rahm, 1970) on *C. (Allenopithecus) nigroviridis* by Verheyen (1963) and Gautier (1985), and on some rare forms on the left bank of the Zaïre/Lualaba River (Thys van den

Audenaerde, 1977). More general information is also given by Schouteden (1947), Dandelot (1965), and Heymans (1975), whereas Colyn (1986, 1987), and Colyn & Verheyen (1987a,b,c) provide new data on species of the lowland rain forest.

The biogeographical importance of the forest area south of the Zaïre/Lualaba River was foreseen by Misonne (1963), Rahm (1966, 1970), Kingdon (1971) and Delany & Happold (1979). Moreover, Laurent (1973), Kingdon (1980), Grubb (1982), Colyn (1987) and Prigogine (personal communication) drew attention to the large number of endemic species in this area, and postulated the existence of a Pleistocene forest refuge. The study of isolated populations of forest cercopithecids on both the left and the right banks of the river led Colyn (1987) to postulate the existence of a 'Major Fluvial Refuge'.

Between 1977 and 1986, the author undertook a number of zoological expeditions in most of the Zaïre forested areas, between 3° N and 3° S. The goal of these field surveys was to study in detail the geographical distribution of Zaïrean mammals: besides his own observations, he also questioned local hunters, and bought animals at village markets (see details in Colyn, 1986; Colyn & Verheyen, 1987a). The present tentative overview of the distribution of the cercopithecine monkeys within the Zaïre lowland rain forests, is based on these observations, and on the study of the available museum material, as well as on a literature survey. The possible causes of the various distribution patterns will be discussed, and the role played by the large river systems in the distribution of guenons will be emphasized. The distribution of other primates, mostly mangabeys and colobus monkeys, will be alluded to when necessary to support our conclusions.

The study area

Our study area, located in the Central Zaïre Basin ('Cuvette Centrale'), ranges from 3° N to 3° S, and from 21° E to 27° E and it embraces both banks of the Zaïre River (Figure 6.1). It was indeed selected because of the key role this geographical area plays in species dispersal. The network of rivers is very dense, as it includes the Zaïre River and many of its tributaries, the Ulindi, Lowa, Maïko, Tshopo, Lindi, Aruwimi and Itimbiri on the right bank, and the Lomami River on the left bank. The relief of the whole area is gently undulating for the most part; further east, however, the ground rises progressively towards the Maniema (Kivu Sub-Region) and the Ituri hills (Upper Zaïre Sub-Region). The forest cover of the whole area is not homogeneous, and two major forest types can be distinguished:

the evergreen rain forests dominated by stands of *Gilbertio-dendron*, with which *Brachystegia* and *Julbernardia* are often associated. They are quite common on the right bank of the Zaïre River.

the semi-deciduous forests, sometimes dominated by *Brachystegia*, *Julbernardia* or *Scorodophloeus*, that are mostly found on the left bank of the river.

Furthermore, 'riparian and swamp forests' are frequent along the Zaïre and Lomami River banks, and enclaves of open wetlands partly covered with grassy vegetation (ferns, grasses, sedges, aroids, Rubiaceae, etc.) and surrounded by the seasonally flooded forests called 'ido' by local Bakumu villagers, are sometimes encountered within the main forest blocks.

Much of the rain forest is now being destroyed by cultivation, especially around Kisangani, and along the major roads.

Our study area overlaps two major zoogeographical regions corresponding to two major forest blocks (Misonne, 1963; Rahm, 1966; Kingdon, 1971; Laurent, 1973; Delany & Happold, 1979; Grubb, 1982; Colyn, 1987; Prigogine, personal communication), which we will call the 'East Central' and the 'South Central' areas, following Grubb

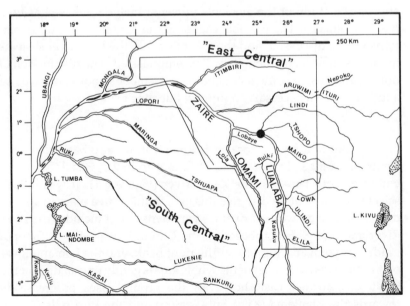

Figure 6.1. The river system of the East Central and South Central forest blocks. The area surveyed is surrounded by a black line, and the location of the city of Kisangani is indicated by a black dot.

(1982). C. 119 species of mammals have so far been recorded in the first area, and more than one hundred in the second.

The guenons and their distribution

Eleven species of guenons were previously known to occur in our study area: *Cercopithecus nictitans, C. mitis, C. ascanius, C. pogonias, C. wolfi, C. neglectus, C. dryas, C. salongo, C. hamlyni, C. lhoesti,* and *C. (A.) nigroviridis* (Schouteden, 1947; Dandelot, 1968; Heymans, 1975; Thys van den Audenaerde, 1977; Colyn, 1986, 1987; Lernould, Chapter 4). The distribution of the various forms, sometimes supplemented by some details on their morphological characteristics, will be given in the following pages. The geographical coordinates of the localities mentioned are given in Appendix 6.1.

Cercopithecus nictitans Linnaeus, 1766

Among the three subspecies recognized by Dandelot (1968), only *C. n. nictitans* is found in the forests of Zaïre. We have seen several of them with approximately the same coat colour between the Mungala and Itimbiri rivers, the lower reaches of the latter apparently acting as a barrier between *C. n. nictitans* and *C. m. stuhlmanni* (Figure 6.2).

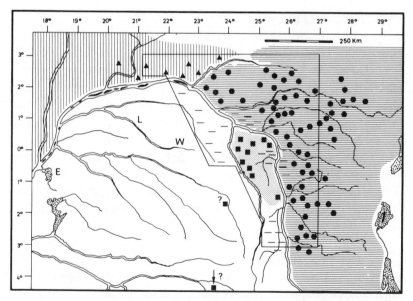

Figure 6.2. Geographical distribution of the *mitis* superspecies. ▲: *C. n. nictitans;* ●: *C. m. stuhlmanni;* ■: *C. m. heymansi.* E: Ekele; L: Lomako; W: Wamba, where *C. m. heymansi* is unknown. ■?: Iligampangu and Piana.

However, in the Musée Royal de l'Afrique Centrale (MRAC) at Tervu-ren, there is the skin of an adult male *C. n. nictitans* (R.G. 8348) from Djamba in the easternmost part of the species' distribution. The two species might therefore come into contact between 25° and 26° E, where the east–west course of the Itimbiri River can no longer act as a barrier.

The *C. n. nictitans* of our study area are distinguished from *C. m. stuhlmanni* populations mostly by the whitish colour of the hairs of their nose patches. The ears are almost hairless, adorned with a few whitish hairs. The top of the head, the neck and the shoulders are bluish-grey, with serially annulated hairs, as on the coat of the back. There is no 'diadem' or superciliary stripe, but a few light-coloured hairs do exist above the eyes. The whiskers do not conceal the ears (Colyn & Verheyen, 1987b). The Zaïre specimens belonging to the MRAC have, on the whole, a less greenish coat colour than those coming from the Sangha River area, west of the Ubangi River.

Cercopithecus mitis Wolf, 1822

Two of the eight subspecies of *C. mitis* are found in our study area.

C. m. stuhlmanni

This subspecies has a broad 'diadem', ochre ear tufts, and a whitish chin. The top of the head, the neck and the shoulders are deep black. The *C. m. stuhlmanni* of Niapu and Akenge are very similar (Allen, 1925). Near the Maïko River, we have seen freshly killed animals whose coat colour was either greenish-grey or greyish-blue, but these colour variants can be observed within the same social group.

C. m. stuhlmanni is common on the right bank of the Zaïre/Lualaba River, from the Itimbiri confluence to 5° S (Figure 6.2). Its distributional limits east of our study area remain imprecise, and animals inter-mediate with the forms *kandti*, *doggetti* and *schoutedeni* are known (Dandelot, 1962; Rahm, 1970; Kingdon, 1971; Lernould, Chapter 4).

C. m. heymansi

This new subspecies has been described recently (Colyn & Verheyen, 1987b), after two adult specimens from Yaenero, in the Lomami/Lualaba forest block. The face is scantily haired, black with a narrow supra-orbital white band with long greyish whiskers extending backwards besides the ears. The top of the head, the neck and the shoulders are bluish-grey, as the rest of the mantle whose hairs are annulated. The hairs of the flanks and the underside of the body are lightly coloured. The head profile looks very much like that of *C. m. opisthostictus*. It is therefore easy to distinguish *C. m. heymansi* from *C. m. mitis, C. m. stuhlmanni* and *C. n. nictitans*.

The pictures of *C. m. maesi* in Schouteden (1947) actually represents *C. m. heymansi*. The *maesi* holotype (Lönnberg, 1919), a subadult individual from Kutu, has a long haired coat typical of 'mountain' forms, and its general colour does not correspond to that of the *C. mitis* of the Central Zaïre Basin. Moreover, its peri-anal area is covered by long reddish hairs, and the black areas of the legs, head, neck and shoulders have brownish or reddish glints. Its origin is uncertain, and its coat colour differs from that of the other adult male *maesi* (Schouteden, 1947). This holotype likely comes from the mountain forests of the eastern side of the Kivu Ridge, as suggested by a comparison with skins from northwest of the Sake Bay that are quite similar to the holotype. Furthermore, its overall coat colour makes it resemble the *mitis* forms intermediate between *stuhlmanni* and *schoutedeni*.

The geographical distribution of *heymansi* (Figure 6.2) raises a real problem. We have observed this form only in the northern part of the forests between the Lualaba and Lomami rivers, which *heymansi* shares with *C. wolfi wolfi* and *Colobus rufomitratus parmentieri* (Colyn & Verheyen, 1987a,c). It is common in the forests between Yaenero and Opala, but was not found near Ubundu, along the Ruiki River, while we did not observe it in the swamp forests that are so common along the Upper Lopori. But among the three skins in the MRAC attributable to *C. m. heymansi*, two come from Iligampangu, on the Upper Tshuapa, and the third is from Piana, near the Sankuru River. More information is therefore needed to clarify the geographical distribution of *C. m. heymansi*; it is, however, worth noticing that *C. mitis* is not known from the Wamba region (Kuroda, Kano & Muhindo, 1985), Lomako (Zeeve, 1985) and Ekele (Gautier, 1985).

Cercopithecus ascanius Audebert, 1799 (Figure 6.3).

The range of this guenon broadly corresponds to the 'South Central' and 'East Central' forest blocks. At least two of the four forms recorded from Zaïre (*C. a. ascanius, C. a. katangae, C. a. whitesidei* and *C. a. schmidti*) are found in our study area.

C. a. schmidti

It is very common throughout the 'East Central' forest block, and it has been observed in each of the interfluvial forest areas (Figure

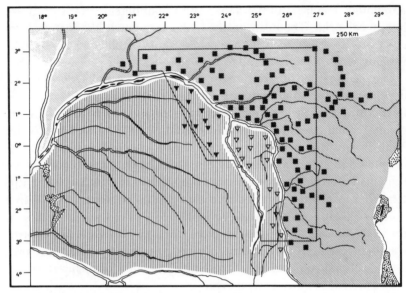

Figure 6.3. Geographical distribution of *C. ascanius*. ■: *C. a. schmidti*;
▼: *C. a. whitesidei*; ▽: *C. a. whitesidei* < *C. a. katangae*.

6.3). The ear lobes are sparsely covered with whitish hairs. The sides of
the face do not have hair whorls as in other *C. ascanius* forms, whereas
the white nose spot is smaller than in *whitesidei*. The colour of the hairs
of the temporal region and of the cheek may vary between individuals,
but that of the mantle or limbs never does so. However, we observed
in the forests around Kalima individuals whose distal part of the hind
legs was bluish-grey to dark slate grey. This colour can even reach the
thigh and the back, whose hairs are distinctly annulated. Around
Kisangani, the back of some individuals is lighter than further east.

C. a. whitesidei.

The skins of specimens collected in the Upper Lopori, near
Simba, closely resemble that of the *whitesidei* holotype described by
Thomas, 1909. The nose spot is sometimes yellowish, sometimes
creamy-white, and seldom white. A small ochre hair tuft is visible on the
ears. The tail, whose dorsal black band nearly reaches the lateral sides,
turns reddish in its distal half. On the contrary, the *C. ascanius* of the
Lomami/Lualaba forest block are intermediates between the *katangae*
and *whitesidei* subspecies. In most individuals, the nose spot is white,
not yellow. The form of the whorl, and especially the colour of the hairs
on each side of the face vary. Some ochre hairs decorate the ears. The
contrasted colours of the tail are like those of *katangae*.

In other respects, downstream from Mbandaka, individuals with the two distal thirds of the tail definitely red, and a bright ochre nose spot, are found (J.-P. Gautier, personal communication; Lernould, Chapter 4).

The analysis of this rich material from the 'South Central' forest block, particularly the comparison of the general colour of the coat, the pigmentation of the hairs of the nose, the sides of the face and the tail, together with the importance of the temporal whorls, leads to the following conclusions:

> The *C. ascanius* populations of the Upper Lopori and Lobaye river systems differ in their colour; the former are clearly similar to the *whitesidei* form, the latter being intermediate between *whitesidei* and *katangae*.
>
> West of the Lomami River, *C. a. whitesidei* populations are liable to erythrism. This is apparently a clinal variation along a southeast–west axis. It is mostly apparent on the nose patch and on the tail.
>
> There is individual variation in the implantation and pigmentation of the face and cheek-hairs; the taxonomic value of this character for distinguishing the subspecies *whitesidei* and *katangae* seems therefore open to question.

Cercopithecus pogonias Bennett, 1833 (Figure 6.4).

    C. p. grayi is the only subspecies known from Zaïre, where it shares the same forest areas as *C. n. nictitans*. We found it from the Ubangi to the Itimbiri River at *c.* 23°30' E.

Cercopithecus wolfi Thomas, 1907 (Figure 6.4).

    Apart from *pyrogaster*, which lives in the forests south of the Kasai River, all three other forms of *wolfi* were found in our study area.

C. w. wolfi

    This subspecies is known from the 'South Central' forest block, south of the Zaïre River. The Lomami River eastwards, and the Kasai River southwards keep it apart from *C. w. elegans* and *C. w. pyrogaster* respectively. Lönnberg (1919), however, records one specimen from Kisangani, and Dandelot (1965) considers *wolfi* as inhabiting both banks of the Lomami River, southwest from Kisangani, without quoting references. *C. w. wolfi* is indeed present east of the Lomami River, and common in the Lobaye and Ruiki river basins (Figure 6.4). The skins from this area are very reddish on the hind limbs and the upper part of the thighs: their reddish flanks contrast strongly with the

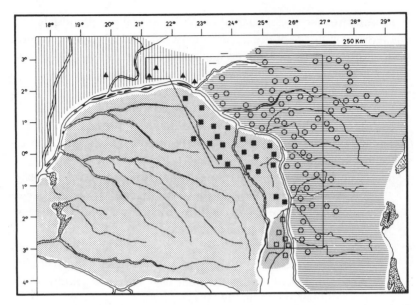

Figure 6.4. Geographical distribution of the *mona* superspecies. ▲: *C. p. grayi*; ○: *C. w. denti*; ■: *C. w. wolfi*; □: *C. w. elegans*.

darker colour of the back. These two characteristics are more marked in specimens from the western bank of the Lomami; the hairs of the throat and of the inner part of the forelimbs of these animals are yellowish, not whitish. In fact, as was the case for *C. a. whitesidei* the pigmentation of *C. w. wolfi* warrants the distinction of several populations within the 'South Central' forest block.

C. w. elegans

The validity of this beautiful form was long questioned (Dandelot, 1968), in so far as the type locality of the holotype and paratype remained imprecise (the middle reaches of the Lomami, quite likely), and as Schwarz (1928a) made reference to three specimens from Kibombo, while *C. w. wolfi* remained unknown in the Lobaye and Ruiki river basins. It was therefore believed that the distribution of *C. w. elegans* stretched throughout the whole forest block between the Lomami and Lualaba rivers (Schouteden, 1947; Booth, 1955; Hill, 1966; Dandelot, 1968; Heymans, 1975; Thys van den Audenaerde, 1977).

C. w. elegans is, however, allopatric with *C. w. wolfi*, which does exist between the two above mentioned rivers, but only south of 2° S (Figure 6.4; Colyn & Verheyen, 1987a). A forest guenon in the north of its range, this subspecies enters the forest galleries further south, in the Kasuku and Lueki river basins. From 3° S, the habitat becomes a true forest–savanna mosaic and the presence of *elegans* there probably depends on the relative extent of savanna vegetation. In the north this form is separated from *C. w. wolfi* by a large area of swamp forest. As access to this region is difficult, we do not know whether the two subspecies actually make contact with each other.

C. w. denti

C. w. denti has a pattern of distribution similar to that of *C. m. stuhlmanni* (Figure 6.4). This subspecies is known from the entire 'East Central' forest block, except for the mountain forests of eastern Zaïre.

C. w. denti is common everywhere in our study area, from the Itimbiri River to the Kindu area, on the right bank of the Zaïre/Lualaba. No significant variation of its coat colour was ever recorded.

Cercopithecus neglectus Schlegel, 1876 (Figure 6.5)

Despite its very large distribution, the de Brazza monkey remains a monotypic species. The forested areas of southeastern Zaïre excepted, it can be found anywhere in the whole Zaïre River complex. The Kasai and Lomami river basins limit its distribution southwards. Northeastwards its dispersal towards Uganda, Kenya and Ethiopia was quite likely facilitated by the Ubangi and Uele rivers, and by the Upper Nile River system.

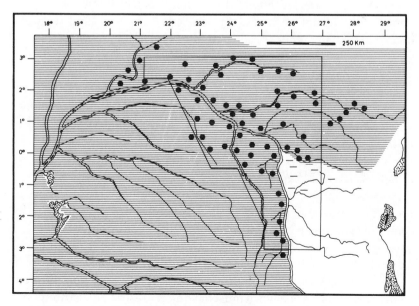

Figure 6.5. Geographical distribution of *C. neglectus*.

In our study area, we did not find it south of the Maïko River (Figure 6.5). Everywhere else, it was common along the river banks, and on the islands of the Lindi and Tshopo, as well as on those of the Zaïre River itself, from Kisangani to Lukolela. It is surprising that the de Brazza monkey, whose distribution is not restricted by the river barriers, does not enter the Lowa, Ulindi and Elila river basins, and that it is not known from eastern Zaïre, from the Semliki forest in the north to the Kahuzi mountain area further south (Gyldenstolpe, 1928; Frechkop, 1938, 1943; Schouteden, 1947; Curry-Lindahl, 1956; Rahm, 1966, 1970; Verschuren, 1972).

At Simba, in the Upper Lopori region, we tried to trace a monkey called 'Ikesse Mpunga' ('Almost a Mpunga') ('Mpunga' = *C. neglectus*) and/or 'Ikesse' by the Ngando, a species not represented in the set of study skins we were carrying with us to help identification. We finally found a skin and skull of a subadult female 'Ikesse-Mpunga', exactly corresponding to the *C. ezrae* of Pocock (1908), considered as a synonym of *C. neglectus* by Schwarz (1928b). Kuroda *et al.* (1985) also reported the same vernacular name in the Wamba region, but wondered if it could not refer to *C. hamlyni*.

Cercopithecus dryas Schwarz, 1932 (Figure 6.6).

This species is only known from a single juvenile individual (skin and skull), and the picture of the captive animal (Schwarz, 1932; Schouteden, 1947). Although the collecting locality is said to be Lomela, we do not know exactly where it came from and we were unable to find any other specimen in the Upper Lopori basin.

Cercopithecus salongo Thys van den Audenaerde, 1977 (Figure 6.6)

C. salongo was described as late as 1977 from an incomplete skin coming from the Wamba region, near Djolu. More recently, Kuroda *et al.* (1985) reported having seen other specimens from the same area, where it is not rare, and mentioned that this guenon was called 'Ekele' by the Ngando.

However, the Ngando of the eastern bank of the Upper Lopori do not know 'Ekele', but a monkey called 'Ikelembe' was reported to us by some villagers near Luale, a name that does not apply to any other species of primate locally identified. Unfortunately, despite our intensive investigations between Mombongo and Simba, we were unable to gather any fresh information on *C. salongo*. If it is a guenon different from *C. dryas*, as we believe, it must have a very restricted distribution.

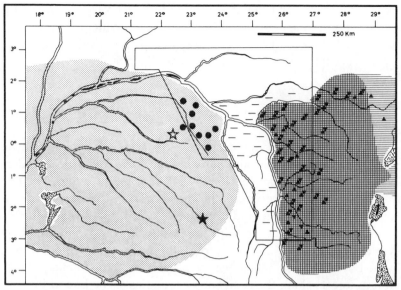

Figure 6.6. Geographical distribution of *C.(A) nigroviridis* (●), *C. hamlyni* (▼), and *C. lhoesti* (▲). The localities where *C. salongo* and *C. dryas* were originally discovered are respectively marked by open and black stars.

Cercopithecus hamlyni Pocock, 1907 (Figure 6.6)

The owl-faced monkey only occurs south of the Lindi River, in our study area. The Epulu and Nepoko river basins mark the northeast limit of its range. It has never been collected in the Upper Nile River system, Lake Mobutu Sese Seko, the Semliki River, the Ruwenzori range, Lake Idi Amin and the Rutshuru River basin in the Virunga National Park (Gyldenstolpe, 1928; Frechkop, 1938, 1943; Schouteden, 1947; Curry-Lyndahl, 1956; Verschuren, 1972).

Rahm (1970) emphasizes the difference in relative length of head–body and tail existing between the mountain forest (Kivu Ridge) and the lowland forest populations. This character, however, has no taxonomic importance, as similar differences were found in a single population around Kisangani; the same applies to the differences in mantle colour, which range from greenish to dull yellow-buff, sometimes tinged with orange.

Cercopithecus lhoesti P. L. Sclater, 1898 (Figure 6.6)

C. lhoesti shares the same range as *C. hamlyni* in our study area, and downstream the Lindi River acts as a barrier. This is a rather common monkey, whose coat colour does not vary much. In the Central Rift, C. *lhoesti* is more often found in mountain areas than C. *hamlyni*, which is confined to the ridge west of Lake Kivu. However, it can also occur in lowland forests, as exemplified by its occurrence near Lualaba.

Cercopithecus (Allenopithecus) nigroviridis Pocock, 1907 (Figure 6.6)

Like C. *neglectus*, this monotypic species is found in the heart of the Central Basin, on both banks of the Zaïre River (Schouteden, 1947; Malbrant & Maclatchy, 1949; Verheyen, 1963; Gautier, 1985). In the 'South Central' forest block this monkey is mostly known from the banks of the Zaïre River, and from the Ruki/Tshuapa River basin. The easternmost collecting locality is Ikela, on the Tshuapa River (Verheyen, 1963). We can now report, for the first time, its occurrence in the Lopori system, where we often observed these monkeys in the swamp forests along the right bank tributaries of the Lopori River. It probably also occurs along the Lilo River, which flows into the Lomami; it might also extend up to the left bank of the Lomami. We were unfortunately unable to explore the swamp forests located between the Ruiki and the Kasuku River basins, but they might represent a suitable habitat for the species.

Contrary to Hill's (1966) statement, C. *nigroviridis* has never been found on the right bank of the Zaïre/Lualaba, around Kisangani (Colyn, 1986), despite our intensive search for it.

Other Cercopithecidae of the area studied

Papio anubis tessellatus Elliot, 1909, and *Cercocebus albigena johnstoni* (Lydekker, 1900) are common in the rain forest of the Zaïre/Lualaba right bank. Their ranges are similar to that of C. *a. schmidti*. *Cercocebus aterrimus aterrimus* (Oudemans, 1890) is peculiar to the 'South Central' forest block. It is also recorded on the Zaïre left bank, up to the Kasai confluence, including the area between the Lomami and Lualaba rivers. *Cercocebus galeritus agilis* Rivière, 1885, common in the north of our study area, is limited southwards by the watershed between the Aruwimi and Lindi rivers. *Colobus guereza occidentalis* (Rochebrune, 1896) has a rather similar distribution pattern. However, it was not found south of the Itimbiri and Aruwimi/Ituri rivers. The last two monkeys are therefore absent from the 'South Central' forest block. The problems raised by the distribution of the colobus monkeys

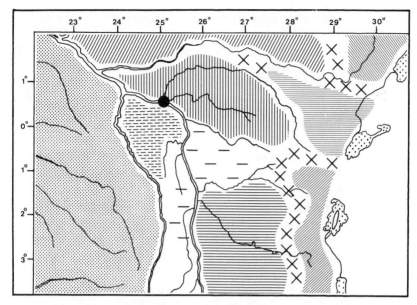

Figure 6.7. The probable role played by the river system in the geographical distribution of the various red colobus forms in Zaïre. Hybridization (X) more often occurs where there are no river barriers, for instance between forms peculiar to the Eastern Mountain Refuge, and between these forms and those of the Fluvial Refuge.

are much more complicated than those raised by *Cercopithecus* and *Cercocebus*. A taxonomic revision of the *Colobus rufomitratus* Peters, 1879 and *Colobus angolensis* P. L. Sclater, 1860 inhabiting the various inter-fluvial forest massifs is now under way. It will emphasize the existence of various population clusters, whose distribution patterns are of two kinds (Colyn, 1987; Colyn & Verheyen, 1987c). The first corresponds to the forms peculiar to the Central Rift (*Colobus angolensis ruwenzori, C. a. prigoginei, C. rufomitratus foai, C. r. ellioti*), and the second to the subspecies limited to the basins of the Zaïre/Lualaba tributaries (*C. a. cordieri, C. r. langi, C. r. parmentieri, C. tholloni*, Figure 6.7).

Discussion and conclusions

Despite its apparent complexity, the present geographical distribution of the 11 species of guenons recorded in the Zaïre/Lualaba/ Lomami River can be accounted for by one of the following six patterns of distribution, each of them shared by several species and subspecies.

1.   The classic pattern 'right bank – left bank', which explains

the distribution of so many land mammals (*Perodicticus, Galagoïdes, Funisciurus, Dendrohyrax, Cephalophus, Pan,* etc.). In the case of guenons, only *C. ascanius* and the mangabeys agree with this pattern of distribution.

2. The *mitis* and *mona* superspecies pattern. Both superspecies have similar geographical ranges: the Itimbiri River separates *C. p. grayi* from *C. w. denti* on the one hand, *C. n. nictitans* from *C. m. stuhlmanni* on the other.

3. The *C. hamlyni* and *C. lhoesti* pattern. These two species are restricted to the 'East Central' forest block, south of 2° N, between the Lualaba and the Central African Rift. A number of forms of Colobus monkeys have a similar distribution: *C. a. cordieri, C. r. langi, C. r. lulindicus* are found in the lowland rain forests of the Lualaba right bank, while *C. a. prigoginei, C. a. ruwenzorii, C. r. foai,* and *C. r. ellioti* live further east.

4. The endemic species of the 'South Central' forest block. Four cercopithecid species share this distribution pattern: *Cercocebus aterrimus* is widespread throughout the area, whereas *Cercopithecus dryas, C. salongo* and *Colobus tholloni* are restricted to the forests located between the Zaïre, Kasai and Lomami rivers.

5. The interfluvial forest pattern. The forest massifs between the Lomami and Lualaba rivers harbour the endemic *Cercopithecus w. elegans* and *Colobus rufomitratus parmentieri.* Furthermore, the local populations of *C. wolfi* and *C. ascanius* are noticeably different from subspecies *C. w. wolfi* and *C. a. whitesidei* living elsewhere in the 'South Central' forest block. Furthermore, *C. m. heymansi* is, to this day, known only from this area.

6. The *C. nigroviridis* and *C. neglectus* pattern. The ranges of these two monotypic guenons are not limited by river barriers. *C. nigroviridis* is restricted to the swamp forests of the Zaïre Central Basin, and occurs on both banks of the river. *C. neglectus* is a riverine species whose distribution is closely dependent on the hydrographic network, except the Lowa, Ulindi, and Elila rivers, south of the Maïko River.

For a better understanding of the problems raised by the distribution of guenons in the rainforests of the Zaïre Central Basin, it is also necessary to consider the distribution of 'peripheral' forms. In the following two paragraphs, we will contrast the 'East Central' sub-species (those of its easternmost part included) with the 'South Central' subspecies, including those living south of the Kasai River and 4° S.

1. On the right bank of the Zaïre River, *C. mitis* is represented by its subspecies *stuhlmanni*, which ranges from the Itimbiri to the eastern part of the 'East Central' block. In the northeast of the country, near the Semliki River and the Ruwenzori range, the local populations are quite different from those found in our study area. Further south, subspecies *C. m. kandti* and *C. m. doggetti* are limited to mountain regions. These isolated forms, although able to interbreed, keep a restricted distribution, limited to mountain forest; they are still distinguishable from the lowland forms.

Our field observations on the distribution of Colobus monkeys in the same area support the conclusions drawn from the guenons' distribution. The ranges of the red colobus subspecies are either limited to certain river basins, or restricted to the eastern mountain forests (Figure 6.7).

Among polytypic species, the following are peculiar to these eastern mountain forests: *C. mitis*, *C. hamlyni* (Colyn & Rahm, unpublished data), *Colobus rufomitratus*, *C. angolensis*, and *Gorilla gorilla beringei* Matschie, 1923. It is surprising that *C. a. schmidti* and *C. w. denti* are not found in most of these mountain forests, despite their wide distribution inside the 'East Central' block. In fact, there are no mountain forms of *C. ascanius* or *C. wolfi*. It is therefore possible that the subspecies of *C. ascanius*, *C. wolfi* and *C. mitis* presently living in the 'East Central' lowland forests did survive the major interpluvial period in the 'Fluvial Refuge'.

A similar situation occurs further west: *C. p. grayi*, *C. n. nictitans* and *Colobus rufomitratus oustaleti* are found on each side of the Ubangi River, ranging to the Congo, Central African Republic, Gabon, Equatorial Guinea, and Cameroon.

2. The large number of 'good' subspecies recorded from the southern part of the 'South Central' block is an excellent argument in favour of another possible Pleistocene refuge. Among these subspecies, two zoogeographical groups can be recognized:

The first includes the forms living south of the Zaïre/Kasai rivers, whose ranges are mostly confined between the Atlantic coast, northern Angola and the Kwango–Kwilu River basins. Here, *C. a. ascanius* and *C. m. mitis* seem to be, on the basis of their external characters, only distantly related to the other subspecies of the 'South Central' block.

It is also in this area that the southern talapoin (*Cercopithecus talapoin*) is to be found (Lernould, Chapter 4).

The second group, of major interest for us here, includes subspecies phylogenetically related to those observed in the Zaïre/Lomami/Kasai River system: *C. w. pyrogaster*, *C. w. elegans*, and *C. a. katangae*. As previously mentioned, the *C. ascanius* populations of the Lobaye basin (Lomami/Lualaba forest block) closely resemble *C. a. katangae*. The maps published by Schouteden (1947), Machado de Barros (1965, 1969), and Kingdon (1980), indicate that *C. a. katangae* is known from 4° S, between the Loange River (*c*. 19° 30′ E) and the Lualaba (Lernould, Chapter 4). There is little chance of *C. a. whitesidei* and *C. a. katangae* being restricted to the two banks of the Lomami (4° S) and east of the Sankuru, whereas no apparent barrier exists for the other species of primates. It is difficult to believe that a hybrid zone between *C. a. whitesidei* and *C. a. katangae* might exist at the same latitude in two different zoogeographic zones (Figure 6.4; Lernould, Figure 4.4, Chapter 4).

Our present studies suggest a clinal variation, west of the Lomami River, along a southeast to northwest axis. Now, if one considers the hydrographic network of the region concerned, it becomes apparent that the rivers cannot play a role in preventing gene flow between the various *C. ascanius* population clusters that survived within the postulated Pleistocene Fluvial Refuge (Figure 6.8). This does not apply, however, to the populations of the Lomami/Lualaba forest block, whose coat colour remains relatively stable.

An increasing erythrism of coat colour occurs in many mammal species along an east to southwest axis, from the area southwest of the Lomami River to the marshy areas between the Kasai River and the Mai-Ndumbe and Tumba lakes. This is particularly obvious among the *C. ascanius* of the Zaïre/Lomami/Kasai forest block. For instance, Amtmann (1966) mentions an increasing darkening of the coat of the squirrel *Funisciurus congicus* Kuhl, as one proceeds towards the more humid areas of the Central Basin, following the well known Gloger's rule.

It is interesting to compare the geographic distribution of *C. a. katangae* and *C. ascanius* from the Lomami/Lualaba forest block, with that of *C. m. opisthostictus* and *C. m. heymansi*. Although their distribution patterns are not obviously related, it is nonetheless possible that the *C. m. opisthostictus* populations were formerly in contact with those

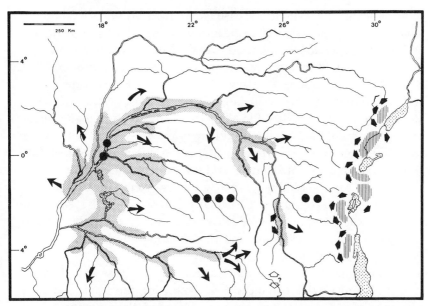

Figure 6.8. Endemic species (black circle) are more numerous in the South Central forest block (4 species) than in the East Central forest block (2 species). *C. neglectus* and *C. (A.) nigroviridis* are considered as endemics of the Fluvial Refuge. Arrows indicate the probable axis of dispersal of the various forms outside the major forest refuges, the Eastern Mountain Refuges (vertical hatching) and the Fluvial Refuge (dotted areas).

of *C. m. heymansi*. Without proposing a formal connection between these two forms, one cannot but be struck by the similarity of their coat colours and of their head profiles.

In the same way, the phylogenetic relationships between *Colobus r. parmentieri* and the *C. r. foai* from the eastern side of the Kivu Ridge (Colyn & Verheyen, 1987c), and the range of the macroscelid *Petrodromus tetradactylus* Peters (Corbet & Hanks, 1968), point towards a former contact between the eastern populations and those of the 'South Central' block, through the Upper Lualaba.

The former location of the postulated Pleistocene Refuge for the subspecies *Cercopithecus w. pyrogaster*, *Cercocebus aterrimus opdenboschi*, *C. galeritus chrysogaster*, *Cercopithecus a. ascanius*, *C. m. mitis*, and *C. talapoin* (southern talapoin) does not raise many problems, if this refuge is located south and west of the 'South Central' block (Laurent, 1973; Kingdon, 1980; Colyn, 1987; Prigogine, personal communication). This is not the case, however, for *C. w. elegans*, *C. a. katangae* and

C. m. opisthostictus, even in the case where their present distributions would not correspond to that of the former refuges (Hamilton, Chapter 2), here supposed to be connected with the Lualaba River system.

The role played by large watercourses in primate distribution has been more fully discussed elsewhere (Colyn, 1987). Except for *C. neglectus* (which swims well), and *C. nigroviridis*, rivers are major barriers, which can explain the allopatric distribution of many cercopithecid forms of the Zaïre/Lualaba/Lomami River system. Wherever a river no longer forms a barrier, because of the layout of its course, two subspecies can come into contact, and even extend their range.

The first possibility is, among others, exemplified by the red colobus monkeys, particularly *C. r. langi* and *C. r. ellioti*, on each side of the Lindi River (Figure 6.8; and Colyn, 1987), and *C. a. whitesidei* and *C. a. katangae* south of the 'South Central' block.

The second possibility is illustrated by the case of *C. hamlyni* and *C. lhoesti* (Figure 6.6). Following the Lindi and Ituri rivers, these two guenons extended their ranges northwards in the forests of both banks. The same mechanism probably explains the occurrence of *C. a. schmidti* north and northeast of the Ubangi River.

On the contrary, in the eastern mountain forests, forms such as *C. m. kandti*, *C. m. doggetti*, and *Gorilla gorilla beringei* are more closely restricted to their mountain habitats, which were only recently isolated from the Kivu Ridge forests.

Our study of the various subspecies of *Colobus rufomitratus* from the eastern Zaïre forests where there is no large watercourse, shows that all forms can interbreed (Figure 6.7).

Endemism within the 'East Central' forest block is of frequent occurrence in many taxonomic groups, but this is not the case for simian primates, with but two exceptions, that of *C. hamlyni* and *C. lhoesti*. On the contrary, primate endemics are numerous in the 'South Central' forest block: *Pan paniscus, Cercocebus aterrimus, Colobus tholloni, Cercopithecus salongo* and *C. dryas*. Furthermore, outside the 'South' and 'East Central' blocks, *C. neglectus* and *C. nigroviridis* can also be considered as endemics of the former Major Fluvial Refuge; during the Pleistocene it is quite likely that these two species were restricted to the Zaïre River basin and its associated marshy areas.

During the past few years, the phyto-zoogeographic influence exerted by the major African rivers may well have been underestimated. The Pleistocene Refuge theory is now accepted by most biogeographers (Hamilton, Chapter 2), but most of these refuges are located in and close to mountain areas. Laurent (1973), Kingdon (1980), Grubb (1982) and Prigogine (personal communication), however, postulated the existence of secondary refuges, such as the Zaïre Refuge located west of the present-day Mai-Ndombe and Tumba residual lakes. Their late recognition is undoubtedly due to our poor knowledge of the flora and fauna of these lowland rain forests.

The large number of species and 'good' subspecies endemic to the basin of the Zaïre River tributaries strongly supports the hypothesis of a Major Fluvial Refuge. The study of guenons provides another good argument in favour of this postulated Fluvial Refuge: the case of *C. nigroviridis*, whose range is restricted to the Central Zaïre Basin, and which is moreover considered as an ancestral form of guenon by Verheyen (1962), and Muleris, Couturier & Dutrillaux (1986).

Acknowledgments

I acknowledge gratitude to the direction of the 'Institut Zaïrois pour la Conservation de la Nature' and to the academic authorities of the 'Faculté des Sciences de l'Université de Kisangani' who facilitated the field research. I also extend thanks to Profs Drs W. N. Verheyen and J. L. J. Hulselmans of the 'Rijksuniversitair Centrum Antwerpen' who supervised my research, as well as to Drs M. Louette and D. Meirte, responsible for the 'Section Vertébrés' in the 'Musée Royal de l'Afrique Centrale'. Last but not least I would like to thank Prof. Dr F. Bourlière who graciously undertook the translation of the manuscript. The field research was made possible thanks to the support ensured by the 'Coopération Technique Universitaire Belge' in the framework of project XIV 'Ecologie et Conservation de la Nature' and thanks to a grant from the 'Fondation pour favoriser les Recherches scientifiques en Afrique'.

Appendix 6.1
Gazetteer

Akenge	02°54' N	26°49' E
Djamba	02°51' N	24°05' E
Djolu	00°35' N	22°27' E
Ekele	00°19' S	18°15' E
Ikela	01°11' S	23°16' E
Iligampangu	01°36' S	23°57' E
Kalima	02°34' S	26°37' E
Kahuzi	02°13' S	28°42' E
Kibombo	03°54' S	25°55' E
Kindu	02°57' S	25°56' E
Kisangani	00°28' N	25°11' E
Kutu	02°44' S	18°08' E
Lomako	00°51' N	21°05' E
Lomela	02°19' S	23°16' E

continued

Appendix 6.1 *continued*

Luale	01°09′ S	23°05′ E
Lukolela	01°05′ S	17°12′ E
Mbandaka	00°03′ N	18°15′ E
Mombongo	01°39′ N	23°09′ E
Niapu	02°25′ N	26°28′ E
Opala	00°39′ S	24°19′ E
Piana	05°04′ S	23°29′ E
Sake	01°34′ S	29°03′ E
Simba	01°25′ N	22°54′ E
Ubundu	00°21′ S	25°29′ E
Wamba	00°01′ N	22°33′ E
Yaenero	00°12′ N	24°47′ E

Part II

Genetic and phenetic characteristics: their use in phylogenetic reconstruction

II.7

Genetic evolution in the African guenons

MARYELLEN RUVOLO

As evolutionary biologists studying living animals, we want to discover the relationships between ecology, behaviour, morphology, social structure, and genetics for certain species groups in order to make some generalizations about overall, broad patterns of evolutionary change. As evolutionary biologists and primatologists, we want ideally to study a primate group that firstly, has many species (so that many comparisons are possible) and secondly, displays greater rather than lesser variability. *Cercopithecus* is perhaps the best choice because it satisfies both criteria, more so even than *Macaca*, for example, another species-rich genus but one not as ecologically varied.

The first step in understanding the patterns linking ecology, behaviour, morphology, social structure, and genetics for a species group is to establish its phylogeny. Knowing a group's evolutionary history is essential for understanding its adaptations. We cannot, after all, talk of living species as 'adapted' without an historical element, because 'adaptation' implies a process of becoming, of changing through time, of moving from one state to another. Furthermore, evolving organisms are not malleable without limit, rather, they are constrained at each point by their previous forms.

Establishing a group's phylogeny provides a baseline for comparisons. It is what allows us to distinguish evolutionary convergence from homology. Without phylogeny, we cannot know, for example, whether two species behave alike because both have independently hit upon the same solution to an ecological problem or whether the behaviours are truly homologous, representing an ancestral innovation common to descendant species.

There are several existing ways to establish phylogeny using genetic techniques. Undoubtedly the most informative for closely related groups is DNA-DNA hybridization because it ideally compares all the single copy genes between two organisms at once. This approach has been most recently applied to hominoids by Sibley & Ahlquist (1984). Comparing sequences of nuclear genes may also be equally effective provided one uses several genes, including their non-protein coding parts and the flanking areas immediately outside the genes. Morris Goodman and co-workers have recently examined a range of primate globin genes by this method (Goodman *et al.*, 1983, 1984; Koop *et al.*, 1986). Mitochondrial sequences are theoretically informative as well. However, because mitochondrial sequences evolve so quickly (up to ten times faster than nuclear ones), there is a much greater chance of homoplasy or 'false' identity at any given site, which makes data analysis trickier. Furthermore, recently discovered examples in mice (Ferris *et al.*, 1983), frogs (Spolsky & Uzzell, 1984) and fruitflies (Powell, 1983) show that mitochondrial genes can 'jump' between species. Because mitochondrial genes can yield different phylogenies from those based on nuclear genes, they therefore might not be best for phylogenetic reconstruction. This could pose a particular problem in a genus such as *Cercopithecus* with its mixed-species groups and hybrids (see Gautier-Hion, Chapter 23; Struhsaker, Butynski & Lwanga, Chapter 24). Another genetic approach is protein electrophoresis, in which proteins are separated by their size and electrical charge. This approach has been more accessible to primatological workers because it is simple and inexpensive. Furthermore, according to a recent calculation by Nei (1985), protein electrophoretic data are not necessarily less informative than DNA sequence data. Nei estimates that the average number of nucleotides examined by protein electrophoresis is 72 base pairs per locus; therefore a study of 14 loci (like the one described here for Cercopithecini) effectively samples 72 × 14, or 1008 base pairs. By comparison, in two recent studies of hominoids, Ferris, Wilson & Brown (1981) examined 280 base pairs by restriction enzyme analysis and Brown *et al.*'s (1982) mitochondrial sequence study looked at 896 base pairs.

Another genetic approach to phylogeny is based on immunological distance as measured by microcomplement fixation, a technique made familiar to most primatologists by Sarich & Wilson (1967). This technique has limits of resolution which make discrimination among closely related species like the Cercopithecini more difficult than with protein electrophoresis. Finally there are cytogenetic analyses, which

Table 7.1. *Proteins studied*

[a]Alb	Albumin
[a]Gc	Group-specific component
[a]PA–1	Prealbumin–1
[a]PA–2	Prealbumin–2
[a]Tf	Transferrin
[a]ADA	Adenosine deaminase
[a]DIA	NADH-Diaphorase
[a]PGD	6-Phosphogluconate dehydrogenase
[a]PGM–1	Phosphoglucomutase–1
[a]PGM–2	Phosphoglucomutase–2
LDH	Lactate dehydrogenase
GPI	Glucose phosphate isomerase
Bf	Properdin factor B
G6PD	Glucose-6-phosphate dehydrogenase

[a] Informative alleles for cladistic analysis

are really in a separate category from other 'genetic' approaches. Chromosomal differences are not the result of one or a few nucleic acid base changes or amino acid differences; they involve new juxtapositions of large pieces of DNA which, for the most part, remain internally unchanged. In practice, chromosomal analysis is often reminiscent of morphological trait analysis. Cytogenetic data reflect chromosome morphology and are complementary rather than equivalent to what one can think of as 'gene-level' molecular data.

I shall describe an electrophoretic study of the Cercopithecini and its phylogenetic implications (Ruvolo, 1983). The 14 genetic loci and 18 species examined are listed in Tables 7.1 and 7.2. What is unusual about this study is that the data have been analysed in several ways, in an attempt to reveal the biases inherent in any one method and thereby to arrive at a hypothesized 'best' consensus evolutionary tree. Figure 7.1 summarizes these methods.

The first distinction to note is between phenetic and cladistic analyses. They differ in their classifying and handling of similarities between taxa. Only the inherited, shared, most recently evolved traits between taxa are considered useful for revealing evolutionary relatedness in a cladistic analysis. In contrast, a phenetic analysis uses all similarities between taxa, without distinction.

Phenetic methods have most commonly been used by geneticists analysing electrophoretic data. After measuring the frequencies of alternative gene forms (known as alleles) within each species, one chooses a distance function to apply to the raw data. For any two

Table 7.2. *Species studied*

Species	Number	Common name
C. (Allenopithecus) nigroviridis	2	Allen's swamp monkey
C. aethiops aethiops	7	Grivet
C. aethiops sabaeus	3	Green monkey
C. aethiops pygerythrus	25	Vervet
C. albogularis	23	Sykes' monkey
C. ascanius	1	Redtail
C. campbelli	4	Campbell's monkey
C. cephus	3	Moustached monkey
C. diana	1	Diana monkey
C. hamlyni	2	Owl-faced monkey
C. lhoesti	1	L'hoest's monkey
C. mitis	8	Blue monkey
C. mona	2	Mona monkey
C. neglectus	13	De Brazza's guenon
C. nictitans	2	Spot-nosed guenon
C. petaurista	2	Lesser spot-nosed guenon
C. pogonias	4	Crowned guenon
C. wolfi	1	Wolf's monkey
C. (Miopithecus) talapoin	7	Mangrove monkey
Erythrocebus patas	17	Patas monkey

species, the function converts the many allelic differences into a single measure of 'distance'. Application to all pairs of species yields a matrix of genetic distances. The many pairwise distances in the matrix are then summarized as a dendrogram (tree diagram) by applying a clustering algorithm. This procedure is very similar to reconstructing a road map of France, for example, beginning only with the distances between Paris and Rennes, Rennes and Poitiers, Paris and Poitiers, and so forth. It is important to note that there are two choices in any given phenetic technique – choice of distance function and choice of clustering technique – and each of these has its own bias. Clustering techniques, for example, may or may not implicitly assume that rates of evolutionary change are constant over time. Fuller discussion of these issues can be found elsewhere (Farris, 1972, 1981), but it suffices for this discussion to remember that, given one data set, more than one phenetic tree or phenogram can be constructed from it.

How much do phenograms of the Cercopithecini differ? Not very much, as we can see by comparing Figures 7.2 and 7.3, both phenograms but constructed using different distance functions and clustering techniques. The biggest difference is in the positions of species *hamlyni* and *lhoesti*. Both phenograms agree on clustering *nigroviridis*

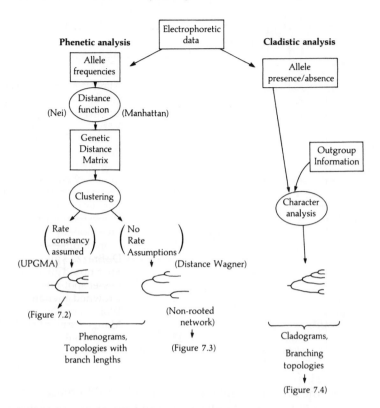

Figure 7.1. Cladistic and phenetic approaches to electrophoretic data.

with *aethiops* and *talapoin* with *patas*. Both phenograms also define two *Cercopithecus* subgroups, the 'diana' subgroup containing *neglectus*, *diana*, *wolfi*, *pogonias*, *mona* and *campbelli*, and the 'mitis' subgroup containing *mitis*, *albogularis*, *ascanius*, *nictitans*, *cephus*, and *petaurista*. Whether or not *hamlyni* and *lhoesti* belong properly within this latter 'mitis' subgroup or are separately placed between the 'mitis' and 'diana' subgroups on the one hand, and the other four species, on the other, is the biggest difference. Keeping this one major difference in mind, namely the placement of *hamlyni* and *lhoesti*, we can turn to the cladistic analysis for a possible resolution of this question.

In the cladistic analysis, the presence or absence of alleles rather than their frequencies are the raw data. Outgroup information, that is, information about taxa outside the group being studied, is what allows an assessment of primitive traits (Wiley, 1981). Very frequently in cladistic analyses, one living species is chosen as the outgroup. One can well imagine that this procedure has biases, because one must

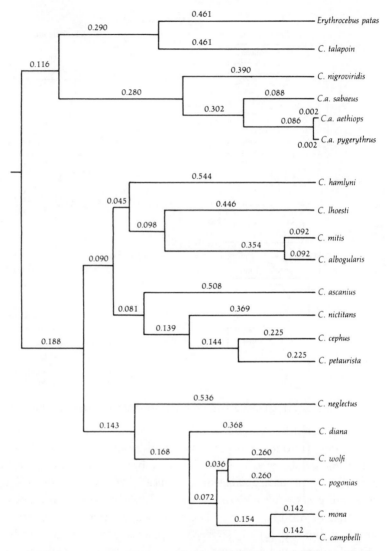

Figure 7.2. A phenogram of the Cercopithecini based on Nei's Standard genetic Distance and UPGMA clustering.

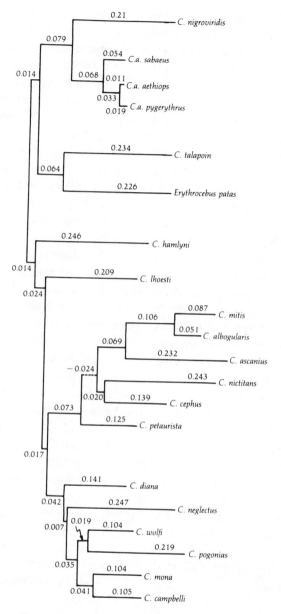

Figure 7.3. A phenogram of the Cercopithecini based on Manhattan Distance and Distance Wagner clustering (no rate constancy assumed).

choose a species which is truly outside the group being studied (but not too distantly related), and because a living species can only approximately represent an ancestral state in retaining some undefined subset of primitive traits. This procedure is also used in analyses which are not strictly cladistic, for example, in cytogenetic studies which begin with the *Macaca* karyotype and generate the various *Cercopithecus* karyotypes by hypothesized inversions, translocations, and other chromosomal changes (Ledbetter, 1981). In general, using several species for outgroup information is preferable to using only one.

As outgroup information for this study, I used allelic presence/ absence data from several species of baboons, macaques, and mangabeys. Those alleles common to baboons, macaques and mangabeys were compared with those for the Cercopithecini. From this comparison, a list of hypothesized primitive alleles was drawn. In cases where either of two alleles was possibly primitive, if one was present in the species *nigroviridis, patas,* or *talapoin,* it was chosen as primitive because cytogenetically these species are considered to be closer to the 42 chromosome ancestral state (Ledbetter, 1981) and because some, although not all, morphologists have concurred. This method of examining several species for outgroup information combined with other knowledge of a group's biology to assess primitiveness has been advocated for electrophoretic cladistic studies by Baverstock *et al.* (1979) and by Arnold *et al.* (1982). Character analysis of the data, which involves grouping species together on the basis of shared derived traits, yields the cladogram Figure 7.4.

Not every genetic locus was informative for this analysis, and this is generally true when constructing cladograms. In some cases, most species have uniquely derived alleles (autapomorphies), and a primitive state is difficult to hypothesize. In others, species either have the primitive allele or uniquely derived (autapomorphic) ones; in other words, no shared derived alleles (synapomorphies) exist for a particular locus. At each node, those alleles which are shared and derived for the descendant species are indicated. It is these synapomorphies which are the basis of any cladistic classification. Each node is therefore defined by a subset of the data, only one or two alleles.

How does the cladogram (Figure 7.4) compare with the previous phenograms (Figures 7.2, 7.3)? More importantly, how can we integrate cladogram and phenogram to arrive at a best current hypothesis for evolutionary relationships among the Cercopithecini? Let us first look at the species of the 'diana' subgroup. The species which comprise this subgroup (*neglectus, diana, wolfi, pogonias, mona* and *campbelli*) are

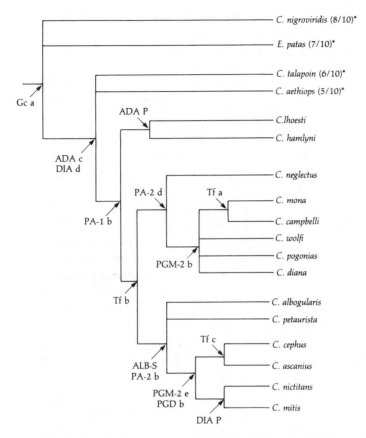

Figure 7.4. A cladogram of the Cercopithecini based on outgroup comparisons and character analysis. Derived alleles are indicated for each node. ★ = number of primitive alleles.

the same for both phenetic and cladistic analyses. What is at issue is their relative placement. Cladistically, *neglectus* is distinct from the other species in the 'diana' group because it lacks allele PGM-2b. On the basis of overall similarity as defined by Nei's distance and UPGMA clustering (the phenogram Figure 7.2), *neglectus* is also the most distant species in the cluster. *Neglectus* and *diana* are reversed positionally in the other phenogram (Figure 7.3), but only barely so (branch length = 0.007). Therefore, since the cladogram and one phenogram agree on *neglectus'* outermost position in the 'diana' subgroup, it appears that way on the consensus tree (Figure 7.5). Concerning the other species in the 'diana' subgroup, *mona* and *campbelli* are closest phenetically and cladistically, and *wolfi* and *pogonias* are phenetically a species pair.

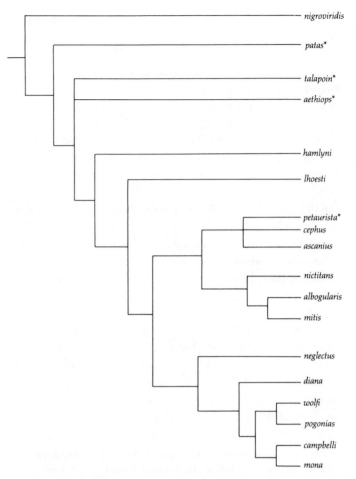

Figure 7.5. Consensus tree of phenetic and cladistic analyses of Cercopithecini protein electrophoretic data.

Cladistically, no one allele associates *wolfi* and *pogonias* more closely than either to *diana*, but this is not contradictory to the phenetic data; rather it reflects lack of cladistic resolution. *Diana* is phenetically distinct from the *wolfi–pogonias* and the *mona–campbelli* species pairs, and at the same time it is cladistically closer than *neglectus* to the pairs. Thus the consensus tree for the 'diana' subgroup places *diana* intermediately between *neglectus* and the two species pairs *wolfi–pogonias* and *mona–campbelli*.

As for the rest of the tree, the first point to note is that there is an unusual genetic trait linking the six species *cephus, ascanius, petaurista, mitis, albogularis* and *nictitans.* That trait is a slowly-migrating albumin variant which in this study shows no variation within these six species and is also monomorphic outside these species for the usual-mobility form.[1] Because albumin variants are rare, one either has to recognize these species as descendants of an ancestral form with this particular mutation, or, less parsimoniously, theorize that the mutation occurred more than once. The former one-mutation case is more likely, and it is also supported by the phenogram (Figure 7.3). Because they lack this albumin variant, *lhoesti* and *hamlyni* appear separately from the 'mitis' subgroup on the consensus tree. *Lhoesti* is positioned closer to the derived 'mitis' and 'diana' subgroups than *hamlyni* in accordance with the phenogram (Figure 7.3).

Within the 'mitis' subgroup, two species pairs are defined cladistically. They are *nictitans* with *mitis* and *cephus* with *ascanius*. *Albogularis* and *petaurista* remained unresolved in the sense that they do not share one or the other derived allele to align them more closely with one group over another. However, non-resolution in the cladogram does not conflict with positioning in the phenogram. Both phenograms have *mitis* and *albogularis* as very similar overall and therefore they appear so in the consensus tree. The really problematic species to place is *petaurista*. One phenogram (Figure 7.2) has it closest to *cephus* and next closest to *nictitans*, but these two neighboring species are in different cladistically defined subgroups. The other phenogram (Figure 7.3) shows it as most divergent of the six species (connected, however, by a negative-length branch). Cladistically, *petaurista* is unresolved. Therefore it is difficult to arrive at a consensus for the placement of *petaurista*. Because overall its greatest similarity is to *cephus* in one phenogram (Figure 7.2), *petaurista* is shown on the consensus tree in an unresolved triad with *cephus* and *ascanius*, but this tentative placement needs more information for better resolution (this is indicated by an asterisk in the consensus tree, Figure 7.5).

Since we have arrived at a consensus for the two *Cercopithecus* subgroups and for the slightly more divergent species *lhoesti* and *hamlyni*, all that remains are *patas*, *talapoin*, *aethiops* and *nigroviridis*. Cladistically, *nigroviridis* and *patas* are equally primitive. However, because *nigroviridis* shares with baboons, macaques and mangabeys a greater number of primitive alleles than *patas* (8 vs. 7 out of 10), it appears as most primitive on the consensus tree. *Talapoin* and *aethiops* are both more derived because they share two alleles with all the other species (ADA c and DIA d). On the phenetic trees, in contrast, *aethiops* is closest to *nigroviridis*, and *patas* is closest to *talapoin*. How can these differences be reconciled? It is important to understand that cladistic and phenetic trees will disagree most on those species which are most primitive. After all, primitive species are primitive because they have a relatively great number of primitive traits; these are precisely the characters that are handled differently between the two types of analyses. In fact, as an operational definition of primitive species within any group of taxa, we can take those species which are most discrepantly positioned by cladistic and phenetic analyses. By this criterion, *talapoin*, *patas*, *nigroviridis* and *aethiops* are primitive. Because of a personal philosophical inclination towards cladism, the consensus tree reflects the cladistic relationships for these four primitive species. In the consensus tree (Figure 7.5), the species *patas*, *talapoin* and *aethiops* have asterisks near them to indicate that their relative positions are currently the best hypothesis given available data and may change with the acquisition of more data.

Integrating the electrophoretic with the cytogenetic phylogenies (Dutrillaux, Couturier & Chauvier, 1980; Ledbetter, 1981; Dutrillaux *et al.*, 1982, Chapter 9) is necessary in order to come to a full understanding of genetic evolution within the Cercopithecini. Such synthesis requires assessing the relative likelihoods of different mutational types. For example, is it less likely that a particular amino acid change rather than a given chromosomal fission could occur twice? Are breaks so frequent at chromosomal fragile sites that identical fissions can occur twice with high probability along different lineages? It is necessary to know these relative likelihoods because if electrophoretic and cytogenetic trees contradict each other, they can be made to agree provided more mutational events are invoked, the particular ones depending on their relative probabilities of occurrence.

Short of knowing these probabilities, can we still attempt to integrate the chromosomal and electrophoretic trees? For most of the derived species, Ledbetter's phylogeny (1981) agrees well with the consensus tree (Figure 7.5) with the exception of the *wolfi–pogonias* species pair. Their placement rests on two shared derived traits (both non-centric fissions of ancestral *Macaca mulatta* chromosomes 4 and 11). If we hypothesize the fissions to have happened twice, this allows the alignment of the species pair with *campbelli* (i.e. with the 'diana' subgroup as it appears on the consensus tree). Invoking two additional non-centric fission events must be balanced against the alternative: that the albumin variant found in this study (which the species pair shares with the 'diana' subgroup) occurred twice. There is little quantitative evidence to judge the relative probabilities of these mutations; however, chromosomal fissions are abundant within the Cercopithecini, while albumin variants are relatively rare within the cercopithecines (Kitchen & Bearn, 1965; Barnicot & Hewett-Emmett, 1971; McDermid & Ananthakrishnan, 1972; Allen & Buettner-Janusch, 1973; McDermid, Vos & Downing, 1973; Smith, 1980; Dracopoli, 1981; Turner, 1981). Dutrillaux *et al.*'s phylogenies (1980, 1982, Chapter 9) are not strictly bifurcating and are therefore difficult to compare.

As for the primitive species, three shared derived characters link *patas, aethiops* and *lhoesti* (non-centric fissions of *Macaca mulatta* chromosomes 2, 5 and 6), while only one shared derived character contradicts that pattern grouping *aethiops* to *talapoin* (centric fission of *Macaca mulatta* chromosome 1, Ledbetter, 1981). Dutrillaux *et al.* (1982) also group *aethiops* and *patas* together, and *lhoesti* is next closest. (*Lhoesti* was not recognized as particularly primitive in the electrophoretic study, but was intermediately placed with *hamlyni*). The chromosomal phylogenies differ at the base of the Cercopithecini tree. Ledbetter (1981) shows *nigroviridis* branching off first, followed by *talapoin*, then by the other primitive species. Dutrillaux *et al.* (1982, Chapter 9) show a bifurcation with *talapoin, lhoesti, aethiops* and *patas* on one side and with *nigroviridis* and the other derived species on the other. An earlier phylogeny (Dutrillaux *et al.*, 1980) showed *nigroviridis* branching off the common ancestral stock while retaining the above bifurcation.

In conclusion, the chromosomal phylogenies conflict with each other as well as with the electrophoretic consensus tree in the placement of the relatively primitive species (see also Gautier, Chapter 23, for phylogeny based upon vocalizations). This non-consensus is in some ways not surprising, because a predictable source of error in cladistic phylogenetic reconstruction which preferentially affects the relatively primitive species rests with outgroup choice, whether using one species (e.g. *Macaca mulatta* as in Ledbetter's study) or a hypothesized slate of ancestral traits based on several outgroups (as in Dutrillaux *et al.*, 1980, 1982, Chapter 9; Ruvolo, 1983; Gautier, Chapter 12).

The trees do agree on the placement of *nigroviridis* as most primitive, and on the inclusion of *patas* and *talapoin* within the group proper. The exact phylogenetic branching order of the most primitive Cercopithecini species still remains a mystery and therefore to a large extent so do their origins. Until the time that single copy DNA-DNA hybridization methods are applied to this problem, they will probably remain cloudy.

Note

[1] Allen and Buettner-Janusch (1973) reported a slow albumin variant with a gene frequency of 2% in a survey of 77 *patas* monkeys; it is unknown whether this is the same variant seen in this study.

II.8

Population differentiation in *Cercopithecus* monkeys

TRUDY R. TURNER, JERALD E. MAIERS and
CAROL S. MOTT

Introduction

To assess population differentiation, biologists have used morphological, behavioural and genetic criteria. The information we have on several populations of vervet monkeys (*Cercopithecus aethiops*) and mitis monkeys (*Cercopithecus mitis*) allows us a unique opportunity to examine the utility of these various measures of population differentiation. The pattern of variation found in our study populations also suggests how differentiation occurred. Different subspecies show little genetic variation; they probably resulted from allopatric separation of populations with subsequent morphological adaptation to local environmental conditions (Templeton, 1981).

Unlike the forest *Cercopithecus* such as the *mitis*, vervets are widely distributed in savanna areas, live in multi-male, multi-female groups and are semi-terrestrial.

We examined two subspecies of vervet monkeys. One hundred and twenty five *C. a. aethiops* were trapped and sampled in Awash National Park, Ethiopia. Three hundred and sixty *C. a. pygerythrus* were trapped at four sites in Kenya (Figure 8.1). The sites were separated by between 90 and 300 km. Three subspecies of *mitis* were sampled, 25 *C. m. albotorquatus*, 14 *C. m. stuhlmanni* and 54 *C. m. kolbi*. Animals were trapped by the Institute of Primate Research, Kenya, in three different sites which differ by altitude, temperature and rainfall. Blood samples were taken from all animals and a series of morphometric measures were taken from all but the Awash vervets. We were thus able to compare local populations of a single subspecies, two subspecies of vervets and three of *mitis*, as well as the two species themselves (Turner, Mott & Maiers, 1986).

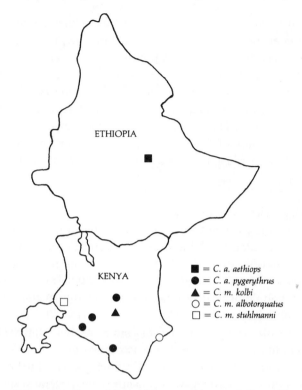

ETHIOPIA

KENYA

■ = *C. a. aethiops*
● = *C. a. pygerythrus*
▲ = *C. m. kolbi*
○ = *C. m. albotorquatus*
□ = *C. m. stuhlmanni*

Figure 8.1. Sites of *Cercopithecus aethiops* and *Cercopithecus mitis* monkeys studied.

Methods

Genetic distance

The blood samples were examined by electrophoresis for polymorphic alleles (Turner, 1981; Mott, Turner & Else, 1984). The *mitis* sample consisted only of adult animals. We have demonstrated (Turner, *et al.* 1986) by comparison with the vervet populations that gene frequencies of adults do not differ significantly from the frequencies of the complete population sample.

Gene frequencies allow us to determine the degree of difference between populations and also the pattern of diversity between populations. The differentiation of populations of a species from each other can be measured by the F_{ST} statistic. The fixation index (Wright, 1965), F_{ST}, is the ratio of the observed variance of allele frequencies to the maximum theoretical variance of these frequencies between populations. It ranges from 0 to 1 with 0 indicating identity.

The data on allele frequencies also allowed us to calculate the genetic distance between populations. Genetic distance expresses gene frequency differences between pairs of populations in a single number. It is the average gene difference per locus. Only groups in which the same loci were examined were compared. No comparisons could be made between the *C. mitis* and the *C. aethiops* species since we are not certain that the observed electrophoretic patterns represent exactly the same alleles. A variety of distance measures are available. We chose Nei's (1975), since it is designed to measure DNA differences from gene frequency data.

Morphological distance

Morphometric data were collected on the adult Kenyan vervet and adult *mitis* populations. Nine measures were obtained: chest girth, body length, hand, lower forelimb, upper forelimb, foot, lower hindlimb, upper hindlimb and tail. These data were examined to determine morphological distance, comparative sexual dimorphism, and comparative shape.

If we were to take a single measure of a single representative of three groups, we could make a simple comparison and determine which pair are closest on that single dimension and which pair are further apart. In a more elaborate situation with measures for several individuals from each of three groups, again on a single dimension, we could use a more elaborate statistical measure which would involve means and standard deviations. This approach could incorporate the variation within each group and determine closeness in terms of overlapping distribution curves.

Groups which might be very close on one morphological dimension might be distant on another. A larger number of dimensions would provide more information on morphological distance than would the single dimension. Statistical approaches to morphological distance use measures of a large number of individuals on a large number of variables. Several statistical approaches to this type of data have been proposed which compute a meaningful measure of morphological distance. One of the simplest measures of morphological distance is the Czekanowski or DD. This assumes that the sum of the absolute difference of the mean of each of the measured morphological dimensions for a species is an adequate measure of morphological distance. The Czekanowski statistic assumes orthogonality and equality of weighting on each morphological dimension (Corruccini, 1975). The Coefficient of Racial Likeness (Pearson, 1926) is a morphological distance statistic which measures the probability of two groups being drawn from the same population. A morphological distance developed by Penrose, CH, is a mean of the sum of the squared standardized difference between two populations (Penrose, 1954; Benoist, 1981). Recently the M statistic approach has been suggested by Cherry *et al.* (1978, 1979, 1982). It is basically a variation on the Czekanowski statistic and has been applied to a wide cross section of organisms.

None of these morphological distance measures corrects for the correlation between the variables. We found that the Pearson's coefficients of correlation on the nine variables were all significantly correlated with one another. So we also calculated the Mahalanobis D (Mahalanobis, 1936) which corrects for the correlations between the variables.

Sexual dimorphism

Body measures and weights were available on the four local populations of *C. a. pygerythrus* and the three subspecies of *C. mitis*. For tabular comparisons, we combined the four local populations of *C. a. pygerythrus* to form a total group. A measure of overall body length was developed by combining the crown to rump, upper forelimb, lower forelimb, hand, upper hindlimb, lower hindlimb and foot measures. A density factor for each individual was computed by dividing the weight in grams by this overall length in centimeters. We then calculated sexual dimorphism for each of the three factors, weight, body length and body density.

T-tests were used to determine whether sexual dimorphism was significant ($P < 0.01$) for each of the seven groups (Table 8.1). To

Table 8.1. *Summary of genetic information*

	C. aethiops aethiops	*C. aethiops pygerythrus*	*C. mitis*
Number of animals	124	364	93
Number of loci	23	18	32
Average heterozygosity ($\bar{H}$)	5.6%	4%	4.2%
Number of polymorphic loci	4	3	4
Proportion of polymorphic loci	17%	17%	12.5%
Polymorphic loci	TF	EST	TF
	EST	GC	CA II
	PGM	ABO blood	DIA
	ADA	groups	PEP D
F_{ST}	0.062	0.112	0.200

determine sexual dimorphism distance between the populations, we followed Relethford & Hodges' (1985) linear regression model.

Since the methodology for assessing morphological distance from shape is less well established, we will consider it separately below.

Results

Genetic distance

The average heterozygosity ($\bar{H}$) and the proportion of polymorphic loci were calculated from allele frequencies. The average heterozygosity for the Ethiopian vervets (0.056), Kenyan vervets (0.04) and the *mitis* (0.042) are similar and are at the low range for vertebrates. The same is true for the proportion of polymorphic loci: vervets (0.17) and *mitis* (0.125).

In our sample the lowest F_{ST} values are found in the Ethiopian vervets (0.062), a single breeding population. Next is the Kenyan vervet population (0.112), a single subspecies divided into four local populations. The *mitis* sample, representing three subspecies, had the highest values (0.20).

The genetic distances between the troops of vervets in Ethiopia and between sites in Kenya are nearly identical (0.0004–0.0162 and 0.0004–0.0159) (Turner *et al.*, 1986). The value for the three subspecies of *mitis*, 0.0147–0.037, is somewhat higher but considerably lower than Nei suggests for subspecies differentiation, or than what is found in populations of *Drosophila willistoni*, 0.20 (Avise, 1976). The figures are, however, in the same range as those found by Kawamoto, Shotake & Nozawa (1982) in their examination of over 2000 baboons and macaques using over 20 loci. Although the authors questioned the idea of subspecific designation in these animals, we believe the small genetic differences may reflect the way in which differentiation occurred.

Morphological distance

Morphological distance is an alternative measure of species differentiation. It can be used as a supplement and support for conclusions based on direct genetic distance data (Sanghvi, 1953; Salzano *et al.*, 1980). Morphological dimensions however, reflect environmental as well as genetic differences (Howells, 1966, 1984; Milton, 1983).

All of the morphological distance measures, the Czekanowski

Table 8.2. *Morphological distance as deferred by the Mahalanobis* D *statistic*

	1	2	3	4	5	6
2	0.94					
3	1.98	0.67				
4	2.39	1.76	1.09			
5	5.22	5.46	4.54	4.67		
6	3.64	4.00	4.20	4.49	4.77	
7	6.02	5.70	4.63	5.31	1.45	4.29

1 = *C. a. pygerythrus* site A
2 = *C. a. pygerythrus* site B
3 = *C. a. pygerythrus* site C
4 = *C. a. pygerythrus* site D
5 = *C. m. stuhlmanni*
6 = *C. m. albotorquatus*
7 = *C. m. kolbi*

distance, the Coefficient of Racial Likeness, Cherry's M statistic, and the Mahalanobis D gave the same results. The four *C. aethiops* groups were all found to be morphologically close. The *C. m. albotorquatus* was close to the four *C. aethiops* groups. The *C. m. stuhlmanni* and *C. m. kolbi* were distant from the *C. aethiops* and morphologically close to each other (Table 8.2).

Sexual dimorphism

The degree of sexual dimorphism, like other morphological traits, can be shown to have a genetic and environmental component (Leutenegger & Kelly, 1977). Most studies of sexual dimorphism on primates and other animal populations have considered only weight (Leutenegger & Kelly, 1977; Leutenegger, 1978, 1982; Lenglet, 1984). In contrast, studies of human dimorphism have dealt primarily with the stature or height (Bielicki & Charzewski, 1977; Hall, 1978; Gray & Wolfe, 1980). Our analysis showed that there was sexual dimorphism in density, body length and weight. Body length and density showed significant differences not revealed by the analysis of weight alone.

The most striking finding is the difference of sexual dimorphism of *C. m. kolbi* from that of the other species and subspecies. The *C. m. stuhlmanni* were closer to the *C. a. pygerythrus* groups than to the other two *C. mitis* subspecies. One population of *pygerythrus* was significantly different from the others in body length dimorphism.

Shape

Method
There have been many attempts to compare body shapes between populations (see Oxnard, 1973, 1975; Webb, 1984). Penrose (1954) proposed that morphological distance has both shape and size components. The size component is reflected by the square of the mean value of all differences between the standardized scores of the variables; shape, according to Penrose, is what remains. We use the seven measures used for body length as well as the chest girth and the tail length to assess the similarity of Penrose's shape components for these animals.

In addition to Penrose, other methods have examined shape including D'Arcy Thompson's use of biocorrelograms (1917). Although this approach is primarily descriptive, it suggests the possibility of viewing shape modification as the way of moving from one form to another. Other approaches have been trend surface analysis by Sneath (1967); tangent analysis by Bookstein (1978); component analysis by Oxnard (Oxnard, 1973, 1975); and Fourier analysis (Lestrel *et al.*, 1976). Shape discussions also focus on the relation of shape and size, allometry (Gould, 1966; Hagaman & Morbeck, 1984).

Shape is a complex notion, including general form, features and proportions. However, in this case, since we wished to analyse animals that share general form and features, we took an approach that could determine a measure of morphological shape distance in the limited sense of proportion. Our method was to factor out body size from each of the nine morphological measures by dividing each measure by the total body length. The result was an average value for each of the nine measures for each of the groups.

The basic notion we used was a form of 'rubber template' pattern recognition (Windrow, 1973; Lee, 1974, 1976) based on the formal mathematics of fuzzy set theory (see Zadeh, 1973). The concept involves measuring the distortion or stretching that would be required to change the one body shape, the mask, to the other, the template. We computed the shape distance as the product of the distortion or dissimilarity between each pair. This is conceptually similar to D'Arcy Thompson's approach, but uses a limited number of measures.

Results
The plot of the Penrose components indicates that *C. m. albotorquatus* is similar to the vervets in both size and shape (Figure 8.2).

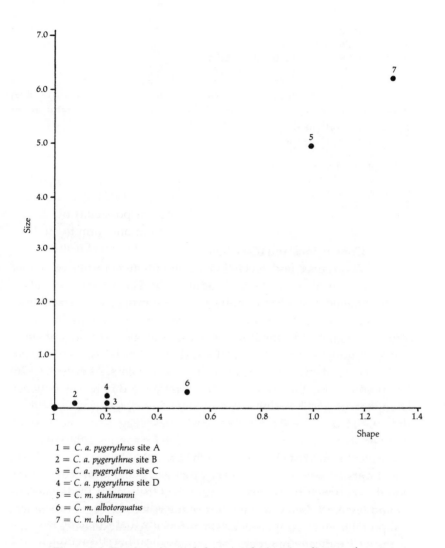

Figure 8.2. Penrose size and shape analysis using distance from troop
No. 1 of *C. a. pygerythrus*.

The rubber template framework indicated that vervet populations'
shapes cluster, as do the *mitis* populations. Even though this technique
indicated the populations of each species cluster together, *C. m.
albotorquatus* was in a somewhat intermediate position (Table 8.3).

Table 8.3. *Fuzzy rubber template cumulative shape distance*

	1	2	3	4	5	6
2	0.22					
3	0.26	0.04				
4	0.29	0.16	0.06			
5	0.39	0.38	0.36	0.38		
6	0.41	0.41	0.39	0.44	0.23	
7	0.49	0.44	0.46	0.48	0.28	0.30

1 = *C. a. pygerythrus* site A
2 = *C. a. pygerythrus* site B
3 = *C. a. pygerythrus* site C
4 = *C. a. pygerythrus* site D
5 = *C. m. stuhlmanni*
6 = *C. m. albotorquatus*
7 = *C. m. kolbi*

Conclusions and discussion

Divergence and speciation are the ultimate results of genetic differentiation of populations. Traditionally, allele frequency differences obtained from electrophoresis of blood proteins have been used to measure differences between closely related populations. While the structural genes examined by this technique have the advantage of being a random sample of the genome (Lewontin, 1974), they may not be the most sensitive markers of evolutionary events. As Wilson (1976) has pointed out, there are few structural gene differences between chimpanzees and humans, even though these organisms are clearly differentiated. Templeton (1980, 1981) has suggested that a small number of major genes may be responsible for species differentiation, while most genes may be neutral to the process. Of importance also in the understanding of evolutionary processes is the knowledge of the population structure and mating systems of the group. These cautions should be remembered when interpreting results of genetic distances computed from electrophoretic assays. Since genetic distance may not be the only indicator of separation, we have elected to supplement the information on allele frequencies with information from other biological systems that may be important to differentiation and adaptation, including morphology, sexual dimorphism and shape.

The genetic information does differentiate between local populations of a single subspecies; and between two or three subspecies of a species. Both F_{ST} values and genetic distance values increase with increasing taxonomic separation of vervet and *mitis* monkeys.

However, the genetic distance between the subspecies is less than expected for this taxonomic rank.

Analysis of morphology also separates the two species as well as the three *mitis* subspecies. The Kenyan vervets are all similar morphologically and show comparable sexual dimorphism and shape. Among the *mitis* monkeys, the *C. m. kolbi* are the most differentiated. *C. m. albotorquatus* is morphologically intermediate to the *C. mitis* and the *C. aethiops*. The similarity of *C. aethiops* and *C. m. albotorquatus* may be the result of an adaptation to similar environmental zones or to an increase in terrestriality of this *mitis* group (J. Oates, personal communication). Morphological distance measures include both size and shape differences. The *C. m. kolbi* are the largest and also live at the highest elevation. When size is factored out, the shape of all three *C. mitis* groups cluster away from the *C. aethiops*. There may be shapes better adapted to savanna environments and terrestriality. Future analysis will attempt to determine how body form differences correlate with the environment.

While the *C. m. kolbi* are the most distinct morphologically from the rest of the *C. mitis*, they are not the most distinct genetically. This may reflect the different rates of change in different biological systems. We suggest that population differentiation among *mitis* monkeys occurred as follows. At one time the *mitis* were more widely distributed across Africa. The populations probably were members of a single extended Mendelian population with clinal variation in morphological traits. As the populations became more isolated from each other due to reduction of forests, they began to diverge. And we believe that both vervet and *mitis* species have differentiated by allopatric separation. The population and mating structure of the *mitis*, with a somewhat smaller effective population size, may have accelerated the process.

It would be of particular interest to compare subspecies and local populations of other species of primates with various mating systems to determine the applicability of our results.

Acknowledgments

This project was supported in part by NSF Grant Nos. SOC 74-24166, BNS 770-3322, and BNS 810-4435. We would like to thank N. C. Dracopoli, J. G. Else and the Institute of Primate Research for their help in Kenya. We would also like to thank the Ministry of Tourism and Wildlife, Kenya for allowing us to conduct this research. Our thanks also go to C. J. Jolly for his help with these projects and to the internal reviewers for their comments.

II.9

Chromosomal evolution of Cercopithecinae

BERNARD DUTRILLAUX, MARTINE MULERIS and
JEROME COUTURIER

Cercopithecine monkeys are probably the primates whose chromosomal evolution is the most elaborate: their chromosome numbers vary from 46 to 72 and, at first glance, there are no possibilities to find out any relationship between so different karyotypes.

We would like to show that in spite of this apparent complexity, this group provides an excellent model to study the chromosomal modifications which took place in the course of the processes of speciation.

This leads us to propose a cercopithecine phylogeny entirely based on the sequence of the chromosomal modifications (Dutrillaux, Couturier & Chauvier, 1980; Dutrillaux et al., 1982). This evolution is characterized by a progressive decrease of chromosome sizes, and an increase of chromosome numbers, due to recurrent fissions of the largest ancestral chromosome.

Material and methods

Most of the karyotypes were studied after fibroblast cultures, following skin or muscle biopsies taken shortly after the natural death of animals. Blood cultures were also performed in a small number of cases but their results were hard to interpret.

Various chromosome banding methods were used: R-bands, Q-bands, C-bands, and different treatments by BrdU (ISCN, 1978).

At least one specimen of each of the following guenons was studied: *Cercopithecus (Allenopithecus) nigroviridis* (ANI); *C. (Miopithecus) talapoin* (MTA); *C. (Erythrocebus) patas* (EPA); three forms of the *aethiops* group: *Cercopithecus aethiops* (CAE), *C. pygerythrus cynosuros* (CPC) and *C. sabaeus* (CSA); three forms of the *lhoesti* group: *C. lhoesti* (CHO), *C. solatus* (CSO) and *C. preussi* (CPR); *C. diana* (CDI); *C. neglectus* (CNE); four forms of the mona group: *C. mona mona* (CMM), *C. m. campbelli*

campbelli (CMCC), *C. m. pogonias nigripes* (CMPG) and *C. m. wolfi* (CMW); five forms of the *cephus* group: *C. cephus* (CCE), two subspecies of *C. ascanius* (CAS), *C. a. ascanius* and *C. a. schmidti*, *C. petaurista* (CPE) and *C. erythrotis* (CEY); three forms of the *nictitans* group: *C. mitis stuhlmanni* (CMS), *C. mitis opisthostictus* (CMO) and *C. nictitans* (CNI); and *C. hamlyni* (CHA).

Most of these animals were kept in captivity at the Museum National d'Histoire Naturelle of Paris, and others at the Station Biologique of Paimpont, the Parc Zoologique of Mulhouse, and the CIRMF at Franceville (Gabon).

Results and discussion

Most of the karyotypes of these species were previously published (Dutrillaux *et al.*, 1978, 1980, 1982; Muleris, Couturier & Dutrillaux, 1981; Muleris *et al.*, 1985), and we will not describe them all once more; only three of them, considered as particularly representative, are shown. Figure 9.1 shows a reconstructed karyotype of the ancestral

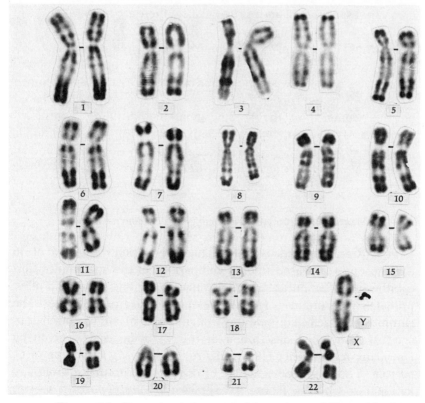

Figure 9.1. Reconstructed ancestral karyotype of Cercopithecinae (from Dutrillaux *et al.*, 1982).

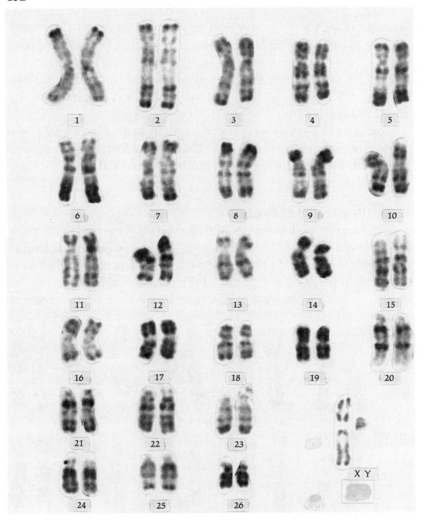

Figure 9.2. Karyotype of *Cercopithecus (Miopithecus) talapoin*.

form of Cercopithecidae (CER). This karyotype, composed of 46 chromosomes, is based upon a comparison of the subfamilies Cercopithecinae, Papioninae and Colobinae on the one hand, and other primates like Pongidae, Hominidae and Cebidae on the other. The chromosomes selected were those shared by a maximum number of species. A detailed description of the methods used is given by Dutrillaux & Couturier (1986). Figure 9.2 shows a karyotype of *C. talapoin*, a species representative of one of the two major groups of karyotypes whereas Figure 9.3 shows the karyotype of a species belonging to the second group of species: *C. nictitans.*

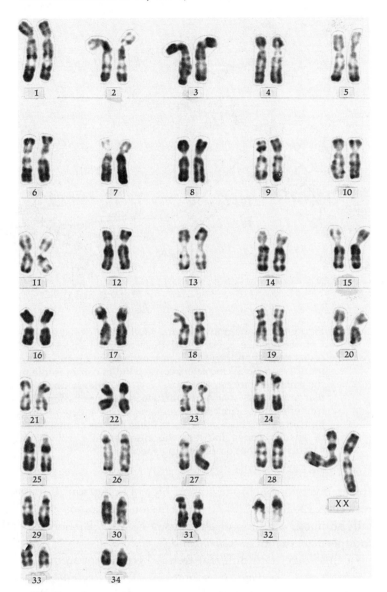

Figure 9.3. Karyotype of *Cercopithecus nictitans*.

Chromosomal phylogeny (Figure 9.4)

Starting from the ancestral karyotype (CER, Figure 9.1) which corresponds to the branching point between Cercopithecinae, Papioninae and Colobinae, three chromosomal rearrangements are shared by all Cercopithecinae and not by the other subfamilies. These

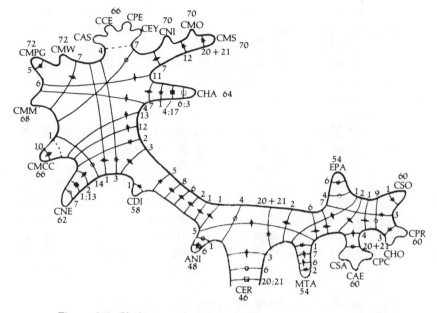

Figure 9.4. Cladogram showing phylogenetic relationships between Cercopithecinae species and their common ancestors (CER). The codes corresponding to species names are given in the text. The numbers in small characters correspond to chromosome numbers of the common ancestor (Figure 9.1) and the symbols indicate the rearrangements which affect them: arrows = fissions, stars = centromeric shifts, circles = inversions, open squares = Robertsonian translocations, black squares = other translocations. Numbers close to species codes indicate the number of chromosomes in the karyotype of the corresponding species.

are: an inversion of chromosome CER 6, a Robertsonian translocation between CER 20 and CER 21, and a fission of chromosome CER 3.

In addition, a fission of chromosome CER 5 is shared by all Cercopithecinae, except *C. talapoin* and *C. nigroviridis*.

At this stage, two different trends become apparent. One is represented by *C. aethiops, C. sabaeus* and *C. pygerythrus cynosurus, C. lhoesti, C. preussi, C. solatus, C. talapoin* and *C. patas*; it is characterized by two rearrangements, namely, a fission of chromosome CER 2, and an inversion or a centromeric shift of the chromosome resulting from the Robertsonian translocation of CER 20 and CER 21 cited above.

On the line leading to *C. talapoin* four rearrangements took place: (1) a fission of chromosome CER 6; (2) a fission + inversion of chromosome CER 7; (3) a fission of chromosome CER 1, (lacking in *C. patas*), and an inversion of one of the two derived chromosomes; (4) an inversion or a centromeric shift of one of the two chromosomes derived from the fission of chromosome CER 2.

Two additional rearrangements are shared by the other species of this group: a fission of chromosome CER 6, and a fission of chromosome CER 7. Although they affect the same chromosomes as in *C. talapoin*, these fissions of CER 6 and CER 7 took place at different breakpoints.

At a later stage, further rearrangements occurred: (1) an inversion, or a centromeric shift, of a chromosome derived from the fission of CER 3, shared by *C. patas* and *C. solatus*; (2) a fission of a chromosome derived from CER 2, shared by *C. aethiops, C. sabaeus, C. pygerythrus, C. lhoesti, C. preussi* and *C. solatus*; (3) a second fission of a chromosome derived from CER 1, and an inversion, or centromeric shift, of chromosome 9; these rearrangements are shared by *C. solatus*, and *C. lhoesti*.

C. aethiops, C. sabaeus and *C. pygerythrus* have a similar karyotype, if we do not take into account heterochromatin variations. All of them display a fission of chromosome CER 4, and a centromeric shift of the chromosome derived from the Robertsonian translocation of CER 20 and CER 21.

C. patas is characterized by the centromeric shift of a derivative of the fission of chromosome CER 6, and an inversion of chromosome CER 4.

In the branch leading to *C. solatus*, a centromeric shift of a derivative from the fission of CER 1 took place, and in the one leading to *C. lhoesti* and *C. preussi*, a centromeric shift of a derivative from CER 3 occurred.

The cladogram of Figure 9.4 is based on these different rearrangements. It must be considered as the most likely interpretation, based on the information available, and no more.

The evolution of the second group of Cercopithecinae is even more complex. From the branching point with the first group, a single rearrangement is shared by all the species: an inversion of chromosome CER 4. A short branch ends up in *C. nigroviridis*, after two chromosome rearrangements, an inversion of chromosome CER 1 and a fission plus an inversion of chromosome CER 6. On the branch common to all the other species, a centromeric shift of a chromosome derived from CER 5, and five fissions affecting chromosomes CER 1 (twice) and CER 2, 6, 8 (once) all occurred before the budding off of the branch leading to *C. diana*. On this branch a single rearrangement affecting a chromosome derived from the fission of CER 1 did occur.

Finally, a single rearrangement, a fission of a derivative from CER 3, took place before a very complex 'populational evolution'[1]. The description of all the rearrangements that occurred at that stage would be too tedious to be reported in detail here. They are all indicated on Figure 9.4, with the relative positions of the different species. From this scheme, it is clear that the group of *C. mona* is relatively homogeneous, except that *C. m. campbelli campbelli* is a little further from the others. *C. erythrotis*, *C. petaurista*, *C. cephus* and *C. ascanius* have all the same karyotype. *C. nictitans* and *C. mitis* must also be closely related.

The branch leading to *C. neglectus* originated relatively early during the evolution of the group, as did that of *C. hamlyni*, although to a lesser degree.

As a matter of fact, all the species of this group are closely related, and share many chromosomal rearrangements in various combinations. This makes it impossible to propose any dichotomic scheme.

The uniqueness of the chromosomal evolution of Cercopithecinae

Among the 21 species studied here, we have identified 17 different karyotypes. Only two groups of species share the same karyotype, the *C. aethiops* group (various subspecies), *C. p. cynosuros* and *C. sabaeus* and the *C. cephus*, *C. ascanius*, *C. petaurista* and *C. erythrotis* group (also *erythrogaster*, unpublished data).

This situation is very different from that found in Papioninae, among which only two chromosomal rearrangements occurred (Dutrillaux *et al.*, 1982). Chromosomal modifications are therefore closely related to the speciation process among Cercopithecinae.

On the whole, we have identified 56 chromosomal rearrangements, i.e. 2.66 per species, on average. However, the majority (32) of these rearrangements took place during phases of populational evolution, and a minority (24) are characteristic of particular species. For this reason, the branches leading to each species are relatively short, and correspond to 1.14 new rearrangements on the average. For the sake of comparison, this average is three times higher among Pongidae and Hominidae (Dutrillaux & Couturier, 1986).

The distribution of the patterns of chromosomal rearrangements is also very peculiar.

1. Fissions, observed 27 times, are the most frequent. They occurred very early during the evolution of Cercopithecinae. They were indeed, responsible for the progressive increase of chromosome number, illustrated in Figure 9.4.

These fissions seem to have followed a simple rule, as they affected the largest chromosomes. For instance, chromosomes 1, 2, 3, 5 and 6 were fissioned before the 'populational evolution' stage in each branch of our cladogram.

In species like *C. m. p. nigripes* or *C. nictitans*, which have 72 and 70 chromosomes respectively, the largest autosome is the equivalent of chromosome CER 10 (Figure 9.1), all the chromosomes larger than CER 10 being fissioned.

This leads to relatively homogeneous karyotypes, composed of a majority of small meta- or submetacentric chromosomes (Figure 9.3).

2. Centromeric shifts, the second most frequent rearrangement, were observed 16 times. This rearrangement, described for the first time in the study of this group of species (Dutrillaux *et al.*, 1980) corresponds to a change in the position of the active centromere. This rearrangement seems to have occurred during the evolution of a few other groups, at the origin of the branches leading to Lemuriformes and to the Atelinae (Rumpler & Dutrillaux, 1986, Dutrillaux, Couturier & Viegas-Péquignot, 1986a). It is always associated with the occurrence of fissions, and is likely due to a similar mechanism.

The fission of a chromosome makes necessary the activation of a new (or latent) centromere. Indeed, this indicates that the species which are able to undergo such chromosomal evolution have acquired either new chromosomal structures, or enzymatic capacities, or both, that most of the other species do not possess. From this point of view, the Cercopithecinae represent a very interesting group of primates for enzymatic and molecular studies.

3. Inversions and translocations, observed 7 and 4 times respectively, are much scarcer in Cercopithecinae than in all the other groups of primates (Dutrillaux *et al.*, 1986b).

Possible consequences of chromosomal evolution in Cercopithecinae

The very complex chromosomal evolution of Cercopithecinae is essentially characterized by a populational evolution. It means that chromosomal rearrangements were collectively acquired in ancestral populations, and that the descendant species have retained some of these various rearrangements in various combinations.

Such a process implies the existence of hybrids which allocate to their progeny the various chromosomal formulae. This pattern of evolution also indicates that the gametic barrier was not very strong when a single chromosomal rearrangement took place. Indeed, this evolution is the opposite of a mere dichotomy and the structure of the karyotypes may be responsible of its occurrence (Dutrillaux, 1986).

The major characteristic of the chromosomal evolution of Cercopithecinae is the increase of chromosome number ($2N$). In their last common ancestors $2N$ was probably equal to 44; in *C. nigroviridis* $2N =$

48; in *C. talapoin* and *C. patas* $2N = 54$ and in *C. diana* $2N = 58$. These *Cercopithecus* are those with the lowest chromosome numbers. In *C. nictitans* and *C. mitis* $2N = 70$, and the species with the highest chromosome numbers are *C. m. wolfi* and *C. m. p. nigripes*, with $2N = 72$.

This increase in chromosome numbers might have had a strong influence on the diversity of the gametes formed by each species. As a matter of fact, the number of different chromosome segregations (NS), at meiosis, is $NS = 2^N$, in which N is the haploid number of chromosomes. The difference between an ancestral form and *C. m. wolfi*, for instance, is $2^{36} - 2^{22} = 48\,515\,282\,432$.

This means that the diversity of the gametes, due to chromosome segregations, is more than 48 thousand million times higher in *C. m. wolfi* than in the last common ancestors of Cercopithecinae.

Since meiotic recombinations by crossing over increase with the number of chromosomes, this means that linkage groups are much more frequently disrupted in *Cercopithecus* than in groups with lower chromosome numbers, like the Papioninae ($2N = 42$).

It is very likely that speciation does not depend on a single gene mutation, but on groups of mutations, responsible for groups of new genetic characters. A dichotomic evolution may eventually occur if such groups of mutant genes are rarely disrupted. In Cercopithecinae, this situation is very unlikely, because of the possible occurrence of multiple recombinations and chromosome segregations discussed above. Thus, dichotomy is replaced by a populational evolution, a process able to lead to new karyotypes by the mixture of different genomes. Here again, the Cercopithecinae provide a very interesting model for the study of the speciation processes.

Phylogenetic relationships with other Cercopithecidae

Papioninae. We have shown that most Papioninae possess the same karyotype, made of 42 chromosomes. Only two chromosomal rearrangements were ever found. The first concerns the karyotype of a single species, *Macaca fascicularis* (Dutrillaux *et al.*, 1979). The second is shared by genera *Cercocebus* and *Mandrillus* (Dutrillaux *et al.*, 1982). Finally, all the *Papio*, *Lophocebus*, *Theropithecus* and *Macaca* species (except *M. fascicularis*) have a similar karyotype, which is likely to represent the ancestral karyotype of the Papioninae. It differs from the ancestral karyotype of the Cercopithecidae by two translocations which took place among Papioninae, and not among Cercopithecinae. This makes it impossible to consider that any of the species of Papioninae is directly or indirectly ancestral to Cercopithecinae.

The chromosomal phylogeny of the colobines remains poorly studied. We have studied three *Colobus* species only, plus one of *Presbytis* and one of *Pygathrix* (Dutrillaux *et al.*, 1984, and unpublished data). However, it is likely that this taxon originated as a third branch of the phyletic tree sprouting between the Papioninae and the Cercopithecinae. This means that we did not find any chromosomal rearrangement shared by two of these taxa and not by the third. However, this does not mean that two of these taxa did not undergo any common evolution after the separation of the third. As a matter of fact, molecular data have shown that the colobines separated first (Cronin & Vincent, 1976). This is in agreement with previous taxonomic schemes based on morphological characters, which merged together the Cercopithecinae and the Papioninae in a family separate from that of the Colobidae.

Finally, it is likely that, at the time of emergence of the three groups of monkeys, gene mutations occurred at a higher rate than did chromosomal rearrangements. This led to the separation of these groups, without modifications of their karyotypes. A low rate of chromosomal evolution has remained characteristic of the papionines, which speciated without displaying chromosomal changes, whereas chromosomal evolution took place at a much higher rate in colobines, and above all in cercopithecines.

Note

[1]Populational ('network') Evolution is a process which takes place when a number of 'incipient species' (forms) happen to coexist in the same or in adjacent habitats. If the reproductive isolation of these forms is weak, they may hybridize. These 'hybrids' may in turn either interbreed, backcross with one parental form, or 'hybridize' again with other 'hybrids' with different karyotypes (chromosomal rearrangements). Such situations lead to a diversity of genetic conditions, as both homozygous and heterozygous individuals differing by two or more chromosomal rearrangements may later build up an efficient genetic barrier between them, and this ends up in true speciation. This hypothesis involving a multi-step process is the more parsimonious to explain the complex distribution of chromosomal rearrangements observed in living species of *Cercopithecus* monkeys.

The proposed model is in agreement with what we observe at present in field conditions. Fertile 'interspecific hybrids' between syntopic species have been observed (Aldrich-Blake, 1968; Struhsaker *et al.*, Chapter 24), as well as between allopatric species along interspecific borderzones (Lernould, Chapter 4; Oates, Chapter 5). Furthermore, forest cercopithecines very often live in semi-stable multispecific troops (Gautier-Hion, Chapter 23). Hybridization has also been frequently observed in captive conditions (Lernould, Chapter 4).

II.10

Quantitative comparisons of the skull and teeth in guenons

ROBERT D. MARTIN and ANN M. MacLARNON

Introduction

Cranial and dental characters have traditionally figured prominently in studies of phylogenetic relationships among primates. However, relatively few studies of morphological features have included quantitative comparisons; the predominant trend has been to rely instead upon qualitative assessment of differences between species. Although publications dealing with quantitative assessments of morphological characters are now appearing in steadily increasing numbers, there has been a somewhat puzzling tendency to move directly into complex multivariate approaches. Multivariate approaches have a considerable appeal because they offer the possibility of summarizing a large number of variables, but they have the drawback that the biologist is rarely able to interpret the results in straightforward biological terms. As in other areas of biology, it is important to ensure that any quantitative approach that may be adopted should take explicit account of known biological principles. In the case of phylogenetic reconstruction based on morphological characters, it is essential both to deal appropriately with the scaling effects of body size and to comply with well established requirements for successful phylogenetic analysis.

One of the major problems encountered in any comparative, quantitative study of primate skulls and teeth is that of the *scaling influence of body size*. In the case of the guenons, it is especially important to take account of body size effects. The talapoin is very small (average adult body weight for males and females combined: 1.25 kg) and the patas is quite large (average adult body weight for males and females combined: 9.46 kg) in comparison to the other guenons considered here (range of average adult body weights for males and females combined:

3.38–6.60 kg). If the scaling effects of body size are not excluded as a first step in quantitative comparisons, the danger exists that species will be grouped largely on the basis of body weight rather than on the basis of more fundamental biological characters.

It is now well established that, in interspecific comparisons, many biological features scale in a regular fashion with body size. The study of scaling effects in primates and other animals has recently become an area of particularly active research (e.g. see McMahon & Bonner, 1983; Peters, 1983; Calder, 1984; Schmidt-Nielsen, 1984; Jungers, 1985). Only in a minority of cases is such scaling found to occur in the form of simple proportionality (*isometry*). In the overwhelming majority of cases, the scaling of individual characters to body size is non-linear (*allometric*) and can usually be described approximately by a simple power function of the form:

$$Y = kX^a$$

(where X = a measure of body size, Y = some character dimension, k = allometric coefficient, a = allometric exponent).

This equation can be converted to linear form through the simple expedient of logarithmic transformation:

$$\log Y = a \log X + \log k$$

Using standard statistical techniques, it is therefore possible to fit a straight line to a logarithmic plot for a given bivariate data set in which each pair of values represents the average condition for a single species. The two constants of the equation (allometric coefficient; allometric exponent) can be inferred from this empirical best-fit line.

With any real biological data set, the correlation between the logarithmic values for the two variables will generally be less than ideal and there will be at least some degree of scatter of points around the best-fit line. Such scatter primarily reflects the special biological adaptations of individual species rather than measurement error or simple random variation. Accordingly, at least in theoretical terms, it can be stated that the best-fit line will indicate the general scaling principle ('allometric law') involved, while the departures of individual points above and below the line (namely, their residual values) will indicate adaptations of single species. A bivariate logarithmic plot thus permits, in principle, a distinction between the 'expected' value for a variable at a given body size and the special quantitative deviation of any individual species. Such deviations indicate particular adaptations relative to the common trend. Shared deviations in the same direction may therefore provide some basis for suggesting relationships between species. This is, for example, likely to be true in cases where an entire taxonomic group of species exhibits an upward or downward shift relative to a best-fit line established for another taxonomic group of species. The term 'allometric grade shift' may be applied here (see Martin, 1980).

Even if it is true that departures in the same direction away from the best-fit line indicate relationships between species, these 'relationships' (in fact, merely *similarities*) can be of two basic kinds. In the first place, it is possible that two species may share positive or negative deviations from a best-fit line because they have independently undergone special adaptation, for example in response to similar functional requirements. In standard evolutionary terminology, such independent acquisition of comparable adaptations represents *convergence*. However, it is also possible that two species may exhibit common adaptations, reflected by departure in the same direction from an allometric best-fit line, because they have retained specific features from some common ancestor. In this case, a similar direction of deviation indicates phylogenetic affinity between species rather than direct functional adaptation. In fact, the shared similarity doubtless reflects some prior functional adaptation in a common ancestor and in standard evolutionary terminology these common deviations may be described as *homologous*. However, it is important to note that the interpretation of homologous similarity in phylogenetic terms is by no means simple. This is because, for any phylogenetic tree, it is necessary to distinguish between features that were present in the initial common ancestral stock (*primitive* homologous features) and features developed in any later stock (*derived* homologous features). The necessity to distinguish between primitive and derived homology in the process of phylogenetic reconstruction was first clearly recognized by Hennig (1950; 1966) and has now become widely accepted and applied (e.g. see Wiley, 1981; Eldredge & Cracraft, 1980; Joysey & Friday, 1982; Duncan & Stuessey, 1984).

The crucial distinction between functional convergence, primitive homology and derived homology is illustrated in Figure 10.1. Clearly, there is an inherent problem in the interpretation of allometric residual values in phylogenetic terms, since deviations in the same direction from a best-fit line could be due to convergence, to primitive homology, to derived homology or to some combination of these. This problem is, indeed, inherent in any quantitative study that simply provides estimates of degrees of *overall similarity* between species. It is not justifiable

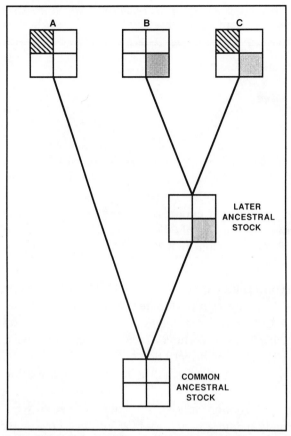

Figure 10.1. Illustration of the distinction between independently-acquired convergent similarities (hatched boxes), primitive homologous similarities (white boxes) and derived homologous similarities (stippled boxes). Although C shares 3 similarities with A and the same number of similarities with B, it is the joint possession of derived homology by B and C that is crucial for the identification of their phylogenetic affinity.

to conclude that more similar species are necessarily more closely related in a phylogenetic sense and the interpretation of diagrams indicating overall degrees of similarity between species is therefore a very complex matter.

Species and methods
The analyses discussed in this paper were carried out on values extracted from two published data sets dealing primarily with morphological variation among Old World monkeys. In both cases, the

Table 10.1. *Guenon species for which morphological data were available for analysis*

Species	Verheyen (1962)	Kay (1978)
Allenopithecus nigroviridis	+	+
Cercopithecus aethiops	+	+
C. ascanius	+	+
C. cephus	+	+
C. diana	+	+
C. lhoesti	+	+
C. mitis	+	+
C. neglectus	+	+
C. nictitans	+	+
C. pogonias	−	+
C. wolfi	+	−
Erythrocebus patas	+	+
Miopithecus talapoin	+	+

studies were concerned with evolution in Old World monkeys generally, whereas the analyses reported here were confined to the guenon group (genus *Cercopithecus*; subgenera *Allenopithecus, Cercopithecus, Erythrocebus, Miopithecus*). Verheyen (1962) provided extensive data on cranial dimensions for a wide selection of Old World monkeys, including 12 guenon species, while Kay (1978) later provided specific data on dental dimensions for a similar range of Old World monkey species, also including 12 guenon species. 11 guenon species were in fact common to the two data sets provided by Verheyen & Kay (see Table 10.1). In both cases, the values given usually combined measurements for adult males and females of each species, thus providing a single figure for each species that takes no account of sexual dimorphism. In some guenon species (e.g. patas monkey; de Brazza monkey), extreme sexual dimorphism is present, with the adult male weighing almost twice as much as the adult female. The influence of sexual dimorphism on the kind of interspecific comparison discussed below is a potential complicating factor that deserves detailed study in its own right but was beyond the scope of the present investigation.

Since the maximum number of variables that could be analysed for any given data set was 18 (see below), it was necessary to exclude 26 variables from the total of 44 cranial dimensions listed by Verheyen (1962). The reduced list of variables that was subjected to allometric analysis was as follows:

1. bieuryal width
2. cranial volume

3. foramen magnum length
4. postorbital constriction
5. external biorbital width
6. anterior interorbital width
7. orbit height
8. orbit width
9. bizygomatic width
10. nasal aperture width
11. palate length
12. upper dental arcade width
13. upper dental arcade length
14. cheek teeth length
15. bicondylar width
16. lower jaw symphysis height
17. ascending ramus height
18. ascending ramus width

In addition, the parameter prosthion-inion length was taken from Verheyen's original list of 44 variables as an indicator of overall skull size. A parameter reflecting skull length (glabella-inion) was used as an indirect indicator of body size by Verheyen and prosthion-inion length was similarly used in the present study in order to maintain broad comparability with Verheyen's analyses. [Verheyen actually analysed his data primarily in the form of simple ratios of individual skull parameters to glabella-inion length. The use of ratios is only appropriate if these parameters scale isometrically to prosthion-inion length. Otherwise, some form of allometric analysis is required (see below).]

Kay (1978) listed only eight dental variables and one of these (crown width of upper second molar) was excluded from the present allometric analysis because of an anomalous value given for *Cercopithecus aethiops* (through a printing error, the value of 3.70 was given instead of 3.20 – R. F. Kay, personal communication). The seven remaining variables included in the analysis were therefore as follows:

1. length of lower second molar (M_2)
2. hypoconid height of M_2
3. metaconid height of M_2
4. crushing surface area of M_2
5. cristid oblique length
6. entocristid length of M_2
7. postmetacristid length of M_2

In this case, actual body weight was used as the reference measure of body size for allometric analysis, in order to retain comparability with the bivariate allometric analyses conducted by Kay. However, body weights were taken from an extensive data-base (largely referring to wild-caught specimens) previously compiled by the authors, rather than from Kay (1978). In fact, Kay only used body weight as an indicator of body size for his preliminary bivariate analyses. When he proceeded to conduct two forms of multivariate analysis (principal components analysis; principal coordinates analysis) on the data set for dental dimensions, he used overall tooth length as an indicator of size in order to adjust the other dental dimensions. However, since this approach may exclude some potentially valuable information (e.g. the size of molar teeth relative to overall body size), molar dimensions were related to body weight in the present study.

A step-wise technique of analysis introduced for the study of scaling of gastrointestinal dimensions in primates and other mammals (Martin *et al.*, 1985; MacLarnon *et al.*, 1986a,b) was applied to the two data sets for guenons (Figure 10.2). In the first instance, logarithmic bivariate analyses were conducted in order to determine empirical patterns of scaling of individual parameters to body size (the latter being indicated by prosthion-inion length for Verheyen's cranial variables and by body weight for Kay's dental variables).

There is still active discussion over the most appropriate best-fit line to use for bivariate allometric analyses. Many authors have used simple least-squares regression for allometric analysis, but uncertainty continues to reign regarding the most appropriate line-fitting technique (e.g. see Harvey & Mace, 1982). Of the commonly used line-fitting procedures, determination of the major axis would seem to be generally preferable for allometric analysis, and this procedure has been used throughout the present study. When correlation coefficient values are relatively high (namely, greater than $r = 0.95$), the choice of line-fitting technique has little influence on the residual values determined for individual species, but as correlation coefficients decline, increasingly divergent results may be obtained with regressions and major axes.

From each bivariate data analysis, residual values were calculated for each guenon species, indicating the *relative* size of the given cranial or dental parameter following 'removal' of the effects of body size. It was hence possible to construct a table of 18 residual values for the 12 guenon species represented in Verheyen's data set for cranial dimensions and a similar table of seven residual values for the 12 guenon species represented in Kay's data set for dental dimensions. Given a table of n residual values for m species, it is possible to construct an

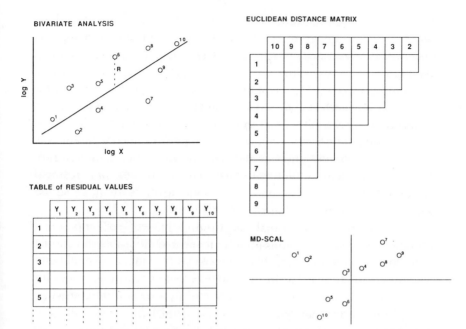

Figure 10.2. Diagram showing the stepwise sequence of analysis employed, for a sample of 10 species. In the first stage of *bivariate analysis*, each parameter (Y) is plotted against some indicator of body size (X) on logarithmic coordinates and a best-fit line (major axis) is determined. It is then possible to determine a residual value (R) for each species, indicating the extent to which it deviates above or below the line. The residual values for all species and all parameters are then tabulated. Subsequently, a half-matrix of Euclidean distances between pairs of species is calculated. In the final stage, various techniques can be used to illustrate the overall pattern of distances between species. Multidimensional scaling (MD-SCAL), which permits a summary of the interspecific distances in the form of an optimized 2-dimensional plot, is the technique favoured here.

($m - 1$) by ($m - 1$) half-matrix of Euclidean distances between pairs of species (Figure 10.2). These distances provide an overall indication of the degree of separation between any two species reflected by the n residual values taken together. Having constructed the distance matrix, it is then possible to apply various techniques to summarize the information provided. One approach is to use some form of hierarchical cluster analysis to generate a dendrogram. Following previous studies of this kind (e.g. Bauchot, 1982; Martin *et al.*, 1985), a clustering technique based on Ward (1963) was employed to generate a dendrogram from each distance matrix. An alternative approach is to apply the technique of multidimensional scaling (Kruskal, 1964a,b), which can be used to generate a two-dimensional plot summarizing the distances between pairs of species in an optimized manner. Previous experience has shown that multidimensional scaling produces more meaningful and consistent results than hierarchical clustering (Martin *et al.*, 1985; MacLarnon *et al.*, 1986a,b) and this is the preferred final step in the sequence of analysis (see Figure 10.2). However, for multidimensional scaling there was a limitation on the total number of parameters included in any one analysis because the available computer programmes permitted simultaneous analysis of a maximum of 18 residual values for each species. For this reason, no more than 18 scaled parameters were included in any one analysis.

Having analysed cranial variables from Verheyen (1962) and dental variables from Kay (1978) separately, it seemed advisable to conduct some form of combined analysis that would provide an overall summary of the data. To achieve this, the seven sets of residual values for relative dental dimensions derived from Kay's data were combined with 11 sets of residual values for relative cranial dimensions derived from Verheyen's data to produce a joint table containing the maximal permissible number of 18 sets of residuals. The 11 variables from Verheyen's data set for cranial variables (see asterisks in Table 10.2) were selected so as to minimise overlap between measures (e.g. bieuryal width was excluded because it is closely related to cranial volume) and in order to retain parameters with the greatest degree of scatter around the best-fit line, thus preserving as much variation as possible despite the reduction in the number of variables. Since only 11 guenon species were common to both original sets of data, it was necessary to exclude *Cercopithecus pogonias* and *C. wolfi* from the combined analysis. In fact, this combined analysis was not ideal because different indicators of body size had been used to derive the cranial and dental residual values (prosthion-inion length and body weight, respectively) and because two of the species used to calculate the original residual values had to be subsequently excluded. However, the combined analysis does provide a useful summary of the overall quantitative similarities and differences between guenon species indicated by the two sets of data from Verheyen (1962) and from Kay (1978).

It should also be emphasized that the residual values were left in logarithmic form for the calculation of Euclidean distances between species. Residual values derived from bivariate allometric analyses are often converted to antilogarithmic form, since this permits a simple statement of their implication. A logarithmic residual value of 0, indicating that a species lies directly on the best-fit line, yields a reference antilogarithmic index value of unity for a 'standard' species exhibiting the 'expected' value at any given body size. Negative or positive departures from the best-fit line are indicated as simple multiples of this reference value. An antilogarithmic residual value of 1.5 for a given character dimension of one species, for instance, indicates that for that species the dimension is 50% larger than 'expected' (from the best-fit line). Conversely, an antilogarithmic residual value of 0.6 for a given character dimension of one species indicates that the dimension is 40% smaller than 'expected'. However, such antilogarithmic conversion of residual values changes an original symmetrical distribution of values into an asymmetrical distribution, since negative residuals can only take antilogarithmic values between 0 and 1 while positive residuals can in principle take any value greater than 1. Accordingly, if antilogarithmic residual values are used to calculate Euclidean distances between species, the effect of positive deviations from the best-fit line will be exaggerated, compared with the effect of negative deviations which will be largely overwhelmed. In a study of gastro-intestinal scaling in primates and other mammals, use of antilogarithmic residual values to calculate distances between species (Martin *et al.*, 1985) obscured some relationships that become apparent only when logarithmic residual values were used (MacLarnon *et al.*, 1986a,b).

In principle, the general method described above and summarized in Figure 10.2 may be expected to produce results similar to those produced by other techniques of multivariate analysis, such as principal components analysis (e.g. see Kay, 1978). However, it is unclear whether the effect of body size is eliminated in such an effective or explicit fashion with these other techniques. Indeed, Kay (1978) scaled his individual dental parameters to overall tooth size before applying principal components analysis, thus avoiding the usual assumption that the first principal component will represent the influence of body size. The advantage of the method advocated here is that it is possible to analyse and interpret individual residual values in biological terms at intermediate stages of analysis, rather than relying entirely upon presentation of the final results in the form of a single plot. Thus, the biological components that are involved in overall assessments of distance between species can be examined as required and the final pattern can be interpreted explicitly in relation to biological adaptations.

Results

The results of the bivariate analyses of 18 individual cranial dimensions, relative to prosthion-inion length, are summarized in Table 10.2. This table lists the individual empirical allometric formulae subsequently used to calculate residual values for each of the cranial variables. It is interesting to note that very few of the cranial dimensions seem to scale in an isometric fashion relative to the prosthion-inion length of the skull. For the 17 simple linear dimensions of the skull, the expected slope value for each logarithmic allometric equation (equivalent to the value of the exponent in the original allometric equation) would be 1 for isometric scaling. For the single volumetric measure (cranial capacity), the expected slope value for isometry would be 3, though in this case it is already well established that brain size typically scales to body size in a negatively allometric fashion. Of the 17 linear skull dimensions, only five (anterior interorbital width; bizygomatic width; nasal aperture width; cheek teeth length and ascending ramus height) exhibit allometric exponent values whose 95% confidence limits include the value of unity. Eight variables (bieuryal width;

Table 10.2. *Allometric equations for 18 cranial parameters relative to prosthion-inion length (data from Verheyen, 1962)*

Parameter	Allometric equation[b]	95% limits	r
Bieuryal width	$y = 0.477x + 0.785$	0.390–0.571	0.966
Cranial volume[a]	$y = 1.476x - 1.137$	1.219–1.814	0.963
Foramen magnum length[a]	$y = 0.422x + 0.293$	0.229–0.646	0.824
Postorbital constriction[a]	$y = 0.493x + 0.638$	0.298–0.723	0.858
External biorbital width	$y = 0.674x + 0.378$	0.567–0.792	0.973
Anterior interorbital width[a]	$y = 1.126x - 1.523$	0.747–1.735	0.866
Orbit height[a]	$y = 0.520x + 0.282$	0.357–0.707	0.905
Orbit width	$y = 0.595x + 0.156$	0.505–0.692	0.976
Bizygomatic width[a]	$y = 0.858x + 0.104$	0.723–1.014	0.973
Nasal aperture width[a]	$y = 1.101x - 1.187$	0.816–1.500	0.921
Palate length[a]	$y = 1.299x - 1.090$	1.145–1.480	0.984
Upper dental arcade width[a]	$y = 0.796x - 0.093$	0.627–0.998	0.951
Upper dental arcade length	$y = 1.150x - 0.700$	1.029–1.287	0.988
Cheek teeth length[a]	$y = 1.077x - 0.759$	0.959–1.210	0.987
Bicondylar width	$y = 0.786x + 0.165$	0.665–0.923	0.974
Symphysis height	$y = 1.465x - 1.671$	1.254–1.728	0.975
Ascending ramus height[a]	$y = 0.990x - 0.538$	0.651–1.501	0.868
Ascending ramus width	$y = 1.217x + 1.109$	1.066–1.395	0.982

[a]Parameters combined with Kay's dental variables for the combined analysis (see text).
[b]$y = \log_{10} Y$; $x = \log_{10} X$.

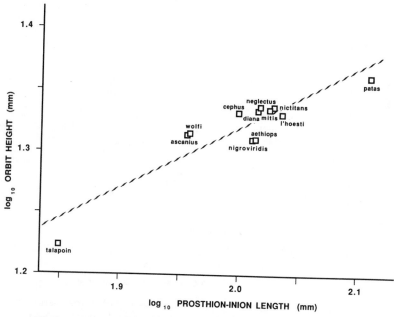

Figure 10.3. Sample logarithmic bivariate plot of orbit height against prosthion-inion length for 12 guenon species (data from Verheyen, 1962). Hatched line is the major axis. See text for discussion.

foramen magnum length; postorbital constriction; external biorbital width; orbit height; orbit width; upper dental arcade width; bicondylar width) clearly scale in a negatively allometric fashion relative to prosthion-inion length, while four (palate length; upper dental arcade length; symphysis height; ascending ramus width) scale in a positively allometric fashion. Hence, allometric analysis is essential for the successful interpretation of the cranial variables analysed in relation to overall skull length. Use of simple proportions (ratios) would only be justifiable if all or most of the cranial variables were found to scale in an approximately isometric fashion.

Sample bivariate plots of individual cranial variables are shown in Figure 10.3 (orbit height) and Figure 10.4 (foramen magnum length). These plots emphasize the extreme positions of the talapoin and the patas monkey with respect to overall body size and show the scatter of individual species relative to the best-fit line (major axis). In Figure 10.3, five species can be seen to lie below the line (*aethiops; lhoesti; nigroviridis; patas; talapoin*). All of these species can be said to have small orbit heights relative to skull size. The remaining species lie above the line to varying degrees and can be said to have relatively tall orbits. In

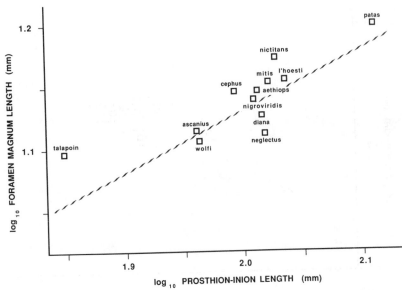

Figure 10.4. Sample logarithmic bivariate plot of foramen magnum length against prosthion-inion length for 12 guenon species (data from Verheyen, 1962). Hatched line is the major axis. See text for discussion.

Figure 10.4, the pattern is more or less reversed, with *aethiops, lhoesti, nigroviridis, patas* and *talapoin* lying above the best-fit line to varying degrees, in this case accompanied by *cephus, mitis* and *nictitans*. It can therefore be said that *aethiops, lhoesti, patas* and *talapoin* share the characteristic of relatively large foramen magnum length in addition to that of relatively small orbit height. Conclusions of this kind can be drawn, with varying degrees of consistency, from the remaining 16 bivariate plots of cranial dimensions against prosthion-inion length.

Results from bivariate analyses of seven dental variables, relative to body weight, are summarized in Table 10.3. In this case, body weight has been used as the indicator of body size and the expected slope (exponent) value for isometry with linear dimensions would be 0.333. The expected exponent value for the scaling of an area (namely, in the single case of M_2 crushing surface area) would be 0.667. It can be seen that the 95% confidence limits on the exponent values include the expected isometric values in all cases. Hence, the use of simple proportions (ratios) would in principle be justifiable in the case of the dental parameters. Nevertheless, for the sake of consistency the empirical allometric equations listed in Table 10.3 have been used for the calculation of residual values in the present study. Examination of

Table 10.3. *Allometric equations for 7 cranial parameters relative to body weight (data from Kay, 1978)*

Parameter	Allometric equation[a]	95% limits	r
M_2 length	$y = 0.301x + 0.579$	0.250–0.353	0.972
M_2 hypoconid height	$y = 0.314x + 0.406$	0.176–0.464	0.840
M_2 metaconid height	$y = 0.314x + 0.364$	0.178–0.461	0.845
M_2 crushing surface area	$y = 0.582x + 0.653$	0.456–0.724	0.951
Cristid oblique length	$y = 0.269x + 0.068$	0.151–0.395	0.842
Entocristid length	$y = 0.362x - 0.024$	0.243–0.491	0.901
M_2 postmetacristid length	$y = 0.425x + 0.016$	0.261–0.612	0.866

[a] $y = \log_{10} Y$; $x = \log_{10} X$.

individual bivariate plots of the dental dimensions permits preliminary interpretation as in the case of the two selected cranial dimensions illustrated in Figures 10.3 and 10.4.

Having established empirical allometric equations for the individual dimensions, it was possible to proceed with the calculation of Euclidean distances between pairs of species and hence to the preparation of overall summary diagrams (dendrograms; multidimensional plots), as indicated in Figure 10.2. The dendrogram based on the residual values for 18 cranial dimensions is shown in Figure 10.5 and the

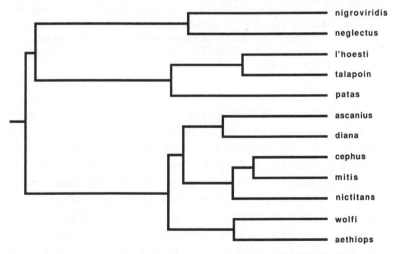

Figure 10.5. Dendrogram showing the results of cluster analysis of the interspecific Euclidean distances based on 18 sets of cranial residual values for 12 guenon species (raw data from Verheyen, 1962). The branch-lengths shown are proportional to the adjusted distances between species.

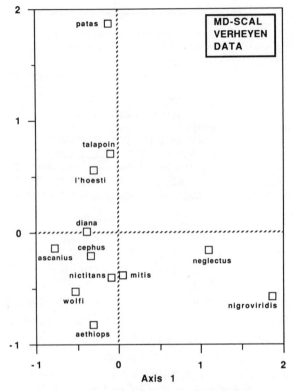

Figure 10.6. Multidimensional scaling plot generated by analysis of the interspecific Euclidean distances based on 18 sets of cranial residual values for 12 guenon species (raw data from Verheyen, 1962). Similarities between species are indicated by both distance and direction in the plot. [NB. The MD-SCAL axes are arbitrary; only the relative positions of the points are of importance.]

corresponding multidimensional scaling plot is shown in Figure 10.6. NB. It is important to note that the axes in a multidimensional scaling plot are arbitrary and have no significance in themselves; only the *relative* positioning of the points conveys relevant information.] The dendrogram suggests that the three species *lhoesti*, *patas* and *talapoin* may be linked fairly closely together on the basis of cranial residual values. Most of the remaining guenons form a second cluster, while *neglectus* and *nigroviridis* are combined together as a third cluster. The multidimensional scaling plot broadly confirms this picture, though it can be seen that *aethiops* is a fairly marked outlier compared to the main group of forest-living guenons (*ascanius; cephus; diana; mitis; nictitans; wolfi*).

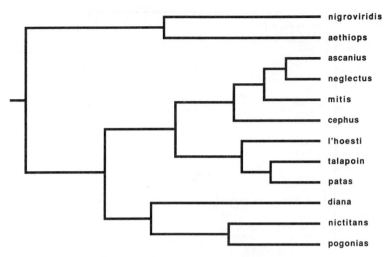

nigroviridis
aethiops
ascanius
neglectus
mitis
cephus
l'hoesti
talapoin
patas
diana
nictitans
pogonias

Figure 10.7. Dendrogram showing the results of cluster analysis of the interspecific Euclidean distances based on 18 sets of dental residual values for 12 guenon species (raw data from Kay, 1978). The branch-lengths shown are proportional to the adjusted distances between species.

The dendrogram obtained with the residual values for seven dental dimensions is shown in Figure 10.7 and the corresponding multi-dimensional scaling plot is shown in Figure 10.8. It can be seen that the dendrogram exhibits several similarities to that obtained with the cranial residual values (Figure 10.5). Once again, the three species *lhoesti*, *patas* and *talapoin* seem to be linked fairly closely together and *nigroviridis* is relegated to a relatively outlying position. However, the forest-living guenons are in this case split into two sub-clusters and *aethiops* is linked (albeit not very closely) with *nigroviridis*. Comparisons with the multidimensional scaling plot for the dental residual values (Figure 10.8) in this case reveals a rather different picture from the dendrogram in certain respects. The diana monkey is seen to be an extreme outlier in comparison with the main group of forest guenons (*ascanius*; *cephus*; *diana*; *mitis*; *nictitans*; *pogonias*). In the dendrogram, it has apparently been linked to *nictitans* and *pogonias* because these two species are moderate outliers in a similar, but by no means identical, direction. Similarly, *aethiops* and *nigroviridis* are quite widely separated outliers in the multidimensional scaling plot (Figure 10.8), whereas they have been linked together in the dendrogram (Figure 10.7). In fact, *aethiops* lies closer to *lhoesti* in the multidimensional scaling plot than it does to *nigroviridis*.

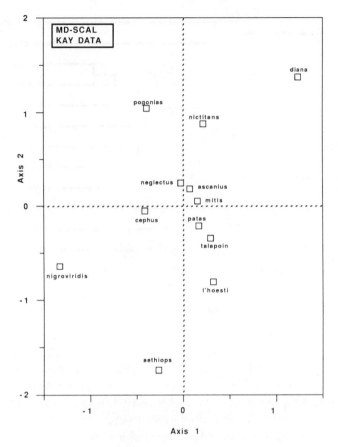

Figure 10.8. Multidimensional scaling plot generated by analysis of the interspecific Euclidean distances based on 18 sets of dental residual values for 12 guenon species (raw data from Kay, 1978). Similarities between species are indicated by both distance and direction in the plot.

The dendrogram for the combined sample of residual values for 11 cranial and seven dental dimensions is shown in Figure 10.9 and the corresponding multidimensional scaling plot is shown in Figure 10.10. In this case, there is fairly close agreement between the two diagrams. In the dendrogram, the now familiar grouping of *lhoesti, talapoin* and *patas* is once again apparent. Further, there is a single main grouping of forest-living guenons (*ascanius, cephus, diana, mitis, neglectus* and *nictitans*), though *diana* can be seen to occupy a relatively peripheral position. Finally, *aethiops* and *nigroviridis* are linked together before being combined with the cluster consisting of *lhoesti, talapoin* and *patas*.

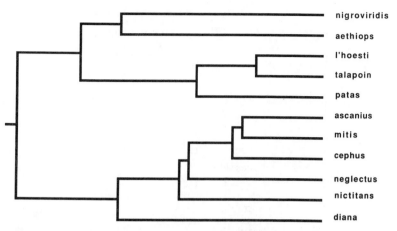

Figure 10.9. Dendrogram showing the results of combined cluster analysis of the interspecific Euclidean distances based on 11 sets of cranial residual values and 7 sets dental residual values for 11 guenon species (raw data from Verheyen, 1962 and Kay, 1978, respectively). The branch-lengths shown are proportional to the adjusted distances between species.

The same general pattern is clearly recognizable in the multidimensional scaling plot. However, the peripheral position of *diana* relative to the main group of forest-living guenons is more marked and it can be seen that the dendrogram seems to have linked *aethiops* and *nigroviridis* merely because they are extreme outliers in a similar, but not identical, direction. In fact, *aethiops* is located somewhat closer to *lhoesti* than it is to *nigroviridis* in the multidimensional scaling plot while *nigroviridis* would seem to be the most isolated species.

Discussion

Despite some variations in detail, a fairly consistent pattern emerges from all of the summary diagrams illustrating distances between guenon species based on cranial and dental residual values (Figures 10.5–10.10). In particular, a grouping of *lhoesti, talapoin* and *patas* emerges from all analyses, suggesting some fundamental similarity between these three species. Whatever specific interpretation may be placed on this similarity, it is particularly encouraging to see that the smallest of the guenon species (*talapoin*) is hence linked to the largest (*patas*). This provides the clearest possible demonstration that the factor of body size has been successfully eliminated in the course of the analyses. Otherwise, any quantitative comparison would automatically tend to separate *talapoin* and *patas*. This proved to be the case in

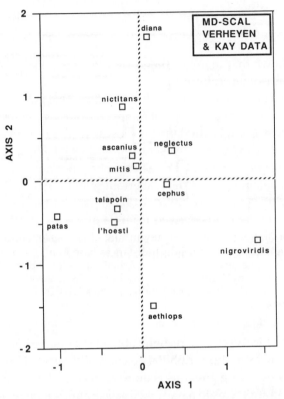

Figure 10.10. Multidimensional scaling plot generated by combined analysis of the interspecific Euclidean distances based on 11 sets of cranial residual values and 7 sets dental residual values for 11 guenon species (raw data from Verheyen, 1962 and Kay, 1978, respectively). Similarities between species are indicated by both distance and direction in the plot.

Verheyen's (1962) original analysis of his data, in which simple ratios of individual skull parameters to prosthion-inion length were used as indicators of phylogenetic relatedness: 'Comme critère des affinités phylogénétiques entre les différentes espèces nous avons adopté la différence moyenne des indices: à l'aide de celle-ci il est possible d'élaborer une phylogénie assez précise' (Verheyen, 1962, p. 10). As has been noted above, the use of ratios as a means of 'removing' size effects is only justifiable where isometric relationships are involved. The use of a ratio between a given parameter and an indicator of body size when these are allometrically related to one another does not 'remove' the effect of size; it merely changes one allometric relationship into another:

if $Y = kX^a$

Then $Y/X = kX^{(a-1)}$

Table 10.2 clearly shows that many of the skull parameters listed by Verheyen do not scale isometrically with skull length. It is therefore not surprising that Verheyen should have concluded that among guenons *patas* (the largest-bodied species) is furthest removed from *talapoin* (the smallest-bodied species), while the latter is closest to the relatively small-bodied *ascanius* and *mona*. It is also understandable why Verheyen should have described *talapoin* as the most aberrant of the guenon species. The talapoin is the most aberrant guenon species in terms of body weight; it therefore follows that it will be the most aberrant in terms of cranial ratios derived from parameters that vary allometrically with skull size. In fact, Verheyen himself recognized the problems posed by allometric relationships and noted that his conclusions regarding *talapoin* were subject to the influence of body size.

It is, incidentally, apparent from Figures 10.3 and 10.4 that there is a potential statistical problem involved in determining best-fit lines because of the extreme body sizes of *talapoin* and *patas*. It is well known that such isolated points at the upper and lower ends of a bivariate distribution may exert a strong bias on the slope of best-fit line, rather like children on a see-saw. A strong bias would only be exerted if, for instance, the point for *talapoin* were located well below the 'real' line and the point for *patas* were located well above it. In this case, the two points would bring about an artificial elevation of the slope. However, it is a fortunate fact that *talapoin* and *patas* tend to diverge *in the same direction* from the best-fit lines in all bivariate plots. As a result, their potentially biasing effects presumably cancel out to a large extent. Of course, it is possible that some biasing effect nevertheless remains and the only way of determining this would be to fit to each bivariate data set a line of slope predetermined on the basis of information about expected patterns of scaling. At present, however, there is no reliable foundation for predicting appropriate scaling exponents for cranial and dental parameters in guenons.

It also emerges quite consistently from the analyses presented above, though more obviously from the multidimensional scaling diagrams than from the dendrograms, that *nigroviridis* is an extreme outlier compared with other guenons. In fact, the case of *nigroviridis* illustrates a weakness of the dendrogram-construction technique, in that this species may be linked to another in a dendrogram because it is also a marked outlier in a vaguely similar direction rather than because of real quantitative proximity (e.g. note the relative positions of *nigroviridis* and *aethiops* in Figures 10.8 and 10.10).

Other features of Figures 10.5–10.10 are less consistent. As a general rule, there is a main cluster of forest-living guenons, but this is more apparent from the multidimensional scaling plots than from the dendrograms. The positions occupied by *aethiops* and by *diana* are inconsistent. In all cases, *aethiops* occupies a somewhat outlying position. However, in terms of cranial residuals it seems to approach the main group of forest-living guenons, whereas dental residual values seem to link it more to the group containing *lhoesti*, *talapoin* and *patas*. With respect to cranial residuals, *diana* is linked quite closely to the main group of forest-living guenons, whereas in terms of dental residuals it is a marked outlier.

As was emphasized in the introductory section, the summary diagrams in Figures 10.5–10.10 provide no more than an overall quantitative indication of degrees of similarity and difference between species. For the reasons explained, it should not be expected that distances between species can be simply equated with degrees of phylogenetic relationship. Indeed, it is entirely to be expected that the distances indicated would represent a complex mixture of convergent, primitive and derived similarities. It is also important to remember that quantitative similarities between species emerging from comparisons of this kind are often interpreted more in terms of the direct adaptations of individual species than in relation to phylogenetic affinity. For instance, in a similar study of gastro-intestinal residual values in primates and other mammals (Martin *et al.*, 1985; MacLarnon *et al.*, 1986a,b), it was found that the resulting dendrograms and multidimensional scaling plots grouped species largely in relation to functionally convergent adaptations for present dietary habits, rather than according to phylogenetic relatedness. In fact, Kay (1978) interpreted the results of his multivariate analysis of dental dimensions in Old World monkeys specifically in relation to the dietary adaptations of individual species, notably with respect to the distinction between frugivory and folivory.

In view of the fact that one would not normally expect simple quantitative 'distances' between species to provide any reliable indication of phylogenetic affinity, it is all the more striking that the patterns indicated in Figures 10.5–10.10 are strongly reminiscent of the phylogenetic tree based on chromosomal evidence (see Dutrillaux *et al.*, Chapter 9). In particular, the chromosomal evidence suggests that there has been a basic dichotomy in the evolution of the guenons, with *aethiops, lhoesti, talapoin* and *patas* forming one subgroup and with typical forest-living guenons (including *ascanius, cephus, diana, mitis, neglectus, nictitans, pogonias* and *wolfi*) constituting the other subgroup (see also Gautier, Chapter 12, for phylogeny based on vocalizations). The position of *nigroviridis* is virtually intermediate between these two main subgroups, though the chromosomal evidence suggests that it may have a slight affinity with the typical forest-living guenons. In the overall summary plot shown in Figure 10.10, it is possible to recognize the same general groupings, though *aethiops* is not as clearly linked with *lhoesti, talapoin* and *patas* as is the case with chromosomal evidence. Overall, however, the cranial and dental residual values for species within the guenon group would seem to be more indicative of phylogenetic affinity than of present adaptations to specific habitat conditions. Of course, the summary diagram in Figure 10.10 allows no distinction between primitive and derived features linking individual pairs of species, so the distances illustrated cannot be interpreted in direct phylogenetic terms. It is, for instance, possible that *nigroviridis* may be more primitive than other guenons in many respects (e.g. see Verheyen, 1962), so the outlying position occupied by this species may be more attributable to the retention of a relatively primitive pattern of residual values than to the development of specialized features. In other words, *nigroviridis* may lie close to the origin (in terms of a specific set of residual values) from which all of the guenon species diverged.

The observation that dental and cranial residual values within the guenon group generally indicate phylogenetic affinity rather than functional similarity is important for a number of reasons. In the first place, Kay (1978) specifically interpreted multivariate distances between Old World monkey species, based on molar dimensions, in terms of dietary adaptations. In Kay's diagram for African monkey species, illustrating the first two coordinates from a principal coordinates analysis, a totally different pattern is seen from that presented in Figures 10.8 and 10.10 above. Extremely wide separation is found between *lhoesti, patas* and *talapoin*, along an axis which Kay interprets as passing from folivory to frugivory. Indeed, Kay concludes from the position of *lhoesti* on his plot that its molar morphology 'suggests that it is by far the most folivorous of all extant cercopithecines'. By contrast, *talapoin* occupies a position that suggests extreme adaptation for frugivory. But limited field observations on members of the *lhoesti* species group indicate that fruits represent a major component of the diet and that a strong emphasis on folivory is most unlikely (J.-P. Gautier and A. Gautier-Hion, personal communication). Further, it is known that *talapoin* consumes a large proportion of arthropods as well as fruits in its diet (Gautier-Hion, 1978). From other studies (e.g. Kay, 1975), it is known that the effect of insectivory on molar morphology in primates is in some respects parallel to that of folivory, so it is difficult to explain why *talapoin* should possess molar teeth that indicate extreme frugivory according to Kay's analysis.

In contrast to the patterns seen in Figures 10.5–10.10 above, Kay's principal coordinates plot for African monkeys indicates no separation of the guenons into the two basic subgroups that may be identified on chromosomal grounds. There is complete overlap in his plot between members of the subgroup represented by *aethiops, lhoesti, patas* and *talapoin* and members of the subgroup containing the typical forest guenons, while *nigroviridis* occupies a central position in the distribution. Even among the typical forest-living guenons there is no good correlation between position in the plot and known dietary habits. For instance, Gautier-Hion (1978) has shown from analysis of stomach contents that *nictitans* is markedly more folivorous than *cephus, neglectus* or *pogonias*, yet its position in Kay's principal coordinates plot indicates that *nictitans* is the most frugivorous of all of these species. Thus, Kay's plot does not reliably indicate either phylogenetic relationships or the distribution of dietary habits among guenons. Without detailed comparison, it is difficult to explain why such divergent results should have been obtained with the analysis presented here, which is in part based on a subset of Kay's original data. It is, of course, important to remember that the present analysis was restricted to species within the guenon group, whereas Kay's analysis was conducted on the entire family Cercopithecidae, including the leaf-monkeys. It is possible that a relatively fine analysis will tend to emphasize phylogenetic relationships among species, whereas a broader analysis will tend to bring out parallel functional adaptations. The phenomenon of 'phylogenetic inertia' among relatively closely related species has been reported in various contexts and several authors (e.g. Harvey & Mace, 1982; Felsenstein, 1985a) have recently emphasized the point that broad quantitative comparisons of species may in fact be biased by the influence of groups of closely-related species. The present study has been consciously focused at the lowest possible taxonomic level for phylogenetic analysis (namely, a collection of species all attributable to the same genus) and it is only to be expected that the direct effects of genetic continuity will be witnessed. However, it should also be noted that the techniques of analysis used in the present study differed from those used by Kay (1978), notably in the methods used to take account of the scaling effects of body size. It remains to be seen whether a re-analysis of Kay's complete data set using the approach advocated here would yield results different from those reported above and closer to those reported by Kay.

The apparent predominance of phylogenetic effects in Figures 10.5–10.10 is also significant in another respect. If it is true that residual values for cranial and dental dimensions of the guenons, following 'removal' of the effects of body size, reflect phylogenetic affinity, then it may be concluded that there is some genetic basis for 'size-free' morphological features. In other words, it would seem that there may be some fundamental distinction at the genetic level between the scaling effects of body size and the specific morphological adaptations of individual species. Accordingly, the results of the present study suggest that the purely empirical approach adopted above to achieve a separation between scaling 'laws' and specific deviations between individual species may reflect an important genetic distinction in development between size and shape factors.

Acknowledgements

Thanks are due to Mr F. Brett for invaluable assistance with the development of the computer programmes required for the analyses reported in this paper. Thanks also go to Dr L. C. Aiello, Dr B. Dutrillaux, Dr A. Gautier-Hion, Dr J.-P. Gautier, Dr R. F. Kay, Mr J. Kingdon, Ms C. Ross and Dr M. Ruvolo for advice and discussion that greatly benefited the study at various stages. The approach applied in this chapter was originally developed with the support of a project grant from the Medical Research Council (London).

II.11

Comparative morphology of hands and feet in the genus *Cercopithecus*

JONATHAN KINGDON

Introduction

On the principle that hands and feet are sensitive indicators of how animals relate to their different substrata (Napier,1960; Prost,1965; Tuttle 1969), the gross proportions of hands and feet have been compared and the detailed topography of one diagnostic bone, the calcaneum, has been examined in representative species of *Cercopithecus*.

The purpose of this very preliminary study is to determine and illustrate the extent or scale of differences and resemblances between species and to see if any features can immediately be identified that suggest primitive or synapomorph conditions (Haines,1958), particularly in relation to the polarities of terrestrial and arboreal locomotion.

Likewise, crude correlations between ecological and anatomical adaptations have been sought. Finally, hands and feet have been examined for features that might indicate genetic proximity in species of differing size and ecological niche but supposedly belonging to common lineages.

Methods and material

Hands and feet of living sedated animals have been X-rayed and supplemented by X-rays of entire dried museum specimens and direct observations and measurement of disarticulated bones. Radiographs have been traced and the ratios and proportions of parts in different species compared.

Because the study is seen as exploratory and, because only one or a very few specimens of some species have been available, only one representative specimen has been studied in many instances. However, a survey of larger samples for several species has encouraged me to believe that, although age and sex have some influence, the characteristics here described as specific are reasonably representative. Variation probably exists within fairly narrow limits in the features selected for study here.

The right foot of each guenon was laid out on the plate with toes and palm as flat as possible (a transparent weight was placed upon the foot) and a vertical X-ray taken. Previous measurements suggested I could expect some individual variation in the length of phalanges and also some slight distortion due to foreshortening of one or more digits. This has reduced the value of comparing phalanges and also overall lengths. Because the relative dimensions of first metatarsals and of the calcaneum seem to be the two most diagnostic features, I standardized all radiographic prints on the basis of a single dimension, i.e. distance between the calcaneal cuboid and metatarsal-phalangeal articulation. This has allowed the relative length of calcaneum and the relative size and position of the first digit to be compared. The relative breadth and development of palm and digits could also be assessed in a more general and approximate way.

Once radiographic prints of the feet had been standardized in this way, prints were made of the hands. This has given an indication of the proportions of the hand relative to the foot in each species.

The calcaneum has been extracted in a number of key species and the following aspects recorded. The lateral aspect indicates the depth of the distal body of the calcaneum as well as the tuber and provides a profile view of the outer edge of the astragalo-calcaneum facet. The superior aspect indicates the overall alignment and breadth of the calcaneum and its facet. The shape and position of the astragalo-calcaneal facet can be charted from this aspect although its actual surface is subject to complex curvature (which can be partially appreciated when this aspect is compared with its profile view).

While it is the astragalus that provides the actual pivot between tibia and foot, the astragalus in turn sits on the calcaneum, principally upon the astragalo-calcaneal facet which lies on or very close to the centrepoint of the calcaneum in the majority of species. The crest of this facet is angled at about 60 to 70 degrees to the axis of the calcaneum. This angle, the width of the facet crest and its position over the calcaneal axis have been measured.

The distal end of the calcaneum articulates with the cuboid and the shape of this articular facet is diagnostic for terrestrial and arboreal locomotion because movement by the cuboid provides an important adjustment of the *pes* to its substrate (Elftman & Manter, 1935; Szalay &

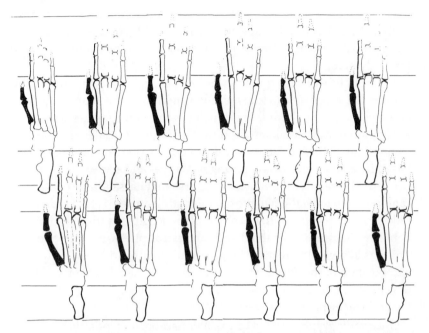

Figure 11.1. Outlines of *Cercopithecus* feet to a standardized measure between calcaneal-cuboid and metatarsal-phalangeal articulations. First digit (thumb) in black. Top row (L to R): *C. (E.) patas, C. aethiops, C. lhoesti lhoesti, C. hamlyni, C. nictitans nictitans, C. cephus ascanius.* Lower row (L to R): *C. (M) talapoin, C. (A.) nigroviridis, C.neglectus, C. diana, C. mona campbelli lowei, C. m. pogonias.*

Decker, 1974). This facet lies above the process that carries the calcaneo-cuboid ligament. This provides some indication of the relative development of this ligament and perhaps of the stability of the articulation at this point. Medial to this ligament lies the groove which channels the tendon that serves the flexors of the toes. The relative position and shape of the facet, process and groove have been recorded.

Twenty species of guenons have been X-rayed, of which 12 are illustrated with radiograph tracings of hands and feet (Figures 11.1 and 11.2). These are *Cercopithecus (Erythrocebus) patas, C. (Miopithecus) talapoin, C. (Allenopithecus) nigroviridis, C. aethiops, C. lhoesti lhoesti, C. hamlyni, C. neglectus, C. diana diana, C. mona campbelli lowei, C. m. pogonias, C. nictitans nictitans, C. cephus ascanius.* The calcaneum was recorded for ten of these and the grey-cheeked mangabey (*Lophocebus albigena*), and the mandrill, *Mandrillus sphinx*, were examined for comparison (Figure 11.3).

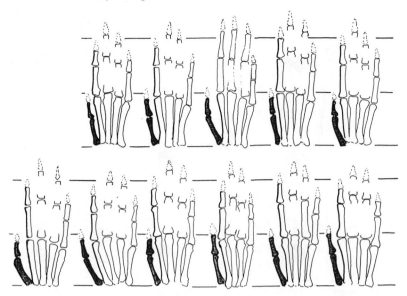

Figure 11.2. Outlines of hands (to same proportions as feet outlines). Top row (L to R): *C. aethiops, C. lhoesti lhoesti, C. hamlyni, C. nictitans nictitans, C. cephus ascanius.* Lower row (L to R): *C. (M.) talapoin, C. (A.) nigroviridis, C. neglectus, C. diana, C. mona campbelli lowei, C. m. pogonias.*

Results

The following observations can be made species by species.

C. (E.) patas

This monkey shows extreme adaptation to a cursorial gait; even so it can still be described as semi-terrestrial. There has been considerable reduction of all digits but elongation of the more proximal bones between calcaneum and metatarsals in the foot. Similar reduction of the digits occurs in the hand. The calcaneum is the longest and narrowest of any guenon's and the astragalo-calcaneal facet is unusual in being behind the mid-point of the calcaneum. This means that the lever arm of the tuber calcanei is shorter than the body of the calcaneum forward of the astragalar pivot. However, this lever arm is actually as long as in any monkey of comparable size because the entire calcaneum is elongated. The only other species with this facet behind the mid-point of the calcaneum is *C. (M.) talapoin.* The calcaneo-cuboid facet has a very shallow cup and is of circular shape with the deepest sections external and vertical. The groove for tendons serving *M. flexor digitorum* is exceptionally shallow. With weak grasping power in short running toes this could be predicted.

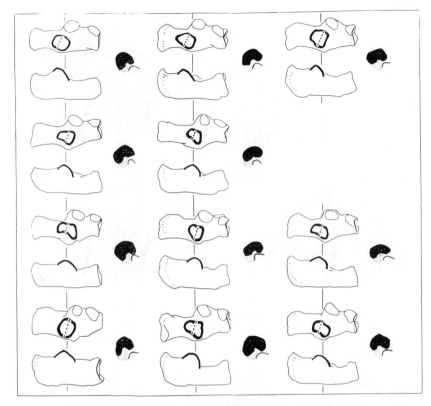

Figure 11.3. Outlines of calcaneum in eleven cercopithecoid monkeys. Superior aspect (above) with astragalo-calcaneal facet in bold outline, lateral aspect. Distal articular facet (right). Tendon groove profile below. Left column: C. (E.) patas, C. aethiops, C. lh. lhoesti, Mandrillus sphinx. Centre column: C. (M.) talapoin, C. (A.) nigroviridis, C. neglectus, C. nictitans nictitans. Right column: Lophocebus albigena, C. mona pogonias, C. cephus ascanius.

C. (M.) talapoin

This miniaturized guenon has surprisingly few signs of allometric disproportion. The first digit on both hand and foot is well developed but not more than in C. (A.) nigroviridis or C. lhoesti. The calcaneum has relatively large articular surfaces and a short robust tuber but is otherwise most like the patas. The kidney-shaped calcaneo-cuboid facet shows an equal balance between its more vertical external half and its more horizontal medial half.

C. (A.) nigroviridis

The Allen's swamp monkey is striking for a powerfully developed first digit of the foot and an exceptionally short, broad hand (the toes are also relatively short). Observations on living animals confirms that the gait is unlike that of most other guenons in that the foot is strongly palmigrade with a well splayed first digit sharing the weight. The hand is predominantly digitigrade but both hand and foot are highly flexible, and this may be relevant to the very horizontal orientation of the calcaneo-cuboid facet (an arboreal adaptation) which contrasts with other rather baboon-like traits.

C. aethiops

The hand and foot of *C. aethiops* exhibit some of the specialised terrestrial traits of the patas in a less extreme form. All digits including the first are slightly reduced, particularly the thumb, and the calcaneum is relatively long and narrow. However the calcaneo-cuboid facet is more oval than in the patas, i.e. more like *C. (A.) nigroviridis* as is the astragalo-calcaneal facet.

C. lhoesti

This monkey is peculiar in having an exceptionally long, narrow foot with a robust first digit and metatarsal. For the length of its foot the calcaneum is the shortest and most lightly built of all guenons suggesting that the lever arm of the Achilles tendon is less developed. The calcaneo-cuboid facet is similar to that of the patas in having a relatively shallow bowl and being nearly circular in contour (a strong indication of its terrestriality). The astragalo-calcaneal facet, on the other hand, has a long, fairly shallow but slightly helical arc and a waist at its apex; its general shape approximates towards that of *C. (A.) nigroviridis* or *C. aethiops* and suggests a degree of flexibility otherwise typical of more arboreal species.

C. neglectus

The de Brazza monkey has the broadest and most robust foot of any guenon and this is powered by a large Achilles tendon attached to a massive tuber calcanei. In other respects the calcaneum resembles that of the swamp monkey and is more arboreally adapted than terrestrial in that the calcaneo-cuboid facet is more horizontal than vertical or circular.

C. diana

A long narrow foot with a well developed first digit and metatarsal characterizes this guenon. Its hand is also relatively narrow and is strongly tapered from the (longest) third digit.

C. mona pogonias

Although members of the *mona* group are similar to one another, *C. m. pogonias* is the most distinctive. This species has a powerful first digit and metatarsal on hand and foot. The second phalanges on hand and foot are particularly short and the hand is one of the smallest and broadest among guenons. The calcaneum is similar to that of *C. neglectus* but is appreciably smaller.

The astragalo-calcaneal facet is elongated backwards and has a long fore-aft traverse, is slightly more helical but narrows at its apex. The calcaneo-cuboid facet is the most horizontally orientated of all guenons, and these two changes in facet shape are unequivocally arboreal adaptations. Bending moments shift medially at the calcaneo-cuboid junction during fast locomotion over very uneven and unpredictably placed supports, because the foot must twist at odd angles and evert as well as invert. The astragalo-calcaneal facet also responds to increased ranges of lateral movements in the *pes* by enlarging its posterior surface.

C. cephus

Precisely the same modifications can be observed in the calcaneum of members of other small-sized arboreal guenons that belong to the *cephus* group. In a preliminary examination of *C. c. cephus*, *C. c. erythrotis*, *C. c. petaurista* and *C. c. ascanius*, no significant differences could be detected. *C. c. ascanius* was taken as typical. It has an enlarged posterior section to the astragalo-calcaneal facet and the calcaneo-cuboid facet is more developed on the medial than on the exterior side. However, here the resemblances stop, because the overall shape and proportion of hand and foot and the general contours of the calcaneum show their greatest resemblance with monkeys of the *C. nictitans–mitis* group. Both have long, tapered hand and feet with relatively slender first digits and metacarpals.

C. hamlyni

This guenon has some resemblance with the *C. nictitans–mitis* group in the proportions of its hand and feet. It is unusual in having very elongated second phalanges. It is not known what this peculiarity signifies in adaptive terms.

L. albigena and *M. sphinx*

Very close correspondence was found in gross overall proportions of the hand and foot between *L. albigena* and *C. n. nictitans*, but the phalanges show marked curvature (in dorso-ventral plane) in *Lophocebus*. In the latter, the calcaneum is the reverse of *C. lhoesti* in showing a horizontal orientation of the calcaneo-cuboid facet and a rather simple circular arc at the astragalo-calcaneal junction. This contrast could imply that the mangabey is flexible with its metatarsals and less so in the ankle, while the *lhoesti* monkey is more flexible in the ankle and less so, within the pes.

The calcaneum of the mandrill is a massive bone with rather a peak to the simple circular astragalo-calcaneal facet and a very deeply vertical component of the calcaneo-cuboid facet. This bone is expressive of almost wholly terrestrial adaptation.

Discussion

As a result of this preliminary survey the following points can be made.

1. Several species show a mosaic of arboreal and terrestrial traits which can be interpreted in more than one way. For example, *C. (A.) nigroviridis* is on several lines of evidence the closest to a generalized ancestor of the group. This does not preclude it being relatively specialised, particularly as it lives in a peculiarly difficult and restricted swamp forest habitat. Its hand is unlike that of any other guenon and, like the foot, is somewhat papionine in its proportions (Burton, 1972). Yet the calcaneum, which is a subtle but decisive indicator of an animal's relationship with its substrate shows that *C. (A.) nigroviridis* has well developed arboreal capacities.

C. aethiops has a somewhat different mosaic of characteristics, the long, narrow calcaneum and shorter digits are less extreme than *C. (E.) patas* but are clearly adaptations towards a more terrestrial and cursorial locomotion. The calcaneum, however, indicates that the vervet also possesses arboreal traits. In this case, the direction of evolution seems clear, the patas-like features are secondarily acquired, the resemblances that it shows with the swamp monkey in the calcaneum are almost certainly retentions.

The combination of narrow foot and very large first digit puts the L'Hoest monkey closest to the swamp guenon but both foot and hand have longer phalanges, an exactly opposite trait to that seen in the vervet and patas. The fact that this species has little use for sustained speed, whether in trees or on the ground, is reflected in the slight dimensions and reduced size of the calcaneum. The combination could be interpreted in three ways: (1) it reflects the ambiguities of a largely terrestrial but slow-paced existence, punctuated by frequent climbing in a heavily wooded environment; (2) derived traits are superimposed on a predominantly arboreal matrix; (3) arboreal modifications are superimposed on older terrestrial traits.

2. Opposite traits are apparent in *C. neglectus*, which has massive bones in hand and foot, a calcaneo-cuboid facet resembling that of *C. (A.) nigroviridis* but a simple arched astragalo-calcaneal facet suggestive of less flexibility at this junction. The robustness of *C. neglectus* seems to be partly a reflection of marked sexual dimorphism; none the less, both sexes may put greater strain upon the foot in particular, as the animal's weight is sustained during slow exploratory locomotion through swamp forest.

3. *C. hamlyni*, in spite of its affinities with *C. neglectus* (see Gautier, Chapter 12), is decidedly different in hand and foot (I have not been able to examine the details of an adult calcaneum except in radiographs).

4. I have already pointed out that the two species which combine arborealism and small size, *C. m. pogonias* and the *C. cephus* group show a combination of lineage characteristics and common arboreal adaptive traits in facets of the calcaneum. The evidence consists of a mosaic of small differences but I consider it is more probable that these monkeys derived from larger (but already arboreal) species than vice versa (see Dutrillaux, Muleris & Couturier, Chapter 9, for karyological evidence).

This tentative conclusion has consequences for an understanding of evolution in the guenons. Fine-branch arborealism is manifested in *C. m. pogonias* and the *C. cephus* group as a derived condition at one extreme and cursive, semi-terrestrialism in the patas at another. The basic nature of guenon locomotion and adaptive niche is likely to lie among the species in between these extremes. Here there is a mosaic of characteristics, each species having a recognizable peculiarity and a picture of the ancestral condition will not emerge by averaging out the differences.

C. diana and the *nictitans–mitis* group are fairly generalized but decisively arboreal species. What could be called the mixed middle ground is occupied by *C. lhoesti* and *C. aethiops* with a terrestrial bias and by *C. (A.) nigroviridis* and *C. neglectus* with more arboreal traits. In spite of its miniaturisation, the talapoin is another species with a mosaic of traits (Rollinson, 1975). What do the mixed group have in common? Allen's swamp monkey, the talapoin, de Brazza's monkey and several populations of vervet are adapted to live in swamps, mangrove thickets and riverine strips. However, other species in the *mona* and *nictitans–mitis* groups also share a proclivity for riverine or gallery forest. It should at least be borne in mind that the ancestral condition could have been correlated with a life in similar habitats that demanded various combinations of arborealism and terrestriality.

My intention here has been limited to seeing if measurable and potentially interpretable differences exist in the manus and pes of guenon species. Further study should include the detailed structure of the wrist and the phalangeal-metatarsal junction in the hand, while more intensive study of the foot should treat the calcaneum and astragalus as an integrated complex. I hope the sketchy outlines of this pilot exercise have advertised the value of applying an intensive, multivariate analysis to the subject in the future.

II.12

Interspecific affinities among guenons as deduced from vocalizations

JEAN-PIERRE GAUTIER

Since the pioneering work of Struhsaker (1970) on cerco-pithecines, a number of primate studies have used vocalizations as phylogenetic criteria. They have mostly dealt with forest dwelling primates such as colobines (e.g. Wilson & Wilson, 1975; Struhsaker, 1981a; Oates & Trocco, 1983) and gibbons (Marshall & Marshall, 1976; Brockelman & Gittins 1984; Creel & Preuschoft, 1984; Creel *et al.*, 1984; Haimoff *et al.*, 1984; Marshall, Sugardjito & Markaya, 1984), while one study involved a comparison between langurs and forest macaques (Herzog & Hohman, 1984) and finally with South American tamarins (Hodun *et al.*, 1981; Snowdon *et al.*, 1986). This is indicative of the supremacy of vocal communication in forest environments, previously emphasized by Waser (1982a).

The use of behavioural traits as taxonomic characters relies primarily on their degree of genetic determination, and Kummer (1970) suggested that the 'best candidate for taxonomic usefulness is the so-called fixed action patterns . . .' such as vocalizations. Ontogenetic studies undertaken in both normal and socially deprived environments, along with neurophysiological studies (e.g. Jurgens, 1979) and structural analyses of hybrid calls (e.g. Gautier & Gautier, 1977; Bouchain, 1985), concur in giving a predominant role to genetic factors in determining the acoustic structure of calls (Newman & Symmes, 1982; Newman 1985). In contrast to the evidence available for many birds, vocal imitation in primates has not been documented. Long-term experiments using captive monkeys living in mixed-species groups (e.g. Bernstein, 1970 and personal observations), as well as the frequent occurrence of polyspecific troops among guenons in the wild (see Gautier-Hion, Chapter 23), show that, while these animals are able to learn the significance of extraspecific signals, they do not change their own vocal patterns.

It has been pointed out for the vocalizations of birds and monkeys that different call types may have been subject to dissimilar kinds of selective pressures and thus have acquired different degrees of species-specificity (Marler, 1957, 1965; Struhsaker, 1970; Becker, 1982). The function of calls, the conditions in which calls are given, and the composition of the community to which species belong may all influence the patterns of selective pressure. The least species-specific calls can be expected to be those in which the emittor's identity is not important (e.g. alarm calls), or calls given at short distance which benefit from additional visual cues (e.g. aggressive calls). On the contrary, calls which serve to enhance inter-individual cohesion or which play a role in the reproductive isolation of a species can be expected to be the most species-specific. This explains why most studies of the phylogenetic implications of calls rely on the comparison of ritualized vocal patterns such as the long calls of tamarins, the songs of hylobatids or the loud calls of Cercopithecinae. In most African species, these highly stereotyped vocal behaviours mainly characterize adult males and require either an enlarged larynx, as in colobines (Hill & Booth, 1957), or extralaryngeal annexes, as in guenons (Gautier, 1971). However, if we suppose that such calls have evolved to prevent interspecific hybridization, they may be expected to diverge more among sympatric than among allopatric species, as has been shown by Marler (1957) for birds. From this perspective, adult male loud calls of guenons may be very good or very poor indicators of interspecific affinities, depending on the degree of sympatry among species. For example, relatively close similarity in the structure and temporal pattern of adult male loud calls is found in two species, *C. n. nictitans* and *C. mitis labiatus*, which are separated by 3000 km (Figure 12.1), while the sympatric *nictitans–cephus* or *mitis–ascanius* pairs of species show very close similarity in most of their repertoire, but adult male loud calls diverge markedly (see Figure 12.8 below).

To compare calls of different species raises a number of methodological problems. Firstly, when can we consider two calls to be structurally homologous? Is it at the point when we cannot distinguish them by ear or is it at the point when statistical analyses comparing a number of their acoustic parameters yield non-significant results? Measurements of such parameters can be made so numerous and precise that a statistical difference is almost inevitably found (see Gautier & Gautier-Hion, 1988), yet the distinctions may not be mean-

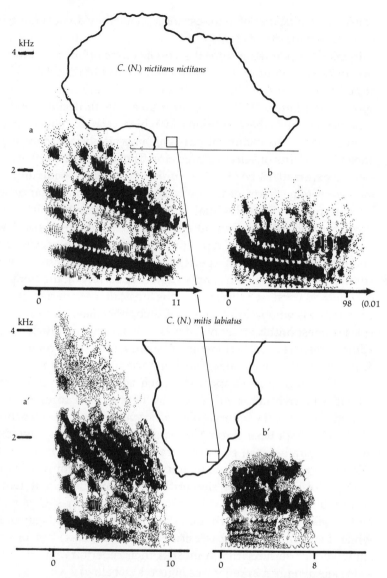

Figure 12.1. Spectrographic analyses of adult male loud calls in *C.n.nictitans* and *C.mitis labiatus* with the geographical locations of species (loud calls of first category: a, a'; loud calls of second category: b, b').

ingful with respect to monkeys as perceivers. Consequently, until more playback experiments are conducted to clarify the level of discrimination in monkeys, it seems wise to compare species on the basis of simple call characters.

Secondly, we have to decide whether two structurally and functionally homologous calls have the same phylogenetic implication as two structurally similar calls given in different functional contexts. In other words, does the function of calls give any additional information in determining interspecific affinities? In agreement with Marler (1957, 1965) and Struhsaker (1970), we believe this to be the case, and call function will be taken into account in the present study.

A third point concerns the problem of the frequency of use. For Kummer (1970), this cannot be used to infer phylogenetic relationships, since *a priori* social and environmental conditions may induce different rates of use of the same call. Thus, depending on population density, territorial vocal battles may or may not occur between *C. cephus* troops (personal observations). Another example is given by Hall (1965a), who thought that the silence of patas monkeys was an adaptive response learned by infants.

It has been frequently claimed that young calls are simply non-elaborated adult calls which cannot be used to deduce interspecific affinities. Indeed in most guenons, the vocal repertoire of immature animals differs notably from that of adults. In the latter, maturational processes change the structure of calls (especially by a decrease in pitch, Gautier & Gautier 1977). Changes in the frequency of use are also found and a number of calls given by immature animals are no longer given by adults, not because they are unable to give them (as can easily be shown in captive animals, personal observations), but because adults are no longer confronted with the usual context in which calls are uttered (Gautier & Gautier-Hion, 1982). Most adult male guenons only retain two or three call types, one of which (at least) has an original structure not derived from young calls.

In a social group, immature monkeys (which remain so for four to seven years) have their own role, and their repertoire of signals depends on and underlies these. Their calls therefore have to be considered as a full part of the species repertoire. The latter cannot be accurately understood by relying on adult calls only. In addition, studies of maturational changes in calls are very useful in determining homology between calls of different species. This is the case, for example, when two structurally different adult calls are shown to derive from the same kind of immature calls (Gautier, 1978).

We will see below that, in some cases, comparisons of adult and immature calls lead to opposite conclusions about interspecific affinities. If we hypothesize that adult calls are more derived than young ones, the latter could be considered as relatively poor indicators of phylogenetic affinities. We might also propose, however, that young calls and adult calls have been submitted to different selective pressures. In any case, resemblances in calls given by young of different species are sometimes so great that it seems difficult not to consider the calls as homologous, and to infer the existence of a common ancestor. Similar examples are given for insects by Hennig (1965), who points out that some phylogenetic affinities can only be shown by comparing immature stages.

Thus, in this study we will search for interspecific affinities by taking into account a number of calls, and examining their structure and temporal pattern, their overall functional context, their presence/absence in different age–sex classes as well as the behaviour which accompanies their emission, and morphological structures which allow it (see Martin, 1978).

Species studied

Twenty-three of the 28 species and about 40% of the 85 forms retained in the revised taxonomic list of Lernould (Chapter 4) are considered in this analysis. For 17 species, I will refer to personal records and observations and to other studies when available (see details in Table 12.1). For others, I will simply refer to the literature. However, great differences exist in the extent to which each species' repertoire has been studied. Data are very limited for about one third of species, and completely lacking for four guenon species: *C. denti*, *C. sclateri*, *C. dryas* and *C. salongo*.

Calls studied

Calls considered in this analysis belong to three functional systems (see the review in Gautier & Gautier, 1977). The first includes calls, given in pacific exchanges between members of the social unit, which serve in intragroup cohesion and pacific inter-individual relationships. The second consists of calls given in warning situations, which tend to spread among troop members through phonoresponses. The third category includes loud calls of the first category which are

Table 12.1. *List of species whose vocal repertoires have been studied*

Genus	Superspecies	Species	Subspecies	L[a]	References
Cercopithecus	*aethiops*	*pygerythrus*	*johnstoni*	3	19, 20
Cercopithecus	*aethiops*	*sabaeus*		3	4
Cercopithecus	*aethiops*	*tantalus*		2	16
Cercopithecus	*cephus*	*ascanius*	*ascanius*	1	11
Cercopithecus	*cephus*	*ascanius*	*katangae*	2	11
Cercopithecus	*cephus*	*ascanius*	*schmidtii*	2	17
Cercopithecus	*cephus*	*ascanius*	*whitesidei*	2	11, 13
Cercopithecus	*cephus*	*cephus*	*cephus*	3	9, 10, 12, 21
Cercopithecus	*cephus*	*cephus*	*cephodes*	1	11
Cercopithecus	*cephus*	*erythrogaster*		1	18
Cercopithecus	*cephus*	*erythrotis*	*camerounensis*	1	11, 21
Cercopithecus	*cephus*	*petaurista*	*petaurista*	2	6, 21
Cercopithecus	*diana*	*diana*	*diana*	2	6
Cercopithecus	*diana*	*diana*	*roloway*	1	11
Cercopithecus	—	*hamlyni*		3	11
Cercopithecus	*lhoesti*	*lhoesti*		2	11
Cercopithecus	*lhoesti*	*preussi*		1	11, 21
Cercopithecus	*lhoesti*	*solatus*		2	11
Cercopithecus	*mona*	*campbelli*	*campbelli*	1	21
Cercopithecus	*mona*	*campbelli*	*lowei*	2	2, 3, 6
Cercopithecus	*mona*	*mona*	*mona*	2	11, 21
Cercopithecus	*mona*	*pogonias*	*nigripes*	3	9, 12
Cercopithecus	*mona*	*pogonias*	*pogonias*	2	21
Cercopithecus	*mona*	*wolfi*	*wolfi*	1	11
Cercopithecus	—	*neglectus*		3	9, 12, 14
Cercopithecus	—	*nigroviridis*		2	11
Cercopithecus	*nictitans*	*mitis*	*stuhlmanni*	2	1, 17
Cercopithecus	*nictitans*	*mitis*	*labiatus*	1	23
Cercopithecus	*nictitans*	*nictitans*	*martini*	1	21
Cercopithecus	*nictitans*	*nictitans*	*nictitans*	3	7, 9, 12
Cercopithecus	—	*patas*		1	11, 15, 22
Cercopithecus	—	*talapoin*	?	3	8, 9

[a]L1: limited data; L2: detailed analyses; L3: exhaustive analyses.
References: 1. Aldrich-Blake, 1970; 2. Bertrand & Gautier, unpublished data; 3. Bourlière *et al.*, 1970; 4. Galat, 1975; 5. Galat, 1983; 6. Galat & Gautier, unpublished data; 7. Gautier, 1969; 8. Gautier, 1974; 9. Gautier, 1975; 10. Gautier, 1978; 11. Gautier, unpublished data; 12. Gautier & Gautier, 1977; 13. Gautier & Gautier-Hion, 1982; 14. Gautier-Hion & Gautier, 1978; 15. Hall *et al.*, 1965; 16. Kavanagh, 1981, 1986; 17. Marler, 1973; 18. Oates, 1985; 19. Seyfarth & Cheney, 1980, 1982, 1984; 20. Struhsaker, 1967e; 21. Struhsaker, 1970; 22. Struhsaker & Gartlan, 1970; 23. Gautier & Girolami, unpublished data.

only given by fully adult males in more or less ritualized sequences and appear to play a role in rallying troop members and spacing out troops within the population and the community.

Calls given in agonistic contexts have been excluded because they are poorly species-specific. Aggressive loud calls proper to adult males (loud calls of the second category, most often a bark type), which have been shown to derive from warning and aggressive calls of adult females and immatures (Gautier, 1974, 1978), have not been considered in detail, despite the fact that their temporal patterns and their associations with loud calls of the first category may be highly organized. Unfortunately, precise studies of such sequences could only be conducted for a few species to date.

Parameters studied

Highly species-specific calls such as adult male loud calls are easily distinguished on the basis of their overall structure. For other call types, whose interspecific similarities are greater, we have relied upon a few simple and stable parameters. These include: presence/absence of the call type; its adherence either to a low- or high-pitched register; the complexity of its acoustic structure (noisy components vs. pure tone, quavering vs. continuous calls); and temporal pattern of emission (unitary vs. multi-unitary calls or calls in series).

Methods of analysis

Our first approach was qualitative and phenetic. The distribution of calls or call characteristics among species permitted hierarchical clustering on the basis of overall similarity. Phylogenetic trends were then inferred from the discrimination between more or less derived forms depending on the vocal characteristics they share with species belonging to 'outgroups' such as Papioninae and Colobinae. In order to rationalize these preliminary inferences, we further applied a cladistic method involving two steps. First, the binary coding of calls and call characteristics allowed us to build a matrix of species by characters (Sneath & Sokal, 1973; Table 12.2). The data were then fed into several computer programs in the 'phylogeny inference package' (Felsenstein, 1982, 1985b).

Having tested a number of these programs, we have retained the Wagner parsimony method because it allows reciprocal changes between ancestral and derived states, as well as no statement about ancestral characters when data do not authorize any statement. Options also allow each character to be weighted differently. Given

Table 12.2. *Matrix of data. Binary coding of the 22 calls characters/15 taxa. Species with the same set of characters are grouped on the same row*

Type of call →	Cohesion-contact calls								Adult males' loud calls, 1st Category								High-pitched warning calls					
Type of character →	Type 2			Type 6			2 + 6															
Number of characters	1 quavered	2 non-quavered	3 + noisy components	4 present	5 quavered	6 non-quavered	7 associated	8 merged	9 present	10 'boom'	11 'pyow'	12 'hack'	13 'gobble'	14 'whoop-roar'	15 'oop-uuh'	16 'hack-roar'	17 present	18 unitary chirp	19 multi-unitary chirp	20 twitters	21 whistle	22 pseudo-chirp
Cercopithecus nigroviridis	1	0	0	1	1	0	0	0	1	0	0	0	1	0	0	0	1	1	0	0	0	0
C. nictitans/mitis	1	0	0	1	1	0	1	0	1	1	1	0	0	0	0	0	1	1	0	0	0	0
C. cephus/ascanius/erythrotis	1	0	0	1	1	0	1	0	1	0	0	1	0	0	0	0	1	1	1	1	0	0
C. petaurista	1	0	0	1	1	0	1	0	1	0	0	1	0	0	0	1	1	0	1	1	1	0
C. erythrogaster	1	0	0	1	1	0	1	0	1	0	0	0	0	0	0	1	1	0	1	1	?	0
C. diana	1	0	0	1	0	1	0	0	1	1	0	1	0	0	0	0	0	0	0	0	1	0
C. hamlyni	1	1	0	1	0	1	1	0	1	1	0	0	0	0	0	0	0	0	0	0	0	0
C. neglectus	0	1	0	1	0	1	1	0	1	1	0	0	0	0	0	0	0	0	0	0	0	0
C. mona/campbelli	0	1	0	1	0	1	1	1	1	1	0	0	0	0	0	0	0	0	0	0	0	0
C. pogonias/wolfi	0	1	0	1	0	1	1	1	1	1	0	0	0	0	0	0	1	0	1	0	0	0
C. (aethiops)	1	0	0	0	0	0	0	0	0	0	0	0	0	0	0	0	1	1	0	0	0	1
C. patas	0	1	0	1	0	1	1	0	0	0	0	0	0	0	1	0	1	1	0	0	0	0
C. talapoin	0	1	1	0	0	0	0	0	0	0	0	0	0	0	0	0	1	1	0	0	0	0
C. lhoesti/'solatus'	1	0	1	0	0	0	0	0	0	0	0	0	0	0	0	0	1	1	0	0	0	0
C. preussi	1	0	?	0	0	0	0	0	1	0	0	0	0	0	0	0	1	1	1	0	0	0

these two possibilities, the program proposes several more or less parsimonious dendrograms which question the author and lead him to consider additional criteria to obtain the most relevant tree of interspecific affinities.

Characterization of calls

Low-pitched cohesion calls

This call system exists in all species and includes a more or less graded series of vocalizations which are only given in full by young animals (Figure 12.2). Within this series, type 2 calls (Gautier, 1975) are the most commonly uttered; they are mostly given without the support of visual clues and enhance intra-troop cohesion. Their frequency of use varies according to species, group size, age class, habitat structure and other ecological factors. Thus the silence of patas (Hall, 1965) as well as that of *C. neglectus* (Gautier & Gautier, 1977) and *C. preussi* (personal observations) may be viewed as adaptive strategies against terrestrial predators, while the frequent use of such calls by talapoins is attributable both to the great size of the troop and to the high structural density of the milieu they colonize, which impedes visual contact.

Type 2 cohesion calls can develop into type 3 lost calls given by young animals, when they lose contact with their mothers or with the social unit.

Gross examination of the structure of these calls allows the separation of species into two categories (Table 12.2). The first includes species which give discontinuous or quavered calls such as *C. nictitans* and *C. cephus* (Figure 12.2); the second includes species whose calls are continuous and non-quavered, such as *C. pogonias*, *C. neglectus* and *C. talapoin*, the latter species having the highest pitch (Figure 12.2). In *C. hamlyni*, both quavered and non-quavered calls are heard, the latter being given by infants, the former by adults (Figure 12.3).

Calls of *C. lhoesti* and *C. solatus*, which share a similar timbre with calls of *C. talapoin* due to the frequent association of high-pitched noisy components with the low register (Figure 12.4), are more difficult to include in one or other category defined above. However, detailed analysis reveals a quavering not easily discriminated by ear because of its very high rhythm and low amplitude (Figure 12.5).

High-pitched contact calls

A high-pitched register is found in all species; it groups a graded series of calls including those, like contact calls, which possess

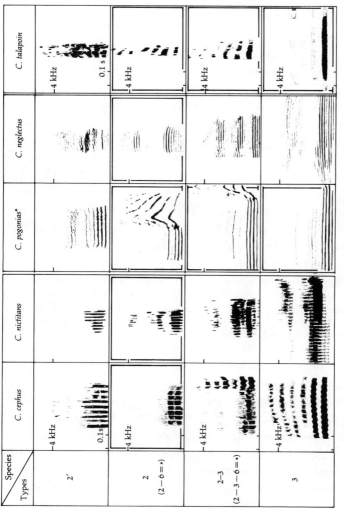

Figure 12.2. Graded series of low-pitched cohesion calls from type 2' (young crouching calls), type 2 (usual cohesion calls) to type 3 (young lost calls). Calls are quavered in *C. cephus* and *C. nictitans*, and non-quavered in *C. pogonias*, *C. neglectus* and *C. talapoin* (from Gautier, 1975). Notice the pitch of talapoin calls which clearly differentiates the species while the structural similarities reveal two species pairs (*nictitans–cephus* and *pogonias–neglectus*). * indicates that type 2 and 6 are given in a single continuous call.

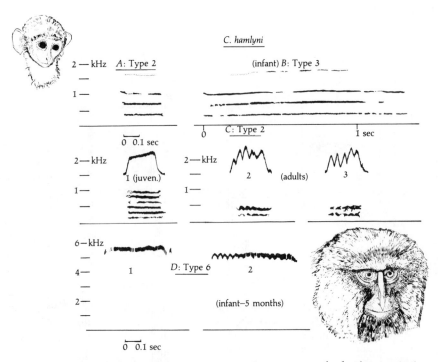

Figure 12.3. Ontogenetic changes in the structure of cohesion-contact calls (type 2 and 6) in *C. hamlyni*. *A* and *C*1: non-quavered type 2 calls of infants and juveniles; *B*: non-quavered type 3 call of infants; *C*2 and *C*3: quavered type 2 calls of adults (the quavering is clearly apparent on the amplitude modulation spectra); *D*1 and *D*2: non-quavered and quavered type 6 calls of a 5-month-old infant.

a well-defined acoustic structure, and other noisy and unstructured whistles and screams which are given in stressful situations and are poorly species-specific.

Together with the low-pitched calls, the high-pitched contact calls (type 6, Gautier, 1975) regulate inter-individual distances, especially between young and mothers. Their occurrence progressively decreases with age and they are not given by adult males of any species. In some species, they remain commonly used by subadults and adult females (e.g. members of the *mona* superspecies and *C. diana*, and to a lesser degree members of the *nictitans* and *cephus* superspecies and *C. nigroviridis*). In others, (e.g. *C. neglectus, hamlyni, patas*) they are only heard in infants. Moreover some species have not evolved these high-pitched contact calls, as in the case of the *aethiops* and *lhoesti* groups and of *C. talapoin*.

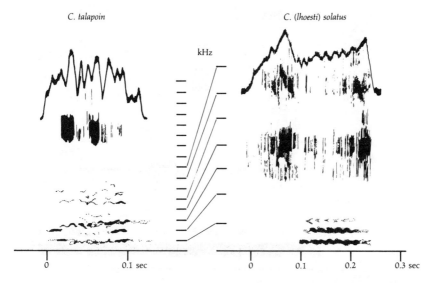

Figure 12.4. Similarity in calls structure of *C. talapoin* and *C. solatus* (here one unit of an agonistic call) resulting from the association of high-pitched noisy components (wide band filter) and low-pitched tonal quavered ones (narrow band filter). Calls of talapoins are 1.5 octave higher than those of *C. solatus*.

Thus if we consider the whole high-pitched register within the guenon group, there is a gradient from species in which only non-structured whistles and screams are heard, to those in which the whole system is given only by infants, to species in which all age classes can perform the whole register.

On the basis of the acoustic structure of their high-pitched contact calls, species can be divided into two clusters (Table 12.2) according to whether their calls are quavered or not (Figure 12.6). In *C. diana*, adult females and immatures are able to give both quavered and unquavered calls (Figure 12.7).

The whole cohesion-contact call system

Within species which possess both type 2 and type 6 calls, another distinction can be made as to whether or not they use the two calls in the same temporal sequences. For example, *C. nigroviridis* and *C. diana* mostly use the high-pitched type 6 call without association with type 2. All other species associate the two call types to a greater or lesser degree; the finest level of association is attained in the *mona* forms which merge the high and low-pitched components into a single continuous vocalization (see Figure 12.2).

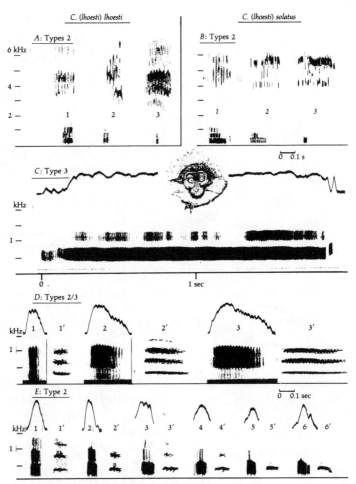

Figure 12.5. Cohesion call system of *C. lhoesti* and *C. solatus*. *A* and *B* 1, 2, 3 show the similarity between type 2 calls of the two species, which associate high-pitched noisy components and low-pitched tonal ones. *C, D* and *E*: detailed analyses of the low-pitched register of cohesion calls in *C. solatus* showing the variability in the duration and energy distribution of calls and the peculiar quavering characterized by a high rhythm and a weak amplitude (D1'–3'; E1'–6': narrow band filter).

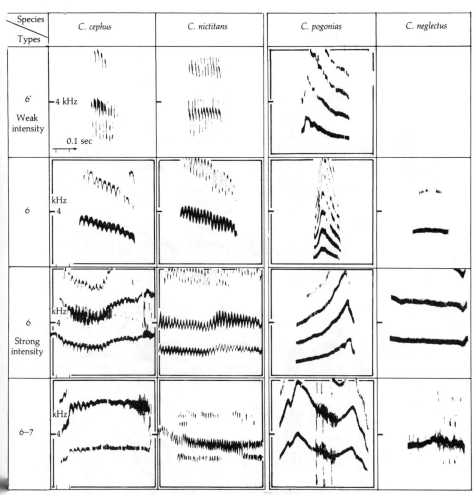

Figure 12.6. Graded series of high-pitched contact calls from type 6′ (weakly voiced sounds), type 6 (usual contact call) to type 6–7 (call evolving towards distress vocalization). Calls are quavered in *C. cephus* and *C. nictitans* and non-quavered in *C. pogonias* and *C. neglectus*. Compare the amplitude of modulation of *C. pogonias* and *C. neglectus*' calls.

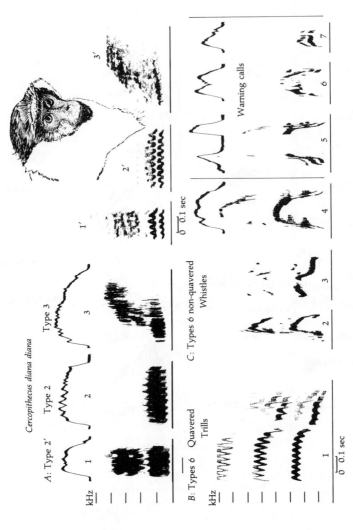

Figure 12.7. Quavered and non-quavered calls of C. *diana*. *A*: low-pitched quavered cohesion calls in young: (1, 2, 3 and 1′, 2′, 3′, respectively wide and narrow band filters) *B*: high-pitched quavered contact call; *C*2, 3, 4: whistling contact calls (given by a highly excited monkey); *C*5, 6, 7: warning calls.

Warning calls

Warning calls are found both in the low-, the medium-, and the high-pitched register. Only the latter will be considered here. The so-called chirps are the most frequent high-pitched alarms. They can be given either in single units (as in *C. nigroviridis*), in multi-units (as in *C. cephus*), or in series (twitters as in *C. petaurista*). A 'pseudo-chirp' which differs in its structure is also given by the *aethiops* forms. *C. diana* does not emit chirps in alarm situations but a very specific whistle. A similar whistling note ends the multi-unit squeak given, together with chirps, by young *petaurista* (Table 12.2). *C. hamlyni, C. neglectus, C. mona* and *C. campbelli* do not utter such high-pitched warning calls.

Loud calls

Loud calls of the first category (Gautier & Gautier, 1977) are present in all species except the talapoin, the *aethiops* forms, *C. lhoesti* and *solatus*. There is no relevant information for *C. patas*. According to J. S. Gartlan (personal communication), this species possesses a loud call very close to the 'oop-uuh' of *C. preussi* (Struhsaker, 1970); how-ever, the 'whoo-wheer' described by Hall, Boelkins & Goswell (1965) and J. Chism (personal communication) does not possess the charac-teristics of loud calls and I have never succeeded in eliciting such calls in a captive group.

Eight species of the ten which possess loud calls utter 'booms', including the *mona* forms, *C. hamlyni, C. neglectus, C. nictitans* and *C. mitis* (Figure 12.8). However, the 'booms' of the two latter species are not really loud; they possess a lower pitch than in other 'boom' emittors and are not given in stereotyped sequences; moreover their function remains obscure. They have nevertheless been considered in the search for phylogenetic affinities because the production of such calls requires the development of large vocal annexes and a very complex behaviour implying a phase of inflating sacs which precedes the emission of the calls (Figure 12.9). Thus, it seems improbable that such a complex phenomenon has evolved independently in several taxa.

Loud calls of species which do not utter 'booms' are clearly distinct from each other (Figure 12.8). Within the same superspecies, however, species distinctiveness is often unclear. This is the case for the 'pyows' of the *nictitans* superspecies (Gautier, 1969; Marler, 1973). Within the *cephus* group, while it is difficult to distinguish the unitary calls of *C. cephus* ('mpaks', Gautier, 1975) and *C. ascanius* ('hacks', Struhsaker, 1970; Marler, 1973), the complex multi-unit hacks given by *C. petaurista* (personal observations) and *erythrogaster* (Oates, 1985) are clearly rec-ognizable.

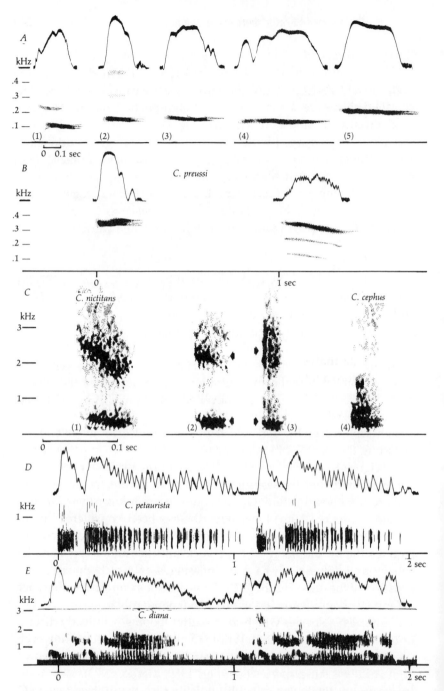

Figure 12.8. Loud calls of adult males. *A*: 'boom' of *C. nictitans* (1), *C. pogonias* (2), *C. campbelli* (3), *C. neglectus* (4), *C. hamlyni* (5); *B*: two-phase 'oop-uuh' of *C. preussi*; *C*: 'pyow' of *C. nictitans* (1) and 'hack' of *C. cephus* (4); transitional sounds given by *nictitans* (2) and *cephus* (3) which reveal interspecific affinities between the two species. *D*: 'hack roar' of *C. petaurista*; *E*: 'whoop roar' of *C.diana*.

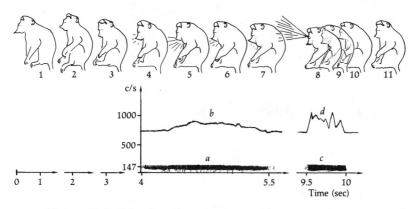

Figure 12.9. Behavioural sequence accompanying the uttering of 'boom' in *C. neglectus*. 1–3: preparatory phases; 4–7: inflation of the vocal sac accompanied by a sound (spectra *a* and *b*); 8–10: uttering of 'boom' (spectra *c* and *d*) and deflation of the vocal sac (from Gautier, 1971).

Data analysis

Before analysing the matrix (Table 12.2), we have to propose for each character the trait which can be considered as the ancestral state. We may assume that the presence of loud calls is a primitive trait because such calls are found in Asian and African colobines, mangabeys, baboons and even some macaques (see a review in Gautier & Gautier, 1977). The 'gobble' given by *C. nigroviridis* is quite probably a primitive trait too, since a similar call structure (given with or without a 'whoop') has been described in langurs (*Presbytis*: Sugiyama, Yoshiba & Parthasarathy, 1965; Ripley, 1967; Poirier, 1968; Horwich, 1976), in mangabeys (Chalmers, 1968; Gautier & Gautier, 1977; Quris, 1980; Waser, 1982a), in one baboon at least ('roar grunt' of *Papio papio*: Byrne, 1981) and in one macaque (*Macaca silenus*: Herzog & Hohman, 1984; Figure 12.10). Loud calls given by all the other guenons show a quite original structure and must be considered as derived traits.

The problem is not so clear for the cohesion-contact call system. In baboons, inter-individual relationships are sustained mainly by low-pitched grunts (Hall & DeVore, 1965; Byrne, 1981). Byrne distinguished four categories of grunts of which at least two (the two-phase grunt and the unitary grunt) are also found in drills and mandrills (Sabater Pi, 1972; Hoshino *et al.*, 1984). Hoshino *et al.* referred to a 'crow' in adult female mandrills, but we cannot decide whether or not this 'crow' is equivalent to the quavered 'crow' described in guenons.

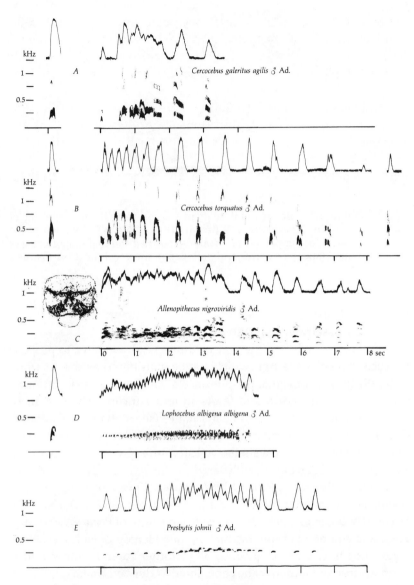

Figure 12.10. 'Gobble' calls in *C. galeritus* (*A*), *C. torquatus* (*B*), *C. nigroviridis* (*C*), *L. albigena* (*D*) and *P. johnii* (*E*). In *A*, *B* and *D*, the 'gobble' is preceded by an advertising 'whoop'.

In both structure and pattern of emission (series and choruses) the grunts of baboons are very close to those of mangabeys (Chalmers, 1968; Quris, 1973; Waser, 1975, 1976; Homewood, 1976; Gautier & Gautier, 1977; Deputte, 1986). The latter are clearly non-quavered. In mangabeys, as in baboons, the high-pitched register is not used in inter-individual relationships. Furthermore, in young baboons there are low-pitched continuous calls ('moans': Andrew, 1963; Ransom, 1976; Byrne, 1981; 'ooer': Hall & Devore, 1965) which are also found in young geladas (Richman, 1976); such 'moans' strongly resemble the non-quavered type 3 calls of *C. patas* and *C. neglectus* (Figure 12.11), *C. mona* forms and *C. hamlyni* (Figure 12.3).

In macaques, inter-individual relationships are regulated by a highly graded system of 'coo' sounds which are non-quavered and belong to a medium-pitched register (e.g. *Macaca mulatta, fuscata, nemestrina, speciosa* and *silenus*: Rowell & Hinde, 1962; Grimm, 1967; Bertrand, 1969; Green, 1975; Hohmann & Herzog, 1985). In these species the high-pitched register is rarely used and poorly structured. Some exceptions do exist, as in the case of the 'food call' of *M. sinica* (Dittus, 1984) which associates low- and high-pitched components as do *mona* forms.

African colobines use well developed high-pitched calls (Marler, 1970, 1972; Struhsaker, 1975, 1981a). In most cases their function remains poorly understood. In *Colobus guereza*, however, they are used in social interactions (Oates & Trocco, 1983). Such calls are highly modulated but do not appear to be systematically quavered. Finally high-pitched quavered calls are uttered by many New World monkeys (see a review in Oppenheimer, 1977).

Consequently, two hypotheses have been envisaged for the cohesion-contact system (Table 12.3). In the first, the presence of a high-pitched register and the presence of non-quavered calls have been considered to be primitive traits. This is also the case for the fusion of continuous calls of type 2 and type 6 found in the *mona* group, since a very similar call has been described in *Macaca sinica* (Dittus, 1984). In the second, we have left the problem unsolved by putting in question marks. In both cases, the presence of loud calls and the presence of a 'gobble' call have been considered as the ancestral state.

The two hypotheses have been tested with and without taking warning calls into account. The latter has been done in order to avoid establishing biased affinities which could result from patterns of convergence (Table 12.3). In the first hypothesis, high-pitched warning

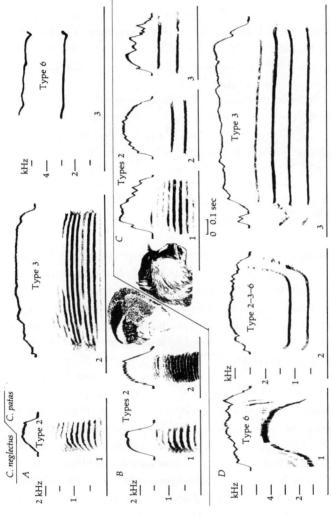

Figure 12.11. Cohesion-contact call system in *C. neglectus*, A, B, and *C. patas*, C, D. In both species, calls are non-quavered and look very similar. However the higher pitch of patas calls give them a timbre quite different from those of *C. neglectus*; in both species, the pitch of Type 2 calls decreases with aging (A_1, C_3: infants; B_1, C_2: juveniles; B_2, C_1: adults) (with the permission of J. Kingdon for the drawings of the heads).

Table 12.3. *Ancestral states of the different characters*

Characters	hypothesis 1 − WC	hypothesis 1 + WC	hypothesis 2 − WC	hypothesis 2 + WC
Loud call	P	P	P	P
'Gobble' call	P	P	P	P
High-pitched type 6 call	P	P	?(P)	?(P)
Quavering of type 2 and 6 calls	A	A	?(P)	?(P)
Highly-pitched noisy components (type 2)	P	P	?(A)	?(A)
Type 2 & 6 association	A	A	?(A)	?(A)
Type 2 & 6 fusion	P	P	?(A)	?(A)
High-pitched warning calls	/	P	/	?(P)
Unitary warning calls	/	P	/	?(P)
Other types of warning calls	/	A	/	?(A)
Number of characters	16	22	16	22
Number of steps	35	44	29	39

(): ancestral state proposed by the program; P: plesiomorphic; A: apomorphic. WC: warning calls. (See text.)

calls have been considered as primitive, except for those given in multi-units or in series; in the second, we did not make a decision.

Finally, we have chosen to give the same weight (i.e. 1) to all characters except the presence of a 'boom' call which is a highly specific feature of many guenon taxa (it has been weighted 3 or 4 respectively when warning calls are considered or not); and the 'gobble' call (weighted 3) which is the only loud call not specifically belonging to the guenon group.

Interspecific affinities

Two different phylogenetic trees are obtained from the four combinations derived from the two hypotheses and from the two sets of characters (Table 12.3). In fact, the same tree is produced from the second hypothesis when taking warning calls into account and when leaving them unweighted, and from the first hypothesis when warning calls are not considered (Figure 12.12).

Furthermore, the second hypothesis (in which we do not make a decision regarding the primitive nature of the cohesion call system or on the warning call system), leads to more parsimonious dendrograms (29 steps instead of 35 for the set of 16 characters and 39 steps instead of 44 for the set of 22 characters, Table 12.3). This second hypothesis

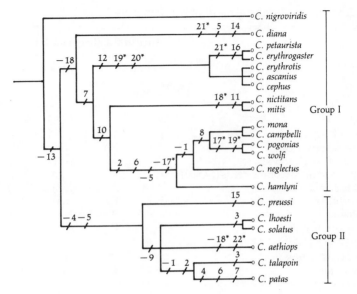

Figure 12.12. Dendrogram resulting from hypothesis 2 and hypothesis 1 (when warning calls are not considered, see text). Numbers refer to the different steps of characters listed in Table 12.2 *: characters referring to warning calls.

agrees with the polarity we proposed in the first hypothesis on two points: the ancestral state of high-pitched contact calls and of high-pitched unitary warning calls, and the derived trait of type 2 and type 6 association and of multi-unitary warning calls (Table 12.3). Conversely, it proposes that the quavering of calls is the ancestral state, suggesting that this character, which is found in New World monkeys, has been retained by guenons, while it has largely disappeared in colobines and papionines, possibly as a result of their larger body weight.

Let us consider first dendrogram 1 (Figure 12.12). *C. nigroviridis* appears as the closest to the ancestral form; this is mostly due to its 'gobble' call. Guenons are then separated into two groups according to whether they possess both loud calls and high-pitched contact calls (group 1, Figure 12.12) or not (group 2, Figure 12.12). The *diana* monkey is the first species to be separated from group 1. It is in fact well distinguished by its whistle alarm, its ability to give both quavered and unquavered contact calls and by its loud call.

The next step separates the *cephus* members from the 'boom' emitters. Within the *cephus* superspecies, two clusters of affinities are found: between *C. cephus*, *C. ascanius* and *C. erythrotis* on the one hand,

and between *C. petaurista* and *C. erythrogaster*, the two westernmost forms, on the other. Among the 'boom' emittors, the two closely related *nictitans* and *mitis* species are easily distinguished from the other taxa. The latter indeed share many derived traits, especially the disappearance of the ability to utter quavered calls and high-pitched alarms. In the latter taxa, the greatest similarity is found within the *mona* superspecies, which is distinguished from *C. neglectus* and *C. hamlyni* by the acquisition of a more graded repertoire while retaining the same timbre for most calls (Gautier, 1975). Two related species pairs are found in the *mona* group, with *pogonias* and *wolfi* on the one hand (which have acquired, probably by convergence, the ability to give high-pitched binary alarms), and *mona* and *campbelli* on the other hand (which share noisy warning sneezes, not considered in the present analysis).

Most calls of *C. neglectus* and *C. hamlyni* are very similar in their overall structure. The only difference observed from the characters used results from the ability of subadult and adult *C. hamlyni* to give quavered type 2 calls which resemble those of *C. nictitans*, while non-quavered calls like those of *C. neglectus* are given by young animals (Figures 12.2, 12.3). This ability confers on *C. hamlyni* an intermediate position between the *nictitans–cephus* group and the *mona–neglectus* group.

Species of group 2 are grouped more by the loss of a number of characters such as high-pitched sounds and loud calls (except for *C. preussi*) than by the presence of shared traits. The most surprising result concerns the relative position of the three members of the *lhoesti* superspecies, which are not grouped together. In fact, detailed observations and recordings made on captive groups of *C. lhoesti* and *C. solatus*, as well as field observations on the latter species (unpublished data), have shown that their repertoires are very similar and it is highly improbable that adult males of these species are able to utter loud calls of the first category as *C. preussi* does.

The repertoire of *C. preussi* is poorly known. Struhsaker (1970) did no more than mention the alarm call and the adult male loud call. After one hundred hours spent in contact with wild troops, we recorded, in addition to these two calls, many unitary barks given by adult males and only 14 cohesion calls given by females and immatures. Loud calls of the first category were found to be quite frequent; they were heard every evening and, on such occasions, up to ten adult males could respond to each other (Galat, Galat-Luong & Gautier, unpublished data). Thus the ability to give loud calls as well as the timbre of their

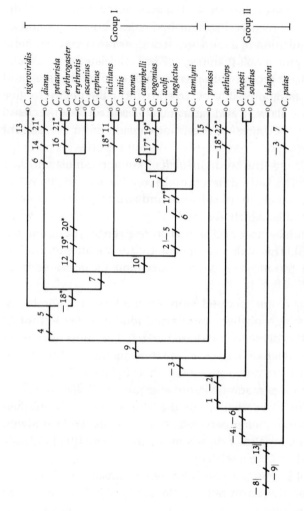

Figure 12.13. Dendrogram resulting from hypothesis 1 when warning calls are considered (see text; legends as in Fig. 12).

cohesion calls, which more closely resemble calls of *nictitans* than of *lhoesti* and *solatus*, makes the repertoire of *C. preussi* clearly different from those of the two other related forms.

The affinities among the other species of group 2 are less clear. In the preliminary treatment of our matrix we observed that a small change of a call character was able to modify the relative branching of these species, while that of the species of group 1 was regularly confirmed. Cohesion calls of *C. lhoesti*, *C. solatus* and *C. talapoin* possess high-pitched noisy components. On the other hand, *C. talapoin* and *C. patas* have in common the absence of quavering in their cohesion calls.

The second dendrogram is obtained under the first hypothesis when taking account of warning calls (Figure 12.13). In fact, the species of group 1, as well as their relative branching, remain the same as in the previous tree, except for *C. nigroviridis*. The latter species becomes the most ancestral form of this group instead of being the most ancestral form of all guenons. All species of group 1 share the ability to give loud calls and are differentiated from species of group 2 by the acquisition of high-pitched contact calls.

Another important difference is that in the second dendrogram, species of group 2 no longer appear as a group of species related to group 1 but as a paraphyletic group which shares primitive traits (synplesiomorphy, Hennig, 1965). In this group, the patas and the talapoin are closest to the ancestral form of guenons and *C. preussi* the closest form to species of group 1 due to its ability to give loud calls. In this group, *C. lhoesti* and *C. solatus* are the first species to acquire quavering in their calls.

Comparison with other phylogenies

Interspecific affinities among guenons obtained in the first dendrogram (Figure 12.12) are in close agreement with those deduced from karyological evolution (Dutrillaux *et al.*, 1982, Chapter 9). Both studies agree on:

 the most ancestral position of *C. nigroviridis*;
 the same two clusters of species, except that in the chromosomal study, *C. nigroviridis* is rooted at the base of group 1 and not at the base of the two groups;
 the primitive position of *C. diana* within group 1;
 the grouping of *C. neglectus* close to the *mona* forms;
 to some extent, the intermediate position of *C. hamlyni* between the *nictitans–cephus* group and the *mona–neglectus* group.

They mostly differ by:

> the affinities of C. *hamlyni*, whose karyotype differs from that of C. *neglectus* by 14 chromosomal rearrangements and the possession of a greater number of chromosomes while the vocal repertoires of the two species are very similar;

> the fact that the same karyotype has been observed for C. *lhoesti* and C. *preussi* while the karyotype of C. *solatus* differs by three chromosomal rearrangements of which one is shared with C. *patas*; conversely, the vocal repertoires group *lhoesti* and *solatus* and separate *preussi*;

> the early departure given to C. *talapoin* in group 2 by the karyological study and to C. *preussi* in our dendrogram.

The results obtained in the second dendrogram (Figure 12.13) also agree with the findings of Dutrillaux *et al.* in that we found exactly the same two clusters of species, with C. *nigroviridis* more closely related to the species of group 1. However, in our study all forms of group 2 appear as more primitive than all guenons of group 1.

Comparing our results with Ruvolo's findings (Chapter 7) based on protein electrophoretic analysis, several points can be made. While Ruvolo also agrees that C. *neglectus* shares a common ancestor with the *mona* forms (the latter being clustered into the same two species pairs as in our study), she finds that C. *diana* is included in this large species group, being either less or more derived than C. *neglectus*, depending on the method of analysis used.

Furthermore, whereas we find close similarities between the repertoires of C. *hamlyni* and C. *neglectus*, none of the three analyses proposed by Ruvolo brings these species together. In addition, electrophoretic data always link *hamlyni* with *lhoesti* (probably C. *lh. lhoesti*).

Ruvolo points out that whatever the method of analysis used (phenetic or cladistic) the branching of C. *petaurista* remains uncertain, being alternately linked with *cephus* or with *albogularis* (a form of the *nictitans* superspecies) or having a primitive position with respect to all *nictitans* and *cephus* forms. Furthermore, *nictitans* and *cephus* are found either to have a common ancestor or to derive from one another, with either a *cephus* form or a *nictitans* form being the most derived. To some extent, we are confronted with the same problem (see below); our analysis also agrees that C. *petaurista* (together with *erythrogaster*) is distinct from *cephus*, *ascanius*, and *erythrotis*, a separation which is not found by karyological analysis.

From outgroup comparisons, Ruvolo concludes that C. *aethiops*, C. *patas* and C. *talapoin*, together with C. *nigroviridis* are 'equally primitive' while the affinities she finds between these species differ according to the type of analysis used. However, in her two phenetic dendrograms, Ruvolo finds affinities between talapoin and patas as shown on our dendrogram 2 (Figure 12.13). Finally, it is worth noting that measurements of skull and teeth in guenons show affinities between *C. talapoin*, *C. patas*, and *C. lhoesti* (Martin & MacLarnon, Chapter 10).

Toward a consensus tree

Comparing our two dendrograms, a consensus tree can be proposed for phylogenetic affinities among guenons by considering additional vocal criteria not used in our matrix and some morphological and behavioural criteria (see Figure 12.15 below). In producing such a tree, we will particularly discuss the controversial points arising from the comparison of our results with those of the genetic studies of Ruvolo and Dutrillaux *et al.*

Our two dendrograms agree in clustering guenons into two groups (Figures 12.12, 12.13). Furthermore, the relative branching of the fourteen species of group 1 is exactly the same in the two dendrograms; only the position of *C. nigroviridis* differs.

The main differences rest on (1) the relative branching of the different species. In the first dendrogram, *C. nigroviridis* is the most primitive form of all guenons; in the second it is most ancestral to group 1, while all forms of group 2 are more ancestral than those of group 1; (2) the relationship between the two groups and within species of group 2, the latter being considered as a paraphyletic group in dendrogram 2.

Let us consider first the species of group 1. Within this cluster, the grouping of *C. diana* with the *neglectus–mona* cluster proposed by Ruvolo, which could be accepted on the basis of some similarities in coat patterns, is not supported by their vocal repertoires. Although *C. diana* calls are well differentiated from those of all other guenons, their closest affinities are found with the cohesion-contact calls of *nictitans* and *cephus* superspecies and especially with *C. petaurista*. It should be noted that both *petaurista* and *diana* give a whistle alarm call and associate a roar with their loud calls. The branching of *diana* in a relatively primitive position is not due to similarities with outgroups but to the fact that this monkey is able to give both quavered and non-quavered high-pitched sounds. A similar 'outlier' position is given to *C. diana* on the basis of skull and teeth measurements (Martin & MacLarnon, Chapter 10).

Similarly, we cannot propose a different grouping for the cluster including the four species of *mona*, *C. neglectus* and *C. hamlyni* and we have no argument to link *hamlyni* with *lhoesti* as Ruvolo does. Must we attribute the close similarity between the repertoires of *hamlyni* and *neglectus* to patterns of convergence? Both species are semi-terrestrial forest monkeys, but the equivalence of their ecological niches remains unknown as *C. hamlyni* has never been studied in the field. The affinity between these two species is also supported by at least two other remarkable features: the very similar coat colour of their young which changes with age, and the existence of a very similar olfactory marking behaviour (see Loireau & Gautier-Hion, chapter 14). However, the two latter traits may also be invoked to link *C. hamlyni* and *C. neglectus* with species of group 2, since a distinct coat colour (although less differentiated than in *C. hamlyni* and *C. neglectus*) is also found in young *C. lhoesti*, whereas olfactory marking behaviour is also present in *C. aethiops* (and in *C. nigroviridis*). In addition, medium-pitched warning calls described in *C. patas* ('huh' calls: Hall *et al.*, 1965) and in *C. aethiops* (Struhsaker, 1967e) are also uttered by *C. hamlyni* (personal observations). Finally, we have seen that very similar 'moans' are given by *C. patas*, *C. hamlyni*, *C. neglectus* and *C. mona* (see Figure 12.11).

The relative branching of the *C. cephus* and *C. nictitans* superspecies which, in our analysis, results only from the presence of a 'boom' call in the latter species is worth mentioning. Detailed studies have shown that the repertoires of *C. mitis stuhlmanni* and *C. ascanius schmidti*, two sympatric species from East Africa (Marler, 1973), as well as of *C. n. nictitans* and *C. c. cephus*, their ecological equivalents in Gabon (Gautier, 1975, 1978), are very difficult to distinguish (see Gautier & Gautier-Hion, 1988). Thus it may be better to propose a common ancestor to the *nictitans* and *cephus* forms with the latter having lost the ability to utter 'booms' (Figure 12.15). This hypothesis is the more plausible as, contrary to many guenons, *cephus* monkeys have undergone a drastic reduction of their vocal annexes which, as demonstrated experimentally, operate in the uttering of 'boom' calls (Gautier, 1971).

The early departure of *C. nigroviridis* found by morphological characters (especially cranial measurements and the presence of a sexual skin: Verheyen, 1962; Hill, 1966; Martin & MacLarnon, Chapter 10) and by genetic studies, is supported in our dendrograms by the ability of this species to give a 'gobble' call. Other vocalizations confirm this, especially the copulation quaver given by females, which looks very much like that given by red colobus females (Struhsaker, 1975). Furthermore, it seems most probable that the swamp monkey is closest to species of group 1 as found in dendrogram 2 (Figure 12.13). Cohesion-contact calls and unitary warning chirps of this species are very similar

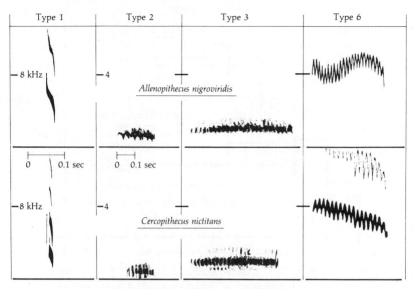

Figure 12.14. Structural similarities of warning calls (type 1), cohesion calls (type 2 and 3), and contact calls (type 6) given by C. (A.) *nigroviridis* and C. *nictitans*.

to those of C. *nictitans* (Figure 12.14) while no obvious resemblance is found between its repertoire and those of any other species of group 2. However, the metallic components of its unitary chirps give them a timbre similar to that of equivalent talapoin calls. In addition, the swamp monkey and the talapoin are the only guenons in which high-pitched chirps are retained in adult males. Both species also have a greenish coat, are the least brightly coloured of all guenons, and possess a sexual skin.

The primitive nature of C. *talapoin* found by the genetic and anatomical studies and in our dendrogram 2 is confirmed by a number of characters of its vocal repertoire. The latter shows astonishing similarities with those of macaques, while there is little analogy in the physical structure of talapoin calls and those of other forest guenons (Gautier, 1974). Thus, the 'clear calls' and the 'coos' of M. *mulatta* (Rowell & Hinde, 1962). the 'coos' and the 'high-pitched musical notes' of M. *nemestrina* (Grimm, 1967), the 'coos' of M. *speciosa* (Bertrand, 1969) and of M. *fuscata* (Green, 1975) as well as the 'whoo' call of M. *silenus* (Hohman & Herzog, 1985) resemble the type 2 cohesion calls of talapoins. All these calls possess a similar timbre and a fundamental tone around 1 kHz. Furthermore, in the talapoin as in some macaques, these calls may be given in choruses. The lost calls of young talapoins,

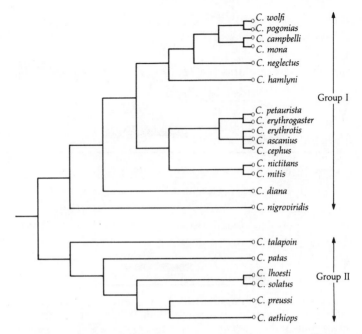

Figure 12.15. Consensus tree of phylogenetic affinities among guenons (see text).

young rhesus and young *nemestrina* all possess an identical structure and are difficult to distinguish. Finally, both in talapoin and rhesus, there is a graded call system arising in approach–avoid behaviours which is structurally similar, and the threat calls (type 4 in talapoin, and 'pant threat' in rhesus) have the same rhythm (4 to 5 units/sec) in both species.

These macaque-like calls confer on talapoin monkeys a position close to the common ancestor of papionines and cercopithecines. Given that we can make no statements about the relative position of talapoins and swamp monkeys, and that while the latter shows affinities with group 1 the former has little in common with those species, it seems justifiable to keep the two clusters of monkeys, with *C. nigroviridis* rooted at the base of group 1 and *C. talapoin* at the base of group 2 (Figure 12.15).

On an acoustic basis, there are also some similarities between *C. talapoin* and young *C. patas*, especially in their cohesion calls which are not quavered (compare Figures 12.2, 12.11). These two species, like *C. nigroviridis* (and also at least *C. solatus*) also give frequent copulatory calls. Such calls are known in colobines (Struhsaker, 1975), in baboons

(Hamilton & Arrowood, 1976), in mangabeys (Gautier & Gautier, 1977), and in macaques (Deputte & Goustard, 1980) and may thus be considered as an ancestral trait.

The early divergence of the patas monkey also relies on the fact that adult males give a two-phase bark which sounds similar to that of baboons (Hall & DeVore, 1965). Thus, data on vocal repertoire, concurring with other studies, suggest that both talapoin and patas are close to the ancestral form of guenons and, as found by Ruvolo in her phenetic analyses and by Martin & MacLarnon from skull and teeth measurements, share several primitive characteristics. Conversely, these two species have few vocal characters in common with the other species of group 2 (see Table 12.2).

We lack arguments to specify the relative position of the *lhoesti* and the *aethiops* superspecies, except that both taxa have a well developed use of the low-pitched register. Furthermore, there is some resemblance in the pattern of emission of cohesion calls given by *C. l. lhoesti* and *C. aethiops*. Finally, we currently have no data to link closely the three *lhoesti* forms: the affinity found between *lhoesti* and *solatus* in vocalizations clearly disagrees with the result of the karyological evidence which links *lhoesti* and *preussi*.

Conclusion

The consensus tree (Figure 12.15) based on characteristics of their vocal repertoires separates guenons into two groups of species which are the same as those found in the chromosomal evolution proposed by Dutrillaux *et al.* Notably, species of group 1 are all forest-dwelling monkeys and, except for three of them (*C. nigroviridis*, *C. neglectus* and *C. hamlyni*) fully arboreal species. Group 2 on the other hand includes the two savanna guenons, the three semi-terrestrial forms of *C. lhoesti* and the talapoin, a monkey which is mainly observed in the lower strata of the riparian forest and which shows several terrestrial traits in its locomotor behaviour (Rollinson, 1975).

All species of the latter group, together with *C. nigroviridis*, show ancestral traits in their vocal repertoires, but their relative affinities are still unclear. As in the study by Ruvolo, there is conflict in the placement of species which are relatively more primitive.

Recently, the repertoires of arboreal forest species have become better known and this knowledge has clearly influenced the choice of the criteria used in our analysis. Though these criteria are appropriate for arboreal forms, they may be less reliable for terrestrial and semi-terrestrial forms which are more primitive (see also Creel & Preuschoft,

1984). One cannot exclude, however, the possibility that these plesiomorphic species are not closely related to one another.

Nevertheless, our conclusions are strengthened by the good agreement among phylogenies drawn from studies in different disciplines. These conclusions also agree with the result of recent studies on fossils, forest history and paleoecology (see Leakey, Chapter 1; Hamilton, Chapter 2; Pickford & Senut, Chapter 3) which all indicate that early cercopithecoids quite likely evolved in semi-arid open woodlands and savannas and were semi-terrestrial forms.

Acknowledgments

I am very grateful to the following colleagues for access to their tape recordings of monkeys' calls and/or unpublished data: M. Bertrand, R. Byrne, J. Chism, S. Gartlan, L. Girolami, M. Herzog, P. Marler, and T. Struhsaker. Special thanks are due to A. and G. Galat who provided me with invaluable information on West African guenons, to R. Quris for assistance with the computer programs and to A. Gautier-Hion for stimulating discussions during the preparation of the manuscript. For helpful comments on the paper, I am grateful to M. Harrison, R. D. Martin and J. Oates.

II.13

What are face patterns and do they contribute to reproductive isolation in guenons?

JONATHAN KINGDON

It is a commonplace to observe that patterns serve to make a species distinctive. It is generally assumed that recognition signs influence or mediate sexual behaviour and, in lower animals, simple visual signals have long been shown to be essential adjuncts to the restriction of mating to members of the same species (Bristowe, 1929; Konishi, 1978). This has not yet been established where behaviour is more complex, as in primates. None the less the visual patterns of several cercopithecid monkeys are so emphatic that there must be a decisive advantage for the evolution of such permanent and easily perceived patterns. Moreover, guenons offer an unusual opportunity to investigate form and function in signals, because some species have virtually no distinctive coloration on the face, others are generally colourful, while some have a concentrated panel of colour and design effectively limited to the face alone. In one species group (that is restricted to the main forest block) fourteen subspecies have been described, of which at least eight have highly distinctive faces while another species shows no variation over a vast and discontinuous range.

So what are face patterns for and do they contribute to reproductive isolation? This enquiry depends upon the assumption that patterning on the face has evolved to communicate (Vine, 1970; Kingdon, 1980) and on the ability of facial features to transmit messages – messages that can be augmented or offset by means of pattern. A consequence is that the variety of faces can be explained in at least two different ways. One explanation is that the messages in face masks differ in content from species to species. Another is that similar messages can be encoded in dissimilar optical signs. I hope to show that both situations may occur.

Some principles in the evolution of visual signals exemplified by the gelada and mandrill

Any signal must evolve from the modification of an already existing structure. Structures that have been totally transformed to serve semiotic purposes are conspicuous in the Cercopithecid genera *Theropithecus* and *Mandrillus*. Examination of the signal structures of these monkeys suggests correlations between species-specific signals and ecologically determined behaviour that is equally specific (Crook, 1966; Jouventin, 1975; Emory, 1976; Dunbar, 1977). They also illustrate some general principles that might govern the evolution of visual signals.

Some general principles

Economy of energy

Conservation of energy is likely to have been a major determinant in this evolution. Messages, especially routine ones, should impinge very little on subsistence activities. Thus frequently recurring movements and gestures are the most likely to provide motor bases for subsequent ritualisation and the coloration of moving parts to serve as optical signals (Cott, 1940; Crane, 1967). Signals will evolve in relation to the distinctive energy budgets that are intrinsic to every evolving species as it comes to exploit a particular ecological niche. Energy costs are likely to modify the siting of signals even if the ultimate determinants of a species' colour and pattern relate to predation or social communication.

Socio-sexual signals

Primate communication and its complexities became more accessible to analysis when Marler (1961, 1968) presented relationships between individuals and classes as a balance between attraction and repulsion or aggregation and dispersion. To reduce the likelihood of attack by conspecifics, primates (in common with many other mammals) have ritualised sexual signals. Female submission and the attractiveness of female genitalia to males provided the behavioural base for the development of 'presentation' as a ritualised appeasement gesture or generalised greeting (Zuckerman, 1932; Maslow, 1936, 1948; Chance, 1956). The mechanism for defusing aggression or attracting therefore tends to polarise at the rear end of an animal (Heinroth-Berger, 1959; Guthrie, 1971). Threat or the capacity to repel focuses instead at the opposite end. The principal weapons, canine teeth, are most developed in adult male monkeys.

Predation

The most widespread colour and pattern of cercopithecine fur is agouti in various shades of brown or grey with dull colouring most in evidence in young animals. Many patterns derive from elaboration of this agouti base. The phenomenon studied here is therefore likely to have evolved out of a predator-related cryptic agouti pattern system.

The case of Theropithecus *and* Mandrillus

Theropithecus gelada (Figure 13.1)

The visual devices evolved by *T. gelada* to mediate its social life are extreme but do illustrate the principle of saving energy. Geladas not only rest on their haunches but spend most of their time feeding on grass in a seated position. Although the habitat is open this mode of foraging effectively removes the female genitalia from view except while walking (Crook, 1966; Dunbar & Dunbar, 1975). Concealment of an important sexual cue can be correlated with the development of vesicles of coloured skin on the chest of females, which mimic a similar arrangement around the perineum. Engorgement of both sets of vesicles during oestrus has an obvious advantage in advertising female condition, yet males too have patches of brilliantly coloured skin on the chest.

Figure 13.1. *Theropithecus gelada.* Right: female genitalia during oestrus; middle: female neck and chest vesicles; left: male neck and chest vesicles.

This shows that functions beyond the advertisement of female condition are served by displacement of what would seem to be a social as well as a sexual signal from one part of the body to another (Wickler, 1967). Such complex mimicry involving extensive physiological rebuilding on an unpromising site and its appearance (in admittedly somewhat different form) in both sexes demonstrates that just as genitals have dropped out of sight in the interest of saving energy, economy of effort in the social order demands their replacement. This imperative derives from the fact that sexual behaviour and the signs of sex and condition are essential not only for reproduction but also for the regulation of social life. Emory (1976) has stressed the mutual visual attraction between male geladas and their harems. Chest patches on geladas reduce the need for cumbersome presentation gestures in both sexes and they balance the adult males' intimidating appearance (in what is probably a subtle equilibrium) with a sign that attracts and reassures perhaps because it simultaneously signifies 'harmless' and also 'gelada'.

Mandrillus sphinx

Many cercopithecines use signs of gender to serve social ends but the best known monkey with brilliant genital and facial colouring is the mandrill, in which a major male advertisement is yawning. The colour scheme of the muzzle is shared with the genitalia, which also signify male adulthood. However, the most visually prominent signal consists of a blue and violet rump with a broad rim of scarlet round the anus. Coloured skin forms a naked disc that extends from tail to scrotum and across the full breadth of the hips. One explanation for common colour-coding of the mandrill's face, rump and genitalia is physiological economy in adapting localised areas of epidermis as the vehicle for signals.

Another view is that selection has favoured an ecologically determined channel for signal transmission (i.e. particular combinations of colour wavelengths travel well in low ambient light). The fact that mandrills, drills and several species of guenons have similar rear-end colouring suggests that the need for effective signal transmission has combined with intrinsic physiological limitations to favour blue and red skin signals.

A species living in a very open habitat, as gelada do, does not use the same colours or sites for its signals as a forest dweller, but there is one respect in which gelada and mandrill males have analogous functions. Because of their assertive behaviour, loud calls and conspicuousness, harem males become the pivot on which other harem members orient their movements (Emory, 1976). In this type of social system males are transformed into social 'beacons'. According to Jouventin (1975), this is a primary role for the mandrills' coloured rump, which is actively followed by the other animals in a group or subgroup.

Rump colour is actually likely to be but one component in a barrage of self-advertising signals (loud calls, scent, exaggerated postures and actions such as yawning, grimacing and branch-shaking). The degree to which these signals are shared by both sexes remains to be determined. Snarls and yawns are especially frequent in the mandrill's repertoire of gestures. The first is a low-intensity, short-range display of canines. Yawns are more conspicuous, more energetic and occur at higher intensities. Both are non-aggressive in function. It is not uncommon among well-armed social mammals for subordinate individuals to display their weapons. Weapon display probably inhibits aggression by advertising its costs whereas 'presenting' serves similar social purposes by deflecting the attention of potential aggressors.

By patterning the blue skin of the nose in a permanent snarl mandrills may inhibit aggression and dispersion with an energy-saving signal that is shared by both sexes. The male muzzle is enlarged by such massive inflation of the muzzle buttresses that it obscures the eyes and even restricts the field of vision in old males. Overshadowing of the eyes may be partly an artefact of the ballooning buttresses but it is also symptomatic of a real evolutionary shift. In both sexes of mandrill the signal code focuses on the mouth and muzzle, whereas other papionines (notably the mangabey, *Cercocebus torquatus*, and the gelada) make an exaggerated use of the eyes and coloured eyelids.

Whereas female genitalia have provided the basis for a signal that is shared by both sexes in gelada, it is a pigmented muzzle that provides the mandrills' single most species-specific signal. Why should an attribute with male connotations be shared by both sexes of mandrills, while geladas share an unequivocally female signal? The most obvious factor common to both sets of signal is that they formulate a locus for social orientation that is wholly species-specific and that they are prominent during routine activities and conspicuous in the monkeys' normal environment.

Thus primary adaptations or ecological specialisation call for changes in behaviour which in turn demand alterations in communication. The ancestral gelada's 'seated foraging' in open grassland and the mandrill 'precursors' entry into forest are examples of ecological shifts which must have demanded change. Subsequent transformations of signals evolve *directly* in relation to what may have started as small but critical innovations, for example, the demand for a new signal focus in seated gelada; improving the visibility and localising the focus of signals in the mandrills' darker, more obstructed environment. Finally, improving the efficiency and specificity of a signal demands that its intrinsic qualities or components are pushed to their limits. In this way, *contrasts, colour, size*, and *geometry* become as exaggerated as other constraints may allow.

The analysis of face patterns in guenons

In choosing the face patterns of *Cercopithecus* for intensive study, several considerations have been uppermost in my mind. One is that the 'face' (effectively the frontal hemisphere of the head) has a relatively simple structure with few moving parts. This reduces the potential for signal elaboration and complication and so removes a major obstacle to analysis (Schlosberg, 1952). Furthermore, this signalling device can only transmit within a narrow arc of movement, because participants generally need to monitor each other. The receptors of signals, eyes and ears, are built into this device, while emitters, voice, facial features and crucial movements of the head (not to say the animal behind it), also emanate from a single concentrated source.

One approach to understanding the role of face pattern is to study the physical and morphological differences between closely related species, extract those elements that are common to all and then construct a single model of those basic elements. This requires a complete inventory of species and an assessment of individual and regional variability.

Where possible the functions of those basic elements should then be considered and behavioural correlates sought. This is problematic, because so little is known about so many species. Notwithstanding the difficulties, there do seem to be some reasonably well understood behaviours that involve the face and head and these appear to be relevant to the analysis of face patterns, whether directly or indirectly.

Facial expression

Facial expression is the most obvious involvement of the head in signalling. Expressive movements of the eyes and mouth combine with head movements, pilo-erection or flattening and minor movements of the ears (Miller, Murphy & Mirsky, 1959; Andrew, 1965). It is possible that refined species-specific facial expressions may occur but these have not been documented (Redican, 1975). The frequency of particular gestures may also vary from species to species but here too the evidence is currently inconclusive (Gautier & Gautier, 1977).

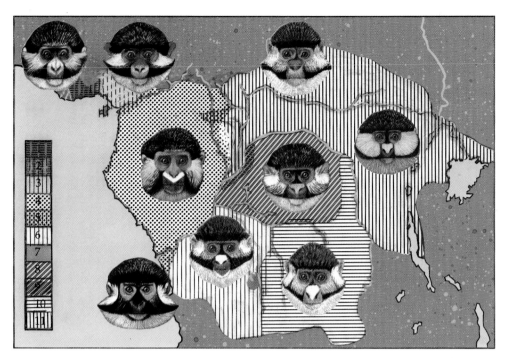

5. Distribution of the *Cercopithecus cephus* superspecies in equatorial Africa: 1 *C. sclateri*; 2 hybrid (*sclateri* × *erythrotis*); 3 *C. erythrotis*; 4 *C. cephus*; 5 *C. cephus* and *C. ascanius ascanius*; 6 *C. ascanius ascanius*; 7 *C. a. atrinasus*; 8 *C. a. whitesidei*; 9 ssp, nov?; 10 *C.a. katangae*; 11 *C.a. schmidti*.

OVERLEAF

6a. Left hand column (from top)
Cercopithecus patas pyrrhonotus
Cercopithecus tantalus budgetti
Cercopithecus pygerythrus johnstoni
Cercopithecus sabaeus
centre column
Cercopithecus talapoin talapoin
Cercopithecus salongo
Cercopithecus l'hoesti l'hoesti
Cercopithecus preussi
right hand column
Cercopithecus nigroviridis
Cercopithecus diana diana
Cercopithecus neglectus
Cercopithecus campbelli lowei

6b. Left hand column (from top)
Cercopithecus pogonias grayi
Cercopithecus wolfi wolfi
Cercopithecus wolfi denti
Cercopithecus mona
centre column
Cercopithecus albogularis albotorquatus
Cercopithecus mitis stuhlmanni
Cercopithecus nictitans nictitans
Cercopithecus hamlyni
right hand column
Cercopithecus cephus cephus
Cercopithecus ascanius s.sp. nov?
Cercopithecus petaurista petaurista
Cercopithecus erythrogaster

7.   *Cercopithecus mitis kandti*
Golden monkey – Virunga volcanoes.

Feature reinforcement

Facial expressions transmit over a more limited range than colour-advertised head movements and an infinitely shorter distance than sound signals. Visual reinforcement for vocal signals could have influenced the evolution of pattern around some guenons' mouths. Movements of the mouth or larger gestures are effectively advertised by eccentric contrasts of tone or colour running across the chin, chest or upper lip in many species. In this way, a specific pattern could acquire a close link with other signals that are equally species-specific. This association would hold even when sound emanates from a closed mouth or the nostrils. It is not possible to assert, however, that visual reinforcement is the primary function of mouth-advertising patterns (Otte, 1974). Where the brows are isolated by a patch of light colour or differentiated hair, this has the effect of increasing the visibility of expressions of the eyes. This advertisement of the brows is widespread, but absent in several species groups (*C. cephus, C. lhoesti, C. nigroviridis*). The ears are prominently coloured in some species (and even local populations) of the *mitis–nictitans, cephus* and *mona* species groups while remaining inconspicuous in others. There is some evidence that the frequency or amplitude of brow and ear movements differ between species. Certainly the effect of colour of brows and ears is to advertise or suppress their visual impact (Kingdon, 1980). The influence of these differences on interspecific communication remains unknown.

Yawning

In *Cercopithecus* yawning is most frequent during periods of tension when both sexes may yawn. By contrast with the multi-male societies of papionines it plays a negligible role in regulating male-to-male relationships except during relatively infrequent intergroup encounters. The white beard of *C. neglectus* and ruff of *C. lhoesti* certainly help to make yawns more conspicuous but this also seems to be an incidental effect.

Eye contact

A rather inaccessible behaviour for the student of primate communication is eye signalling. Not only are eye movements difficult to observe but the observer's own eyes or binoculars can influence behaviour (Perret *et al.*, 1984). None the less, several students of bird, mammal and primate behaviour have recognised their importance (Chance, 1962; Kleiman & Eisenberg, 1973; Kingdon, 1982). Eye-avoidance serves both subdominant and dominant individuals. Since a sustained stare is threatening in almost all mammals, the general response is flight. Social cohesion is served by 'cut-off', because it is a very economic means of controlling flight. Animals that avoid eye contact remain together and when the behaviour is reciprocal it seems to dampen aggression.

When the animal seeking to approach another is an adult male attracted to an oestrus female, there are high reproductive costs if the female's tendency to flee is not offset by some friendly sign by the male. There is evidence that eye-avoidance has been ritualised into stereotyped movements by some species of the *cephus* group and that these may occur during courtship (Kingdon, 1980). Even during violent interactions between highly aroused animals individuals can be seen to make frequent 'cut-off' gestures. In between bouts of attack and retreat the members of two interacting groups will make ostentatious post-ures, turn about, lie down or groom one another. I have observed this in group encounters between troops of *C. lh. lhoesti*, *C. cephus ascanius* and *C. mona pogonias*. In such circumstances it is difficult to assess whether visual devices play any role.

Presenting

The restriction of bright blues and reds to the genitals and faces of guenons must beg the question whether there is any connection. As with the mandrill considerations of physiological economy or optical efficiency may apply. Whereas terrestrial guenons tend to have bright genital signals, the arboreal *cephus* and *mona* lineages have facial instead of genital signal foci (Kingdon, 1980). Further discussion of this topic follows.

Suggested functions for pattern as exemplified by selected guenon species

Cercopithecus (Allenopithecus) nigroviridis

Allen's swamp monkey is a drab khaki-coloured guenon with the classic countershading of a predominantly cryptic animal, but attention can be drawn to several features which may have a general significance for the face patterns of other guenon species. These relate, in my opinion, to the importance of regulating eye-to-eye contacts.

Fur on the cheeks is exceptionally long, giving the male in particular a somewhat leonine appearance. This forms a broad ruff and its primary function must be to enlarge the visual impact of the facial disk. In common with a lot of other guenons, macaques and at least one

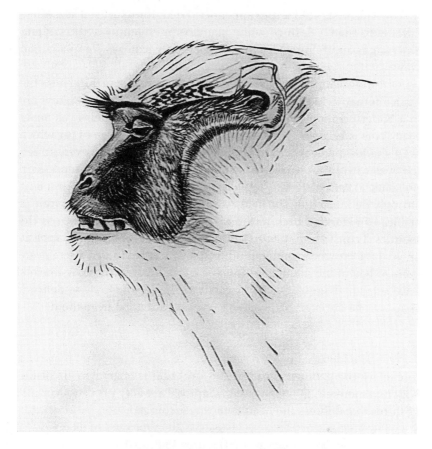

Figure 13.2. Side view of *C. (A.) nigroviridis.*

mangabey species, there is also a cow-lick where the white chin and throat fur thrusts up against the agouti ruff. This ridge of elevated fur describes a sinuous curve over the cheek and, from a side view, pulls the area of white nearer the eye (Figure 13.2). When the angle is sufficiently oblique, the top of the fur ridge actually provides a thin screen between the viewer and the viewed monkey's eyes. For the monkey transmittor of signals this screen not only allows it to monitor another animal with its head averted but its glances are sufficiently screened to reduce the likelihood that it will precipitate an attack in retaliation for staring. So much for a lateral view.

Apart from full frontal, two other positions may be considered. The head is sometimes raised by captive swamp monkeys in a fast and rarely seen gesture, as if the monkeys were following an overhead rocket. This first rapid swing appears to be highly ritualised and offers little more than a flash of white like that provided by a rabbit's tail. Yawning is another gesture that momentarily exposes some white to view.

For the regulation of eye-to-eye contacts depression of the head is the main alternative to eversion and, from this position, most monkeys are in a better position to monitor one another without their eyes being seen. This is because they can pull forward a brow ridge of fur which both shades the eyes and, if sufficiently depressed, may even screen the eyes with a thin barrage of long hairs. In *C. nigroviridis* depression presents a flat, plate-like top of the head, its forward edge being strongly delineated by the heavily shaded eye sockets. This margin is artificially extended back to the ears by a black line of fur across the temples. I consider that its function may be to alter the visual appearance of the crown, turning an otherwise amorphous area into a more defined geometric shape. In this way an extension of the brow-line not only helps to disguise eye glances but the line, by helping to define a shape above the eyes, also serves to distract attention from them.

Cercopithecus erythrogaster

The idea that pattern can evolve to *distract* attention may help to explain the geometric pattern or 'cap' that is characteristic of the Nigerian monkey, *C. erythrogaster*. Captives are very prone to 'cut-off' gestures that depress the head and expose this cap.

Cercopithecus mona group (Figures 13.3, 13.4)

The brow and temple line occur in many guenon species but the *mona* group exemplifies how these two structures can be major structural elements in the evolution of face patterns (Figure 13.3).

The most generalised pattern is that of *C. campbelli lowei*. When this species is compared with its relative it can be seen that the temple line and brow patch provide foci of coloured fur which, when they become darker or lighter in the tonal scale or warmer or cooler in the chromatic scale, generate highly distinctive patterns. When tones, colour contrasts and the actual margins of colour pattern shift, the visual focus also changes, even moving from one area of the head to another. For example, either the cheek or the crown can receive emphasis by small alterations along the upper or lower margins of the temple line. Intensification of colour can lift the light ear or brow patch into much greater visual prominence (especially if the temple focus 'creeps' around the margins and therefore frames it). Whereas some face

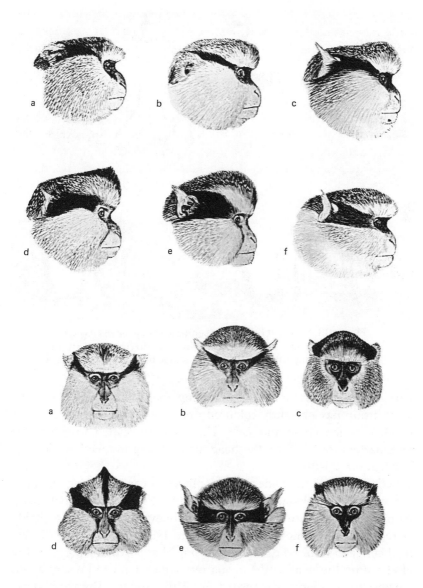

Figure 13.3. Profile and front views of mona monkeys. *a*: *C. m. campbelli lowei*, *b*: *C. m. mona*, *c*: *C. m. denti*, *d*: *C. m. pogonias pogonias*, *e*: *C. m. p. grayi*, *f*: *C. m. wolfi*.

Figure 13.4. Threat behaviour sequences in *C. mona pogonias* (above), *C. m. mona* (below). In each case (from left to right), *a*: 'cut-off' phase, *b*, advertising phase, *c*: threat + vocalisation, *d*: 'cut-off' phase.

patterns are entities that are separate from the body, all members of the *mona* group have brightly-coloured undersides and there is no clear break between chest and face. This may be related to the entire forequarters (rather than the head alone) being invoked in 'display' and 'cut-off' postures.

In *C. mona* the underside and inner linings of arms and legs are pure white, while cheeks and brow are also very pale. Adult male *mona* monkeys respond to a hand-make monkey puppet with a highly stereotyped set of movements (Kingdon, 1980) and the same display has been seen in the wild. The advertising phase of this display consists of elevating the head and forequarters which tend to present a tall, narrow strip or stripes of white to view. This is followed by a hard stare, lunge and/or vocalisations in a lower bunched position. Such an aggressive bout is followed by 'cut-offs' which present the dullest areas of the animal to view: a drab crown, an agouti back or a hunched side view (Figure 13.3).

When this display is made by *C. pogonias* there is a greater preponderance of lateral movements (Figure 13.4). These expose more side-view, less underside, more head, less body (and the movements appear to use less energy that in *C. mona*. A major visual difference between *pogonias* and *mona* is the development in the former of a crest with white or golden bands interposed between black temple bands and a tall black ridge along the top of the crest. The ears are plumed with tufts of golden hairs which remain prominent from most angles. The lateral movements made during aggressive displays present the flattened and enlarged crown to view. Tilting of the head is conspicuous for its disturbance of the linear field but any major movement of the head by this species will direct the eye towards the crown. This coronet is so visually assertive that it is as much the badge of the species as chest patches in gelada or the blue muzzle of the mandrill. It is possible that the ecological origins of a greater frequency in lateral turning lie in scanning behaviour. For a small omnivorous monkey the need for wide-angle optical searching might predispose these monkeys to movements that have been further ritualised in display. Intensified pattern on the head corresponds here with more reliance on head than body movements. It may therefore also represent an overall saving of energy expenditure.

Cercopithecus neglectus

Like *mona* monkeys the cryptic de Brazza monkey has converted pale countershading on the underside into very distinct species-specific patterns which can be correlated with equally specific behaviour. White undercoating has become localised at opposite poles of the body. It surrounds the mouth and the genitalia but can be hidden from view by crouching. In addition to providing a visual focus on the mouth, the white beard does for the *neglectus* chin what the crest does for the *pogonias* crown – it creates a longitudinal white signal from in front but its flattening creates a broad white flag when the head is turned. In conformity with this trait *neglectus* does not have broad cheek ruffs; instead, fur on the sides of its face is distinctly adpressed.

I have observed two distinct forms of flagging of the beard. One resembles *C. m. mona* in employing simple up and down jerks. It appears in aggression by individuals of either sex. The second mainly concerns male behaviour after group dispersal when side views of the beard panel are exposed to view. Movements over a wide arc are associated with silent searching behaviour, booming and with the broadcast of loud calling by the adult male. The last performance may also serve to space males but the 'booms' have the immediate and unequivocal effect of rallying group members (Gautier-Hion & Gautier, 1978). In any event, white beard and buttocks are the main visual beacons advertising an otherwise unobtrusive and quiet monkey. The white beard is also a frame for the mouth. This should help a vocaliser's audience to identify the source of a call but it also frames and advertises silent yawns (which are more frequent in males than females).

Cercopithecus diana

Advertisement with colour and pattern reaches its apogee with the Diana monkey, where colour, tone and refined linear design are used to focus attention on the thighs and rump. *C. diana* is a fully arboreal species but lives in the canopy and emergents of mature forest where its greater weight inhibits its exploitation of thinner branches. Intensive observation on captives has suggested that this exceptional colour focus can be correlated with the development of a species-specific code whereby exaggerated ritual postures provide clues as to sex, age, status and perhaps intentions (Mörike, 1973). Postures of dominant individuals are energetic and emphatic. The pattern may serve to convert individual differences in muscle-tone into a geometric semaphore. Tail postures are a part of this semaphore.

Cercopithecus diana, C. aethiops, C. lhoesti

C. diana, *C. aethiops* and *C. lhoesti* are all groups in which a dark face and muzzle is more or less framed by a white collar or bib. The evolutionary origins of this pattern type would appear to derive from an upward extension of underside countershading and an intensification of tonal contrast. All three species project semaphores of bold body postures. It is possible that a visual signal system that puts most reliance on bold body movements may evolve a particular range of sizes for its components. Therefore the scale of these units could conceivably favour the obliteration of facial features. Maximising contrast around the face has three effects. The head is turned into a sort of focal target. Expressions are blacked out except at very close quarters. A tonal mask overlies and suppresses variability and individuality in faces and their features. It is possible that any sources of ambiguity in the faces of other group members are potentially energy-expensive distractions leading to unnecessary confrontations. Blacking out of faces may therefore be a device for reducing the scope for distraction and ambiguity during routine existence. It may be a way of simplifying social order.

This mode of suppressing facial expression and individuality could correspond with strong lighting in exposed habitats. This cannot apply to *C. lhoesti*'s habitat today (although there could be some carry-over if *C. lhoesti* was once less forest-adapted). However, uniformity of facial colour is very much less in *C. lhoesti* than in *C. aethiops*. An intensely black oval does enclose the mouth and nose of *C. lh. lhoesti*, but between these and a wholly dark crown lie broad bare cheeks. This prognathous muzzle is wider than that of any other guenon (particularly in the male). It is significant, therefore, that the bare skin not only gives visual prominence to an anatomical detail that is species-specific but also a bright shade of violet seems to be linked with health or status in adult males. The signal that emanates from the vastly inflated muzzle of a mandrill finds here a sort of small-scale equivalent. Among the questions this raises is a phylogenetic one: is this the vestige of a sexual dimorphism in muzzle and facial colour that was more emphatic in guenon ancestors? Or are there correlations between more terrestrial habits and greater sexual dimorphism (Gautier-Hion & Gautier, 1985) that are independent of phylogeny?

C. cephus and C. nictitans/mitis species groups

These two arboreal species groups contain the largest number of species and subspecies within the genus and they also embrace the greatest variety of faces. Some members of both groups have nose spots, others do not. It is a feature that is likely to have been independently evolved (Booth, 1956b; Hill, 1966). Nose spots are an example of an eccentric area of the face being isolated by altered tonal values. I consider that their principal function is to serve as a visual distraction from the eyes. It can be correlated with head-flagging in both *cephus* and *nictitans* (Kingdon, 1980). Its general effect, of course, is to provide a distinctive cue for species recognition that is especially conspicuous when the head moves.

When I first began to study the structure and evolution of guenon colour and pattern, I devised various tests to measure degrees of difference between species. Every method of scoring suggested that there was greater diversity and distinctness of face patterns within the *cephus* group than between *cephus* and most other species (Kingdon, 1980). This fact alone calls for some explanation, particularly in the light of the group's almost wholly allopatric distribution and the apparent lack of any significant difference in gross anatomy, size, ecological niche, voice or social behaviour. Furthermore, this group's closest affinities in karyology, anatomy and voice are with the larger-sized *nictitans–mitis* group (Dutrillaux *et al.*, 1982, Chapter 9; Gautier, 1975, 1978, Chapter 12) which generally lacks bright colour or patterning (especially on the face). The lack of highly developed face patterns in the more generalised, and what appear to be the more conservative of the two related species groups (Kingdon, 1971) may provide an impor-

Table 13.1. *A selection of significant differences between* C. mitis *and* C. ascanius *(from Struhsaker & Leland, 1979)*

	C. mitis stuhlmanni	C. ascanius schmidti
% time spent in visual scanning	4	20.5
% time spent foraging	5	0.6
% time spent resting	30	10
Intra-group spread, metres	0–120	6–230
Group size	24	30–35
Home range surface area, ha	61	24

tant clue as to why and how the *cephus* group acquired its extraordinary diversity.

The basic differences between monkeys belonging to the *nictitans–mitis* and *cephus* complexes are well known for several representative species. Adult male *nictitans–mitis* weigh about 7 kg, females 4 kg. Adult *cephus* males are 4 kg, their females about 3 kg. That smaller body sizes require higher energy budgets is broadly confirmed by the food differences that have been recorded for members of the two species groups (Struhsaker, 1969, 1978; Gautier-Hion & Gautier, 1979; Gautier-Hion, 1980). From the point of view of the present discussion it is the techniques by which food is found that is more significant than its exact composition. The differences are most dramatically illustrated by the comparisons, of blue monkeys and redtails in Kibale Forest (Table 13.1; Struhsaker & Leland, 1979).

The various parameters given for the larger monkeys (notably body and group sizes and spatial dispersion) are closer to those of other guenons and in general are likely to be closer to the common ancestral condition. The five-fold greater frequency in visual scanning is an objective measure of significant specialisation in the *cephus* group. The capacity to find and catch flying insects calls for greater sensitivity to visual cues and improved coordination; both are likely to be based on some physical and neural refinements. Notwithstanding this, advance must be very subtle – their larger relative is scarcely less coordinated and visually acute.

None the less, specialisation in *cephus* has been more fundamental than a mere improvement in food-getting techniques. Struhsaker & Leland's figures demonstrate an overall intensification in the use of one sensory channel, vision. Furthermore, the combination of being smaller and more numerous would tend to increase predator pressure (especially from eagles), while greater agility and greater numbers of alert monkeys would put a higher premium on rapid mutual alarm systems. The trend towards smaller bodies, wider dispersion and larger numbers would in one-male groups have changed relations between and within sexes, and so altered the social behaviour of ancestral *cephus*. Such changes could in turn be served by appropriate alterations in communication. A bias or heightened sensitivity towards the visual might therefore be expected to favour the evolution of visual signals and in fact *cephus* are strongly patterned, especially their faces.

Face patterns and head-flagging

The correlation between face patterns and head-flagging in the *cephus* group is now well established (Kingdon, 1980) but it probably represents the elaboration of a relatively minor behaviour that is observable in other species, i.e. *C. n. nictitans* and *C. (A.) nigroviridis*; both flick their heads, apparently as a part of cut-off behaviour. Head-flagging may be employed by other species, notably *C. m. pogonias*, but the social contexts and behavioural sources of ritualisation appear to be different.

It is significant that *pogonias*, which had independently taken insectivory and fine branch arborealism even further than *cephus*, should also have elaborated a colourful face mask and that its most stereotyped movements also seem to relate to cut-off behaviour. My own observations of several *cephus* species, including *erythrotis*, *ascanius*, *cephus* and *petaurista*, suggest that there can be an unbroken gradient between simple eye-avoidance gestures, and fast, rhythmic head-flagging. An individual monkey can pass from the former to the latter, which is an unmistakably ritualised performance.

I pointed out that all semi-terrestrial guenons exhibited male genital colouring and that the main colour focus was at the front end of arboreal species (Kingdon, 1980). I suggested that a phylogenetic switch might have transferred an appeasing signal from back to front by means of a direct association between 'presenting' and eye-avoidance. The evidence lies in contrived body-twisting displays (especially common in young guenons) where the two types of appeasing gestures are prominently and simultaneously projected.

Subsequent observations and reflection have persuaded me that eye-avoidance or cut-off gestures are as potent mediators of social behaviour and as subject to ritualisation as 'presenting'. I now consider that exaggerated eye-avoidance or cut-off by highly arboreal monkeys have provided the basis for an *alternative* type of signal with origins that are probably independent from 'presenting'. A functional switch *has* occurred but it first evolved in *cephus* by increasing the speed, amplitude and frequency of movements that were part of a pre-existent tension-defusing mechanism common to all monkeys. Subsequently these movements were advertised by making them more visible by means of a concentrated coloured 'flag' at the front end of the animal.

Just as the gelada's habitat and feeding techniques reduce the effectiveness of genital signals and favour frontal ones, fine branches and the dispersion of individuals produce similar dilemmas for *cephus*. Like the gelada, *cephus* have a signal device which is in 'open storage' and can be activated with minimal expenditure of energy and with the simplest of movements.

Are head-flags independent of facial features?

I have already examined the details of fur and skin colour that allow face patterning to be developed, and it was shown that very small alterations of tone and colour can generate radically different patterns (Kingdon, 1980). The design principle involved in *cephus* face patterns appears to differ from previous examples in that colouring is more independent of particular features. Eccentric geometries are imposed on non-expressive areas such as nose, cheeks, chin and ears rather than designed to advertise mouth, eyes or genitals. This corresponds with use of the entire head for head-flagging, but it might still be expected that some designs serve such a purpose better than others. Were any one pattern to possess intrinsic advantages, its owners should be more successful.

It is true that some *cephus* species have much more extensive ranges but their distribution also accords well with the likely extent of past refugia and the patterns of climatic history. *C. c. sclateri* in the Niger delta has by far the smallest range of any *C. cephus* species. It has the least geometric and contrasting face pattern of the entire species complex and is closest to the theoretical ancestral model (Kingdon, 1980). This species could therefore be a relic from an early stage of the group's evolution. By the same token, *C. c. cephus* and *C. c. ascanius* could have progressed further in adapting to a common *cephus* niche. Symptoms of their more advanced condition could lie in more highly developed face patterns and the occupation of huge distribution ranges.

Do *C. cephus* face patterns contribute to reproductive isolation?

The present moment could be one of peak diversity in the evolutionary history of the *cephus* group, but it is not clear what this diversity signifies. Different types of face pattern and degree of development in five species and at least ten recognisable racial populations suggests that the face-flags were elaborated during relatively recent periods of time when the forests were repeatedly fragmented by climatic fluctuations. Today many of these ecologically equivalent and allopatric populations maintain long continuous boundaries between species and races without evidence of large-scale genetic exchange. The existence of numerous, relatively discrete populations within the main forest block is perhaps the strongest argument for face patterns playing a part in reproductive isolation.

C. cephus face patterns appear to differ fundamentally from others in that different patterns have evolved to serve the same function. This distinction is a crucial one. *Cephus* populations seem to have become isolated during very early stages of pattern innovation and it may be because of this timing that their faces look like experiments in signal geometry. In each population new patterns were generated out of interaction between animal eyes (their search for pattern) and the limited physical components from which pattern could be built.

The special importance of these patterns is that their elaboration may have been subject less to extraneous biological constraints than to rules of physiology, visual perception and signal clarity. If the ultimate effect of a face pattern is to make a monkey look totally different to both taxonomists and fellow guenons, that difference, however superficial, can act as an obstacle to communication. It can become a barrier no less than a physical obstacle. *Cephus* patterns could therefore be prime movers in speciation.

Further study of the process of visual elaboration to understand better how signals evolve may now be as important for our understanding of speciation as defining more obvious adaptive specialisations.

II.14

Olfactory marking behaviour in guenons and its implications

JEAN-NOËL LOIREAU and ANNIE GAUTIER-HION

Studies on primate olfactory communication have been mainly limited to prosimians and platyrrhines, among which many species display more or less ritualized behaviours associated with the release of odours produced by specialized glands (e.g. Moynihan, 1967; Schilling, 1980; Epple *et al.*, 1981).

In catarrhines, such behaviour is rare although there is evidence that chemical signals play a role in sexual relationships (Keverne, 1980). However, in 1968, Gartlan & Brain described a marking behaviour implying mouth and chest rubbings in a high density *Cercopithecus aethiops pygerythrus* population living on Lolui Island. Since then, despite numerous studies on this monkey, such behaviour has not been reported. However, G. Galat (personal communication) observed it in captive *C. a. sabaeus*.

Having observed a very similar behaviour in captive de Brazza monkeys, *C. neglectus* (Gautier & Gautier, 1977), we carried out a further study on captive groups and we searched systematically for similar behaviours in several other guenon species.

Marking behaviour and gland structure in *C. neglectus*

The study was carried out on two groups, including respectively three animals (one adult male, one adult female and their six-month-old infant female) and six animals (one adult male, one adult female, one subadult male, two juvenile and one six-month-old males). Their home cages included an indoor compartment and an outdoor enclosure. Experiments were conducted in these latter whose surface areas were of 25 m² for the small group and 100 m² for the large one. These enclosures were 45 m apart from each other and could be connected by a corridor.

Description

Each sequence of marking included a series of rubbings against inanimate substrates (trunks, branches, pieces of metal, wire-netting, etc.). 77.6% of the 85 sequences analysed on videotapes included rubbing series involving the muzzle, throat and chest, with or without palm rubbing and licking (Table 14.1). Other sequences were limited to the rubbing of muzzle and/or throat and/or palms.

Each sequence began with marked nose contact against the object, followed by a forward thrust of the body, bringing the muzzle, throat and chest into serial contact with the object. After this, the animal relaxed and resumed its original position before initiating another rubbing (Figure 14.1). Between three to 41 series of rubbings could be observed in any one marking sequence lasting from 1 minute to 3 min 30 sec.

Individuals concerned

Marking sequences have been observed in all age and sex classes except infants less than one year old. Adults marked more frequently than juveniles; furthermore, marking behaviours of adult males lasted longer and included more rubbings than those of adult females (respectively 3 min 30 sec and 1 min; 41 and 11 rubbings).

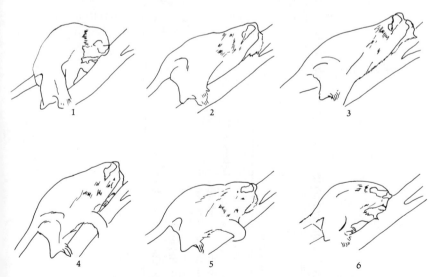

Figure 14.1. Marking behaviour by an adult male *C. neglectus*. From 1 to 6, details of a rubbing sequence against a branch (drawings from a 16 mm film).

Table 14.1. *Different series of rubbings included in marking sequences according to age and sex*

Kind of rubbings	% observed	% per age and sex classes			
		Adult ♂ (2)	Subadult ♂ (1)	Adult ♀ (2)	Juvenile ♂ (2)
Muzzle + throat + chest	54.1	48.6	57.7	66.8	44.5
Muzzle + throat + chest + lickings	15.3	–	30.8	20.0	22.1
Muzzle or muzzle + throat	11.8	20.0	7.7	6.6	–
Muzzle + throat + chest + palms	8.2	17.3	–	6.6	–
Palms	7.0	14.3	–	–	11.1
Muzzle + lickings	3.5	–	3.8	–	22.2
Number of sequences observed	85	35	26	15	9

Table 14.2. *Kinds of experiments conducted to release markings and percentage of responses*

Tests	N	% with markings	Number of sequences/test	Max. number of series/test
Non-social experiments				
Re-introduction of one group in its familiar cage	14	21.4	2.7	15 (ad. ♂)
Introduction of a new branch in the cage	14	42.8	2.3	15 (sad. ♂)
Introduction of one group in a new cage	21	58.6	2.2	41 (ad. ♀)
Social experiments				
Intergroup encounters	33	3.0	2	5 (sa. ♂)
Introduction of a branch marked by the other group	8	37.5	1.33	30 (ad. ♂)

Situations

The marking behaviour of the de Brazza monkeys was rarely observed in routine conditions (a total of 38 sequences for 540 h; 0.07 sequences/animal/h). Out of these 38 markings, 46% appeared 'spontaneously', 23% after aggressive bouts with monkeys living in neighbouring cages and 31% following the visit of unfamiliar persons.

Two types of experiments have been conducted to release this behaviour; the first consisted of non-social stimulations leading to changes in the usual environment of monkeys; the second included direct or indirect intergroup confrontation. 53% of non-social stimuli and less than 10% of social ones induced marking behaviours (Table 14.2). The introduction of a new branch in the usual enclosure and the introduction of the group to a new enclosure were the most efficient stimuli. On the other hand, markings were only observed in 3% of cases following intergroup confrontations and the introduction of a branch previously marked by one individual of the other group did not provoke more responses than an unmarked new branch.

Intragroup responses

67% of marking sequences induced a response among other members of the same group including infants. However, markings made by adult and subadult males provoked a response more frequently than those of adult females and immature animals (Table 14.3).

Table 14.3. *Different kinds of responses elicited by marking behaviour according to the category of the first marking animal*

1st marking monkey	Number of markings	% releasing responses	Kinds of responses, %					
			Look	Approach	Sniffing	Pushing	Marking	N
Ad. ♂♂ (2)	35	71.5	7	23.5	47.5	8.5	13.5	59
Sad. ♂ (1)	26	77.0	13	5.0	72.0	7.5	2.5	39
Ad. ♀♀ (2)	15	53.0	—	—	63.5	—	36.5	11
Juv. ♂ (2)	9	44.5	—	—	80.0	—	20.0	5
Total	85	67	8	14	59	7	12	114

Table 14.4. *Different kinds of responses elicited by marking behaviour according to age and sex*

| Responding monkey | Kinds of responses, % | | | | | N |
	Look	Approach	Sniffing	Pushing	Marking	
Ad. ♂♂ (2)	14.3	—	50.0	—	35.7	14
Sad. ♂ (1)	9.0	36.4	27.3	—	27.3	11
Ad. ♀♀ (2)	21.0	29.0	37.5	—	12.5	24
Juv. ♂ (1)	—	—	80.5	13	6.5	46
Inf. ♀♂ (2)	5.3	26.3	58.0	10.4	—	19

Five kinds of responses were observed (Table 14.4). Nearly 70% consisted of sniffing the marked area. Sniffing predominated among juveniles which were also the only animals that harassed marking monkeys. In contrast, marking the same or a different place was the characteristic response of adult and subadult males.

Gland structure

In between two marking sequences, it was not infrequent for the marker to rub its chest with hands or feet or for another group member to come and sniff at it. A histological examination of a piece of chest skin was carried out and compared with a piece of dorsal skin, taken from a dead adult male. Results showed that on the chest skin: (1) sebaceous glands are less numerous due to sparser hair; (2) sudoriparous glands with apocrine secretion have a larger cross-section and form larger clusters; (3) furthermore, frequent absence of any conspicuous secretion within the gland, the uniform aspect of the epidermal surface and the dense structure of the dermis suggest a more intense glandular activity.

Thus, in de Brazza monkeys, as in numerous primates (Geissmann, 1987), the sternal region shows a specialized cutaneous glandular area.

Functional aspects

The form of the behaviour, the existence of a specialized glandular area and the response given by conspecifics, all suggest that the marking behaviour leads to the production of a substance having an olfactory signal function. It must be noted, in addition, that both nose secretions and saliva and possibly secretions from mouth-corner glands appear to be stimulated during markings and might also be implied in the olfactory signal.

The fact that, on the one hand, non-social stimuli are the most efficient in triggering the behaviour and, on the other, that intragroup responses are more conspicuous than intergroup ones, tends to suggest that the marking behaviour may play a role in the overall control of the environment (which could be similar to the 'confidence giving effect' described by Mykytowycz, 1974, in rabbits); it might also play a role in intragroup social relationships, the more so as the behaviour seems to be correlated to social hierarchy (as observed for extra-markings performed by males which looked like challenging responses).

Marking behaviour among other species of guenons

Species concerned

Long-term observations and experiments on members of *C. nictitans, cephus, mona* and *lhoesti* superspecies in our colony and at the Zoological Garden of Mulhouse (a total of 12 species) failed to disclose this behaviour. By contrast, marking sequences, quite similar to that described for *C. neglectus*, have been observed and experimentally induced on both *C. hamlyni* and *C. (Allenopithecus) nigroviridis*.

Ecological and behavioral correlates

The four guenon species which display marking behaviour are all semi-terrestrial and live either in open areas (*C. aethiops*) or in forest. The use of olfactory signals by the de Brazza monkeys can be related to several other traits: the low development of their visual and vocal signalling (Gautier & Gautier, 1977), their small group size, their cryptic behaviour and their small home range (Gautier-Hion & Gautier, 1978); this last characteristic has indeed been found to enhance olfactory communication (Harrington, 1974). It must also be pointed out that a similar cryptic behaviour has been reported in *Colobus verus*, while olfactory marking is also suspected to occur (Galat & Galat-Luong, 1985). Data are lacking about the ecology of *C. hamlyni* which is known to share a number of call characteristics with *C. neglectus* (Gautier, Chapter 12).

However, none of these characteristics are typical of *C. nigroviridis*, which lives in large troops and has a well-developed vocal repertoire (Gautier, 1985 and unpublished data) or of *C. aethiops* (see Fedigan & Fedigan, Chapter 20). Furthermore, the semi-terrestrial *lhoesti* superspecies does not appear to use olfactory markings. Thus, no consistent socio-ecological correlates can be found among species which display olfactory markings.

Phylogenetic correlates

The results given in this book, deriving from anatomical, karyological and paleontological approaches, suggest that the semi-terrestrial guenons belong to a more ancient radiation than the arboreal forms. It may therefore be suggested that those guenons which are closest to the ancestral stock may have retained the anatomical and behavioural potential to communicate by olfactory signals, a pattern generally considered as primitive. These signals should be used more or less according to the habitat and to specific socio-ecological features. They can be considered as almost vestigial, and only expressed in extreme situations, in *C. aethiops*, and should be better developed in species with small ranges such as *C. neglectus* (and possibly *C. hamlyni*). The total absence of such chemical signalling in arboreal guenons makes a further argument in favour of their more recent evolution.

Part III

Ecology and social behaviour

III.15

The diet and dietary habits of forest guenons

ANNIE GAUTIER-HION

In reviewing the literature on forest guenons, it is somewhat disappointing to notice the small number of species that have been studied so far. As concerns their diets, detailed data relate only to four different sites: the Kibale Forest, Uganda, the Kakamega Forest, Kenya, the Makokou Forest, Gabon and the Taï Forest, Ivory Coast. *Cercopithecus mitis* and *C. ascanius* diets were studied both at Kibale (Rudran, 1978a; Struhsaker, 1978) and at Kakamega (Cords, 1987b); at Taï, Galat & Galat-Luong (1985) worked on a community including *C. diana*, *C. campbelli* and *C. petaurista*, while at Makokou the studies concerned *C. talapoin*, *C. neglectus*, *C. nictitans*, *C. pogonias* and *C. cephus* (Gautier-Hion, 1971b, 1978, 1980; Gautier-Hion & Gautier, 1978).

For some other species, our information is limited to chance observations of plant species, plant parts or prey eaten by monkeys and/or to short observation periods (Haddow, 1952; Jones, 1966; Bourlière, Hunkeler & Bertrand, 1970; Gartlan & Struhsaker, 1972; Schlichte, 1978; Gautier, 1985; Oates, 1985). Our knowledge is therefore scanty, particularly for the semi-terrestrial guenons, though it is most important to know what extent these monkeys collect food on the ground, and whether their semi-terrestrial habits are related to dietary adaptations or not. In this review I will refer almost exclusively to the four communities listed above: to avoid repetitions, authors will not be repeatedly acknowledged when there is no ambiguity in their identity.

In addition to an obvious bias in favour of arboreal species, the difficulty of identifying common dietary trends within species or among guenons is magnified by the fact that most species have been studied at only one study site (*C. mitis* and *C. ascanius* are the only exceptions) and, furthermore, that data frequently concern a single troop at each location. Moreover, although all the species 'normally' present in these regions are represented in the communities studied, poaching and/or selective habitat destruction may have changed their absolute and respective densities.

On the other hand, even when diets have been well analysed, data about feeding strategies are extremely scarce. This mostly stems from the paucity of information on the food available – a kind of work which requires large research teams because of the high diversity of the flora and the complexity of phenological patterns in tropical forests – and from our ignorance of the chemical composition of tropical fruit in contrast with our increasing knowledge on leaf material (see Waterman, 1984).

The last difficulty in a comparative review of guenon diets is a methodological one. Several techniques have been used to quantify diets: the analysis of stomach contents and the direct observation of foods taken and measurement of time spent feeding on different food items. Hladik (1977a) showed that different methods give different results but failed to find coefficients of correction that could be applied to different species. Thus I did not attempt to use such coefficients and the extent to which the comparison of results obtained at different sites is justified may be questioned.

Morphological and physiological constraints on food choice

The analysis of morphological and physiological traits able to set constraints on the capacity animals have to localize, acquire and assimilate food has been recently reviewed (e.g. Rodman & Cant, 1984; Chivers, Wood & Bilsborough, 1984; Jungers, 1985). Among these traits, body size, food processing apparatus (dentition and gut morphology) and locomotory patterns, are the most important.

The interest in studying their relationships with foraging strategies is to establish predictive correlations with gross dietary characteristics of both extinct and living species (Kay & Covert, 1984; Fleagle & Mittermier, 1980). Furthermore, an interspecific comparative approach to these problems may generate hypotheses on the evolution of adaptive traits linked with food acquisition, once characteristics due to 'phylogenetic inertia' are identified (see Waser, 1987). Such a perspective has to avoid spurious correlations where allometric relations are not considered (Martin, 1984 and Chapter 10).

Body size

Energy needs are negatively correlated with body weight with a coefficient (0.75) close to that defined by Kleiber (1961). The same is

Table 15.1. *Body weights, canine size and their sexual dimorphisms (given by the ratio male/female calculated from linear dimensions) for arboreal species (above) and semi-terrestrial ones (below)*

Species (number)	Body weight (g) Male	Female	Canine size (cm) Male	Female	Sexual dimorphism Weight	Canine
C. talapoin[a] (9)	1380	1120	1.06	0.65	1.07	1.63
C. cephus[a] (9)	4100	2900	1.45	0.75	1.12	1.99
C. pogonias[a] (7)	4500	3030	1.47	0.84	1.14	1.75
C. nictitans[a] (17)	6600	4220	1.78	0.85	1.16	2.09
C. nigroviridis[b] (6)	5950	3700	1.80	0.80	1.17	2.25
C. neglectus[a] (8)	7000	4000	2.00	0.75	1.21	2.67
C. lhoesti[b] (4)	7500	4300	2.20	0.90	1.20	2.44

[a]Body weights recorded from wild specimens of the same community (Gautier-Hion & Gautier, 1985).
[b]*C. nigroviridis*: wild and captive specimens; *C. lh. lhoesti* captive specimens (J.-P. Gautier, personal communication).

true for the weight of ingested food (Martin, 1984; Martin *et al.*, 1985). *C. talapoin* excluded, the mean body weight of guenons ranges from 3 to 7.5 kg (Table 15.1) and there exists a broad interspecific overlap, due to sexual dimorphism in weight. The latter is all the more important the heavier a species, and is enhanced in semi-terrestrial guenons (Gautier-Hion & Gautier, 1985). In one community, the adult male of a given species may have a body weight more similar to that of an adult of either sex of another species than to that of his own adult female (Table 15.1).

Food processing apparatus

As for energy requirements, the surface of gut compartments increases in an allometric way with body size (Martin *et al.*, 1985). However, Chivers & Hladik (1980) were able to differentiate three gross dietary categories, on the basis of the coefficients of gut differentiation: faunivorous, frugivorous and folivorous monkeys. The four guenon species envisaged in their study display typical frugivorous traits, with a greater differentiation of the stomach and large intestine than that found in faunivorous species; as for the smallest relative surface of small intestine, found in talapoin monkeys, it probably results from their small body size.

However, many points remain to be clarified. For example, Hill (1966) found that the stomach of the swamp monkey (*C. nigroviridis*) differs from that of other guenons in its 'incipient' sacculation, which recalls that of colobines. Taking into account the ancestral position of this species, this trait could well be considered as an ancestral rather than a derived dietary adaptation.

Tooth morphology hardly differentiates guenon species as they are all characterized by large incisors and bilophodont molars (Hylander, 1975). Here again, Hill (1966) has drawn attention to the upper cheek of the swamp monkey, which differs radically from those of other guenons and brings it closer to the genera *Cercocebus* and *Papio*, while the relatively small size of their molar lophs brings *C. talapoin* and *C. nigroviridis* together. To what extent these differences reflect phylogeny or dietary adaptations remains to be established. However, it is worth noting that Martin & MacLarnon (Chapter 10) found that dental measurements of guenons indicate more phylogenetic affinities than dietary habits as suggested by Kay (1978).

It must be added that the sexual dimorphism of canine size also increases with body size and does so more in semi-terrestrial forest species than in arboreal ones (Gautier-Hion & Gautier, 1985). Such a trend seems better interpreted as a defensive adaptation rather than as an adaptation to a peculiar diet.

Motor patterns and positional behaviour

The physical accessibility of food items depends largely upon the locomotor abilities and the animals' means of prehension (Andrews & Aiello, 1984). All guenons are 'above-branch feeders' (Rose, 1973) characterized by a relatively long trunk and relatively high brachial index. Verheyen (1962) has pointed out how difficult it is to differentiate guenon species on the basis of their skeletal measurements, especially the arboreal forms. Detailed analyses conducted on five sympatric guenons have shown to what extent differences in skeletal material are small and that most of the traits considered as characteristics of a terrestrial tendency in baboons, such as the increased size of the forelimbs, were not 'diagnostic features of habitat in semi-terrestrial guenons' (Rollinson, 1975). However, like terrestrial papionines, *C. neglectus* and *C. talapoin* used mostly walking and galloping gaits while trotting characterized mainly arboreal *Cercopithecus*. Furthermore, like the monkey species which collect food on the ground, *C. neglectus* was the only guenon endowed with 'refined prehension'. On the other hand, all arboreal species were characterized by 'cupping prehension', which was particularly well developed in *C. pogonias* and *C. talapoin* and might enhance their ability to capture mobile prey.

The studies of Kingdon (Chapter 11) also show that subtle indication of species–substrate relationships (ground, trees, small or large branches) can be evidenced by comparing the morphology of their hands and feet.

Conclusion

To sum up, it is clear that the forest guenons are very homogeneous as far as their energy requirements, food-processing methods and locomotor abilities are concerned; the biological constraints on foraging behaviour are probably very similar for the different species. However, so few species have been adequately studied that it would be premature to draw any general conclusion on the meaning of interspecific differences, if any (see Fleagle & Mittermier, 1980), all the more so since semi-terrestrial forms have been so poorly studied.

Dietary characteristics

Quantitative features of the diets

The relative contribution to diet of the major food categories (average yearly figures) is given for eleven species, in relation to their mean body weight (Figure 15.1).

Fruit and seeds

Their mean relative abundance varies from 43 to 85% without any obvious relation to body weight. Moreover, great differences occur within the same species or between equivalent species, according to habitat (Figure 15.1a). The reduced fruit intake of talapoins is due to the fact that most of the published data come from 'commensal' troops, which consume a number of manihot tubers stolen at soaking sites; while, when only 'wild' troops are concerned, the mean fruit intake reaches about 60% (personal observations).

Semi-terrestrial guenons (*C. neglectus* and *C. nigroviridis*) eat a lot of fruit in riparian forests, this habitat being potentially as rich in fruit as *terra firma* (see Emmons, Gautier-Hion & Dubost, 1983).

Leaves and other plant material

Leaf consumption generally increases with increasing body weight (Figure 15.1c). Whatever the habitat, the same folivorous tendency characterizes species belonging to a same superspecies whatever may be their rate of consumption, very low as in the *mona*

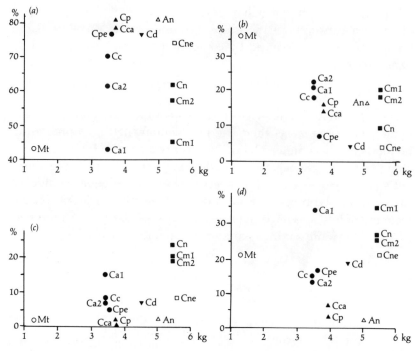

Figure 15.1. Relative abundance of the main food categories ingested by different guenon species (according to their mean body weight: adult male weight + adult female weight/2. (*a*) fruit-seeds; (*b*) animal matter; (*c*) foliar material; (*d*) all plant food except fruit. Mt = *C. talapoin*; Cc = *C. cephus*; Cp = *C. pogonias*; Cn = *C. nictitans*; Cne = *C. neglectus* (Makokou); Ca1 = *C. ascanius*; Cm1 = *C. mitis* (Kibale); Ca2 = *C. ascanius*; Cm2 = *C. mitis* (Kakamega); Cd = *C. diana*; Cca = *C. campbelli*; Cpe = *C. petaurista* (Taï Forest); An = *C. nigroviridis* (Zaïre).

superspecies (<5%), moderate as in the *cephus* superspecies (5–10%) or high as in the *nictitans* superspecies (about 20%). As for *C. diana*, which has no equivalent species in other guenon communities, its folivorous diet is in accordance with body weight, as for talapoins which seldom eat leaves.

Correlations with body weight are lowest when all plant food (fruit excluded) is taken into consideration, due to an additional rate of plant consumption by guenons of Kibale, and to the high rate of manihot tubers taken by talapoins (Figure 15.1d).

Animal matter

Percentages of prey consumption range from less than 5% (*C. diana* and *C. neglectus*) to 35% in *C. talapoin*. This tiny monkey excluded,

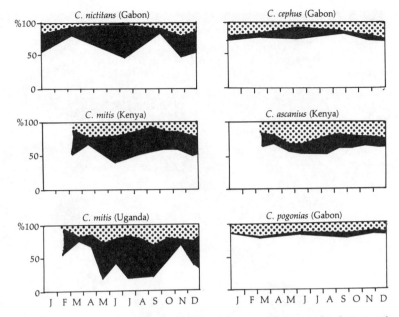

Figure 15.2. Seasonal variations of the three main food categories ingested by five guenon species in three habitats. Fruit and seeds (white); other plant material (black); animal matter (dotted). Adapted from Rudran (1978a), Uganda, Gautier-Hion (1980) Gabon, and Cords (1987b) Kenya.

no overall correlation is found between prey intake and body size; however, within a given community, the heavier a species, the less insectivorous it is (Fig. 15.1b).

Seasonal variations

The amplitude of seasonal variations of gross food type intake varies with the species; it is larger in the heaviest *nictitans* superspecies, at all three study sites (Figure 15.2). Furthermore, guenons display everywhere a bimodal pattern of seasonal consumption of fruit. Monthly percentage abundance of fruit may be as low as 10% (as is the case for *C. mitis* at Kibale) or reach up to 90% (in *nictitans* at Makokou, Figure 15.3). Variations are smaller in the two *cephus* members and are particularly small in *C. pogonias* (range between 76 and 88%).

The decrease in fruit consumption is compensated for by an increased consumption of other plant food in the heaviest species *nictitans* and *mitis*; the same is found for *C. ascanius* at Kibale, while *C. ascanius* at Kakamega, like *C. cephus* at Makokou, compensate by increasing their intake of leaves and/or prey. At the same time, *C. pogonias* monkeys increase their intake of animal matter.

Figure 15.3. On the left, *C. nictitans* feeding in a *Drypetes* sp. tree overloaded with fruit; on the right, the same animal with its cheek-pouches filled up.

Sex-related differences

The four species for which sex differences in diet have been investigated show the same trends: adult females eat less fruit, more leaves and more insects than adult males (Figure 15.4). This trend is even more marked in young individuals, as shown by Rudran (1978a) in *C. mitis* and by Quris *et al.* (1981) in *C. cephus*. Juvenile *C. mitis* eat 1.7 times less fruit than adult males and five times more insects.

However, results from the Makokou community have shown that sex-related differences vary depending on species: they are most obvious in *C. nictitans*, and far less so in *C. pogonias*. These sex-related differences underline the difficulty of defining species-specific dietary characteristics (Gautier-Hion, 1980).

Diurnal variations

The average intake of different food types varies according to the time of the day, following a similar pattern at all the study sites. Fruit and seeds are mainly eaten in early morning and late afternoon, while prey are most frequently taken between 0900 and 1500 h

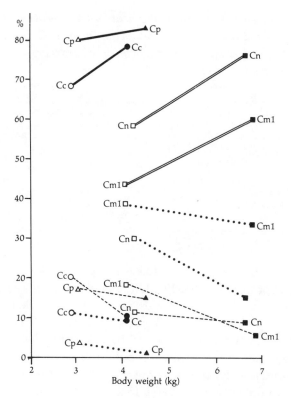

Figure 15.4. Relative abundance of the main food categories ingested by males and females of four guenon species, according to body weight. Females are indicated by open symbols. Fruit-seeds: black line; other plant material: dotted line; animal matter: pecked line. Symbols as in Figure 15.1 (from Rudran, 1978a, for Cm1 and from Gautier-Hion, 1980, for Cn, Cc, and Cp).

(Gautier-Hion, Gautier & Quris, 1981; Sourd, 1983, for *C. cephus*; Cords, 1987b, for *C. mitis* and *C. ascanius*).

Qualitative features of the diets

Plant material

A variety of plant parts is included in the guenons' diets: fruit, young and mature leaves, leaf and flower buds, blossoms, stems, bark, etc. and they belong to many species. A grand total of about one hundred plant species is frequent (*C. cephus*, 107 – Gautier-Hion, 1980; Sourd & Gautier-Hion, 1986; *C. ascanius*, 101, *C. mitis*, 104 – Cords, 1987b). These lists can be made longer by increasing observation time

Table 15.2. *Contribution of top species-specific plant food species to diets of different guenon species and diversity of plant food species used (for* C. cephus, *only one year-round study has been considered in order to have comparable methods to those used in other sites: see text)*

Sites	Kibale		Kakamega		Makokou
Species	*C. mitis*	*C. ascanius*	*C. mitis*	*C. ascanius*	*C. cephus*
Number of plant food species	59	80	98	104	57
% diet in top 5 species	34.7	36.5			42
% diet in top 10 species	65		55	55	63.8
% diet in top 20 species	87		79	82	81.4
H (diversity index)	3.20	3.33	3.55	3.50	3.62

(Rudran, 1978a) but, on the whole, monkeys sample new food items only infrequently. For instance, the 17 additional plant species added by Sourd & Gautier-Hion (1986) after one extra year of field work, to the previously given list of plant items taken by *C. cephus* at Makokou (Gautier-Hion, 1980) accounted only for less than 7% of the diet.

All studies show that the relative contribution of plant species to monkeys' diets varies greatly, and that a few plant species make up the staple diet of each monkey species (Table 15.2); as a consequence the diversity index of plant species consumed yearly is low, and very similar at different sites (range: 3.20–3.62); in all species studied, about 80% of the diet is acquired from about 30 plant species.

Occasional food items have also been mentioned. Gums are taken by *C. ascanius* and *C. nictitans*; mushrooms are frequently eaten by the semi-terrestrial *C. neglectus*.

Animal prey

It is difficult to identify prey taken by arboreal monkeys by direct observation and most observers rely on indirect cues, for instance, the plant substrates on which prey are caught and the motor patterns used to approach and catch them. Insects make up the bulk of the animals taken, while predation upon vertebrates is quite rare, including galagos in *mitis*, eggs and fledglings in *cephus*, eggs and freshwater shrimps in *talapoin* and reptiles in *neglectus* (in Butynski, 1982a). All the observations made on *C. mitis* and *C. ascanius* (Rudran, 1978a; Struhsaker, 1978; Cords, 1987b) agree that the kind of substrates explored and the modes of capture used differ between species; *C. mitis*

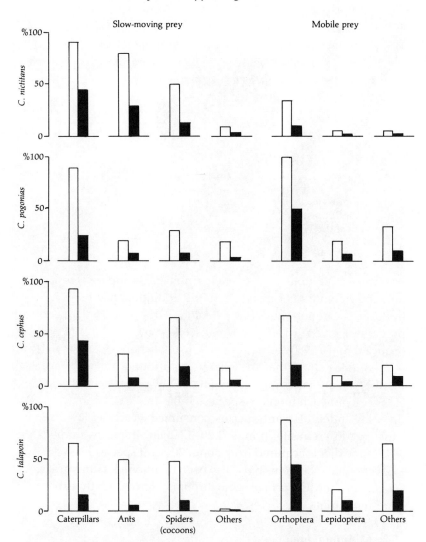

Figure 15.5. Percentage of presence (open bars) and relative abundance (black bars) of different prey types collected by four species of the Makokou community (Gautier-Hion, 1978, 1980).

appears to concentrate more on slow-moving or immobile prey than *C. ascanius*.

These data agree with the analyses of stomach contents (Figure 15.5). *C. nictitans*, like *C. neglectus*, mostly preys on caterpillars and ants, while *C. cephus* is more opportunist, eating both slow-moving caterpillars and fast grasshoppers; *C. pogonias* concentrates first on

Figure 15.6. Young *C. talapoin* eating a cricket (photo: A. R. Devez).

grasshoppers and other Orthopteroids. On the other hand, *C. talapoin* catches grasshoppers, caterpillars, moths and butterflies, beetles and spiders, concentrating mainly upon mobile prey (Figure 15.6). The well-developed 'cupping' prehension described for the last two monkeys (Rollinson, 1975) may well be related to catching fast-moving arthropods.

Sex-related differences

Differences have been found between the prey categories taken by male and female arboreal guenons at Makokou (Gautier-Hion, 1980). Generally, males prey on sedentary animals more than females do. These differences are significant in *C. cephus* and *C. pogonias*, but very small in the heaviest *C. nictitans*.

Interpopulation and intergroup differences

Differences between populations of the same species are documented only for *C. mitis* and *C. ascanius*. At Kibale, these two species are more folivorous and less frugivorous than at Kakamega (Rudran, 1978b; Struhsaker, 1978; Cords, 1987b). Comparing two troops of *C. mitis* living 500 m apart from each other, Rudran also found clear differences in the total number of food plants used by each troop, in their diet diversity and in their pattern and frequency of use of common plants and food items (mean overlap 45%). These differences have been accounted for by differences in tree species density and phenological patterns.

Conclusion

All the guenons so far studied are mostly frugivorous. Depending on their body size, they supplement their diet either with animal matter or leaves, or with a combination of both. The absence of relationships between fruit intake and body size is probably due to the small range of weights within the group, and is in agreement with the results of a study of the whole community of 66 mammalian primary consumers of Gabon, in which frugivorous rates of 50% were still found in species reaching 50 kg of body weight (Emmons, Gautier-Hion & Dubost, 1983). On the other hand, the same study pointed out that leaf consumption rapidly increased as soon as body weight reached 4 kg; such a trend is found in guenons (Figure 15.1) as in other primates (Clutton-Brock, 1977a). As to prey consumption, an inverse trend related to body weight is found within each community.

Despite methodological discrepancies, species-specific dietary characteristics are very similar in the study sites for the members of each superspecies: *mona, cephus* and *nictitans*, especially as far as leaf and prey consumption are concerned (*C. mitis* and *C. ascanius* from Kibale appear as exceptions). This homogeneity may point either to similar genetic constraints upon dietary adaptations and/or to a similarity of ecological constraints upon species food niche partitioning. The differences in fruit consumption within the same superspecies may be due to the fact that fruit is the most variable food resource. Here again, however, differences mainly involve the guenons of Kibale.

Age–sex related differences in diet have been observed in some primate species but there is no single rule; in some species, males are more frugivorous than females and the reverse is true in others (see Clutton-Brock, 1977a). In guenons, most of these differences can be explained by body weight and related nutritional requirements as differences are more apparent in the heaviest and the most sex-dimorphic species. However, the variability in extent of such differences between species of similar body weight (for example *C. pogonias* and *C. cephus*) is probably induced by other factors such as competition.

Similarly, diurnal variations in food choice have been related to energy requirements, energy-rich food like fruit being preferentially sampled in the morning to restore energy balance (Clutton-Brock, 1977a): this does not, however, explain the second peak of fruit intake observed in the afternoon. This reasoning might possibly be reversed, fruit being taken when insects are not: in tropical forests, poor light inhibits both the activity of many insects and an efficient prey search; the latter would be maximized during the sunshine hours.

Food partitioning

Mean annual overlap between specific diets

A great overlap in resource use is obvious among species belonging to the same community when one compares the relative abundance of gross dietary categories in their diets. This can be made more meaningful by qualitative analysis.

Plant material

At Makokou, the three arboreal guenons (*C. nictitans, cephus* and *pogonias*) and the semi-terrestrial *C. neglectus* share 75% of the fruit species they eat (Figure 15.7). Moreover, the few plant species which appear to be species-specific are only occasionally exploited, while the most frequently taken fruits are usually shared by all monkeys (and are also the most popular plant foods for many other fruit-eating mammals, Gautier-Hion *et al.*, 1985a). Thus the food plant most commonly eaten by *C. cephus*, which accounts for 12.5% of its annual feeding scores at Makokou, was found in 52% of *C. cephus* stomachs, 62% of those of *C. pogonias* and 38% of those of *C. nictitans*. As a matter of fact, the top ten species eaten by *C. cephus* are also the most popular foods for the two other arboreal sympatric guenons (Gautier-Hion, 1980; Sourd & Gautier-Hion, 1986).

Figure 15.7. The semi-terrestrial de Brazza monkeys feed in trees. Here an adult male climbs down from an emergent after feeding on fruits of an Apocynaceae liana of the tree canopy (radio-tag can be seen on his back).

An equivalent trend was found for *C. mitis* and *C. ascanius* at Kakamega (Cords, 1987b), with 70.5% of species-specific food items (including substrates used for prey captures) shared yearly by both species of monkeys, and with five to eight plant species among the ten most frequently sampled shared each month. A different situation was observed at Kibale, where the annual overlap in plant food items between *C. mitis* and *C. ascanius* was less than 34%, whereas there exists a smaller overlap for the most important foods (Struhsaker, 1978).

Animal matter

Despite the differences shown above, the overlap index of insectivorous diets was high between *C. nictitans* and *C. cephus* (0.88), *C. cephus* and *C. pogonias* (0.78) and less between *C. nictitans* and *C. pogonias* (0.55). In parallel, Cords (1987b) found that differences between *C. mitis* and *C. ascanius* were small, with most prey being caught on leaves (respectively 71 and 68%) and on trunks (19 and 20%). Moreover, modes of capture did not differ much in speed of movements, with respectively 19.8 and 20.5% for slow ones, 72.9 and 68.5% for medium speed and 7.3 and 11% for fast ones. At Kibale, percentage overlap for microhabitats exploited by *C. mitis* and *C. ascanius* was lower (23.6%, Struhsaker, 1978), and more significant differences occur for use of substrates and movements, *C. ascanius* moving quickly in 21.3% of the cases and *C. mitis* in 6.5%.

Seasonal variations of interspecific diets overlap

Both Struhsaker (1978) and Cords (1987b) found that overlap in plant species use increased as the proportion of fruit in the monkeys' diets grew larger. Similarly, at Makokou, diets of the three arboreal guenons were very similar during the two periods of the year when the total number of fruiting species available was high (Gautier-Hion, 1980). In this area, diets changed on two occasions (Figure 15.8): (1) in the dry season when fruit availability decreased: then *C. nictitans* and, to a lesser extent, *C. cephus* switched to leaves, leading to a decrease in interspecific overlap; (2) around November, despite the fact that no significant decrease in fruit availability did occur; at this time of the year, male diets remained relatively similar for all species, while females switched to young leaves and/or insects. This decrease in inter-sex overlap was explained as resulting from the greater protein requirements of pregnant females.

The patterns described are indeed very complex, interspecific differences being, in some communities, inextricably mingled with intersexual ones. As a matter of fact, at Makokou, it was found that the diet of *cephus* adult males varied seasonally in much the same way as that of *nictitans* of both sexes, while the diet of *cephus* females remained more or less the same throughout the year, as did that of *pogonias* of both sexes.

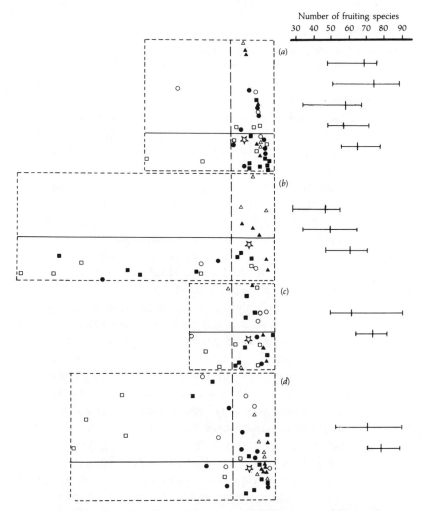

Figure 15.8. Factorial analysis of correlations summing up the seasonal trophic patterns of the guenon community of Makokou, on the left, (from Gautier-Hion, 1980), and the monthly variations of the number of fruiting species (mean and interannual variations for 5 years, (from Gautier-Hion *et al.*, 1985b)), on the right. (*a*) January–May; (*b*) June–August; (*c*) September–October; (*d*) November–December. On the horizontal axis, leaves are the dispersion factor; on the vertical one, animal matter (respectively 46% and 18% of the total variance). Fruit, indicated by a star, contribute little to the dispersion. Black symbols = males; open symbols = females. Squares = *C. nictitans*; circles = *C. cephus*; triangles = *C. pogonias*.

Conclusion

Dietary overlap in guenons is exceptionally high and raises the problem of interspecific competition. The evidence for competition and its role in structuring primate communities has recently been reviewed by Waser (1987). Some arguments play in favour of such a role in guenons:

>   the fact that at all sites, diets tend to differ the most when fruit production is lowest may indicate that fruit is a limiting resource and that competition for fruit is the prime factor leading to dietary shifts;

>   the fact that *mitis* and *ascanius* are more frugivorous at Kakamega than at Kibale, may result from a 'competitive release', due to the absence of *Lophocebus albigena* and other fruit-eaters at the former site (Cords, 1987b);

>   the fact that both population densities and folivorous tendencies are higher at Kibale than at Makokou, may result from a higher competitive pressure at the most densely populated site, where fruit might be temporarily in short supply;

>   the fact that at Makokou, the greatest dietary differences occur between *nictitans* and *pogonias* may offset their quasi-total overlap in foraging heights (Gautier-Hion, Quris & Gautier, 1983). However, the differences in body weight of the two species account also for dietary differences. In the same way, the close similarity between the diets of *cephus* females and *pogonias* of both sexes may result from their very similar body weights or from their shift in foraging heights, *cephus* entering the lowest forest strata (Gautier-Hion *et al.*, 1981);

>   the fact that in polyspecific troops, most interspecific aggressive interactions occur in feeding situations (Struhsaker, 1978; Cords, 1987b) may indicate competition for food. As pointed out by Waser (1987), however, such aggressive interactions in large fruiting trees do not appear to exclude any visiting monkey.

However, there is no reason for guenons to compete only for fruits; animal prey might well be as important, if not more, for niche separation, as suggested by Struhsaker (1978). Unfortunately, we have no information on the kinds and numbers of invertebrate prey available at different times of the year, in the different forest layers. However, it is worth noticing here that inter-individual foraging distances were found to be significantly greater when guenons were foraging for insects than when they were searching for fruit (Quris *et al.*, 1981).

As already stressed (Gautier-Hion & Gautier, 1979), the regular spacing of body weights and sizes of feeding apparatus, which has often been described in other communities and is supposed to contribute to niche separation, is not observed in guenon communities. Furthermore, the differences in weight are often larger between males and females of the same species than between different species, and the same can also occur for dietary overlap in some species pairs. Thus in this case, competition between the sexes may have played a more important role than interspecific competition. In addition, it has been shown elsewhere (Gautier-Hion *et al.*, 1983 and Chapter 23) that polyspecific associations tend to increase the overlap of foraging heights, ranging patterns and diets of participating species, rather than making them more dissimilar. Waser's viewpoint (1987), which states that the niches of guenons are not regularly spaced along important niche dimensions but appear as 'nested niches', probably reflects the actual situation better. However, this view does not generate clear hypotheses on the evolution of such ecological partitioning.

Food selection

Evidence for food choice

At all study sites, the consumption of different plant foods by guenons varies a great deal. To assess whether this variability reflects a selective feeding, it is necessary to establish that the differential use of food does not result from differences in availability and abundance of the various kinds of food items. Among the ten species which made up the major food items of *C. mitis* at Kibale, Rudran (1978a) found that the frequency of use of five of them was related to their production, while the five remaining species were eaten more than could be expected from their availability. At Makokou, out of the 200 fruit species collected twice a week for a year, along a forest transect, only 25% were sampled by a *C. cephus* troop. A significant correlation was found between the percentage abundance of fruiting species and the percentage of sampling by monkeys, indicating that the density of fruiting trees of different species was one of the factors influencing their consumption. However, among the top twenty *cephus* plant food species, only three were included in the list of the 20 most abundant species at this site (Sourd & Gautier-Hion, 1986).

Furthermore, for both *C. mitis* and *C. cephus*, it has been shown that while the sampling of some fruit species by monkeys did correlate with their fruiting patterns, this was not the case for all species, some being collected more, and others much less, than expected from their availability. Therefore, besides food density and availability, food preferences must also be taken into consideration as determinants of guenon diets.

Food choice criteria
Two categories of cues may influence an animal's decision to eat or reject a given food type: immediate chemical and physical signals, and nutritional criteria (Richard, 1985).

External criteria
The role of colour in guiding food choice in monkeys has recently been investigated in South America (Janson, 1983), southern Africa (Knight & Siegfried, 1983) and Gabon (Gautier-Hion *et al.*, 1985a). The last study examined the manner in which morphological characters of 122 fruit species were involved in the choice of 39 vertebrate fruit-eaters (including six monkeys). It was shown that consumer groups were first of all influenced by the weight of the fruit (on the first axis: 52% of the total inertia, Figure 15.9), and then by its colour and flesh characteristics (on the second axis: 21% of the total inertia). Guenons were found to select significantly the most brightly coloured fruits (orange, red, yellow, $p < 0.001$), the succulent fleshy fruits, the dehiscent fruits with arillate seeds ($p < 0.001$), and, less obviously, the fruits weighing between 5 and 50 g ($p < 0.05$). Furthermore, it was found that these external fruit characters were interrelated; this was especially the case for the kind of flesh which is not independent of fruit colour: very juicy fruit and arillate fruit are brightly-coloured, while fruits with a dry pulp as well as dehiscent fruits with no arillate seeds tend to be dull-coloured. Consequently, by relying only on colour, monkeys may be able to determine fruit-flesh as well. Among the fruit species that a *C. cephus* troop sampled more than expected from their availability (on an annual basis), it was shown that 67% were orange or yellow, and 86% possessed a succulent pulp; while 70% of species sampled less than expected were green or brown and 40% had a dry pulp (Sourd & Gautier-Hion, 1986).

Nutritional criteria
There is some experimental evidence that monkeys, like rats, are able to control the nature and quantity of nutrients ingested (see in Hladik, 1981). Dietary shifts towards protein rich pellets have been induced in captive talapoin monkeys by removing arthropods from

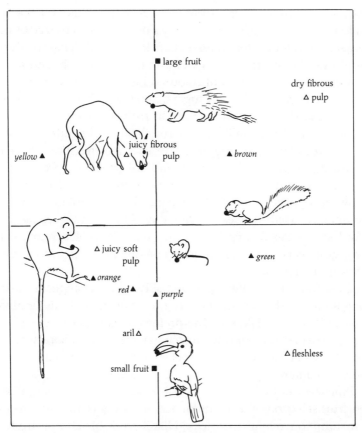

Figure 15.9. Multifactorial analysis showing the choice patterns of four consumer groups (large canopy birds; small and large rodents; squirrels, ruminants and monkeys; a total of 39 species from the Makokou community), according to 7 fruit characters (colour, protective coat of flesh, type of edible flesh, seed protection and number, fruit and seed weight). Fruit weight and kind of flesh contribute most to the vertical axis (52% of the total inertia); colour and type of flesh contribute most to the second axis (21% of the total inertia). Monkeys select fruit by colour ($p < 0.001$), kind of flesh ($p < 0.001$) and fruit and seed weight ($p < 0.05$). (From Gautier-Hion *et al.*, 1985a).

their daily diet (unpublished data). Indirect evidence from the field is plentiful, but most studies concern leaf-eaters, and mainly New World species. Their results tend to indicate that foods rich in proteins and with a low fibre content are selected first (e.g. Milton, 1979; Glander, 1981). However, recent studies have also shown a correlation between selectivity and overall digestibility of the food (Waterman, 1984).

Nutrient content of 23 plant species eaten by *C. cephus* (i.e. more than 75% of those sampled by monkeys on an annual basis) were analysed

(Sourd & Gautier-Hion, 1986). It was found that: (1) the pulp of indehiscent brightly coloured fleshy fruit was characterized by a higher sugar and water content than that of all other fruit types; (2) the arils of brightly-coloured fruit possessed the highest protein and fatty acids content; (3) the pulp of dull-coloured indehiscent fleshy fruit had the lowest content in water and sugars, and the highest content in fibres.

Consequently, by selecting bright fruit, monkeys are informed to some extent of the quality of the fruit flesh and its nutrient value (due to the correlation among these three characteristics). By eating a mixture of succulent fleshy fruit and arils, they are provided in turn with a large amount of sugars, proteins and lipids. This may explain why, in order to attain such a balanced diet, monkeys can discard a commonly used species when other fruit types become available (Rudran, 1978a; Gautier-Hion *et al.*, 1981). This also accounts for the change in diet of female guenons when they are pregnant and therefore in need of more protein.

Therefore food preferences probably reflect changes in nutritional requirements. However, the latter are not the only determinants of selectivity as, at Makokou, the preferred fruit species of *cephus* monkeys (up to 37% of monthly samplings) was mostly characterized by its acidity, while the top eighth species was characterized by its high K and Mg content.

Practically nothing is known of the negative role of chemical defences on fruit selection and the influence of alkaloids and tannins on fruit consumption remains unclear (Hladik, 1977b). Unripe fruit, which contains more tannins than ripe ones (Waterman, 1984), seem to be infrequently eaten by guenons. Taking into account the major role played by guenons in seed dispersal, and the fact that plants have apparently developed bright colours and succulent pulp to attract them, one can expect chemical defence to concern mainly seeds; clearly, guenons are not seed-eaters, and when the seeds are swallowed, they are very seldom crushed (Gautier-Hion, 1984).

Conclusion

As far as we know, forest guenons appear to be selective feeders whose staple diet, at least its plant component, is made up of a relatively small number of species among all those available. This selectivity could ensue from the fact that, for most of the year, fruit constitutes an abundant resource.

Guenons' choices rely partly upon both external and nutritional criteria. It remains, however, to be established whether the Makokou results can be generalized to include all guenons and to the whole of Africa, and, especially, whether, depending on the fruit spectrum available, monkeys at other sites select fruit by using the same cues. It is, however, interesting to note here that both Janson (1983) and Terborgh (1983) reported that the five primate species they studied in Peru also tended to select yellow, orange and red fruits.

The situation found at Kibale, where guenons are significantly more folivorous, appears as an exception. McKey (1978) found that on the poor soils of Douala reserve, where leaf material has a high rate of toxic secondary compounds and a poor digestibility, the black colobus (*Colobus satanas*) was mainly a seed-eater. Conversely, the folivorous tendency of the Kibale guenons might be a response to the low level of chemical defence of leaves in this site (Gartlan *et al.*, 1980) and to their higher overall digestibility (Waterman & Choo, 1981). However, a recent study in the La Lopé Reserve (Gabon) has shown that, despite the fact that the overall toxicity of plant material is lower than in Douala Reserve, the black colobus remained a seed-eater, as in Cameroon (Harrison & Hladik, 1986). In addition, *Colobus guereza* in northeast Gabon was found to ingest a large amount of fruit (Gautier-Hion, 1983) as is the case for the lowland gorilla (Tutin & Fernandez, 1985). Thus the breadth of the frugivorous niche in West Africa, as compared with East Africa, remains an open question (Gautier-Hion, 1983) until comparative measurements of plant food availability can be made.

Feeding and ranging

Almost all locomotor activities of monkeys are related to their food requirements (Fleagle, 1985). This explains why correlations between ranging behaviour and metabolic needs have been so frequently looked for. However, no general trend has yet been convincingly established except, to some extent, when considering gross categories such as folivorous vs. frugivorous monkeys (Hladik, 1981). Reviewing the problem, Martin *et al.* (1985) found that most authors agreed that 'supplying areas' increased in excess of what can be predicted from the metabolic hypothesis; no satisfying explanation has yet been proposed to explain this pattern. Results concerning related changes in feeding and ranging behaviours are also confusing; in some species, home range sizes and the day range lengths increase with lower food availability, whereas the reverse is found in others (Clutton-Brock, 1977b).

In fact the energy expenditure is not necessarily related to the mean surface area of home range (Struhsaker, 1978) and the concept of home range tells us almost nothing about the tactics of food acquisition: daily distance travelled, surface areas visited daily, ranging diversity, etc. The data provided by Struhsaker (1978) are particularly significant: while the mean annual surface area covered by a *C. mitis* troop at Kibale was about 60 ha, and that of a *C. ascanius* troop of 24 ha, the latter troop travelled slightly more than the former each day (1450 m vs. 1300 m), and both species visited approximately the same surface area daily (6.75 ha vs. 7 ha). In fact, available data on ranging behaviour in guenons make evident that the various parts of the yearly home range are differently used, just as plant food species may be sampled differently: usually, only a small part of the home range is regularly visited. Thus at Kakamega, redtails and blues spent 77% and 83% of their time respectively in 55% of their annual home ranges (Cords, 1987b) while, at Makokou, a *C. cephus* troop spent 50% of its time on 20% of its annual home range. For the latter species, seasonal changes in these percentages were correlated neither with overall fruit availability nor with the number of fruit species eaten (Sourd, 1983; Figure 15.10).

However, better correlations between feeding and ranging are found when one considers shorter time periods. For example, Rudran (1978a) found a direct relationship between the abundance of foods within given quadrats and their frequency of use by *C. mitis*, and a less diversified ranging when food was more concentrated, with fruiting sources having the main effect. In the same way, ranging diversity increased with plant food diversity for *C. ascanius* (Struhsaker, 1978).

The relative influence of fruit availability and of forest structure on the ranging behaviours of a *C. cephus* troop was tested in Gabon (Gautier-Hion *et al.*, 1981). It was shown that (1) there was a different use of the various forest types described, not clearly related to their fruit-feeding value; (2) despite significant differences both in the number of fruiting trees and in the number of fruiting species between the dry and the rainy season, the overall ranging patterns of the troop did not differ (with 90% of time spent in 44% of the home range in both seasons), while the number of hectares visited daily remained the same (10.9 ± 2.6 in the dry season vs. 11.0 ± 1.8 in the rainy season); however, a more precise analysis showed that the tactics used for locating fruit differed according to the season. During the dry season, when fruiting species were scarce, feeding tactics relied on the densest fruiting species, at the expense of dietary diversity. During the rains, when the diversity of the fruiting species was twice as great, monkeys were more selective and concentrated upon areas which contained several highly attractive fruit species.

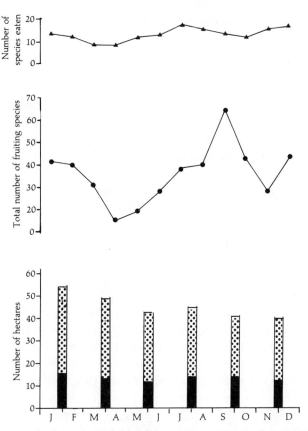

Figure 15.10. Home range size, fruit availability and fruit diet for a C. *cephus* troop along one annual cycle. Number of hectares used is given for 50% of time (black bars) and 95% of time (black bars + dotted bars); a total of 1550 hours of observation). Number of fruiting species was measured on 6 km trail, every two weeks (from Sourd, 1983).

As a matter of fact, a simple model based upon the average values of fruit availability in the home range demonstrated that, visiting 11 ha every day gave the guenons a 90% chance of encountering the three most numerous species during the dry season, and of finding at least eight out of the 14 species available during the rains. In both cases, to encounter more species, monkeys would have to increase considerably their searching area. Thus, in such cases, 11 ha appeared to be the best compromise for both seasons, without expending too much energy on travel. In order to maximize their search for food, monkeys find solutions which are more clever than a mere increase of foraging area or a reduction of their level of activity (as can be expected from broad correlations between home range size, daily travels and food availability).

The guenons, at least the species so far studied, are therefore selective rangers, a characteristic that is clearly not unrelated to their being selective feeders. The way they manage their ranging patterns, alternately choosing differently-structured forest types, relying in turn upon the most abundant food species and on a few highly appetitive ones, suggests that they have an excellent knowledge of their home ranges, and are able, to some extent, to monitor the timing of fruit production, at least of their major plant foods.

Further detailed studies on guenons' ranging behaviours, together with reliable estimates of food availability, are badly needed. They are, I feel, as important as studies on social behaviour to gain some insight on the cognitive abilities of these species living in a highly diversified, complex, and changing environment.

Discussion

To place dietary characteristics of guenons into a phylogenetic perspective is presently quite impossible, as data are lacking for too many species, mainly on the semi-terrestrial ones. As far as we know, these semi-terrestrial monkeys do not appear to be less frugivorous than the arboreal forms, and they also collect most of their food in trees.

As a matter of fact, a relatively homogeneous dietary pattern is found among all forest guenons studied, fruit being their staple diet. The relative percentage of non-fruit material and of animal matter mostly depends on fruit availability, and also on body weight, the lightest species in a community, and the thinner of the two sexes within a species, consuming more 'insect' prey.

To what extent this 'average' overall dietary pattern is likely to change with changing habitats remains little known. Significant changes in diets have, however, been observed in commensal talapoin troops, with manihot tubers making up to 17% of the diet, a tradition that is widespread in West Africa and relatively recent, since manihot was only introduced in the nineteenth century.

This contrasts with the extraordinary flexibility and opportunism which characterize vervets, which are able to face both changing habitats and changing resource levels, a behaviour which greatly contributes to the success of the species. This is well illustrated by the opposite destinies of the successful *aethiops* introduced on St Kitts and Barbados Islands and of the *C. mona* introduced on Grenada, which quickly faded out (see Fedigan & Fedigan, Chapter 20).

This makes forest guenons very prone to extinction with the rapidly dwindling 'old' rain forests, some canopy species, such as *C. diana*, being even more threatened than *cephus* or *mona* forms, which inhabit lower strata and second growth forests, or even a few riparian semi-terrestrial species such as *C. neglectus*.

III.16

Guenon birth seasons and correlates with rainfall and food

THOMAS M. BUTYNSKI

Introduction

Discrete mating and birth seasons are being reported for an increasingly large number of primate species in a wide variety of habitats. Since the 1930s there has been considerable laboratory and field research on seasonal reproduction in primates, its adaptive value, and its proximate and ultimate causative factors (see reviews in Lancaster & Lee, 1965; Schapiro, 1985; Lindburg, 1987).

Seasonal reproduction is part of a species' reproductive strategy. In theory, natural selection favors individuals who time reproductive events to coincide with varying environmental conditions in such a way as to maximize reproductive success. Births, therefore, occur at the optimal time for the survival of mother and offspring. Where reproductive seasons and peaks occur, they are believed to be direct or indirect responses to critical, more or less predictable seasonal changes in the environment. In the tropics, where marked annual variations in daylength and temperature are absent, annual changes in rainfall and nutrition are generally thought to be most important in determining the timing of reproductive events (Sadleir, 1969; Bourlière & Hadley, 1970; Sinclair, 1978; Rowell & Richards, 1979; Nash, 1983; van Schaik & van Noordwijk, 1985). For at least some species, however, the timing of reproduction may represent a compromise among these and several other selective forces – for example, predation and climatic factors other than rainfall. Developmental and phylogenetic constraints may also be involved.

Given the large differences among primate life history strategies, phylogenies, and the ecological complexity within which each species lives, it is not surprising that correlates between rainfall, the availability of nutrients and the timing of reproductive events have not been recognized across the Order Primates. This is, however, not to say that such correlates might not exist at a lower taxonomic level, such as within a genus or among closely related genera where the life history characteristics of the species being compared are less variable and different (e.g. body size, growth rates, food habits, social organization, age at first reproduction, gestation and longevity).

This paper reviews the reproductive patterns of guenon monkeys. It uses a comparative approach to examine the relationship between rainfall and nutrition and the effects of rainfall and nutrition on the distribution and timing of guenon mating and birth seasons. In particular, it focuses on the timing of gestation, birth and weaning relative to seasonal changes in rainfall and the availability of nutrients. The primary questions addressed are: (1) When do guenons mate and bear young? (2) What similarities and differences exist among the annual reproductive cycles of guenon species living in the same and different areas? (3) What are the correlates between the timing and length of guenon mating and birth seasons and the timing, distribution and amount of annual rainfall? More specifically, can rainfall be used to predict when guenon mating and birth seasons occur? (4) What are the relationships between rainfall and nutrition, and between nutrition and guenon mating and birth seasons?

This chapter is mainly concerned with the occurrence and adaptive function of birth seasonality. The possible proximate cues in the physical and social environments which initiate and limit mating seasons in guenons have been little studied. What is known and hypothesized concerning such proximate cues has been reviewed several times elsewhere (Lancaster & Lee, 1965; Klein, 1978; Rowell & Richards, 1979; Schapiro, 1985; Lindburg, 1987) and, therefore, will not be re-examined here.

The guenons are a particularly appropriate group of primates on which to undertake comparisons of seasonal reproductive patterns and of the possible effects of rainfall and nutrition on these patterns. This is because: (1) The guenons include a large number of closely related species (18–26 species) (Lernould, Chapter 4). As such, comparisons of their reproductive patterns should provide clues to the more important selective forces imposed by the environment within which each species lives. (2) Guenons occur under a wide spectrum of rainfall conditions and, therefore, are found in an array of biotic environments. (3) Several species have been the focus of long-term studies in widely spaced study sites.

The guenons are restricted in their geographical distributions to Africa. This group is taken to include four genera: *Cercopithecus,*

Miopithecus, Allenopithecus and *Erythrocebus* or only one, *Cercopithecus,* including the three other taxa as subgenera (Lernould, Chapter 4).

In this paper the term 'forest *Cercopithecus*' refers to all species of *Cercopithecus* except for *Cercopithecus aethiops.* With one exception, the forest *Cercopithecus* are all equatorial in distribution (within 20 degrees of the Equator) and all inhabit primary or secondary rain forest. The one exception, *C. mitis,* lives also in riverine and dry forest and is the only forest *Cercopithecus* whose geographic distribution extends into the subtropics. All are intermediate in size with adult males weighing 2.2–7 kg.

C. aethiops is a medium sized guenon – adult males weigh about 5 kg. It has the largest distributional range of the African non-human primates and, together with *C. mitis,* is one of only two guenon species to occur in subtropical Africa. Unlike the forest *Cercopithecus* its typical habitats are savanna and woodland. It is, however, also frequently found in riparian forest and on the edges of rain forest.

Miopithecus talapoin occurs mainly in flooded primary or secondary forest, and in riverine forest. It is, by far, the smallest of the guenons with adult males weighing about 1.4 kg. Its distribution is confined to the Equatorial Region.

With adult males weighing approximately 12 kg, *E. patas* is the largest guenon. Like *C. aethiops* it is exceptional among guenons in that its typical habitats are savanna and woodland. The distribution of *E. patas* is mostly confined to the Equatorial Region.

Terminology, data base and methods

Several terms used in this paper are defined here. Some definitions have been adopted from Lancaster & Lee (1965).

Mating season A distinct period of the year to which fertile copulations are confined.

Birth season A discrete period of the year to which most or all births are confined.

Birth peak The period of the year in which a high proportion of births is concentrated.

Weaning period The period during which the neonate is both suckling and taking foods other than milk.

Weaned No longer suckling.

All areas occupied by guenons are characterized by 'dry seasons' and 'wet seasons'. Where two wet and two dry seasons occur, the 'long wet season' is defined as the wet season having the more months with rainfall greater than the mean monthly rainfall. Likewise, the 'long dry season' is defined as the dry season having the more months with rainfall less than the mean monthly rainfall.

This review is based on data from free-ranging, unprovisioned populations of guenons in their natural habitats. In most instances the studies cover more than one year and in several cases data were collected over more than four years.

The tables below provide summaries of the data base used in this paper. The details on sample size and duration of the study provided in Table 16.1 give some basis for assessing the confidence which can be placed on each data set. The 'study period' should be viewed with caution, however, as there were numerous times during some of the longer studies when no data were collected for several months or more.

The studies vary considerably in the detail, quantity, quality and kinds of data obtained. For example, only a few provide information on more than one social group over several years. Some studies, while probably delineating most of the period over which births take place, tell us little about if and when birth peaks occur.

Probably all of the sets of data on reproduction presented here are somewhat biased by differential monthly sampling intensity – some months were sampled more often than others and/or some months had larger total numbers of females monitored for new infants. Except for *C. mitis* in W Uganda, this bias is not corrected since the necessary data are not available. To correct for differential monthly sampling intensity in *C. mitis* in W Uganda the total number of births seen each month was divided by the total number of adult females monitored during that month. This yields the number of births per adult female per month. For the remaining data sets it is only possible to report the number of births per month, or simply that at least one birth was observed during particular months.

Annual rainfall and the monthly distribution of rainfall are based on data from a few to many years depending on the information available (Table 16.1). The monthly distribution of rainfall is expressed as a percentage of total annual rainfall.

Faced with these and other differences among data sets, only a broad examination of the data is warranted at this time. To provide a better foundation for comparisons, the data are presented in a simple but uniform manner. Because of the usually small yearly sample sizes, reproductive data from studies which spanned more than one year are lumped by month (Table 16.1). Where the number of observed births is fairly large (i.e. $\geqslant 26$ births), and information on the monthly distribution of births is available, the monthly birth data are presented here as a percentage of total births.

Table 16.1. *Summary of guenon birth seasons, the study sites from which these data were obtained, and primary sources providing information on rainfall and guenon reproductive seasonality*

Source	Location	Species	Birth season[a] (J F M A M J J A S O N D)
Bourlière *et al.*, 1970; Galat-Luong & Galat, 1979	Adiopodoume, SE Ivory Coast	*C. campbelli*	— (J); + + + (A S O); — O (N D)
Gautier-Hion, 1968, pers. comm.	Makokou, NE Gabon	*C. cephus*	O (J); + + + (A S O); — (D)
Gautier-Hion, 1968, pers. comm.	Makokou, NE Gabon	*C. nictitans*	O (F); — (D)
Gautier-Hion, 1968, pers. comm.	Makokou, NE Gabon	*C. pogonias*	O _ _ _ (F M); — (N D)
Gautier-Hion & Gautier, 1978	Makokou, NE Gabon	*C. neglectus*	O (J); — (D)
Butynski, Struhsaker, Leland, Skorupa, Oates Kalina, Waser, Wallis Ghiglieri, Isbell, unpubl. data	Kibale, W Uganda	*C. lhoesti*	O (J); — (D)
McMahon, 1977	Karkloof, SE South Africa	*C. mitis*	O (O)
Scorer, 1980	Cyprus, NE South Africa	*C. mitis*	O (M)
Lawes, unpubl. data	Ngoye, E South Africa	*C. mitis*	— (M); C (D)
Omar & DeVos, 1971; Rowell, 1970	Aberdares, Central Kenya	*C. mitis*	O (M); — (D)
Cords, 1987b	Kakamega, W Kenya	*C. mitis*	O (J); — — – (O N)
Rudran, 1978a; Butynski, 1982b, unpubl. data; Lwanga, unpubl. data	Kibale, W Uganda	*C. mitis*	O (M) —————
Aldrich-Blake, 1970	Budongo, NW Uganda	*C. mitis*	—— —————— —
Cords, 1984, unpubl. data	Kakamega, W Kenya	*C. ascanius*	— (F); — C (N D)
Haddow, 1952	Uganda, several sites	*C. ascanius*	O_ _ _ – (J); — —–––
Struhsaker & Butynski, unpubl. data	Kibale, W Uganda	*C. ascanius*	O (A) ——————
Gautier-Hion, 1970, 1973	Makokou, NE Gabon	*M. talapoin*	O (J) ———— —
Rowell & Dixson, 1975	Mbalmayo, S Cameroun	*M. talapoin*	O (S)
Kavanagh, 1983a	Bakossi, SW Cameroun	*C. aethiops*	O (O)

Sample size[b]	Habitat	Study period (yrs)	No. wet seasons/ yr	Mean annual rainfall (cm) and source
16	Rain forest	6	2	220 Bourlière *et al.*, 1970
15	Rain forest	10	2	162 Gautier-Hion *et al.*, 1985b
18	Rain forest	10	2	162 Gautier-Hion *et al.*, 1985b
5	Rain forest	10	2	162 Gautier-Hion *et al.*, 1985b
10	Rain forest	10	2	162 Gautier-Hion *et al.*, 1985b
46	Rain forest	13	2	154 Struhsaker & Butynski, unpubl. data
?	Dry forest	3.8	1	114 McMahon, 1977
5	Dry forest	1.7	1	98 Scorer, 1980
≥11	Coastal scarp forest	1	1	132 Lawes, unpubl. data
26	Rain forest	1.2	2	93 Nat'l Atlas Kenya, 1970
15	Rain forest	2.2	2	192 Nat'l Atlas Kenya, 1970
122	Rain forest	9	2	154 Struhsaker & Butynski, unpubl. data
16	Rain forest	1.3	2	150 Aldrich-Blake, 1970
>10	Rain forest	2.2	2	192 Nat'l Atlas Kenya, 1970
21	Rain forest	13	2	150 Atlas of Uganda, 1967
185	Rain forest	10	2	154 Struhsaker & Butynski, unpubl. data
57	Inundated rain forest	4	2	162 Gautier-Hion *et al.*, 1985b
?	Rain forest	≥0.8	2	149 Rowell & Dixson, 1975
5	Secondary rain forest	1.4	1	358 Kavanagh, 1983a

Continued

Table 16.1. *Continued*

Source	Location	Species	J	F	M	A	M	J	J	A	S	O	N	D
Kavanagh, 1983a	Buffle Noir, Central Cameroun	*C. aethiops*	—	O										
McMahon, 1977	Karkloof, SE South Africa	*C. aethiops*											O	
Dunbar, 1974	Badi, W Senegal	*C. aethiops*	—	O										
Henzi, unpubl. data	Mzuki, ENE South Africa	*C. aethiops*											O	
Basckin & Krige, 1973; Krige & Lucas, 1974, 1975	Durban, E South Africa	*C. aethiops*	—											O
Lancaster, 1971	Livingston, S Zambia	*C. aethiops*	—											O
Struhsaker, 1967b, 1971, 1973; Klein, 1978; Lee, 1984a; Lee *et al.*, 1986; Isbell, unpubl. data	Amboseli, S Kenya	*C. aethiops*	—	……										O
Whitten, 1982	Isiolo, Central Kenya	*C. aethiops*	—				+	+						O
Harrison, 1982	Mt Assirik, SE Senegal	*C. aethiops*				O	—							
Kavanagh, 1983a	Kalamaloue, N Cameroun	*C. aethiops*								O				
Boulière *et al.*, 1976; Galat & Galat-Luong, 1977	N'Dioum, N Senegal	*C. aethiops*	+	+							O		+	+
Gartlan, 1969	Lolui Island, SE Uganda	*C. aethiops*									O			
Struhsaker & Gartlan, 1970	Waza, N Cameroun	*E. patas*	—											C
Hall, 1965a	Murchison, NW Uganda	*E. patas*	O											—
Chism *et al.*, 1984; Olson pers. comm. in Lindburg, 1987	Mutara, N Kenya	*E. patas*	O				+			+				—

[a] A line below the month indicates that the month is probably part of the primary birth season, a plus denotes that a few births occur but that the month is likely not part of the primary birth season, and dots indicate that the month is probably part of the primary birth season but that more data are needed to show this. The circles denote the approximate middle of the primary birth season.

[b] Number of births and/or foetuses. The sample for *C. lhoesti* is based on sightings of clinging infants.

Sample size[b]	Habitat	Study period (yrs)	No. wet seasons/ yr	Mean annual rainfall (cm) and source
3	Savanna/ woodland	1.4	1	145 Kavanagh, 1983a
?	Fringe of dry forest	3.8	1	114 McMahon, 1977
?	Gallery forest	0.3	1	150 Dunbar, 1974
?	Savanna/ woodland	?	1	63 Henzi, unpubl. data
>22	Savanna/ woodland	2.2	1	101 Henzi, unpubl. data
14	Gallery forest	0.7	1	73 Rhodesia, 1965
>166	Savanna/ woodland	>5.2	2	35 Western & Lindsey, 1984
39	Gallery forest & grassland	2.2	2	45 Whitten, 1982
20	Savanna/ woodland	2	1	95 McGrew *et al.*, 1981
13	Savanna/ woodland	1.4	1	65 Kavanagh, 1983a
17	Savanna/ woodland	1.6	1	21 Atlas Senegal, 1977; Galat & Galat-Luong, 1977
Many	Bush, forest fringe	1	2	157 Jackson & Gartlan, 1965
≤19	Savanna/ woodland	0.2	1	66 Struhsaker & Gartlan, 1970; Atlas Cameroun, 1979
?	Savanna/ woodland	1	2	94 Atlas of Uganda, 1967
85	Savanna/ woodland	2.6	2	61 Chism *et al.*, 1984; Chism, unpubl. data

Table 16.2. *Summary of guenon, Cercocebus and Colobus diets and reproductive parameters*

Species	Body wt (kg) adult ♂/adult ♀	Interbirth interval (yr)	Gestation (mos)	Age when weaned (mos)	Diet (%)				Source
					Fruit & seed	Arthropods	Leaves	Flower	
C. campbelli	4.3/2.2	1.0		12					Hunkeler et al., 1972
C. diana	5.0/?				41	24	6	—	Curtin, unpubl. data in Cords, 1987a
C. cephus	4.1/2.9		5.6		78	12	6	0	Gautier-Hion, 1980, unpubl. data; Gautier-Hion & Gautier, 1985
C. pogonias	4.5/3.0		5.6		82	16	1	0	Gautier-Hion, 1980, unpubl. data; Gautier-Hion & Gautier, 1985
C. nictitans	6.6/4.2		5.6		70	9	17	1	Gautier-Hion, 1980, unpubl. data; Gautier-Hion & Gautier, 1985
C. neglectus	7.0/4.0	1.0	5.6	12	74	5	9	3	Gautier-Hion & Gautier, 1978, 1985, unpubl. data

Species									Reference
C. mitis	6.9/4.2	3.8[a]	4.7	30	17	57	19	4	Cords, 1987b, Chapter17; Rowell, 1970
C. mitis					?	60[b]	23[b]	5[b]	Scorer, 1980
C. mitis					?	46[b]	33[b]	21[b]	Harcourt, unpubl. data in Homewood, 1976
C. mitis					?	65	?	?	Aldrich-Blake, 1970
C. mitis					1	51	25	9	Breytenbach, unpubl. data
C. mitis					19	45	21	9	Rudran, 1978a
C. mitis					5	66	24	3	Lawes, unpubl. data
C. mitis		2.4–5.0		16	37	23	31	7	Butynski, unpubl. data
C. mitis					13	45	18	24	Schlichte, 1978
C. ascanius	4.2/2.9	4.3[a]		24	25	62	7	2	Cords, 1987b, unpubl. data
C. ascanius		1.5		18–20	22	44	16	15	Struhsaker, 1978, Chapter18, unpubl. data
M. talapoin	1.4/1.1	1.0	5.5	12	36	43	2	2	Gautier-Hion, 1968, 1973, 1978, unpubl. data; Gautier-Hion & Gautier, 1985; Rowell, 1977, unpubl. data

Continued

Table 16.2 (*cont.*)

Species	Body wt (kg) adult ♂/adult ♀	Interbirth interval (yr)	Gestation (mos)	Age when weaned (mos)	Diet (%)					Source
					Fruit & seed	Arthropods	Leaves	Flower		
C. aethiops	5.1/3.5	1.2			63	13	7	13		Harrison, 1982
C. aethiops		1.7		9–18						Whitten, 1983
C. aethiops		1.4	5.4							Cheney et al., 1981; Bramblett et al., 1975
C. aethiops		1.2–1.4			51	7	19	18		Struhsaker, 1976; Dunbar & Dunbar, 1974a; Klein, 1978
C. aethiops		≥1.7		10–18						Lee, 1984b
E. patas	12.0/6.0	1.0	5.6	12	high	high	?	?		Sly et al., 1983; Chism et al., 1984, unpubl. data; Chism & Rowell, Chapter 21; Koster, unpubl. data

Species									Reference
C. badius	10.5/7.0	2.0–2.2	6.0–7.0	>26	6	0	72	2	Struhsaker, 1978; Struhsaker & Leland, 1985
C. albigena	10.0/7.0				59	11	5	3	Waser, 1974
C. albigena					58	25	0	0	Freeland, 1979
C. albigena					69	12	12	5	Wallis, 1978
C. albigena			5.8		81	6	6	5	Gautier-Hion, 1978, unpubl. data
C. galeritus	10.2/5.4				78	1	11	1	Homewood, 1978
C. galeritus					73	3	14	1	Quris, 1975

[a] Interbirth interval when previous infant survives.
[b] Per cent of plant portion of diet only (i.e. arthropods not included in percentage).

Mating seasons are estimated by back-dating from the time of birth using the known or an approximate mean gestation period for each species. The length of gestation is established for eight species of guenons (Table 16.2). For all other species of guenon a gestation of 5.5 months is used here.

Concerning the statistics, all probability values are two-tailed.

Results and discussion

Rainfall and the availability of guenon foods

This section reviews the relationship between rainfall and the abundance of guenon foods. If the pattern of rainfall can be used to predict periods of relative food abundance it would be important for two reasons. First, since seasonal reproduction in guenons appears to be related to the pattern of rainfall (see below), a good case could be made for nutrition as a primary ultimate factor determining the timing of reproductive events. Second, information on reproductive seasonality in guenons is often not accompanied by data on food availability. If periods of food abundance are related to rainfall then it would be possible to use rainfall data to predict the timing of nutrient abundance in sites where such data have not been collected.

Table 16.2 reviews the data available on guenon diets (also see Gautier-Hion, Chapter 15; Chism & Rowell, Chapter 21). It indicates that guenons are frugivore–omnivores and that the typical guenon diet is comprised of about 50–60% fruits and seeds (range 23–82%), 10–20% arthropods (range 1–37%), 10–20% leaves (range 0–31%), and 5–10% flowers (range 0–24%).

Fruits are the most prominent foods in the guenon diet. They are the foods over which most of the overt aggression during feeding occurs and lists of preferred plant foods usually give fruits a disproportionately large number of high ratings. Although they are often low in protein, fruits are high in energy and generally easily digested.

Guenons spend a large portion of their foraging time searching for arthropods and these appear to be relished when captured. Arthropods are easy to digest and the most nutritious of the items eaten by guenons, providing high levels of protein and energy.

Young leaves comprise a significant part of the diet of many guenons. They often contain considerable protein of high biological value. In Table 16.2, 'young leaves' are not differentiated from 'mature leaves'. It should be noted, however, that where the data are provided, young leaves are a much more common item in the diet of guenons than are mature leaves.

Unlike studies of other guenons, those of *C. aethiops* suggest that the seasonal abundance of flowers may play a critical role in the timing of matings and births (Klein, 1978; Harrison, 1982; Whitten, 1983). Many flowers, particularly those of *Acacia* spp., are easy to digest and high in protein and energy.

Given the obvious importance of fruits and arthropods (and in the case of *C. aethiops*, flowers), it is not surprising that primatologists have most often focused on these components of the diet when examining the question of seasonal variation in nutrient availability and its effect on the behaviour and ecology of guenons.

Data on annual changes in the abundance of guenon foods are available for five savanna/woodland sites and five forest sites (Table 16.3). In all sites there appears to be one period each year when guenon foods are usually relatively abundant. This is from approximately November–March in eight of the ten sites, including all five forest sites. In the other two sites, both in savanna/woodland habitats, food is most common from about April–June.

In nine of the ten sites, including all five forest sites, food for guenons is most abundant during the wettest time of the year and/or first half of the subsequent dry season. This result is not surprising, at least not for guenons in savanna/woodland habitats. Numerous quantitative studies of food availability in the savannas and woodlands of Africa document predictable and dramatic increases in plant growth (Rosenzweig, 1968; Bourlière & Hadley, 1970; Coe, Cumming & Phillipson, 1976) and arthropod abundance (Dingle & Khamala, 1972; Kemp, 1973; Sinclair, 1978; Lack, 1986) during and soon after the wettest period. As such, rainfall has frequently been cited as the primary ultimate factor determining the availability of nutrients for primates and many other animals living in the drier areas of Africa.

In complex forest habitats, where rainfall is higher and more evenly distributed than in savanna/woodland, the influence of rainfall on the food supply is far less obvious and more difficult to substantiate. In fact, until recently it was often assumed that there were no significant seasonal changes in the supply of food in tropical forests – favourable conditions for reproduction were thought to be present on a more or less continuous basis. With long-term, detailed studies this view has changed considerably (Charles-Dominique, 1977; Milton, 1982; Gautier-Hion *et al.*, 1985b; Smythe, Glanz & Leigh, 1982; Smythe, 1986). For example, in three rain forest sites on three continents the rate of annual leaf-fall changes 2- to 4-fold from season to season while fruit-fall varies 5- to 6-fold (Bourlière, 1979).

Table 16.3. *Summary of relationships among guenon birth seasons, wettest periods, and months when fruit, arthropods and other foods are most abundant. Some data are also provided for* Cercocebus

Location & Source	Species	Variable	Month[a] (J F M A M J J A S O N D)
Makokou, NE Gabon			
Gautier-Hion, 1970, 1973	*M. talapoin*	births	
Gautier-Hion, 1968	*C. cephus*	births	
Gautier-Hion, 1968	*C. nictitans*	births	
Gautier-Hion, 1968	*C. pogonias*	births	
Gautier-Hion & Gautier, 1978	*C. neglectus*	births	
Gautier-Hion, unpubl. data	*C. galeritus*	births	
Gautier-Hion, unpubl. data	*C. albigena*	births	
Gautier-Hion, 1980; Gautier-Hion et al., 1985b		fruit	
Gautier-Hion, 1980		arthropods	
Hladik, 1978		young leaves	
Gautier-Hion et al., 1985b		wettest mos	
Kibale, W Uganda			
Butynski, Struhsaker, Leland, Skorupa, Oates, Kal-ina, Waser, Wallis, Ghiglieri, Isbell, unpubl. data	*C. lhoesti*	births	
Rudran, 1978a; Butynski, 1982b, unpubl. data; Lwanga, unpubl. data	*C. mitis*	births	
Struhsaker & Butynski, unpubl. data	*C. ascanius*	births	
Waser, 1974; Wallis, 1978; Leland, unpubl. data	*C. albigena*	births	
Butynski, unpubl. data		fruit	
Ghiglieri, 1984		chimp fruit	
Nummelin, 1986		arthropods	
Struhsaker & Butynski, unpubl. data		wettest mos	

Location / Reference	Species		
Cyprus, NE South Africa			
Scorer, 1980	*C. mitis*	births	
Scorer, 1980		fruit	
Scorer, 1980		wettest mos	
Ngoye, E South Africa			
Lawes, unpubl. data	*C. mitis*	births	
Lawes, unpubl. data		fruit	
Lawes, unpubl. data		wettest mos	
Durban, E South Africa			
Krige & Lucas, 1974, 1975	*C. aethiops*	births	
Basckin & Krige, 1973		food	
Henzi, unpubl. data		wettest mos	
Isiolo, Central Kenya			
Whitten, 1983	*C. aethiops*	births	
Whitten, 1983		flowers	
Whitten, 1982		wettest mos	
Amboseli, S Kenya			
Struhsaker, 1967b, 1971, 1973; Klein, 1978; Lee, 1984a; Lee et al., 1986; Isbell, unpubl. data	*C. aethiops*	births	
Lee, 1984b		diet quality	
Klein, 1978		flowers	
Post, 1982		baboon food	
Western & Lindsey, 1984		wettest mos	
Mt Assirik, SE Senegal			
Harrison, 1982	*C. aethiops*	births	
Harrison, 1982		flowers & fruit	
McGrew et al., 1981		wettest mos	

Continued

Table 16.3 (*cont.*)

Location & Source	Species	Variable	Month[a]											
			J	F	M	A	M	J	J	A	S	O	N	D
Mutara, N Kenya														
Chism et al., 1984	*E. patas*	births	○										—	
Chism et al., 1984		food					●	- - -						
Chism et al., 1984		wettest mos						×··	··········					
Tana River, E Kenya														
Homewood, 1976	*C. galeritus*	births	○									—		
Homewood, 1978; Marsh, 1978		fruit & young leaves	●	- - -	- - -	- - -	- - -					- - -	- - -	
Kenya Atlas, 1962		wettest mos					······							··×······

[a]The solid line below the month indicates that the month is probably part of the primary birth season, a dashed line denotes that these months have a relative abundance of food, and a dotted line indicates the wettest months. Open circles denote the approximate middle of the primary birth season, closed circles represent the approximate middle of the season of relative food abundance, and × indicates the wettest month of the year.

The data summarized in Table 16.3 not only suggest that the abundance of guenon foods in the forest environment varies with the time of year, but also that the period of abundance is fairly predictable – occurring from about November to March. A relative abundance of food for primates (particularly fruit) during this time of the year has also been noted in other forest sites in Africa (e.g. Hoshino, 1985). Interestingly, as in savanna/woodland areas, food is relatively abundant during the wettest time of the year and in the following dry season.

In summary, the data available suggest that, both in forest and savanna/woodland habitats, the abundance of guenon foods varies considerably through the year and that the period of greatest abundance is related to rainfall in a predictable way. Specifically, guenon foods seem to be most abundant during the wettest time of the year and within one to three months thereafter.

Mating seasons, birth seasons and rainfall

Forest *Cercopithecus*

Sixteen data sets on eight species of forest *Cercopithecus* provide information on the guenon reproductive cycle in nine rain forest and dry forest sites. *Cercopithecus ascanius* in W Uganda is the only guenon known to give birth during all months although a distinct birth peak is obvious with only 3% (5/185) of the births occurring during June to August (Figure 16.1(*a*)). The other seven forest *Cercopithecus* studied to date appear to be seasonal breeders although additional observations may show that *C. mitis* (Figure 16.1(*a*)), and perhaps others, bear young throughout the year in some localities (Table 16.1). One other guenon, *Cercopithecus mona*, is reported to be a seasonal breeder but details are not available (J. S. Gartlan, personal communication in Rowell & Richards, 1979). These seven species have birth seasons which range from at least 3 to 10 months in length. With the possible exception of *C. mitis* in NW Uganda, all birth seasons center on November, December, January or February, especially January or February. These rather consistent patterns are exhibited by populations in East, West and South Africa. Where ample data are available, forest *Cercopithecus* exhibit an annual peak in births (Figure 16.1).

The rain forests of Equatorial Africa experience moderate to high annual rainfall and two wet and two dry seasons each year. Thirteen sets of data on the timing of the birth season of eight species of forest *Cercopithecus* were obtained from seven such areas (Table 16.1). Mean

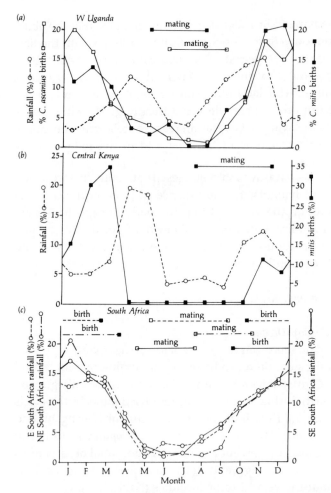

Figure 16.1. C. *ascanius* and C. *mitis* mating and birth seasons as they relate to the annual distribution of rainfall. (*a*) C. *ascanius* and C. *mitis* in W Uganda; (*b*) C. *mitis* in Central Kenya; (*c*) C. *mitis* in three South African sites.

annual rainfall among areas differs 2.4-fold (93–220 cm) and the timing of the long and short wet seasons is often reversed, as are the long and short dry seasons. Although there is considerable variation in the months over which the wet and dry seasons occur, all seven areas have a dry season (either a long or short dry season) from December to February. These are the central months of the forest *Cercopithecus* birth season. The annual distributions of rainfall for W Uganda and Central Kenya (Figures 16.1(*a*),(*b*)) are typical for these seven areas.

During which months are forest *Cercopithecus* born in areas which have only one wet season? Three studies of *C. mitis* living in forest in subtropical Africa (E, NE and SE South Africa) provide some insight (Table 16.1, Figure 16.1(c)). They show that births occur primarily from November to February. This is similar to the timing of the middle part of the birth season of forest *Cercopithecus* in areas with two wet seasons. In southern Africa, however, these are the wettest, not the driest, months.

Using 5.5 months as the approximate mean gestation indicates that forest *Cercopithecus* are mating mainly in July–October. In areas with two wet seasons, this period includes either a long or short dry season and the beginning of either a long or a short wet season. In areas with one wet season, this is the dry season or start of the wet season (Figure 16.1).

Miopithecus talapoin

Like most of the forest *Cercopithecus*, *M. talapoin* lives in areas with two wet seasons and relatively high rainfall (149–162 cm/yr) (Table 16.1, Figure 16.2). It might, therefore, be expected that this species would exhibit a seasonal pattern of mating and births similar to the forest *Cercopithecus*. This is the case for *M. talapoin* in NW Gabon, where births occur from November to April. This expectation was not met, however, by the S Cameroun population where births were observed in June–August. Rowell & Dixson (1975) note that the long and short dry seasons are switched in the two study areas and that *M. talapoin* gives birth during the short dry season in both sites. None the less, the annual rainfall and its distribution are quite similar in the two areas, making it surprising to find such a large difference in the timing of the birth season. As indicated above, most of the data on forest *Cercopithecus* show birth seasons centered on the November–February dry period regardless of whether this represents the long or short dry season. This generalization would appear not to fit *M. talapoin*. Among the forest guenons, the S Cameroun population of *M. talapoin* is the only one reported to have a birth season centered outside the November–February period. Further study of the ecology of this population could be most useful in helping us understand the ultimate factors determining the timing of the birth season in forest guenons.

Erythrocebus patas

E. patas has been studied in three sites, all of which receive similar amounts of annual rainfall (61–94 cm) (Table 16.1, Figure 16.3).

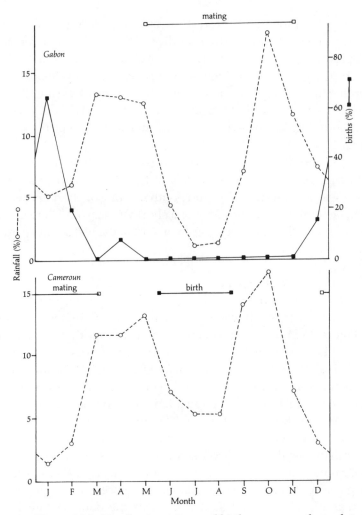

Figure 16.2. *M. talapoin* mating and birth seasons as they relate to the annual distribution of rainfall in NE Gabon and S Cameroun. Sample sizes and the sources of these data are provided in Table 16.1.

These three populations have birth seasons restricted to 3–4 months within a 3–7 months long dry season. Two of the sites have a bimodal rainfall but the driest season, and thus the birth season, occurs during about the same time of the year in all three sites (November–February). Two other populations of *E. patas*, one in SW Niger (Koster, personal communication) and one in N Ghana (Chism, Rowell & Olson, 1984; Chism & Rowell, Chapter 21) also appear to have a birth season at this time of year, again during the driest period.

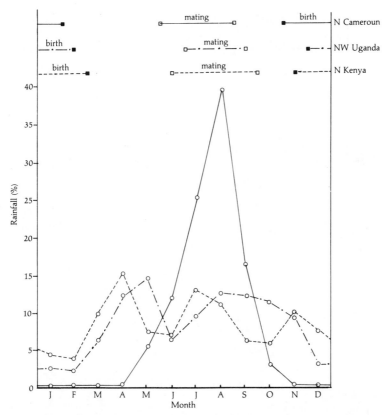

Figure 16.3. *E. patas* mating and birth seasons as they relate to the annual distribution of rainfall in N Cameroun, NW Uganda and N Kenya. Sample sizes and the sources of these data are given in Table 16.1.

The mating season for *E. patas* is mainly in July–August. This is about the wettest time of the year in NW Uganda and N Cameroun but is the short dry season in N Kenya.

Cercopithecus aethiops

One wet season with low rainfall. Four studies of *C. aethiops* have been undertaken in areas where rainfall is low (21–73 cm/yr) and where there is but one wet season (N Cameroun, N Senegal, S Zambia, ENE South Africa) (Table 16.1, Figures 16.4, 16.5). In all four sites there is a severe dry period of about 5–7 months. The *C. aethiops* birth season in these areas is 2–4 months long and occurs during the first half or middle of the wet season. This is the case even though the wet season in

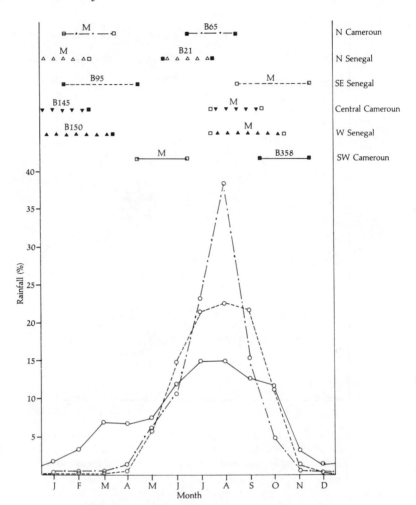

Figure 16.4. Months included in the *C. aethiops* mating and birth seasons and the annual distribution of rainfall in six sites with but one wet season. The monthly patterns of rainfall for W Senegal, Central Cameroun and N Senegal have not been plotted here since they are almost identical to that for N Cameroun. The horizontal lines with 'M' above them indicate the time of the mating season and those with 'B' denote the birth season. The number after the 'B' is the mean total annual rainfall in centimetres. Sample sizes and the sources of these data are given in Table 16.1. Note that all six sites have the wet season centered on July–September and that, in general, the higher the annual rainfall the more the birth season precedes the month with peak rainfall.

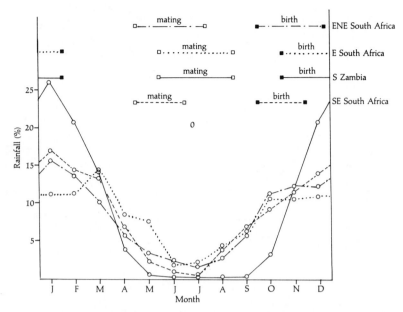

Figure 16.5. *C. aethiops* mating and birth seasons as they relate to the annual distribution of rainfall in four sites with but one wet season. Sample sizes and the sources of these data are given in Table 16.1.

N Senegal and N Cameroun centers on August while in S Zambia and ENE South Africa it centers on January. Mating occurs during the middle of the dry season in all four sites.

One wet season with moderate rainfall. *C. aethiops* has been studied in five sites where there is moderate rainfall (95–150 cm/yr) and one wet season (SE Senegal, W Senegal, Central Cameroun, SE South Africa, E South Africa) (Table 16.1, Figures 16.4, 16.5). Under these conditions the birth season is 2–3 months long and occurs during the middle of the 4–6 months dry season and first half of the wet season. Mating takes place during or at the end of the wet season. The wet season centers on July–August in Equatorial Africa and on January–February in southern Africa. It is, therefore, important to note that the mating and birth seasons in these areas also differ by about six months.

One wet season with high rainfall. In SW Cameroun, where rainfall is very high (358 cm/yr) and unimodal, births occur during the two months at the end of the wet season (Table 16.1, Figure 16.4). The mating season is at the end of the dry season.

Two wet seasons. Where there are two wet seasons each year (S Kenya, Central Kenya, SE Uganda) the birth season of *C. aethiops* extends over 4–6 months (Table 16.1, Figure 16.6). It begins during the

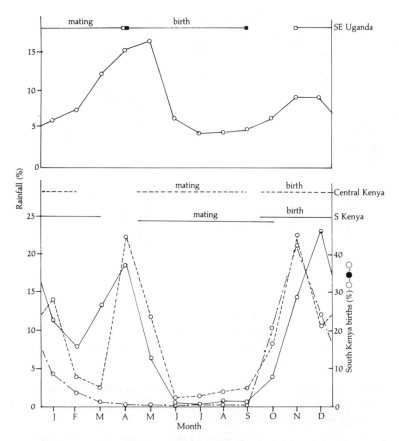

Figure 16.6. *C. aethiops* mating and birth seasons as they relate to the annual distribution of rainfall in three sites with two wet seasons. Sample sizes and the sources of these data are given in Table 16.1.

first half of the long wet season and continues into or through the subsequent dry season. The two wet seasons occur during approximately the same time of the year in all three sites but the long wet season in Uganda occurs at the time of the short wet season in Kenya and the timing of the births differs accordingly.

Although the long wet season in the two Kenya sites provides only about 5–10% more precipitation than the short wet season, the births are still centered on the long wet season. This may be related to the fact that the long wet season is followed by the relatively mild short dry season rather than by the severe long dry season when little or no rain falls for about four months.

In all three localities mating begins during the middle of the short wet season and continues throughout most of the subsequent dry season.

Cercopithecus aethiops reproductive patterns

1. *C. aethiops* in all 13 study sites has but one mating and one birth season per year. This is the case even though annual rainfall in these sites varies by a factor of about seventeen (21–358 cm) and its monthly distribution differs greatly.

2. *C. aethiops* in 10 sites have birth seasons centered on one of the months from November to March, while in three sites it is centered on June, July or August – about six months later. Relatively few births occur during April–May or September–October. Back-dating 5.4 months from the birth dates places the main mating season from June to October or from January to March.

3. The length of the mating and birth seasons differs significantly between areas with one and two wet seasons (Mann-Whitney U-test, $p < 0.01$, $n_1 = 10$, $n_2 = 3$). The mating and birth seasons each cover 2–3 months (mean = 2.6 mo) where rainfall is restricted to one wet season. Where rainfall is bimodal the mating and birth seasons each cover 4–6 months (mean = 5.0 mo). This is the case even though areas with but one wet season usually received considerably more precipitation (mean = 118 cm/yr) than do areas with two wet seasons (mean = 75 cm/yr) (Table 16.1). Given the apparent complicating effect of the distribution of rainfall on the length of the birth season, it is not surprising to find only a weak association between total annual rainfall and length of the mating or birth season ($r = -0.28$, $n = 13$, $p > 0.10$).

4. Where there is but one wet season it appears that we can estimate the approximate timing of the middle of the *C. aethiops* birth season if we know the total annual rainfall in the site and which month is, on average, the wettest (Figure 16.7). This is because there is a highly significant positive correlation between total annual rainfall (X) and the number of months by which the middle of the birth season precedes the wettest month of the year (Y) ($r = 0.86$, $n = 10$, $p < 0.002$). The equation is:

$$Y = 0.33 + 0.03X$$

Since the length of the gestation period is known to be 5.4 months (Table 16.2), this equation can, of course, also be used to estimate the approximate center of the mating season. The data available further suggest that these mating and birth seasons will each be 2–3 months

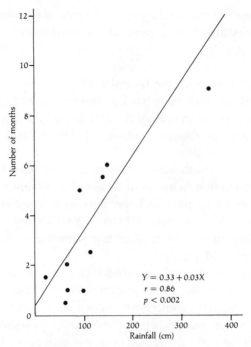

Figure 16.7. In areas with one wet season there is a positive relationship between mean annual rainfall and the number of months by which the middle of the *C. aethiops* birth season precedes the wettest month of the year. Sample sizes and the sources of these data are given in Table 16.1.

long, since this is their length wherever *C. aethiops* has been studied in areas with one wet season.

Out-of-season matings and births

In at least some populations of guenons, births occur outside what is obviously the 'primary' birth season (Table 16.1). 'Out-of-season' births have been recognized for two of 13 *C. aethiops* populations, one of three *E. patas* populations, and probably occur in some populations of forest guenons as suggested by the data for *Cercopithecus cephus* and *C. campbelli* (Table 16.1). In the two *C. aethiops* and one *E. patas* populations in which out-of-season births have been observed, roughly 3–12% of the annual births occurred outside the primary birth season.

Most of the out-of-season births of *C. aethiops* and *E. patas* occur about 5–6 months after the primary birth season. This is approximately equivalent to the gestation period and indicates that some matings occur during the primary birth season. P. Whitten (personal communication) suggests that *C. aethiops* females mating at this time may be low-ranking. Chism *et al.* (1984) provide some evidence that, in *E. patas*, these are females whose pregnancies failed or who lost infants soon after giving birth during the primary birth season.

Mating seasons, birth seasons and time of year

The data now available, although biased towards a few species and research sites (Table 16.1), indicate that some months are more likely to be part of the guenon mating and birth seasons than are others ($\chi^2 = 50.3$, d.f. = 11, $p < 0.001$) (Table 16.4). November–February are most often included in the birth season while May–September are least often included. Eighty-five per cent (28/33) of the birth seasons are centered on December–February.

Using a 5.5 month gestation period and back-dating from the above information, I estimate that June–September are most often included in the guenon mating season while December–April are least often included.

Rainfall and the length of mating and birth seasons

As mentioned earlier, *C. aethiops* in areas with one wet season have significantly shorter mating and birth seasons than in areas with two wet seasons. This applies also to guenons when examined as a group (Figure 16.8). There is a highly significant difference in the length of guenon mating and birth seasons where there is one wet season (mean = 2.8 mo) and where there are two wet seasons (mean = 5.8 mo) (Mann-Whitney U-test, $p < 0.0005$, $n_1 = 14$, $n_2 = 20$). With only one exception, the guenon mating and birth seasons are shorter than 3 months in areas with one wet season. *C. mitis* in E South Africa is the exception with 4–5 month mating and birth seasons. In contrast, the mating and birth seasons are 3 months or longer in all areas with two wet seasons.

Mating and birth seasons are longer in areas with two wet seasons even though several areas with one wet season receive considerably more rainfall than some with two wet seasons. The distribution of rainfall, therefore, appears to be a significant factor affecting the length of guenon mating and birth seasons – apparently more so than total annual rainfall. This may reflect the fact that for primary production, the distribution of rainfall is generally much more important than the annual total (McGrew, Baldwin & Tutin, 1981).

Table 16.4. Frequency of occurrence of each month in the mating and birth seasons of guenons as determined from the 34 data sets reviewed in Table 16.1. The mating season is estimated by back-dating 5.5 mos from the time of birth. Values in parentheses are percentages

Mating season months	Aug	Sept	Oct	Nov	Dec	Jan	Feb	Mar	Apr	May	June	July
Birth season months	Jan	Feb	Mar	Apr	May	June	July	Aug	Sept	Oct	Nov	Dec
Month is part of season	25	22	15	12	5	6	6	5	7	12	21	25
	(16)	(14)	(9)	(7)	(3)	(4)	(4)	(3)	(4)	(7)	(13)	(16)
Month is middle of season	7.5	8.0	1.0	0	0	1.0	2.5	0.5	0	0.5	3.5	8.5
	(23)	(24)	(3)	(0)	(0)	(3)	(8)	(2)	(0)	(2)	(11)	(26)

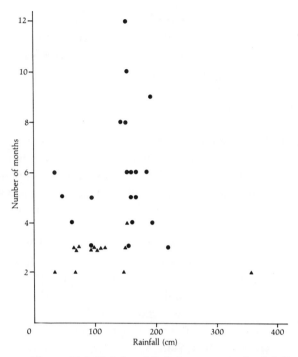

Figure 16.8. Relationship between annual rainfall and the number of months in the guenon mating season (or birth season). Note that, in general, guenons in sites with one wet season (▲) have shorter mating and birth seasons than guenons in sites with two wet seasons (●). Sample sizes and the sources of these data are given in Table 16.1.

As can be seen in Figure 16.8, there appears to be no association between annual rainfall and the length of the mating and birth seasons ($r = 0.20$, $n = 34$, $p > 0.10$). None the less, Figure 16.8 may demonstrate the approximate upper limit for the length of the guenon mating and birth seasons as they relate to annual rainfall. It suggests that this limit is about six months at 50 cm annual rainfall and about eight months at 100 cm of rain. Guenons probably do not have year round mating or birth in areas receiving less than 150 cm of rain.

Rainfall and the timing of mating and birth seasons

The mating seasons of the two studied populations of *M. talapoin* occur during the long dry season and long wet season. In the three populations of *C. aethiops* which have been studied in localities with two wet seasons, mating begins during the middle of the short wet season and continues through most of the following dry season.

Such apparent patterns between mating seasons and conditions of rainfall are, however, the exception rather than the rule for guenons. Unlike the birth season, the guenon mating season seems less consistently associated with a particular set of rainfall conditions. For example, the three studies of E. *patas* place the mating season in the first half of the wet season in N Cameroun, during the short dry season in N Kenya, and in the short dry season and first half of the long wet season in NW Uganda. None the less, matings occur at the same time of year in all three of these populations. This implies that if rainfall is a proximate cue used by guenons to initiate and terminate mating it is used differently by guenons in different localities. It seems more likely that some factor other than rainfall is the primary proximate cue for mating in guenons.

Forest *Cercopithecus* in areas with two wet seasons, as well as *M. talapoin* and *E. patas*, all have birth seasons centered on either the short or (most often) the long dry season. Forest *Cercopithecus* in places with one wet season give birth at the same time of year as other forest *Cercopithecus* but for them this is the wet rather than the dry season.

In contrast to the other guenons, it is possible to generalize about the timing of the *C. aethiops* birth season only after the amount and distribution of the annual rainfall are considered (see above). This is probably because this species is found under such a great range of rainfall (21–358 cm/yr) and because this rain may occur during one or two wet seasons.

Timing of rainfall, food abundance, births and weaning

Savanna and woodland

As indicated above, there appears to be a relative abundance of food for guenons during the wettest month of the year and during the 1–3 months which follow. That this is a period of food abundance also applies to many of the other species of animals living in Africa's forest, savanna and woodland areas – a large proportion of which produce young at this time (Sadleir, 1969; Bourlière & Hadley, 1970; Dingle & Khamala, 1972; Sinclair, 1978; Lack, 1986). Several studies of primates living in dry areas have found that, as the wet season wanes and the dry season progresses, nutritional stress is common, reproduction ceases, and mortality increases (Bearder & Martin, 1980; Cheney, Lee & Seyfarth, 1981; Chism *et al.*, 1984).

Four studies of *C. aethiops* and one of *E. patas* provide information on rainfall, food availability and birth seasonality in areas receiving 35–101 cm of rain annually. As can be seen in Table 16.3, the time when the availability of food is highest in savanna/woodland habitats is during at least the latter part of the wet season and soon thereafter. SE Senegal appears to be an exception. There, food for *C. aethiops* may be most abundant just before and during the first half of the wet season.

In areas with 35–101 cm of rain per year the *C. aethiops* birth season begins 1–3 months prior to the period of abundant food and ends 1–3 months before the period of abundant food is over. Thus, the middle of the birth season occurs about 0.5–2 months prior to the middle of the period of relative food abundance. This suggests that *C. aethiops* are matching the time of maximum lactation, and the nutritional stress this brings to the mother (Sadleir, 1969; Klein, 1978), to about the first half of the period when food is most plentiful. This is also the time of most rapid infant growth (Lee, 1988).

C. aethiops infants begin to take solid foods at about two months of age (Klein, 1978; Whitten, 1983). The average length of nursing bouts rapidly declines after infants are approximately 3–4 months old (Lee, 1988). Although infants are not weaned until 9–18 months of age, they apparently receive little milk after they are five months old and are capable of sustaining themselves through independent feeding when six months old (Klein, 1978). Thus, *C. aethiops* infants are rapidly switching over to a vegetable–arthropod diet during the second half of the period when nutrients are most available. It seems likely that milk constitutes a minor but important part of the neonate's diet during the subsequent dry season when food is relatively scarce. Infants are weaned primarily during the following year's period of food abundance.

C. aethiops appears to be rather flexible as to the age at which suckling stops. Presumably, if conditions for the growth of the infant are favorable, suckling ends as early as nine months after birth. The mother may then produce another infant during the coming birth season. On the other hand, if conditions allow for only a slow rate of infant growth, suckling may continue until the infant is 17 or more months old (Lee, 1988).

There are no data on food availability on sites where *C. aethiops* occur under conditions of moderate or high rainfall. Where rainfall is moderate (e.g. Central Cameroun with 145 cm per year – woodland and gallery forest) food is likely to be more abundant over more months.

Under these circumstances the birth season occurs in the dry season, 5–6 months before the time of maximum rainfall. Infants are largely on a vegetable–arthropod diet by the height of the wet season when food is probably most abundant.

Where rainfall is extremely high (e.g. SW Cameroun with 358 cm per year – mosaic of farmland and secondary forest in the rain forest zone) the annual supply of food for *C. aethiops* is likely to be far less seasonal. With so much precipitation during the wet season (250 cm in 5 mo) it is reasonable to suggest that, as in the rain forest, food production is highest during the wet season and for 1–3 months thereafter. This might explain why *C. aethiops* births in SW Cameroun occur at the end of the wet season and beginning of the dry season. This is similar to what is observed for the forest *Cercopithecus*.

Newborn infants in SW Cameroun have about seven months of relatively dry weather before the heavy rains begin. While births may be timed to take advantage of favourable nutritional conditions, it may also be that births are timed so that infants and their stressed mothers are not subjected to five months of extremely wet conditions when they might be particularly susceptible to diseases associated with damp conditions (Gautier-Hion, 1968). A similar situation has been described in Ethiopia where *Theropithecus gelada* experience a severe annual wet season of frequent soaking rains (Dunbar, 1980). Although the wet season is the time of maximum food availability, post-natal death increases at this time. Not surprisingly, *T. gelada* births occur primarily during the dry season.

The initial impression upon looking at the earlier data for *C. aethiops* was that no pattern exists between the timing of births and rainfall or the supply of food (Else *et al.*, 1986). As such, it was hypothesized that birth synchrony *per se*, not synchrony with rainfall or food availability, has been selected for and is responsible for birth clumping in this species (Bramblett, DeLuca Pejaver & Drickman, 1975; Rowell & Richards, 1979; Kavanagh, 1983a). The data presented above do not support this hypothesis. The findings of this review do not, however, refute the possibility that synchrony of births may also provide social and survival advantages which are independent of rainfall and food conditions, such as those obtained through allomothering and improved predator detection (see below).

We might predict that *E. patas* would have a birth season which is related to rainfall and food abundance in a manner similar to that observed for *C. aethiops* in areas of low rainfall. This is, however, not the case. Births occur during the driest period in all four sites where

E. patas have been studied. In N Kenya, and probably also in the other three sites, food is most available during the middle of the wet season, 2–6 months after births occur (Table 16.3). Based upon our knowledge of the relationships between the *C. aethiops* birth season, rainfall and food supply, the *E. patas* births occur far in advance of the time that *C. aethiops* births would be expected, given the same distribution and amount of rainfall (for example, compare *E. patas* in N Cameroun with *C. aethiops* in N Cameroun or S Zambia).

Although *E. patas* are not weaned until about 12 months of age they shift rapidly from dependence on milk to independent foraging when approximately 4–5 months old and can survive the death of their mother when seven months old. This transition period corresponds to the wet season when 'weaning foods' are most available (Chism *et al.*, 1984; J. Chism, personal communication), suggesting that this may be a particularly critical period for infants. This is probably more so than for other guenons since *E. patas*, although the largest of the guenons, are weaned at a relatively young age and live in the harshest of the habitats used by guenons. As such, *E. patas* neonates presumably require the most favourable conditions the environment has to offer when switching from a largely milk diet to mainly independent foraging.

Forest

Seasonal changes in plant productivity can have a measured effect on the ecology and behaviour of mammals living in rain forest. It appears that there is a cause–effect relationship between the low availability of fruit (Otis, Froehlich & Thorington, 1981; Milton, 1982; Smythe *et al.*, 1982) and young leaves (Montgomery & Sunquist, 1978) and increases in the mortality of some species of primates and other mammals in rain forest. It should not be surprising, therefore, that changes in the supply of food for primates living in rain forest appear to affect the timing of reproductive events.

Data sets on rainfall, food availability and birth season are available for five species of forest guenons in one site in NE Gabon (162 cm rain/yr) and three species in one site in W Uganda (154 cm rain/yr) (Table 16.3). In both areas, fruits and arthropods are most abundant during November–March, particularly December–February. This is during the middle of the dry season and follows the wettest month of the year by about 2–3 months. Thus, late gestation and early lactation tend to occur near the middle of the period when nutrients are probably most available. In addition, many of the neonates born at this time are spared prolonged wet weather for the first few months after birth (i.e. January–February).

C. *mitis* in NE South Africa lives in dry forest. Here there is one wet season and the annual rainfall is about 98 cm (Table 16.3). In E South Africa, C. *mitis* occurs in forest which also experiences one wet season but which receives more rain (132 cm/yr). Here the birth season is two months longer, extending from October to February. In both sites the birth season roughly coincides with the periods of most rainfall and highest fruit production. It appears that the period of food abundance and the middle of the birth season for C. *mitis* in the subtropics are similar to those observed for C. *mitis* and other guenons in the tropical rain forests. This is the case even though rainfall differs considerably both in quantity and distribution (Figures 16.1, 16.2).

The eight species of forest guenons considered in Table 16.3 represent the range of body sizes, and probably the basic life histories and ecologies, of forest guenons as a whole. As such, these findings suggest that the general relationships among rainfall, food availability, and birth season exhibited in Table 16.3 are widespread among forest guenons.

The data presented above suggest that there are correlates between the distribution and amount of rainfall, the period over which nutrients are most easy to obtain, and the timing of guenon reproductive events. I conclude (1) that the distribution and amount of rainfall are associated with the temporal supply of nutrients for guenons, and (2) that seasonal differences in the availability of nutrients, in turn, influence the timing of reproductive events in guenons – particularly late gestation, early lactation and the period when infants are rapidly switching from milk to a largely plant–arthropod diet.

In summary, it appears that rainfall is a primary ultimate factor determining the seasonal availability of guenon foods, while food is an important ultimate determinant of the timing of guenon reproductive events.

Interbirth intervals and the age guenons are weaned

There is considerable interspecific variation among guenons as to the length of the interbirth interval and age at which weaning ends. It is interesting to note that the smallest and the largest species of guenon have the briefest interbirth intervals and are weaned at the youngest age (Table 16.2). M. *talapoin*, E. *patas*, C. *campbelli* and C. *neglectus* have interbirth intervals of about one year and are weaned at approximately 12 months of age. C. *aethiops* have a fairly short interbirth interval of 12–20 months and are weaned at 9–18 months. C. *ascanius* have an interbirth interval of 1.5–4.3 years and young are weaned at 18–24 months of age. *C. mitis* have an interbirth interval of 2.4–5 years (Table 16.2). *C. mitis* infants begin to take solid food at about three months (Butynski, unpublished data) but suckling continues until they are 16–30 months old.

Interspecific synchrony of mating and birth seasons

Table 16.3 clearly shows that birth seasons occur at about the same time of year for forest guenons living in the same area. *M. talapoin* and the four species of forest *Cercopithecus* at Makokou, NE Gabon, all have birth seasons centered on late January or February. This places the middle of the mating period for all five species in July or August. Similarly, all three species of forest *Cercopithecus* at Kibale, W Uganda, appear to have birth peaks during December or January. The same is the case for *C. ascanius* and *C. mitis* in W Kenya.

Does this interspecific synchrony of mating and birth seasons extend to primates other than guenons? Although *Cercocebus albigena* is about twice the size of the typical guenon, its ecology is similar to that of the forest guenons in that it is a diurnal, arboreal, frugivore–omnivore with a diet of about 65% fruit and seeds, and 15% arthropods (Table 16.2). *C. albigena* at Kibale has a birth season of at least nine months (Table 16.3). In a sample of 29 births only one infant was born during April–July. These data suggest that *C. albigena* at Kibale has a birth season similar to that of the three guenons in that forest. The 5.8 months gestation of *C. albigena* places the primary mating period in the short wet season. Likewise, *C. albigena* in NE Gabon also appears to have a birth season which coincides with that of the five guenons in the area (Tables 16.2, 16.3).

Cercocebus galeritus, like *C. albigena* and the guenons, has a diet in which fruit is the major component (75%) (Table 16.2). *C. galeritus* at the Tana River, E Kenya, has a November–February birth season (Table 16.3). In NE Gabon this species seems to also have a birth season centered on these months (A. Gautier-Hion, unpublished data). Thus, both *C. albigena* and *C. galeritus* appear to follow the general forest *Cercopithecus* strategy of birth seasons centered on November–February when nutrients are probably most available.

Colobus badius is diurnal, arboreal and about the size of *C. albigena*. Unlike the guenons or *Cercocebus*, *C. badius* is a folivore. At Kibale its diet consists of about 72% foliage and 6% fruit and seeds. Arthropods are rarely eaten (Table 16.2). A sample of 68 births collected over 13 years by Struhsaker (unpublished data) shows that *C. badius* infants are born during all months. The distribution of births indicates a primary birth peak around April–June, when 46% of the births occur, and a distinct secondary peak about November. The birth peaks are, there-fore, during the second halves of the two wet seasons. The wet seasons in Kibale may be the periods when young leaves are most abundant as is the case at Makokou (Hladik, 1978). An assumed gestation of 6.6 months (Struhsaker & Leland, 1985) suggests that the two peaks in mating activity are centered on the two wet seasons with the primary mating peak occurring during the short wet season.

It appears that, in Kibale, *C. badius* has an annual distribution of mating and births which is distinctly different from those of guenons or *Cercocebus* – no guenon has been found to have a strong bimodal pattern of births and only *C. ascanius* is known to have births during all months. In addition, a birth peak located in April–June has not been reported for any guenon.

In general, fruits and arthropods seem considerably more seasonal and limited in their availability than foliage. Temporal and spatial differences in the quality and availability of foliage, fruits and arthropods may be largely responsible for the observed contrasts between the birth patterns of the folivorous *C. badius* and those species relying largely on fruits and arthropods.

At least part of the adaptive significance of birth seasonality in primates may lie not in when births occur, but simply in the fact that births are clumped in time (see above) (Bramblett *et al.*, 1975; Kavanagh, 1983a). This synchrony of births may reduce the risk of predation to mothers and their infants. There are at least two ways in which this might be achieved. First, it may produce a predator-swamping effect (Rowell & Richards, 1979; Schapiro, 1985). Second, pregnant females and mothers with clinging infants are less mobile than other group members (Rowell & Dixson, 1975; Chism, Olson & Rowell, 1983) and, therefore, probably more susceptible to predation. Predation on these individuals may be reduced if births are syn-chronized and slow animals are simultaneously handicapped. With many similarly slow animals around them, mothers and infants are less likely to be left behind without the benefits of predator detection and avoidance that being within the group affords.

That synchronous births might have a significant effect on reducing predation on primates has been called into question by the fact that primates often live in small groups into which few infants are born each year (Schapiro, 1985). I suggest that this point may be less relevant where the synchrony of births extends to neighbouring groups, to populations and, especially, to populations of both closely related and distantly related species in the same community. In particular, it seems that interspecific synchrony of births within primate communities could have a significant swamping effect on predators in areas where polyspecific associations are large and common – for example, in W Uganda and NE Gabon (Gautier-Hion, Chapter 23).

Data from captive studies

Data on the timing of reproductive events of guenons living under captive conditions are obviously of considerable value. It is concluded here, however, that such studies are of little or no value in helping us predict the following for wild populations of guenons living in their natural habitats:
1. whether mating and birth seasons occur;
2. how long mating and birth seasons last;
3. during which months matings and births take place.

For example, compare the results for wild populations summarized here with those from captive studies by Rowell (1970), Bramblett *et al.* (1975), Rowell & Richards (1979), Sly *et al.* (1983), Schapiro (1985), Else (1985) and Else *et al.* (1986). This conclusion also calls into question whether studies of captive colonies, where variables such as nutrition, photoperiod, rainfall and temperature are not experimentally manipulated, can tell us much about the proximate and ultimate factors affecting guenon reproductive events in the wild.

Conclusions and summary

1. This review suggests several generalizations concerning the guenon reproductive cycle. These should be useful for predicting the timing and length of mating and birth seasons of guenons in areas where research has yet to be undertaken, and of species of guenon which have yet to be studied. These generalizations are listed here. It will be interesting to see how they fare as additional information accumulates on guenon reproductive cycles and on the environments in which guenons live.

    (a)   All guenons have mating and birth seasons and mating and birth peaks. However, in areas of high rainfall (>150 cm/yr) some populations exhibit year round mating and birth.

    (b)   The majority of guenon populations (28 of the 33 reviewed here) have mating seasons centered on July,

August or September, and birth seasons centered on December, January or February.

(c)   Mating and birth seasons usually last three months or less where there is one wet season and three months or longer where there are two wet seasons each year.

(d)   Areas with less than 50 cm rain/yr do not have guenon mating or birth seasons which are longer than six months. Where rainfall is less than 100 cm/yr the mating and birth seasons are not longer than eight months. Year round mating or births do not occur where rainfall is less than 150 cm/yr.

2. There is considerable interspecific synchrony of mating and birth seasons among forest guenons living in the same area.

3. There are correlates between the distribution and amount of annual rainfall, the period over which nutrients are most available, and the timing of gestation, birth, lactation and the rapid transition from a mainly milk diet to a mainly plant–arthropod diet. *E. patas* young undertake much of the transition from milk to other foods when nutrients are most available. For other guenons the time when food is most abundant coincides with late gestation, birth and early lactation. It appears that, for guenons, rainfall plays a dominant role in determining the timing of food abundance. Food, in turn, is a primary ultimate factor determining the timing of reproductive events. In other words, rainfall, through its effect on the food supply, plays a major, but indirect, role in determining the timing of guenon reproductive events.

Acknowledgments

I thank the Uganda National Research Council, President's Office and Uganda Forest Department for permission to work in Kibale Forest. Annie Gautier-Hion, Jan Kalina, Janice Chism, Tom Struhsaker, Marina Cords and Thelma Rowell made valuable comments on the manuscript. Their help is much appreciated. I particularly wish to thank the following field workers for generously providing valuable unpublished data: Tom Struhsaker, Jeremiah Lwanga, Mike Lawes, Annie Gautier-Hion, Peter Henzi, Stan Koster, Jan Kalina, Pat Whitten, Lysa Leland, Joe Skorupa, John Oates, Peter Waser, Michael Ghiglieri, Lynne Isbell, Thelma Rowell, Marina Cords, Simon Wallis, and Janice Chism. The financial support provided by Wildlife Conservation International and World Wildlife Fund – US is gratefully acknowledged.

III.17

Mating systems of forest guenons: a preliminary review

MARINA CORDS

Animal mating systems are a major proximate determinant of gene flow. On one hand, they may influence patterns of breeding dispersal, and hence genetic mixing within one generation. On the other hand, they can filter genes passed through time to future generations. Understanding mating systems is thus prerequisite to understanding gene flow within the deme.

The term 'mating system' refers to two aspects of animal reproduction: (1) the pattern of actual sexual behaviour (i.e. who mates with whom), and (2) the genetic consequences of such behaviour (i.e. how many copies of whose genes reach the next generation). Usually, we measure the former and infer the latter, assuming a straightforward correspondence between mating and production of offspring. As we shall see, however, that correspondence is not straightforward in forest guenons, and though we may know little about patterns of actual mating, we understand even less about their genetic consequences.

Recent field studies of forest *Cercopithecus* monkeys (*C. aethiops*, *Erythrocebus*, *Miopithecus* and *Allenopithecus* are not included here) indicate that a species' mating patterns may vary between populations, between social groups in one population, and from year to year in one group. Little is known about interspecific variation, however, since mating patterns in most species have not even been described. This paper therefore focuses on within-species variation in mating system, and the reader should remember that even this discussion derives mainly from a few studies of only two species, *Cercopithecus ascanius* and *C. mitis*. In each of these species, mating varies from a 'female defence polygyny' pattern, in which one male monopolizes reproductive access to a group of receptive females by aggressively excluding other males, to 'promiscuity' during multi-male influxes, when members of both sexes mate with multiple partners. We know, therefore, two extremes: but the interesting questions remain, (1) what is the actual distribution of mating patterns within a group over time, between groups, and between populations? (2) What causes that distribution of mating patterns? (3) What are the evolutionary consequences? Although complete answers to these questions await further study, existing data suggest some preliminary conclusions and working hypotheses.

The biological background: female reproductive parameters

The limits of possible variation in mating systems depend partly on basic reproductive biology. Four aspects of female reproductive biology are reviewed here, all of which can limit the number of fertile and receptive females available to males. Unfortunately, the variability of these aspects of female reproduction is unknown due to insufficient data.

First is the age at first reproduction (given as age at first birth in Table 17.1(a)). All figures from wild populations (except *C. campbelli*) are *estimates* based on observed changes in body size over time. Females seem to give birth for the first time at about 4–5 years. Data on captive animals indicate that there can be considerable variation among individuals held in identical (but unnatural) conditions.

Interbirth intervals are given in Table 17.1(b), excluding cases in which the first of two infants died. Note that there are few data from wild populations, in which few individuals are usually recognized, and that short studies (the majority) will tend to find short interbirth intervals. It is safe to conclude, however, that most forest guenons are *not* annual breeders. In a given year, some fraction (on average) of all breeding females will conceive. An unanswered question is how variable this fraction is, i.e. how synchronized – among females – are births? The size of annual cohorts of surviving infants has varied three-fold (2–7 individuals) in a group of *C. ascanius* and six-fold (1–6 individuals) in a group of *C. mitis* in Kakamega.

All wild studied populations show some birth seasonality, although births occur in most or all months of the year in some cases (Table 17.1(c); also Butynski, Chapter 16). The environmental cues triggering successful breeding are unknown. The timing of birth peaks relative to rainfall is also somewhat variable between populations (Butynski, Chapter 16). The importance of social factors in stimulating mating (but

Table 17.1. *Age at first birth, interbirth interval, and birth seasonality for female forest guenons*[a,b]

Species and study site	(a) Estimated age at first birth	(b) Interbirth interval (months)[c]	(c) Birth seasonality
C. ascanius Kakamega, Kenya	—	49, 52, 60 (n = 3 int, 3 ♀♀)	Sep–Feb, peak Dec–Jan
C. ascanius Kibale, Uganda	4–5 yr (n = 1)	15 (11.5–28)* bimodal: 12 and 24 (n = 15 int, 7 ♀♀)	All months, peak Nov–Feb
C. ascanius Bangui, CAR	—	—	Aug–Sep
C. campbelli Adiopodoume, Ivory Coast	3 yr (n = 1)	12 (n = 5 int, 3 ♀♀)	Peak Dec–Jan, some Aug, Sep
C. cephus Makokou, Gabon	—	—	Peak Dec–Jan
C. cephus, captive Paimpont, France	5, 5 yr (n = 2)	27.4 (n = 1 int, 1 ♀)	—
C. lhoesti Kibale, Uganda	—	—	Peak Dec–Feb
C. mitis Kakamega, Kenya	—	47 (24–54) (n = 10 int, 10 ♀♀)	Most months, peak Jan–Mar
C. mitis Kibale, Uganda	—	Approx. 29–60[d]	Peak Dec–Jan
C. mitis Aberdares, Kenya	—	—	Nov–Mar, peak Feb–Mar
C. mitis Budongo, Uganda	—	—	Most months, peak Jan, Jul–Aug
C. mitis, captive Limuru, Kenya	65 (53–97) mo (n = 7)	18 (8–63) (n = 25 int, ? ♀♀)	All months
C. mitis, captive Limuru, Kenya	—	Mean: 19.2 (14.4–25.8)* (n = ? int, 10 ♀♀)	All months
C. neglectus Makokou, Gabon	—	—	Nov–Apr
C. neglectus, captive Paimpont, France	3.5, 3.8 yr (n = 2)	14.5, 12.2, 16.1 (n = 3 int, 1 ♀)	—

Continued

Table 17.1. *Continued*

Species and study site	(a) Estimated age at first birth	(b) Interbirth interval (months)[c]	(c) Birth seasonality
C. neglectus, captive Limuru, Kenya	60 (48–94) mo ($n = 6$)	23 (11–51) ($n = 13$ int, ? ♀ ♀)	Most months
C. nictitans Makokou, Gabon	—	—	Peak Dec–Jan
C. nictitans, captive Paimpont, France	5 yr ($n = 1$)	22.3 ($n = 1$ int, 1 ♀)	—
C. pogonias Makokou, Gabon	—	—	Dec–Feb (one in Apr)
C. pogonias, captive Paimpont, France	5 yr ($n = 1$)	21.2 ($n = 1$ int, 1 ♀)	—

[a] Data from Aldrich-Blake, 1970; Omar & DeVos, 1971; Hunkeler *et al.*, 1972; Galat-Luong, 1975; Gautier-Hion & Gautier, 1976, 1978; Rudran, 1978a; Galat-Luong & Galat, 1979; Rowell & Richards, 1979; Else *et al.*, 1985; Cords & Rowell, 1987, and unpublished records of T. Butynski, M. Cords, A. Gautier-Hion and T. Struhsaker.

[b] For $n < 4$, actual values are given. Otherwise, medians and ranges (in parentheses) are given. Means were substituted for medians only if the latter were unavailable.

[c] Interbirth intervals were calculated excluding cases where the first of two infants died. An asterisk indicates that this information was not available.

[d] This figure was calculated from group birth rates (births/female/year, $n = 122$ births), not from records of individual females. All births were counted, regardless whether the infant survived one year. Inasmuch as some adult females are post-reproductive, this method may overestimate the interbirth interval.

not necessarily conception) is suggested by the coincidence of multi-male influxes and increased sexual activity (see below). In captive *C. mitis*, however, neither Rowell & Richards (1979) nor Schapiro (1985) found evidence for social facilitation of successful breeding, in that births were not clumped. Similarly, *within* a limited mating period of *C. ascanius* in Kakamega, the estrus periods of females in one group were not more synchronous than would be expected by chance (Cords, 1984). Social stimulation apparently may only loosely synchronize breeding among individual females.

A final aspect of female reproductive biology is that receptivity (estrus) is not restricted to times when females are physiologically ready to conceive. Rowell (1970) studied a captive group of *C. mitis* in Uganda, in which behaviour was monitored concurrently with physiological cycles measured with vaginal lavages. All five females showed mating periodicity, but the correspondence with the menstrual cycle was highly variable between individuals: two females showed mating peaks in the second half of the cycle, one in the first half, one had no peak, and the fifth mated infrequently, but mostly in the first half. Copulations also occurred, though relatively rarely, during pregnancy. Girolami (1985) found that mating occurred throughout the menstrual cycle in captive *C. mitis* in South Africa, though males and females in her study were allowed contact only 15 minutes per day. During a 2.5 month study of *C. diana* in the Frankfurt Zoo, a pregnant female was seen to copulate 23 times, and 18 of the copulations occurred within a week of the infant's birth (Mörike, 1973).

In wild guenons too, receptivity and fertility are evidently not closely linked, either over the course of a menstrual cycle or over the year. In *C. ascanius* and *C. mitis* in Kakamega, where daily monitoring during the breeding season allowed us to determine estrus period durations, some females were receptive for a month or more (Cords, 1984, Cords *et al.*, 1986). In *C. ascanius*, sexual activity continued into pregnancy (Cords, 1984). On an annual time scale, mating sometimes started earlier and/or continued longer than the period when conceptions occurred (Tsingalia & Rowell, 1984; Cords *et al.*, 1986). Observations like these call for caution in interpreting behavioural data. The significance of sexual activity in these monkeys may not be as obvious as it appears.

Mating patterns: a review

Female defence polygynous mating
The association of guenons with female defence polygyny goes back to the first review of their social structure (Struhsaker, 1969). There are three kinds of evidence for female defence polygynous mating in guenons, two of which are indirect, in that mating was not necessarily observed (Cords, 1987a).

First, most group counts include only one adult male. There are about 90 independent group counts for 12 species combined, and they derive from censuses (Struhsaker, 1969; Quris, 1976), longer studies where individuals were not recognized (Aldrich-Blake, 1970; Gautier-Hion & Gautier, 1974; Galat & Galat-Luong 1985), and extended studies in which recognized males are known to have persisted in groups for up to several years (Hunkeler, Bourlière & Bertrand, 1972; Struhsaker & Leland, 1979; Butynski, 1982b; Cords, 1984; Tsingalia & Rowell, 1984; Cords & Rowell, 1986).

Although such a consistent pattern of group counts provides fairly strong evidence that guenon groups usually contain only one adult male, there are problems in interpreting these results as evidence of female defence polygyny (Cords, 1987a). We know from long-term studies of identified groups that complete group counts are seldom achieved on the first attempt; it is therefore especially likely that census counts – which make up the majority – have overlooked individuals. The behaviour of males mating in groups of females during multi-male influxes, when mating is promiscuous, suggests that such males may be especially easily overlooked because they can be peripheral and cryptic. Furthermore, group counts have been made throughout the year, whereas multi-male influxes – and promiscuous mating – occur most often during limited times when females are receptive. Finally, group counts in fact indicate social structure, not mating patterns, and so constitute only circumstantial evidence for the identity of the latter (see also Gautier-Hion & Gautier, 1985).

The intolerance shown by males resident in groups of females toward non-resident males (Galat-Luong & Galat, 1979; Struhsaker & Leland, 1979; Cords, 1984; Tsingalia & Rowell, 1984; but see Butynski, 1982b) is further indirect evidence for female defence polygyny. Resident males spend noticeably more time scanning their surroundings than other group members, or non-resident males, and when intruders are detected, they are usually chased away. Observations of the Kakamega *C. mitis* suggest that intolerance does not characterize all male–male relationships, however: in this population, some non-resident males regularly associate with one another over a period of several months. Overt friendly interactions are rare, but the males stay within sight of each other, often feeding and resting in the same tree. A telling case is that of a male who regularly chased others while a temporary resident during a multi-male influx, but who was seen a year later, as a non-resident, associating peacefully with some of the same individuals. We also know of one previously non-resident male who, after acquiring a newly formed group of females, tolerated the occasional visits of his former close associate. During these visits, only one or two females were receptive, and the former associate, who was not seen to

Table 17.2. *Tenures of recognized males who resided in groups of females for at least one year[a]*

Species and study site	Tenure lengths (months)[b]	Reference
C. ascanius Kibale, Uganda	39, 28, 24, 23, 23*, 21*, 17*, 15*, 13*	Struhsaker, unpub.
C. ascanius Kakamega, Kenya	39	Cords, unpub.
C. campbelli Adiopodoume, Ivory Coast	34	Hunkeler et al., 1972
C. mitis[c] Kibale, Uganda	94, 92, 58, 42, 37, 35, 24, 18, 12	Butynski, unpub.
C. mitis Kakamega, Kenya	49*, 46*, 26*, 26*, 26, 22, 14	Cords, unpub.

[a]Data on all species, except C. campbelli, come from studies in which observations were, at least in part, intermittent. Adapted from Cords, 1987a.
[b]Values reported are minimum values since many males were still in their groups when observations ceased.
[c]Information required for assigning asterisks (see below) was not available for these data.
*Tenures of males known to have been joined by other males during some periods when females were either seen to mate, or inferred to have mated because of births that occurred subsequently. The timing, duration, and probable significance of these periods vary.

interact with females, apparently posed no 'threat' to the resident, who could monitor the receptive female(s) closely.

The strongest evidence for female defence polygynous mating systems comes from long-term studies of particular social groups that individual males have monopolized when offspring were conceived. Table 17.2 gives tenure lengths for selected males who persisted in groups for at least 12 months, i.e. a time including at least one breeding period. There are other reports of males who stayed in groups less than a year, but who were the only males seen in their groups when offspring were conceived (Butynski, 1982b; Cords, unpublished data; Struhsaker, unpublished data). Many of the males whose tenures are given in Table 17.2 were observed to mate. Several were joined by other males, some of whom also copulated, during certain mating periods. In most mating periods, however, there is no evidence that residents were joined by other males; if one considers additionally the timing of multi-male mating periods relative to births, it is reasonable to conclude that most of the males whose tenure lengths are given in Table 17.2 did monopolize females reproductively and probably sired most of the offspring conceived during their stay in a group.

This conclusion holds despite the fact that all of these groups were monitored intermittently to some degree: the pattern of events in recognized multi-male influxes suggests that they would be difficult to overlook if they occurred, even with intermittent sampling. One cannot be certain, however, that females did not sometimes mate with 'sneaking' extra-group males, even in the absence of a *bona fide* multi-male influx. Intermittent observations are very likely to miss such copulations. Continuous tracking of receptive *C. mitis* females at Kakamega has shown that 'sneaked' copulations outside multi-male influxes may account for 5–20% of a group's observed copulations (Cords & Rowell, 1986).

Promiscuous mating

Promiscuous mating in guenons is associated with the break-down of the one-male group structure. It has been directly observed in *C. ascanius* (Struhsaker, 1977, Chapter 18; Cords, 1984) and in *C. mitis* (Tsingalia & Rowell, 1984; Cords *et al.*, 1986; Butynski, unpublished data; Henzi & Lawes, 1987) in three study areas. Aside from these few detailed studies, which are discussed below, multi-male influxes are known from other groups in the same populations, and the presence of more than one adult male in a group of females has been confirmed for several other populations and species (Bourlière, Hunkeler & Bertrand, 1970, *C. campbelli*; Galat-Luong, 1975, *C. ascanius*; Gautier-Hion & Gautier, 1985, *C. cephus, pogonias* and *nictitans*; Butynski, unpublished data, *C. lhoesti*; Curtin, unpublished data, *C. diana*). In some of these cases, more than one of the adult males present was observed to mate.

During multi-male influxes at Kakamega and Kibale, an average of 3–6 males (but as many as 11 in Kakamega) attended a group of females on any one day. The identity of those males changed, however, over the course of 2–3 months. As many as 19 males joined the focal group of *C.mitis* at Kakamega during a single mating period. Henzi & Lawes (1987) report at least 25 males joining their study group of *C. mitis* at Cape Vidal over a similar period. Many of the incoming males at Kakamega were observed as non-resident 'floaters' in the months preceding the mating period. Even males regularly resident in groups of females, however, sometimes leave their own groups for a few hours to court and mate females in a neighboring group. In all three study areas, any one male stayed with the group from a few hours to several months, and sometimes these new arrivals persisted as lone residents long after other newcomers and the original resident left. In other cases, the previously resident male persisted throughout and after a multi-male influx.

The presence of several males in the group coincides with elevated levels of sexual activity: *C. mitis* females in Kakamega were seen to copulate several times per day significantly more often during a multi-male influx year than in the next one-male polygynous year (Cords, unpublished data). Male copulation rates per day (with focal group females) were similar in both years, but more males mated in the focal group during the multi-male influx.

During multi-male influxes, members of both sexes may mate with multiple partners over the course of the entire mating period, a single female oestrus period, and even a single day (Cords, 1984; Tsingalia & Rowell, 1984; Cords *et al.*, 1986; Henzi, personal communication). Both aggression and active choice influence who mates with whom. Agonistic interactions are especially frequent among males, and often indicate competition for mates (e.g. Cords, 1984). Some male–male aggression seems to be incited by females: *C. mitis* females at Cape Vidal and Kakamega appear to lead their consorts directly towards other males, where a male–male conflict typically erupts. Among *C. ascanius* males in Kakamega, agonistic rank was correlated with measures of sexual activity over short periods of constant male membership during a multi-male influx (Cords, 1984). In two promiscuous mating periods in sympatric *C. mitis*, however, rank and sexual activity were not related (Tsingalia & Rowell, 1984; Cords *et al.*, 1986). In one of these periods only, exclusion of the resident male (who ranked highest but never copulated) rendered the correlation significant. The association between copulation and 'winning' agonistic interactions just prior to copulation further indicates that intermale aggression can affect mating patterns within the group, even in *C. mitis*. Aggressive competition for mates among females is less common, but it does sometimes prevent copulations (Cords *et al.*, 1986).

Active choice of mates by males and females is suggested by persistent following and solicitation behaviour and by the lack of acknowledgment of such advances when made by certain individuals but not others. In one promiscuous mating period among Kakamega *C. mitis*, it was found that males approached females more than vice versa overall, but that for those approaches ending in copulation, females were much more likely to have been the initiators (Cords *et al.*, 1986; also Struhsaker, personal communication, for Kibale *C. ascanius*). Henzi & Lawes (1987) similarly found that *C. mitis* mountings were more likely to lead to ejaculation if initiated by the female. Females also play a major role in forming and maintaining 'consortships', but both males and females could successfully evade persistent followers of the opposite sex (Tsingalia & Rowell, 1984; Cords *et al.*, 1986).

Although behavioural mechanisms for choice clearly exist, the criteria on which it is based remain obscure. The observations that preferences by one individual do not remain constant, and that individuals frequently diverge in their choice of partner, suggest that each monkey is choosing to be promiscuous, within some preference set peculiar to the individual. In so far as members of both sexes are exercising a choice, the availability of mutually attracted partners also brings about intra- and interindividual variation in apparent preferences.

The mating success of a male participating in a multi-male influx can also depend on how long he attends the group. Positive correlations between a male's length of stay and sexual activity were found for incoming males not resident in neighbouring groups in *C. ascanius* (Cords, 1984) and *C. mitis* (Tsingalia & Rowell, 1984; Cords *et al.*, 1986) in Kakamega. A long stay, like high agonistic rank, however, did not guarantee high mating success during a multi-male influx.

The reproductive consequences for males participating in an influx, and jockeying for mates via aggression, choice and mere presence, cannot be measured directly because paternity is unknown. Paternity might be estimated by evaluating observed sexual behaviour one gestation length prior to the birth of offspring. Stern & Smith (1984) found this method unsatisfactory for rhesus monkeys, but their behavioural sample was smaller than those of the Kakamega guenons. Perhaps more limiting in the present case is the fact that females may mate with more than one male over short periods (hours, days) and, at a more practical level, that birth dates and gestation lengths are often unknown. Furthermore, we know that males may mate in more than one group in a given year (Cords *et al.*, 1986, Henzi & Lawes, 1987), so an estimate of their annual reproductive success should include observations made in groups other than the focal one.

In spite of these difficulties, it seems that (1) in some cases, males resident in social groups before and after multi-male influxes do not have a reproductive advantage relative to those who join groups temporarily *in years when such influxes occur*, and (2) in the remaining cases, there is no basis for concluding that previous and subsequent residents are likely to have sired more offspring than other males who mated concurrently, again in years when multi-male influxes occur. In the Kakamega *C. mitis* group, the long-term resident male probably sired few if any of the offspring conceived during two multi-male influxes, as he was seen to mate very infrequently (Tsingalia & Rowell, 1984) or not at all (Cords *et al.*, 1986), despite intensive monitoring. In Kakamega and Kibale *C. ascanius*, males resident in a group before and/or after multi-male influxes may have sired some of the offspring conceived during the influx, but other males mated concurrently (and sometimes more frequently) and so are also possible fathers (Struhsaker, 1977; Cords, 1984). The above conclusions stand pending further information on the extent (i.e. how many groups) of a male's sexual activity, and the exact relationship between mating and conception.

A final point is that multi-male influxes and coincident promiscuous mating do not always correspond to peak conception times (Cords *et al.*, 1986; Struhsaker, unpublished data). This may simply reflect 'slop' in a system of reciprocal behavioural stimulation among animals that do not perceive fertility (Cords *et al.*, 1986). Another possibility, however, is that these intense bouts of socio-sexual activity do not have primarily a direct reproductive function, but rather serve to develop social relationships among males and between males and groups of females. These relationships in turn may enhance *survival* by increasing familiarity and the exchange of information, and such indirect effects on lifetime reproductive success may be as important as any conceptions that happen to occur at the time.

Causes of variation

On theoretical and empirical grounds, it seems reasonable to assume that a resident male would 'prefer' not to share reproductive access to his group of females. His success in monopolizing his group reproductively should depend both on his intrinsic ability to control access to the group, and on the amount of competition he faces. A male's intrinsic ability to control access to his group probably depends, at least in part, on the number and dispersion of receptive females it contains, as well as the forest density which can affect his monitoring ability. The number of receptive females may reflect the female reproductive parameters reviewed previously, as well as ecologically determined population parameters like group size and age structure. Dispersion of females similarly may reflect the cost and benefits of social foraging and predator avoidance.

Data relevant to these factors are somewhat equivocal. In *C. mitis*, multi-male influxes and promiscuous mating seem to have occurred at a higher rate in Kakamega than in Kibale, though sample sizes are small. Female reproductive parameters appear similar in the two study areas, but groups are larger on average and individuals are more dispersed in Kakamega. However, a series of male replacements and multi-male influxes has been observed in one Kakamega group whose size (21 members) is similar to that of groups in Kibale. In *C. ascanius*, the incidence of multiple males in groups in three study populations was not related to group size (Cords, 1984). These interpopulational comparisons are not very convincing, but case histories of individual resident males do support the hypothesis that the number of simultaneously receptive females limits a male's ability to monitor those females, and hence to control who mates with them. In the Kakamega *C. mitis* study group, only one female at a time was seen to be estrous when there was no multi-male influx (Cords *et al.*, 1986). When this group split into two (smaller) one-male units, extra-group males were seen to mate in either new group only when more than two of its females were simultaneously receptive (Cords & Rowell, 1986). At these times, a resident was unable to keep a close watch on all receptive females, since they were often over 100 m apart.

A male's intrinsic ability to control access to his group probably also depends on certain individual characteristics, such as his previous experience, confidence, and enduring relationships with other males in the population. Very long-term studies of individual males are needed to illustrate this point, but the possibility is suggested by our emerging awareness that males have a network of social relationships transcending the bounds of female groups (see Rowell, Chapter 22).

The amount of competition a male faces should also affect his success in excluding rivals from his group, whatever his intrinsic ability to do so. Given that females come in groups, an operational measure of competition is the ratio of adult males to groups of females in a population. This ratio is independent of population density, and depends only on group size and sex ratio of adults. The latter is extremely difficult to estimate accurately in the field, not only because extra-group males are cryptic, but also because the density of males appears to vary considerably on a local spatial and temporal scale. Thus large-scale censuses are apt to overlook individuals, while estimates of adult male density made during limited times in a few group ranges may provide misleading figures for the population as a whole.

The ratio of males to groups was estimated roughly for *C. ascanius* in three populations, assumed to have identical sex ratios and age structures, but it did not correlate with the incidence of multiple males in groups (Cords, 1984). This result seems to speak against the importance of male–male competition, but it may well be a spurious result: the assumption of similar sex ratios and age structures may be incorrect, and the population estimates of the incidence of multi-male groups, based on few data, may show only chance differences.

For *C. mitis*, there are also few firm figures. Butynski (1982b) found that male *replacements* in groups occurred more frequently in a part of the Kibale Forest where the males:groups ratio was estimated to be seven times higher than elsewhere. Group density in the Kakamega study area is 1.6 times higher than in Kanyawara, Kibale (Butynski, unpublished data; Cords, 1987a): the density of extra-group males in Kakamega (at least on our study area during promiscuous mating bouts) is almost certainly at least this much greater than at Kanyawara, where such males are almost never seen (Butynski, 1982b and personal communication). Thus, the males:groups ratio is very probably higher in Kakamega than in Kanyawara, corresponding to the apparent lower frequency of multi-male influxes and promiscuity at the latter site. Again, these conclusions must be considered preliminary, pending better estimates of male density and of the incidence of promiscuous mating bouts.

Evolutionary consequences for males

To discuss evolutionary consequences of mating system variability, one must know how male behaviour translates into producing descendants, and one must consider lifetime reproductive success. Presently existing data do not allow us to do either of these things with great confidence. These data suggest a few working hypotheses, however, in so far as (1) mating behaviour during bouts of promiscuous mating *has* genetic consequences for non-residents, and (2) the only male *seen* in a one-male group sires the offspring produced during his stay. The basis for preliminary acceptance of these premises has been discussed above.

First, male–male competition does not only involve defending a group of females by aggressively excluding intruders. Although this strategy is feasible for some males in some years, the presence of more than a few simultaneously receptive females and the availability of extra-group males seems to make effective defence impossible. When the one-male group structure breaks down, aggression between males may influence their mating success, but choice and mere persistence also play important roles.

One might still argue that the alternative competitive mechanisms used in bouts of promiscuous mating are relatively unimportant if such bouts are uncommon. Long-term residents might make up the losses incurred during promiscuous mating bouts by persisting unchallenged in other years. The male resident in the Kakamega *C. mitis* group, who copulated rarely or not at all during the two multi-male influxes he weathered in a five-year tenure, was nevertheless the only male seen in the group when 14 of the 23 infants born during his tenure were conceived. Using data on 39 tenures by 33 males in seven groups of Kibale *C. ascanius*, in which multi-male influxes sometimes occurred, Struhsaker (Chapter 18) similarly demonstrates a positive correlation between tenure length and the number of offspring probably sired *in a particular group*. These last words are critical, however, for we know that non-resident males may attend more than one group in a single year, and the question is still wide open as to what they do in other years, when the observer no longer sees them. It is at least possible that non-resident males range widely (indeed, most 'disappear' after a time) and attend one or more groups every year. Fair and comparable estimates of their reproductive success should consider offspring produced in all groups attended over a time interval identical to tenure lengths of residents; but because field studies have focused on groups, and not individual males, such estimates are not possible. For this reason, presently existing data cannot address the difference in reproductive pay-off of these two male strategies, no matter what assumption one makes about the connection between observed mating and paternity.

A further complication in estimating reproductive advantages of alternative male strategies is that an individual male may adopt more than one strategy in his reproductive lifetime. A few *C. mitis* and one *C. ascanius* male at Kakamega have switched between resident and non-resident status over periods up to several years. The meaningful comparison may therefore not only be between males spending variable lengths of time as residents or as non-resident participants in bouts of promiscuous mating, but also between non-resident males that do participate effectively in such bouts and those that do not. If participation in such mating bouts affects survival as well as immediate reproduction, the difference between these latter two classes of males may be even more important.

Evolutionary consequences for females?

As described above, variation in guenon mating patterns corresponds with dramatic variation in male behaviour, the genetic consequences of which are naturally of interest. Females too, however, behave differently when one or more males are present: in the latter case, they may mate more often and with more partners. We should at least consider whether mating system variation has reproductive consequences for females.

To some degree, of course, the options available to females depend on what the males are doing: if a resident male successfully rebuffs other males, or there are no (or few) non-resident males in the area, females will not have the opportunity to mate with males other than the one resident in their group. On the other hand, if this opportunity exists, females need not take it, but it seems that they usually do. Moreover, they do not choose to mate with only one of the available males, but with several. Why?

Schwagmeyer (1985) reviews adaptive functions of mating by females with multiple partners. Because male guenons do not provide material benefits to their mates, because females produce single offspring, and because females can resist unwanted male attention, all but two of the hypotheses reviewed are inapplicable to these animals. These two are that mating with more than one male (1) ensures fertilization, and (2) promotes male tolerance or care of young by obscuring certainty of paternity.

Dewsbury (1982a) has called attention to the limited nature of sperm production in mammals generally, though data on polygynous non-human primates are lacking. Precise figures would be needed to determine whether female guenons could benefit from taking multiple partners by increasing the chances of fertilization: although females copulate at higher rates during bouts of promiscuous mating (at Kakamega), at least some males do so also (especially if one were to include their sexual activity in other groups) and these males *may* consequently have reduced sperm counts per ejaculate. Females do conceive during periods when they mate with one or with several males, but because the correspondence between receptivity and ovulation is unknown, it is impossible to determine whether the rate of successful breeding is higher in the latter case.

Male tolerance of offspring translates roughly into males not perpetrating infanticide (Hrdy, 1981). Tolerance is expected to depend on males' association of a young infant with their mating 5–6 months earlier with the mother. Infanticide has been reported for *C. ascanius* (Struhsaker, 1977) and *C. mitis* (Butynski, 1982b), and is suspected for *C. campbelli* (Galat-Luong & Galat, 1979), though it appears to occur rarely. In all cases the marauding male was a newcomer to the group presumed not to have mated with the mothers of the infants that were killed, at least not at the time of conception. To test directly whether mating with multiple partners decreases the likelihood of infanticide, one should show that the infants of females that did mate with a particular male during a bout of promiscuous mating are less likely to be harmed by him than the infants of females that were not his partners during that bout. If there is insufficient variance in female promiscuity, however, one would be limited to the less elegant comparison of the fates of infants conceived during promiscuous mating bouts at the hands of resident males that were or were not participants in such bouts. At present, the data allow neither of these comparisons.

Paternal care in forest guenons consists mainly of thwarting predator attacks by alarm calling and mobbing. Such behavior would benefit all infants in a group, including those whose mothers had not mated with the male. Therefore, within-group comparisons of females that did or did not mate with a resident male cannot demonstrate that females trade copulations for paternal care. Instead, one needs to compare care-giving tendencies in males who have and have not mated with at least some of the females in their groups. Such data are not available, and would be difficult to obtain, since the observer must assess not only whether a male provides care, but also how often he has the opportunity to do so.

Schwagmeyer's discussion of multiple mating by female ground squirrels includes only functional explanations of this phenomenon, perhaps because these animals suffer an increased vulnerability to predators by mating multiply. Among forest guenons, however, mating with multiple rather than single partners does not impose any such obvious costs on females, and so the possibility should be considered that it has no *function*, but only a *cause* (*sensu* Tinbergen, 1951) and thus does not affect female reproductive success. Vandenbergh & Drickamer (1974) showed that induced sexual receptivity in only two female rhesus monkeys stimulated sexual behavior in males and other females in their group, even outside the normal breeding season. A similar 'voyeur' effect may operate in guenons, and when many males are present, females simply have partners that are unavailable at other times. In so far as male availability to females is influenced by heritable social and competitive skills among males, females avoid mating with the very 'poorest quality' males, while differences in 'quality' among those males taken as mates may be minimal.

Priorities for future research

The unanswered questions in the above discussion of evolutionary consequences of mating system variation for males and females indicate most clearly that new methods are required in future studies. The paternity of offspring must be determined and related to detailed observations of behaviour during mating periods. Further studies of female behaviour in relation to physiological cycles would be interesting, but it seems likely that no neat relationship will be found, and that such studies will not help us to determine paternity with confidence from behavioural data alone.

Further observational data are needed, however, to arrive at reasonable estimates of variation in mating systems within groups over time, within populations, between populations and between species. Mating patterns in most forest *Cercopithecus* species have not yet been studied. Estimates of variation are needed, in turn, to clarify the causes of variation, both at the level of individual animals' abilities and histories, and at the level of ecologically determined population parameters, which must, of course, be known as well.

Acknowledgments

My thanks go to Tom Butynski, Sheila Curtin, Annie Gautier-Hion, Peter Henzi, Mike Lawes, Thelma Rowell and Tom Struhsaker for their comments and/or access to their unpublished data.

III.18

Male tenure, multi-male influxes, and reproductive success in redtail monkeys (*Cercopithecus ascanius*)

THOMAS T. STRUHSAKER

Introduction

Social organization in the forest *Cercopithecus* species has generally been considered to have as its basis the one-male group, a social group containing only one fully adult male with several adult females and their young (Struhsaker, 1969; Aldrich-Blake, 1970; Hunkeler, Bourlière & Bertrand, 1972; Gautier-Hion and Gautier, 1974; Quris, 1976; Struhsaker & Leland, 1979). This group structure appears to be maintained largely through extreme intolerance among the adult males. Furthermore, it has been assumed that the *Cercopithecus* one-male group also reflects the mating system, namely, that the resident male generally has more or less exclusive access to estrous females. Exceptions, of course, occur. The short-term presence of more than one adult male in a particular social group has been reported from several studies: for redtails (*C. ascanius*) in Kibale, Uganda (Struhsaker, 1977), Kakamega, Kenya (Cords, 1984), and Central African Republic (Galat-Luong, 1975); for blues (*C. mitis*) in Kibale (Rudran, 1978a; Butynski, 1982b and unpublished data) and Kakamega (Tsingalia & Rowell, 1984): also for *C. cephus* and *C. pogonias* in Gabon (A. Gautier-Hion, personal communication).

Recently, however, the basic concept of the one-male group, and particularly of the one-male breeding unit, has been questioned (Cords, 1984; Tsingalia & Rowell, 1984). Central to the discussion are observations of two or more adult males simultaneously present and copulating in the same social group; and particularly, cases in which short-tenure males copulated more frequently than the long-tenure, resident male. Published accounts of such cases, however, are based on only one breeding period (Cords, 1984; Tsingalia & Rowell, 1984).

This chapter is intended to contribute to a more complete understanding of the problem by presenting data spanning 10 years of study in the Kibale Forest, W Uganda where seven groups of redtails were observed, some for as long as five years. The results demonstrate the extreme variation both within and between groups in male tenure length, the number of males in association with a group, the frequency of multi-male influxes, and possible reproductive success by individual males within the focal study groups. Despite this variation, the data demonstrate the utility of the one-male group concept and support the hypothesis that resident males of long tenure (so called harem males) have definite reproductive advantages over short-tenure or peripheral males within the focal study groups.

Study site and methods

Observations were made in the Kibale Forest of W Uganda. This is a medium-altitude rain forest, transitional in form between lowland and montane forest. Redtails were observed in two study areas within Kibale, Kanyawara and Ngogo, which are separated by approximately 11 km. Details of the habitat and sample methods are given in Struhsaker (1975, 1980) and Ghiglieri (1984).

Data reported here were collected from seven groups of redtails from 1973 to 1983 (Table 18.1). Sampling varied from a few hours to more than 100 hours per month. Two groups were sampled more intensively than others: TTK group (1973–5; $\bar{x} = 5.6$ days per month, SD $\pm$ 1.899) and S group (1975–6; $\bar{x} = 4.9$ days per month, SD $\pm$ 2.205). After 1976 the S group and all other groups were sampled on average only 1.4 to 2.9 days per month: S 1977–80 ($\bar{x} = 2.3$); Sn ($\bar{x} = 1.7$); Ss ($\bar{x} = 1.8$); SWK ($\bar{x} = 2.9$); RAT ($\bar{x} = 1.4$); BTP ($\bar{x} = 1.8$). Observations resumed on some of the groups in 1984 provide additional qualitative data.

Two groups were excluded from some of the analysis: SWK group because samples of it were usually short and infrequent; and RAT group because of exceptional circumstances (see p. 349). Group counts were infrequent and could rarely, if ever, be stated with certainty to be entirely accurate. Counts, however, were generally consistent, giving confidence in the approximate group size as indicated by the ranges in Table 18.1.

At Kanyawara, TTK and SWK groups were contiguous with one another. At Ngogo the BTP group's range bordered both the RAT group and S group, and later the two new groups, Ss and Sn, which formed upon the fission of the S group (see Struhsaker & Leland, Chapter 19).

Table 18.1. *Redtail sample data, Kibale Forest, Uganda*

Group name	Group size	No. adult ♀	Site	Sample time		Span of sample		No. months
				Hours	Months	Dates		
TTK	c. 35	9–10	K	609.7	16	Mar 1973–Jan 1975		23
SWK	c. 28–35	≥7–9	K	54.8	8	Nov 1973–Jan 1975		15
S	c. 35–50	15–16	N	1011.0	41	Jul 1975–Mar 1980		57
Sn	14–20	5	N	187.5	25	Jun 1980–Jun 1983		37
Ss	c. 30–35	10–12	N	248.1	25	Jun 1980–Jun 1983		37
RAT	c. 20–30	NA	N	188.5	22	Jan 1978–Dec 1981		48
BTP	c. 25–35	9–10	N	377.9	40	Apr 1978–Jun 1983		63
Totals				2677.5	177			

Results

Number of adult males simultaneously present in a social group

The number of adult males simultaneously in association with a group varied both within and between social groups (Table 18.2). Some groups, such as TTK and Sn, never had more than one male, whereas others often had two or three males (SWK). When data from all groups are combined, however, only one male was present during 70% of the 177 group-months sampled (Table 18.2).

Deviation from this weighted mean value could be explained for some groups. For example, attraction to an unusually large number of females may account for the relatively large number of extra-group males which associated with S group (see p. 349). The RAT group, on the other hand, was exceptional in having a large hybrid resident male (cross between redtail and blue and about 25–33% larger than redtail males; see Struhsaker, Butynski & Lwanga, Chapter 24) who persistently and successfully chased off adult male redtails. As a consequence, there was a nearly constant change in male membership. In at least 55% of the sample months of RAT group, there was more than one male redtail. No male was resident for more than two months until the hybrid male left (see Figure 18.5 below). If one excludes data for this group, the average and probably more typical percentage of months in which a group can be expected to have only one adult male redtail increases from 70% to 74%.

Table 18.2. *Frequency (number of months) with which each troop of* C. ascanius *was seen to include one or more adult males*

Group	None seen	1	2	3	4	5	6	Months sampled
	No. months with × No. ♂♂							
TTK		16						16
SWK	1		4	3				8
S		26	10		3	1	1	41
Sn		25						25
Ss		17	4	2	2			25
RAT	2	8	9	2		1		22
BTP		31	6	2	1			40
Totals	3	123	33	9	6	2	1	177
%	1.7	69.5	18.6	5.1	3.4	1.1	0.56	

Table 18.3. *Number of different adult males which associated with each troop of* C. ascanius

Group	No. A ♂♂	Sample months	Spread of sample (months)
TTK	1	16	23
SWK	4	8	15
S	16–21	41	57
Sn	1	25	37
Ss	7–8	25	37
BTP	8–12	40	63
RAT	13–20	22	48

Total number of different adult males per social group

The number of individually distinctive adult males associating with any particular social group in the entire sample varied from one to perhaps as many as 21 (Table 18.3). The range in numbers of males given in Table 18.3 for groups S, Ss, and BTP reflects an uncertainty in recognition of particular males by the observer. In other words, a male that was not recognizable by the observer could have been scored more than once.

The number of different males associated with a group was neither a function of sample size nor spread of the sample over time. Compare, for example, groups Sn and Ss, or BTP and RAT. Expressed as a rate (number of different males per sample month), the variation between groups is from 0.04 to perhaps as high as 0.91, or an average of 0.34, which represents a new and different male in association with a group every three months (3.7 months if RAT group is excluded).

Problems in locating adult males within social groups can be important when dealing with relatively small sample sizes. However, they cannot account for the variation in numbers of males in association with particular groups which were sampled with similar intensity. For example, the SWK group was sampled less than any of the groups and some samples were of only a few hours' duration. Nevertheless, at least two adult males were located in all but one sample.

Multi-male influx

The simultaneous association of two or more adult males in one-male groups has been referred to as a multi-male influx (Struhsaker, 1977). Combining the data for the five groups of Ngogo clearly suggests a seasonal trend in multi-male influxes (Figure 18.1).

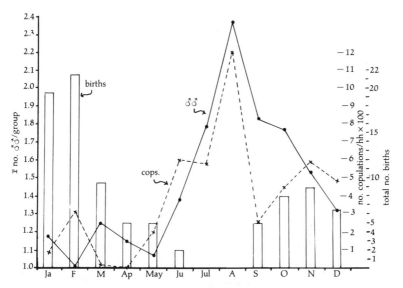

Figure 18.1. *C. ascanius*. Correlations between number of males per group (solid line), number of copulation bouts per hour × 100 (broken line), and total number of births (histogram). Data combined from 5 different social groups at Ngogo, Kibale during 8 years (1975–83). See Figures 18.4–18.7 for sample periods for each group. Data for the two groups at Kanyawara were excluded because their monthly distribution of births was significantly different from that at Ngogo. Furthermore, TTK group had only one male and SWK was inadequately sampled (Figure 18.3).

These influxes are positively correlated with copulation bouts ($r = 0.78$, $p < 0.01$, $n = 12$) and negatively correlated with births ($r = -0.60$, $0.05 > p > 0.01$, $n = 12$).

This analysis, however, fails to consider the extreme variation between groups in all three of these variables. The apparent correlation between multi-male influxes and probable conceptions implied in Figure 18.1 is more likely the result of multi-male influxes occurring in some groups, while conceptions were developing in others; it does not reflect a causal relationship (see p. 353 and Table 18.6 below).

Male tenure length

Minimum tenure length was determined in 30 cases for 24 different males in five groups (SWK and RAT groups excluded: see Methods). On average, a particular male remained with a specific social group for 8.1 months, with a mode of three months (Figure 18.2). The utility of an average figure for tenure is limited not only by the very

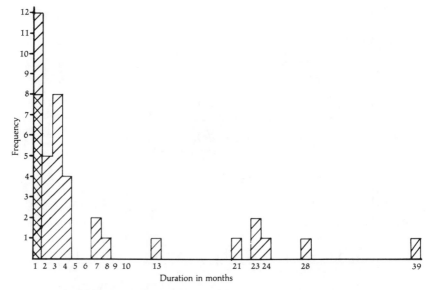

Figure 18.2. Frequency distribution of male tenure length in *C. ascanius* groups (39 samplings for 33 males; mean duration of male tenure = 6.5 months). The 8 cases indicated by cross hatching for 1-month tenure were all from the RAT group (see text). Tenure lengths are minimal estimates especially for those of 7 months or more.

high variance, but also in some cases by minimal estimates of tenure length. This was because some males included in this analysis were in the group when study of it began, while others were still present when studies were terminated.[1]

In the distribution of tenure lengths in Figure 18.2, there is a clear cut-off point in frequency for tenures greater than four months. Two classes can be recognized: the many *short-tenure* males remaining for four months or less; and the fewer *long-tenure* males staying for more than six months. Depending on whether or not the data from RAT group are included, 25.6% to 33.3% of the males had long tenure.

Temporal patterns of male tenure

The chronology of male tenure varied considerably between and even within those social groups studied for sufficiently long duration (Figures 18.3 to 18.7). Neither site location nor gross differences in habitat seemed to be important variables influencing temporal patterns of male association. For example, groups TTK and SWK were adjacent to one another in similar habitat types at the Kanyawara site,

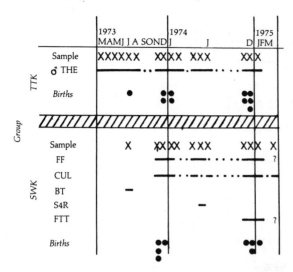

Figure 18.3. Chronology of male tenure in TTK and SWK groups. An × indicates sample months. A small dot means the male listed in the left column was presumably present. Each large black circle represents one birth.

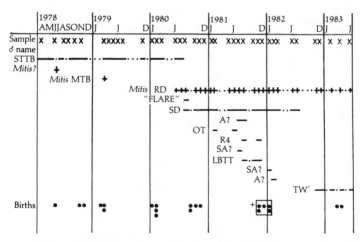

Figure 18.4. Chronology of male tenure in BTP group. Symbols as in Figure 18.3. *Mitis* (+) refers to a male blue monkey with the group. Births enclosed in a box were conceived when more than one male was present in the group. The small + indicates the birth of a back-cross between a female hybrid and a male blue.

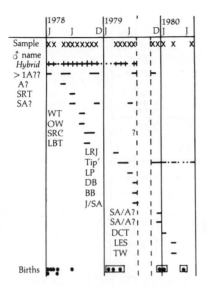

Figure 18.5. Chronology of male tenure in RAT group. Symbols as in Figure 18.3.

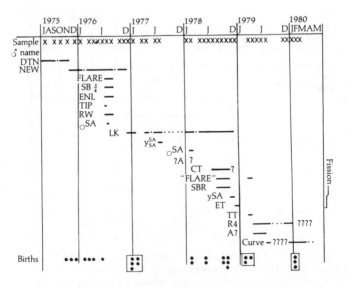

Figure 18.6. Chronology of male tenure in S group. Symbols as in Figure 18.3.

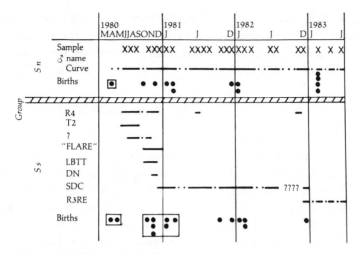

Figure 18.7. Chronology of male tenure in Sn and Ss groups. Symbols as in Figure 18.3.

and yet had very different patterns of male association. The TTK group had only one male throughout the entire 16 months of study, whereas the SWK group always had two or three males. Likewise, the five groups at Ngogo showed a similar diversity of male tenure patterns despite being contiguous with one another.

The patterns of male association in groups BTP, S, and Ss (Figures 18.4, 18.6, 18.7) may reflect the more typical long-term pattern. In these cases, there was an apparent oscillation between exclusive residency and influxes of one or more males who remained for relatively brief periods. Replacement of the long-tenure resident male often followed these influxes. Surprisingly, however, the new resident male was not usually involved during the peak of male influx. Rather, it seemed that he became associated with the group after most of the other males had left.

Several exceptional groups at Ngogo are worthy of mention. The RAT group, as described earlier, had a string of short-tenure males until the hybrid male transferred to a neighboring group (Figure 18.5). The S group at Ngogo, which subsequently divided, had 50 members including 16 adult females. This group had a high rate of male influx and turnover (0.45 different males/month). After the fission, the small segment Sn had only five adult females. This small number of females may have made it relatively less attractive to extra-group males and may account for the very long and exclusive tenure of male Curve

(Figure 18.7). In contrast, the large segment resulting from the fission (Ss) remained with some 10–12 adult females and, for the first seven months, continued to have an oscillating pattern of male association much like that of the original parent group (Figures 18.6, 18.7).

Correlates of male tenure length

Copulation bouts

There was a significant correlation between a male's tenure length in months and the number of copulation bouts he had in a specific group (Table 18.4, $r = 0.40$, $0.05 > p > 0.01$, $n = 30$ tenures for 24 males). A copulation bout was defined as a series of sexual mounts between the same male and female occurring within a 10-minute period. Intervals between bouts were greater than one hour.

Because long-tenure males were inevitably sampled more, one might expect to observe them copulate more by chance alone. However, conversion of the data to mounting rates (number/time) is also biased because in most months long-tenure males were present when no females were in estrus, while short-tenure males tended to visit groups when females were in estrus (although it is not always clear whether estrus preceded or was precipitated by male influxes).

Copulation in long-tenure and short-tenure males

Another way of examining the relation between tenure length and copulation success is to compare the performance of long- and short-tenure males. Eleven long-tenure males performed 78.2% of 87 mounting bouts, while 19 short-tenure males did only 21.8%. The difference in the average number of copulation bouts performed was 6.2 times greater in favor of long-tenure males ($\bar{x} = 6.2 \pm 5.5$ vs. $\bar{x} = 1 \pm 1.21$) and is highly significant ($U_s = 191$, $p < 0.001$).

These results support those in the preceding section, namely that long-tenure males have greater copulatory success than short-tenure males. It must be emphasized that this refers only to copulations within the focal study groups. Short-tenure males could have been copulating in other groups as well.

Tenure and possible number of offspring sired

Supporting the preceding analyses is a highly significant correlation between tenure length and the possible number of offspring sired in the focal study groups (Figure 18.8, $r = 0.91$, $p < 0.001$, $n = 38$ cases for 32 males from six groups).

Table 18.4. *Tenure length and number of copulation bouts (30 cases for 24 different males)*

Name of male	Tenure (months)	No. copulation bouts
Curve	⩾39	4
STTB	⩾28	1
SD	⩾24	7
THE	⩾23	4
LK	23	9
SDC	⩾21	5
NEW	13	22
TW	⩾8	2
R4	7.5	7
R4	7	5
Curve	7	2
DTN	⩾4	3
CT	4	2
LBTT	4	0
SBR	3	2
T^2	3	1
'Flare'	3	0
LBTT	2.75	2
Flare	2.5	4
SB$\frac{3}{4}$	2.5	2
ENL	2.5	0
RW	2.5	0
'Flare'	2	0
ET	2	0
R4	2	0
OT	2	2
DN	1	0
TT	1	0
'Flare'	1	0
Tip	<1	1

Paternity determination is beyond the scope of this study. At best we can only discuss possible paternity based on backdating from birth dates to the time of conception. The assumptions involved in this analysis are: (1) a gestation period of 5–6 months (even 4 months does not alter the correlation); (2) equal probability of paternity when more than one male was present in the group at the estimated time of conception (e.g. if two males were present, $p = 0.5$ for each conception); (3) that males did not leave and then rejoin the focal study group between samples (where doubt exists, then a mean of maximum and minimum estimates of tenure length was used). There is also the

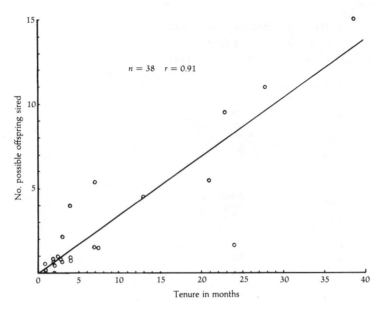

Figure 18.8. Significant correlation between male tenure length and number of probable offspring sired (see text).

possibility that the observer overlooked peripheral males engaging in kleptogamy ('stolen' copulations) away from the core of the group. However, the sample is thought to be sufficiently large and representative to compensate for these possible exceptions.

It should be emphasized that the assumption of equal probability of paternity when more than one male was present in the month of conception may not be valid. Earlier analysis has shown that in most cases long-tenure males copulated more than short-tenure males and that resident males can have clear copulatory advantages over transient males (Struhsaker, 1977). If this proves to be the general case, then the correlation between tenure length and the possible number of offspring sired within focal study groups is even stronger.

Conceptions and number of males present in group

Another way of describing the mating system in redtails is to consider the number of adult males present in a group at the time of conceptions. Despite the variation between and within groups, more than 69% of all infants were conceived when only one male was present (Table 18.5). There was a trend toward more conceptions when only one male was present than when more than one male was in the group

Table 18.5. *Conceptions and number of males present*

| Group (years) | Resident male | No. conceptions (%) | | Total |
		With only resident ♂	With more than one ♂	
TTK	THE	8–10 (80–100%)	2? (20%?)	10
1973–4				
S	DTN	4 (100)	0	4
1975-80				
	NEW	4 (57.1)	3 (42.9)	7
	LK	7 (70)	3 (30)	10
	R4 & Curve	0	6 (100)	6
Sn	Curve	12 (100)	0	12
1981–3				
Ss	R4	0	7 (100)	7
1981–3	SDC	6 (100)	0	6
BTP	STTB	13 (100)	0	13
1978–83	SD	0	5 (100)	5
	TW	1 (100)	0	1
RAT	none			
1978–80	>2 mo	1 (16.7)	5 (83.3)	6
Totals		57 (65.5)	30 (34.5)	87
Totals (exclude RAT)		56 (69.1)	25 (30.9)	81

($U_s = 82$, $0.10 > p > 0.05$). In all cases, the resident males were also long-tenure males. These data suggest that long-tenure or resident males have at least a two-fold advantage in terms of conceptions over the short-tenure males within the focal study groups. Because long-tenure males may also have sired some or even all of the infants conceived when other males were present, it must be emphasized that this is a conservative estimate of reproductive success for these males.

Effects of male tenure length and multi-male influx on birth-rate

Although the sample is small (five groups), there is a significant inverse correlation between birth rate and the number of males present per sample month ($r_s = -0.9$, $p = 0.05$) and a weakly significant positive correlation between birth rate and percentage of sample months in which only one adult male was present ($r_s = 0.88$, $0.10 > p > 0.05$, Table 18.6). These results suggest that short tenure of adult males, highly dynamic male membership, and/or multi-male influxes may be deleterious to female reproduction (see Struhsaker & Leland, Chapter 19).

Table 18.6. *Birth rate and number of males associated with group*

Group (size)	Number months sampled	No. males associated	% months only one male	No. males per sample month	Annual birth rate
Sn (14–20)	25	one	100	0.04	0.74
TTK (c. 35)	16	one	100	0.06	0.55
BTP (c. 25–35)	40	8–12	77.5	0.2–0.3	0.42
Ss (c. 30–35)	25	7–8	68	0.28–0.3	0.36
S (c. 35–50)	41	16–21	63	0.39–0.51	0.39

Patterns of movement between groups by individual males

The association of a male with a particular social group was not necessarily continuous over time. For example, some males were known to return to a particular group after an absence of 17 and 22 months (R4 and 'Flare', Figures 18.6, 18.7). Likewise, male TW' was resident in the BTP group for approximately 23.5 months before being replaced by male IV. After an absence of about eight months, TW' returned and stayed in the BTP group with male IV for less than one month before disappearing again in July 1985. One year later TW' again returned briefly to the BTP group in August 1986 when he was seen to copulate twice (one complete), even though the resident male IV was also in the group. TW' had a fresh wrist wound at this time, suggesting he had fought with IV. TW' was next seen as a solitary 3.5 months later in December 1986.

At least three males, 'Flare', R4, and LBTT (Figures 18.4, 18.6, 18.7) moved between different groups.

Although we can document movements of males within and between focal study groups, we know virtually nothing about their activities outside these groups. In spite of this gap, two sets of observations provide preliminary insight into the problem. First, incidental observations were made of one distinctive solitary male (BT) during 19 days that were spread over 40 months at Kanyawara. No systematic observations were made of this male, and the sample of his movements was clearly biased in favor of areas used by the focal study groups. None the less, a minimal estimate of his range during this time was 37 ha (convex polygon), or about 32–85% larger than that of the TTK group (20–28 ha) living in the same part of the forest. Of greater significance, however, is that this male's range overlapped in part the ranges of at least five social groups (including TTK). He was occasionally seen peripheral to some of these groups, but never interacting with them.

Secondly, the longest studies in Kibale are of three groups at Ngogo whose ranges were either contiguous and partially overlapping, or very near to one another (S group which split in two, BTP, and RAT). In spite of this, only three of 37 males that associated with these three groups were seen in more than one of the groups (Figures 18.4 to 18.7), suggesting that males did not move frequently between adjacent groups.

Discussion

Comparison with other studies

In the Kakamega Forest of W Kenya, a study of redtails yielded results which fall within the range of variation described for the Kibale redtails (Cords, 1984, 1987a). Male tenure was variable (less than one to more than 39 months). In one group studied for two years, there was a positive correlation between measures of mating success and the time a male spent in the group. Contrary to the general trend, but still within the short-term variation of some Kibale groups, was the finding that of nine births, five were conceived when more than one male was present. The remaining four cases were equivocal. Depending on the exact dates of birth and gestation length, one infant was conceived with only one male in the group, and another one when one or perhaps more males were present. The number of males in the group was not known for two of the conceptions (from Figure 1 and Table 4 in Cords, 1984). In this group there was no obvious long-term resident male until after the six-month period when most copulations were observed.

Studies of the closely related blue monkey (*C. mitis*) in Kibale show ranges of variation in male tenure similar to those of the Kibale redtails. During a five-year study at Kanyawara, the average minimal tenure length for adult males in seven groups was 41.1 months (±25.3, range 13–72 months), while at Ngogo tenure was seven times shorter in the one group of this subpopulation ($\bar{x} = 5.8 \pm 5.8$, range 0.4–17 months, Butynski, 1982b and unpublished data). Consistent with this are the results from the earlier and shorter study of two groups of blues at Kanyawara where Rudran (1978a) reported tenure lengths of at least 16 and 22 months.

There were also differences between the two study sites in the amount of time more than one adult male blue was with a particular group: 2% of the sample months at Kanyawara, and 11% at Ngogo (Butynski, unpublished data). In both areas, however, it appears that multi-male influxes were considerably less frequent and comprised less of a group's time in blues than among the Kibale redtails (see also Rudran, 1978a).

Most, if not all, of the 35 births recorded for blue monkey groups at Kanyawara were conceived when only one male was present, whereas paternity of 23 blues born at Ngogo was less certain because of the shorter male tenure there (from Butynski, unpublished data). Similarly, during Rudran's (1978a) study of Kanyawara blues, all 10 infants born in his two study groups were conceived when only the resident male was known to be present. In general, it appears that among the blues in Kibale, there was a close correlation between tenure length and probable number of offspring sired at Kanyawara, but much less so for the study group at Ngogo (from Butynski, 1982b and unpublished data).

Tsingalia & Rowell (1984) provide details on a multi-male influx in a group of blues during a six-month period in the Kakamega Forest. They question the reproductive advantages accruing to the resident male of a one-male group. A re-evaluation of their paper, however, demonstrates overwhelming advantages to him. The study group numbered some 45 to 50 individuals, larger than any group ever recorded for the species (Cords, 1987a), and it eventually divided in 1984 (Cords & Rowell, 1986). During the six-month period reported by Tsingalia & Rowell (1984), 16 different adult males associated with the group, including the male TA who was the resident and only male with the group during the preceding two years. Nine of the 16 males were seen to copulate, including the resident male TA. Of the 4 to 6 females thought to be in estrus, however, only three conceived during this period of multi-male influx. In contrast, of the 15 total births recorded for this group, 12 were conceived when only one male (TA, the resident) was present. Thus, like the redtail and blue studies in Kibale, long-tenure or resident male blues at Kakamega probably had a pronounced reproductive advantage over short-tenure males within the focal study group. Depending on the assumptions made regarding the probability of paternity during the multi-male influx at Kakamega, this difference in reproductive advantage could be as great as 16- to 36-fold in favor of the long-tenure male.

Less detailed observations on four other groups of blues at Kakamega suggest that, in intergroup variation in male membership, they were more similar to the Kibale redtails than blues. Two groups had relatively rapid turnover in male membership, while each of two other groups had only one resident male for at least two years (Tsingalia & Rowell, 1984).

The results from both Kibale and Kakamega demonstrate great variation in male tenure length and the frequency of multi-male influxes in redtails and blues. In general, these studies also show a clear reproductive advantage to the long-tenure males over those of shorter tenure within the focal study group. Most conceptions occurred when only one male was present in the group. Combining all available evidence for either redtails or blues supports the hypothesis that the one-male group reflects not only a distinct social organization, but also a polygynous mating system in which the long-tenure or resident male generally, but not invariably, has reproductive advantages within his group over those of transient or short-tenure males. The major differences in interpretation of results from Kibale and Kakamega apparently stem largely from differences in sample size and duration of the study reported, although demographic variables may also play a role.

Cords (1984, p. 314, and Chapter 17) suggests that the redtail social structure 'fluctuates between one- and multi-male groups'. This description is misleading, however, because the multi-male influxes do not resemble multi-male groups of other cercopithecids, such as mangabeys, baboons, macaques and red colobus in terms of the degree of male–male tolerance, affiliative behaviors, agonistic coalitions and duration of tenure in a social group.

Situations inevitably arise in which a resident or harem male is unable to exclude competitors and maintain exclusive access to estrous females. Exceptions to exclusive access to estrous females by the harem male have been described for a number of other species, including gelada baboons (Dunbar, 1979, 1984), elephant seals (Le Boeuf, 1974), red deer (Clutton-Brock, Guinness & Albon, 1982) and horses (Rubenstein, 1980). Some view these examples of kleptogamy or mating with females on the periphery of the harem as a fairly regular, though less common, form of mating strategy alternative to the more typical harem system (Rubenstein, 1980). The fact that kleptogamy and multi-male influxes occur is a reflection of male competition for mates. These alternative male behaviours neither negate nor replace the basic concept of the one-male group and polygynous mating system.

Determinants of tenure and multi-male influxes

The number of extra-group males per one-male social group will likely influence the intensity of competition among males for mates and, correspondingly, the frequency of multi-male influxes and the duration of male tenure. Variation between males in tenure length may, in turn, be influenced by individual behavioural characteristics (Cords, 1984, and Chapter 17). Differences between groups in the frequency of multi-male influxes and tenure of the resident male may also be influenced by the number of estrous females they contain.

Number of males per social group

Determining the ratio of males per social group is made difficult because solitary males are often inconspicuous. As yet, we do not have accurate estimates of adult sex ratios or density of males for any forest guenon. Cords' (1984 and Chapter 17) estimates, based on the assumption of a unitary adult sex ratio, are questionable and require verification.

Despite these problems of determining the density of males, the two subpopulations of blues in Kibale were sufficiently different so as to allow relative comparisons and provide insight into the effects of differing male densities. At the Ngogo site, where there was an unusually low population and social group density, the density of solitary males was estimated to be seven times greater than at the Kanyawara site, where group and population density were high. Furthermore, there were approximately 6.7 times more females per male at Kanyawara than at Ngogo. These differences in sex ratio, density, and ratio of adult males per social group correlate well with the shorter tenure of males and the more frequent occurrence of multi-male influxes at Ngogo as compared to Kanyawara (Butynski, 1982b and unpublished data).

Number of estrous females available

Groups having many estrous females should offer greater potential attractiveness to solitary males than groups with few or no estrous females. Evaluation of this variable in guenons is complicated by both cryptic and postconception estrus. Furthermore, the degree of synchrony in estrus among females of a particular group can vary annually because of broad birth seasons (e.g. nine month span for some redtail groups) and differing interbirth intervals for individual females (e.g. redtails, 11.5 to 28 months) (see Butynski, Chapter 16).

In the absence of detailed data on estrus in female redtails, we must resort to an analysis of the relation between the number of adult females in a group and its attractiveness to extra-group males. A strong correlation between these two variables was shown for the S group and its subsequent daughter groups that formed after the group fission (see p. 349). A similar situation occurred in the unusually large group of blue monkeys at Kakamega (Tsingalia & Rowell, 1984).

Another line of evidence supporting this hypothesis that the number of females in a group is an important determinant of male tenure and influx is the significant correlation between the number of adult females in a group and the number of different adult males associated with that group per sample month (see Tables 18.1 and 18.7; $r_s = 1.0$, $p = 0.01$, $n = 5$ groups). Similar correlations have been reported for gelada baboons. Dunbar (1979) observed that the number of males in association with harems and the frequency of takeovers by males increased when similar numbers of females were compressed from 25 harems into 17. On the other hand, Cords (1984 and Chapter 17) concluded there was no correlation between the incidence of multiple males and the size of a group in her comparison of three study populations of redtails. Although group size can provide a rough index of attractiveness to extra-group males, the critical factor is the number of females per group. Furthermore, comparisons of the incidence of multiple males and average group size *between* populations will not be meaningful unless other variables, such as the ratio of extragroup males per one-male social group, are taken into account.

The likelihood that any particular group will experience a multi-male influx or male replacement may depend not only on the ratio of adult males per social group, but also the degree of estrous synchrony between females of neighboring groups. The more closely synchronized they are, such as in annual and highly seasonal breeders, the less likely any particular group may be expected to experience male influxes or replacement.

In summary, male tenure and the frequency of multi-male influxes in redtails and blues are probably determined by a complex interplay of these variables related to male competition for mates: density and ratio of adult males per one-male social group; number of estrous females within and between groups; and behavioural characteristics of individual males (see also Cords, Chapter 17; Dunbar, 1979, 1987). Improved predictability is likely to hinge on more refined long-term data and samples from populations of varying demographic structure.

Birth rate and male association

Data were presented for redtails suggesting that short tenure, rapid turnover, and/or frequent multi-male influxes might be detrimental to female reproduction (p. 353). How these factors influence birth rates is not known. Social stress resulting from increased levels of aggression among males and by males toward females during influxes and takeovers may be a possible factor contributing to reduced birth rates (see Chapter 16 and Dunbar, 1979). Stress may also adversely affect fecundity of males. Copulations with more than one male within a relatively brief time may also block pregnancy, as demonstrated in deer mice (Dewsbury, 1982b) and suggested for chimpanzees (Tutin, 1980).

Males can also have an effect on birth rates in terms of infant survivorship. Following infanticide, for example, interbirth intervals can be significantly shortened (Leland, Struhsaker & Butynski, 1984; Struhsaker & Leland, 1985). Infanticide may help explain some of the pronounced differences in birth rates between groups of blues similar in size and composition (Butynski, 1982b and unpublished data), but not for redtails where survivorship seemed to be similar between groups having different birth rates.

Outstanding issues and speculation

There are several issues relevant to the theme of this chapter for which there are either no data or only a few qualitative observations. It seems important, however, at least to raise these points as a means of indicating gaps in our knowledge and to stimulate more research.

Female choice of mates or paternity deception?

The distinction between female choice and deception among redtails and blues is confounded by cases of infertile estrus. During male influxes into groups of redtails (S group, Struhsaker, 1977) and blues (Tsingalia & Rowell, 1984) several females copulated, sometimes with more than one male, but few conceived. Was this due to stress-induced infertility or paternity deception?

Females may not only be promiscuous during multi-male influxes, but they have also been observed moving far to the periphery of their group to copulate with males from other groups (for blues, Cords, 1987a; Struhsaker, unpublished observations). Are these examples of mate choice, or of paternity deception as a contingency behavior against possible male replacement and infanticide, i.e. 'bet-hedging' (Hrdy, 1974; Struhsaker, 1977; Struhsaker & Leland, 1985).

To what extent is female choice dependent on familiarity with her mate? In terms of fitness, one would expect ovulating females to copulate with males having advantageous qualities, as evaluated through observation, and thus familiarity. If this is true, why do females copulate with what appear to be strangers on the periphery during multi-male influxes? Paternity deception may explain some of these cases. However, in others it is likely that the males were not strangers at all since they are known to return to a group after long absences (p. 354; and also in Kibale blues, T. M. Butynski, personal communication).

Long-term reproductive success of males

Although we have reasonable indications of possible reproductive success by males within focal study groups, we know virtually nothing about their activities elsewhere. Why do short-tenure males leave a group when they are not obviously forced out? What becomes of long-tenure males once they leave a group? The alternative reproductive behaviours of long vs. short tenure are probably not genetically determined because we know of at least two males (R4 and TW') who were both long- and short-tenure males. These different patterns may reflect ontogenetic stages with, for example, old and young males having short tenure, and middle-aged males having long tenure. Alternatively, this variation in male behaviour may simply reflect opportunism depending on the interplay of variables discussed earlier (see also Cords, Chapter 17 for male qualities). Among the most important of these may be a male's relationship with specific males encountered while attempting to enter or defend a group.

In order to understand more fully male reproductive success, we need to follow males once they leave a focal group and to study solitary males. If the available information from our focal study groups is representative, however, then all indications are that solitaries and short-tenure males do not achieve higher reproductive success than long-tenure males even outside the focal study groups.

Genetic implications

Inbreeding is more likely to occur in some groups than in others. For instance, in the Sn group of redtails at Ngogo and some of the blue groups at Kanyawara (Butynski, unpublished data), the same resident males have remained for five to six years without any evidence of peripheral males interacting with females. The possibility of father–daughter matings creates the potential for deleterious inbreeding. We do not know to what extent this might account for low birth rates and/or high infant mortality in some of these groups. The answers to these and many of the others raised throughout this paper will ultimately hinge largely on genetic determination of paternity.

Summary of results and conclusions

1. The number of adult males per redtail social group was variable, but only one male was present during 70% of the sample time.
2. Tenure length of males in any one group ranged from less than one month to at least 72 months. Two categories could be distinguished: short-tenure ($\leq$4 months) and long-tenure ($\geq$7 months) males.
3. Long-tenure males copulated more and had a greater likelihood of siring more offspring than short-tenure males within the focal study groups.
4. Although multi-male influxes correlate with the peak of copulations in the population, these influxes do not necessarily lead to a high incidence of conceptions within a particular group.
5. The advantages to a male of maintaining long tenure in a particular group lie not only in achieving copulations with females who are in estrus outside the peak season, but also in protecting his offspring against possible infanticidal attacks by intruding males.
6. Multi-male influxes in one-male redtail groups are temporary states. Relations between males remain strongly antagonistic and intolerant. At no time do they bear any resemblance to the multi-male groups of other cercopithecids where males often interact in friendly behaviour and form coalitions.
7. The fact that multi-male influxes occur in the one-male groups of redtails and blues is a reflection of male competition for mates and does not nullify the utility of the concepts either of the one-male group or the polygynous mating system in these species.

Acknowledgments

I am grateful to the following colleagues for useful comments and discussion relevant to the preparation of the manuscript: Ms L. Leland, and Drs T. M. Butynski, J. Chism, M. Cords, A. Gautier-Hion, and T. Rowell. Major funding for the field work, data analysis, and writing was from the New York Zoological Society. Some of the data analysis was done while I was a Fellow at the Center for Advanced Study in the Behavioral Sciences, funded in part by National Science Foundation BNS 76-22943 and the Alfred P. Sloan Foundation 82-2-10. Additional data were collected, analysed and written up during the tenure of grants from the H. F. Guggenheim Foundation and the National Geographic Society (No. 2929-84) for studies of redtail monkey behavioral ecology. Permission to study in Kibale was granted by the President's Office of Uganda, Uganda National Research Council, and the Uganda Forest Department. The Department of Zoology, Makerere University provided my local affiliation in Uganda. The assistance of all these parties is gratefully acknowledged.

Note

[1]Subsequent to completing the analysis for this chapter, longer tenures were established for two males: R3RE, at least 3.75 years; and Curve, 6 to 6.5 years.

III.19

Group fission in redtail monkeys (*Cercopithecus ascanius*) in the Kibale Forest, Uganda

THOMAS T. STRUHSAKER and LYSA LELAND

Introduction

Group fission, or the permanent separation of members of a social group into two or more new social groups, has been described for only seven species of Old World monkeys; *Colobus guereza* (Dunbar & Dunbar, 1974b); *Macaca fuscata* (Sugiyama, 1960; Furuya, 1969); *M. mulatta* (Southwick, Beg & Siddiqui, 1965; Chepko-Sade & Sade, 1979; Melnick & Kidd, 1983; Malik, Seth & Southwick, 1985); *Papio anubis* (Nash, 1976); *Theropithecus gelada* (Dunbar, 1979, 1984); *Cercopithecus aethiops* (Hauser, Cheney & Seyfarth, 1988); and *C. mitis* (Cords & Rowell, 1986, T. M. Butynski, personal communication). Group fission, as defined here, is distinct from the temporary separation of group members, e.g. vervets (Struhsaker, 1967a) and baboons (DeVore & Washburn, 1963; Stoltz, 1972), and the fusion–fission societies of hamadryas baboons (Kummer, 1968) and chimpanzees (van Lawick-Goodall, 1968).

In the Kibale Forest of W Uganda, three cases of group fission have been observed in three of the six species studied there: redtails (*Cercopithecus ascanius*), blues (*C. mitis*, T. M. Butynski, unpublished data), and mangabeys (*Cercocebus albigena*, Leland, unpublished data). This chapter describes a case of group fission in redtails, where the group was studied for five years prior to and five years following division. We examine the ecological and behavioural (particularly reproductive) consequences of group fission; consider the adaptive significance of group fission in redtails; and compare these results and conclusions with case studies for other species.

Methods and study site

Observations of redtails were made in the Kibale Forest of W Uganda beginning in 1973 and continuing through 1986, with a break of 15 months in 1983–4. Detailed studies were made of eight groups in two study sites within Kibale (Ngogo and Kanyawara, see Table 19.1 and Figures 19.1 and 19.2; see also Struhsaker, Chapter 18, Figures 18.3–18.7 for details of sample months and hours). Incidental observations were made of numerous other groups. Further details on the habitat and sample methods are given in Struhsaker (1975, 1980, Chapter 18) and Ghiglieri (1984).

Results

Background information: Kibale redtail monkeys live typically in one-male social groups numbering about 30–35 individuals, but ranging in size from 10 to 50 (Cords, 1987a; Struhsaker, Chapter 18). Females and their young constitute the permanent core of the group, while adult males have relatively short tenure. Old juveniles and subadult males emigrate from their natal groups. Females and juveniles are the most active defenders of territories; while resident males participate moderately or not at all in territorial encounters. Multi-male influxes do occur, although there is considerable intra- and intergroup variation in their frequency and duration (Struhsaker, Chapter 18; Cords, Chapter 17). Adult and subadult males are generally aggressive and intolerant of one another. No all-male groups have been seen in Kibale nor have they been described elsewhere.

The fission and subsequent group dynamics

The fission occurred in the S group of redtails at the Ngogo study site. Observations began on this group in July 1975, when it numbered some 35 individuals including about 15 adult females (Struhsaker, 1977, and Chapter 18). By January 1980 the S group had increased to nearly 50, with 16 adult females. At this time, the group was clearly undergoing temporary divisions during much of the day, although the two parties then rejoined late in the afternoon to sleep together. This temporary division was initiated as early as November 1979. By January and February 1980 the daily splitting or fragmentation was both frequent and obvious. No samples of the group were made in April and May 1980, but by late June 1980 the S group had clearly divided into two new groups who engaged one another in territorial conflicts. Thus, the group fission was a gradual process spanning at least 4–6 months.

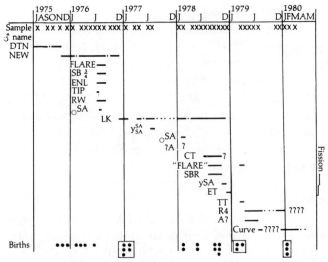

Figure 19.1. Chronology of male tenure in the original S group of redtails. An × indicates sample months. A small dot means the male listed in the left column was presumably present. Each large black circle represents one birth (see Struhsaker, Chapter 18); those in boxes represent birth conceived when more than one male was in group.

Adult name membership in the S group during 1979 prior to fission was marked by the simultaneous presence of two males (R4 and Curve) for an unusually long period of time (up to six months). In 1978 male membership was even more dynamic (Figure 19.1). Immediately before and during the division, only male Curve was seen. After fission, he remained with the smaller splinter group (called Sn) which numbered about 15, including five adult females. Male R4 must have returned during our absence in April and May, for he was seen in June 1980 with the larger new group (called Ss) along with another male (Figure 19.2). The Ss group numbered 32–35 with 11 adult females. Because few of the adult females and juveniles could be individually recognized and agonistic interactions were rarely observed, we do not know if the original S group divided on the basis of genealogical lines or dominance relationships.

Although accurate counts of the groups were rare, it appeared that the two new groups remained relatively stable in size and composition at least through June 1983, when observations were interrupted until September 1984. By January 1986 the Ss group had decreased to about 20–25 individuals, or an estimated reduction of 22–43%. By January 1987 the Sn group had only 10 individuals (including two or three adult females), a 33.3% reduction. Four of the original five adult females in the group disappeared and presumably died.

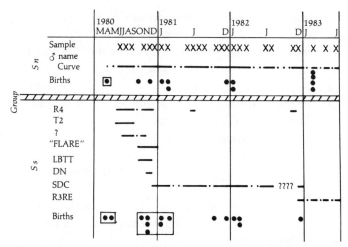

Figure 19.2. Chronology of male tenure in the two new groups of redtails resulting from the group fission. Symbols as in Figure 19.1.

Following the high influx and rapid turnover of at least seven adult males in the Ss group during 1980, there was usually one adult male present in the group through January 1987; one male R3RE was the resident male from December 1982 through at least August 1986. In Sn group adult male Curve was the only male seen for six years. He disappeared during our absence between June and December 1986, and a new male was with the group on 1 January 1987. Thus, both groups ultimately had greater stability in male membership.

Rates of group fission

Group fission has been verified only once in the study of eight different social groups of Kibale redtails and is clearly a rare event. If the sample of 435 group-months spanned by these studies is representative, the rate of group fission can be estimated roughly as: 0.028 per group per year; one fission every 36.5 years; or once every 6 to 7.3 generations, assuming a generation time of at least 5–6 years (estimated age of first parturition, unpublished data and Cords, 1987a).

Ecological consequences

Home range

Because adult females are the principal defenders of territories among Kibale redtails, it might be expected that the smaller Sn group would be at a disadvantage, having only half as many adult females as the larger Ss group and most other groups in the area. After group fission, ranging patterns might change as follows:

1. The small group might be forced out of the original home range into an entirely new area.
2. As presumed close kin, the two new groups might use the original home range on a time-share basis, but with the larger group dominant over the smaller. This could pose potential problems with optimal foraging, particularly in regard to insect foods.
3. The original range might be divided between the two new groups.

The actual ranging patterns which developed represent a mix of the second and third alternatives listed above, plus unexpected range extensions by the new large group.

The larger Ss group attained a home range 16.3% smaller than the original group, while Sn's range was half that of Ss (Table 19.1 and Figure 19.3). Thus, in this case at least, home range size appears to

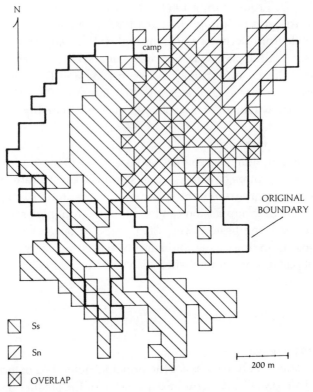

Figure 19.3. Home range of original and two new groups.

Table 19.1. *Quadrat use, home range, core area and density in redtail groups*

	S (original)	Ss (large)	Sn (small)	Sn/Ss
1. 0.25 ha quadrats used per day (11.5 h) (% of home range)	31.4 (11.6%)	24.8 (11.2%)	23.7 (21.2%)	0.96
2. New 0.25 ha quadrats used per day (11.5 h) (% of home range)	6.8 (2.5%)	11.75 (5.2%)	6.8 (6.1%)	0.58
3. Home range size (ha)	67.5	56.5	28.0	0.50
4. Proportion of range overlapping original		71.7%	87.5%	1.22
5. New area added (ha) (% of home range)		16.0 (28.3%)	3.5 (12.5%)	0.21
6. Shared range (ha) (% of home range)		17.8 (31.5%)	17.8 (63.6%)	
7. Core area size (ha) (% of home range)	17.3 (25.6%)	17.5 (31%)	7.5 (26.8%)	0.43
8. Overlap with original core (ha) (% of core)		6.25 (36%)	4.75 (64%)	0.76
9. Shared % of use of original core area[a]		18.5%	17.1%	0.93
10. Shared core area (ha) (% of core)		2.75 (16%)	2.75 (37%)	
11. Shared % of use of shared core area[a]		9.4%	9.4%	
12. Density: ha/indiv.[b]	1.35–1.44	1.36	1.27	0.93
13. Combined density of Ss and Sn	1.33–1.42			
14. Mapping sample time (h) (days): full[c] & partial ()	451 37 (3)	271.75 9 (28)	188.25 10 (21)	

[a] Shared percentage of use was derived from summing the shared percentages of daily entries for each quadrat that constituted part of the core area for both of the two groups being compared.
[b] The area used in density estimates equals the sum of exclusively used area plus half of the area overlapped by both of the new groups. These are minimal density estimates because they do not take into account overlap with neighbouring groups. Specifically: Ss = 38.7 + 8.9 = 47.6 ha/35 monkeys; Sn = 10.2 + 8.9 = 19.1 ha/15 monkeys.
[c] Full day $\geqslant$ 11.5 h.

reflect group size. Surprisingly, however, a greater proportion of the small group's range (87.7%) than the large group's range (71.7%) overlapped with the original range. This means that Ss added nearly 4.6 times (16 ha) more new area to its range than Sn (3.5 ha).

This range extension is also reflected in the average number of new 0.25 ha quadrats used per day of mapping (Table 19.1). Compared with the original group, it remained unchanged for the Sn group (6.8 quadrats) but increased 1.73-fold (11.8 quadrats) for the Ss group. While Sn's range was restricted to the north and east by grassland and colonizing bush and appeared to contain areas of suboptimal habitat, Ss unexpectedly extended its range greatly to the south and southwest (Figure 19.3).

The area shared by the two new groups (17.8 ha) represented 31.5% of the larger group's entire range, but twice that of the smaller group's range (63.6%). Sharing this area may have had an adverse effect on foraging efficiency, especially for Sn group (see also below for use of home range by other groups). Nevertheless, a possible benefit of fission was that by partitioning the original home range, fewer individuals were foraging in the same areas at any one time, i.e. spatial and temporal overlap between individuals was reduced.

Daily travel distance

Larger groups may be expected to travel further to forage in the same habitat than smaller groups (e.g. Waser, 1977). If food competition within the original S group was increasing as a consequence of increasing group size, then one might expect the S group to have traveled further each day; and after the fission, the two new groups should have traveled less each day.

Contrary to expectations, the original group decreased its average daily travel distance from 2019 m when the study began in 1975, to 1720 m a year before fusion occurred, as shown by the running means on a monthly basis ($0.01 > p > 0.001$, two-tailed, Table 19.2). In fact, the average daily distance for the last five daily travel samples of the S group in May and June 1979 was only 1198 m and significantly shorter than any five-day sample in 1975 and 1976. This decrease was not due

Table 19.2. *Original S group of redtails: running mean of daily travel distance on monthly basis*

Year	Months (no days)	Running mean (m)
1975	July (2) + Sept (3)	2019
	+ Oct (5)	2079
	+ Nov (5)	1961
	+ Dec (4)	1917
1976	+ Jan (2)	1883
	+ March (4)	1877
1979	+ May (3) & June (2)	1720

Table 19.3. *Group size and daily travel distance in redtail groups*

Group	Approx. size	Distance (m) $\bar{x}$ & SD	Range (m)	N
S (original) Total: 1975, 1976, 1979	35–50	1766 ± 390	932.5–2685	30
S (original) 1979 only	45–50	1198 ± 163.5	932.5–1410	5
Ss (large)	30–35	1546 ± 287	1117.5–2022.5	9
Sn (small)	15	1595 ± 312	1277.5–2300	10
TTK (Kanyawara, Kibale)	35	1447 ± 249	1085–2028	34
Kakamega, Kenya (Cords, 1987)	26	1543 ± 296	920–2385	84

to increased habituation of the group over time because they were well habituated from the start of the study.

If one compares the average distance for the entire sample of the original group, then the two new groups moved less each day than the original group. However, a comparison of the sample of the original group during the year preceding fission (1979) with those of the two new groups shows that the new groups actually moved further (Table 19.3, $U_s = 37$; $0.05 > p > 0.025$). Despite a two-fold difference in group size, the two new groups moved similar distances each day (Table 19.3, $U_s = 46$, $p > 0.10$). Comparison with two other groups, one from the Kanyawara study site in Kibale and one from the Kakamega Forest of Kenya (Cords, 1987a), show similar daily travel distances despite differences in habitat and group size.

It appears that there is no obvious correlation between group size and daily travel distance. If increasing competition for food within the group was a factor contributing to fission, then it is not reflected in an increase in daily travel distance. Relevant to this finding are similar results from Kibale for blue monkeys (Butynski, unpublished data) and red colobus (Struhsaker & Leland, 1987) which also show no correlation between group size and daily travel distance.

Quadrats used per day

Related to daily travel distance is the mean number of different 0.25 ha quadrats used per day. For both new groups this number decreased from 31.4 prior to fission to 24.8 (Ss) and 23.7 (Sn) after fission (Table 19.1). The larger number for the original group could reflect a larger group spread. It is interesting, however, that both Ss and Sn entered a similar number of different quadrats per day. This means that Sn, with half the group size and half the home range size of Ss, still moved similar distances and covered a similar amount of area on a daily basis as Ss. The area covered, however, represented only 11.2% of Ss's range, but nearly twice that (21.2%) of Sn's range. In other words, on subsequent days Sn reused areas of its range more frequently. The question remains as to whether or not Sn was utilizing its habitat on a sustainable basis.

Core area (Figure 19.4)

We examined this by analysing the most frequently used 0.25 ha quadrats whose combined entries equalled 55–58% of the total quadrat days. (One quadrat day is an entry by group members into a specific quadrat on one day.) These quadrats are referred to as the core area.

The size of Ss group's core area was essentially the same size as the original group's. The small group's core area, although less than half the size of the original or large group's core, represents a similar proportion of its total range (26.8%, Table 19.1 and Figure 19.4).

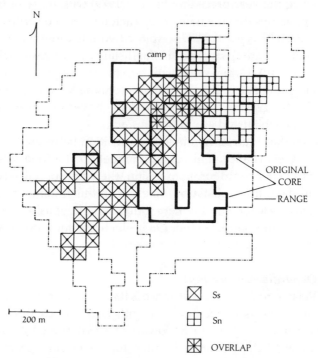

Figure 19.4. Core area of original and two new groups.

Of the 17.3 ha making up the original group's core area, only half were used as core areas by either Ss or Sn. Only 35.7% of the larger group's core contained quadrats of the original core area, but this proportion nearly doubled to 63.3% for the smaller group. If one can assume that the core area of the original group was relatively prime habitat, then the smaller Sn group benefited as much as, if not more so, than the larger Ss group. Both the large and small groups had similar shared percentages of use of the original core area (18.45% and 17.05% respectively, Table 19.1).

The Ss and Sn groups shared eleven of their core area quadrats (2.75 ha) representing 15.7% and 36.7% of their entire core areas respectively. The shared percentage of use for these quadrats was 9.4%. Sn, however, used this area twice as intensively (18.3%) as did Ss (9.6%).

Density

In terms of minimal density estimates for these two groups, the Ss group had 1.36 ha per individual while Sn had only slightly less (1.27 ha, Table 19.1). The overall density of the two groups combined is essentially the same as the density of the original group prior to fission (1.35–1.44 ha per individual). Thus, considering the density for these groups only, fission did not lead to a decrease in density.

Density, of course, increases when neighbouring groups are considered. At least five other groups used parts of Ss's range, in particular the BTP group, which was essentially the only group to also penetrate into Sn's range. As these deep incursions by BTP into both Ss and Sn's ranges did not occur to such an extent before the fission, it is possible that the BTP group took advantage of the presumed weakening of territorial defense following fission. In addition, the BTP group was joined by a large adult male blue monkey (*C. mitis*) at the time of fission in the S group. He was apparently attracted to the BTP group by a hybrid female (redtail × blue cross) who was reaching maturity at this time (see Struhsaker, Chapter 18). Both the male blue and hybrid female, who were considerably larger than redtails, were instrumental in winning territorial conflicts; and they too may have influenced the deep incursions into Ss's range.

Intergroup conflicts

The qualitative nature of intergroup conflicts was variable. Violent physical contact between members of different groups was rare. Most encounters involved chases and vocalizations specific to intergroup conflicts. Occasionally one group would supplant another without any overt interaction.

Table 19.4. *Rates of intergroup conflicts in redtails:[a] pre- and post-group fission*

Focal group	Observ. hours	Contestant group[b]				
		BTP	Other	Sn	Ss	Total
S (original)	1011	12 (1.2)	19 (1.9)	—	—	31 (3.1)
Ss (large)	248.1	9 (3.6)	10 (4.0)	3^a $(1.2)^c$	—	22 (8.9)
Sn (small)	187.5	1 (0.5)	1 (0.5)	—	$7 (3.7)^c$	9^d (4.8)

[a]Considers only those encounters occurring during samples of focal study groups.
[b]First number is number of encounters and second is the rate (no. per observation hour × 100).
[c]Rates differ for Ss and Sn groups because a greater proportion of Sn's range is overlapped by Ss than vice versa. Consequently, the probability of an inter-group encounter between these two groups is greater during samples of Sn.
[d]Does not include 3–4 major intrusions by BTP and one by N group: they were in close proximity but there was no interaction.

The frequency of intergroup encounters, expressed as the number per observation hour, underwent dramatic changes following the group fission (Table 19.4). Rates increased three-fold for the large Ss group due, apparently, to at least three factors. First, extensive intrusions into Ss's range by the neighbouring BTP group increased conflicts with this group by more than three-fold. Second, the range extension of the Ss group to the south led to contact with more groups, accounting for the two-fold increase in encounter rate with other groups. Finally, formation of the Sn group not only added another group to its territorial boundaries, but it also resulted in more frequent conflicts in the course of dividing up the original range.

Sn's conflict rate with Ss group was slightly higher than the total rate of the original group with all other groups (Table 19.4). This high rate may be partially related to the process of establishing and maintaining two new territories within the area of the original range.

Sn's interaction rate with other groups decreased nearly three-fold. For one thing, Sn's new home range was restricted to an area where contact was possible only with Ss, BTP, and very rarely with one other group. The conflict rate with BTP group also decreased (by 2.25-fold) because Sn tended to avoid interaction with this group whenever BTP made its frequent and extensive intrusions into Sn's range. On at least five occasions the Sn group sat and watched BTP group intrude and feed within its range (once when only 25 m apart) without showing any

Table 19.5. *Intergroup conflicts in redtails: trends in outcome[a] pre- and post-group fission*

Group	Wins	Losses	Draw or uncertain	Total
I. vs BTP group				
S (original)	4 + 1 = 5 (38.5%)	3 + 0 = 3 (23.1%)	5 + 0 = 5 (38.5%)	13
Ss (large)	3 + 5 = 8 (50.0%)	3 + 1 = 4 (25)	3 + 1 = 4 (25.0%)	16
Sn (small)	0 + 1 = 1 (33.3%)	1 + 0 = 1 (33.3%)	0 + 1 = 1 (33.3%)	3[b]
II. vs other groups[c]				
S (original)	3 + 0 = 3 (15%)	8 + 0 = 8 (40%)	8 + 1 = 9 (45%)	20
Ss (large)	5 + 1 = 6 (40%)	3 + 2 = 5 (33.3%)	2 + 2 = 4 (26.7%)	15
Sn (small)	0 + 0 = 0	1 + 0 = 1 (100%)	0 + 0 = 0	1[d]
III. Totals				
S (original)	8 (24.2%)	11 (33.3%)	14 (42.4%)	33
Ss (large)	14 (45.2%)	9 (29%)	8 (25.8%)	31
Sn (small)	1 (25%)	2 (50%)	1 (25%)	4

[a]Includes all data from main study groups as well as other groups and incidental observations. Consequently, samples are larger than those used to compute rates of occurrence. First entry represents score when main group in left column was the focal study group, and the second entry is the score when the combatant group was being studied or when incidental observations were made.
[b]Does not include 3 + 3 = 6 major intrusions by BTP, with proximity but no defence or other interaction.
[c]Excluding encounters between Ss and Sn.
[d]Does not include one major intrusion with proximity but no interaction.

response. It seemed that a new type of relationship had developed between them: no longer territorial, but more closely resembling a dominance relationship independent of spatial parameters.

Pronounced differences in the outcomes of intergroup encounters followed group fission (Table 19.5). The proportion of encounters won by Ss group compared with the original group increased, while losses remained about the same, and draws or uncertain results decreased. These results are only weakly significant ($\chi^2 = 3.41$, d.f. $= 2$; $0.10 > p > 0.05$, one-tailed). None the less, the suggested trend is counter-intuitive. One might have anticipated fewer wins and more losses following a reduction in group size.

The Sn group had too few encounters with groups other than Ss to allow statistical analysis. As noted above, except for the Ss group, the Sn group avoided interactions with other groups.

It is not surprising that Ss, with twice as many adult females, was able to win six out of 10 encounters with the smaller Sn group. What is surprising is that they won only 60%; that on one occasion Sn was able to win; and that 30% of the encounters were draws or had ambiguous outcomes. Thus, in spite of once having defended this area as a united group, and presumably being relatively closely related, the Ss and Sn groups established and maintained territories against one another. It is of interest that Sn's relationship with Ss was aggressive rather than tolerant, as it was with BTP group.

Reproduction

Male reproductive success
Adult male membership in the original group prior to fission was generally dynamic. A total of 16 to 21 different males associated with it during the five years prior to fission. Although two males, New and LK, had relatively long tenures, most did not. Multi-male influxes were common and in the two years immediately preceding group fission, no less than 10 males associated with the group (Figure 19.1).

Following group fission, male Curve was the only adult male in association with the small group. In contrast, the new large group continued to experience an influx of at least seven males during the first seven months post-fission, but eventually stabilized with only one male predominating from 1981 to 1986 (Figure 19.2 and Chapter 18).

One can estimate the possible reproductive success of males by assuming a gestation of 5–6 months, and equal probability of paternity when more than one male was simultaneously present during months of estimated conception (Struhsaker, Chapter 18). For the original group in the pre-fission period, estimates were made for the 16 recognizable males. The average number of possible offspring sired per male was estimated as only 1.7 (SD ± 2.38), being highly variable with a c.v. of 140%.

Following the group fission, male reproductive success within the focal study groups improved considerably. In the large Ss group the mean for six recognizable males was estimated as 2.5 (SD ± 2.19), though still quite variable (c.v. 88.3%). Male Curve, the only male with the small Sn group, is believed to have sired all 15 infants born in this group from 1980 to 1983. His success was clearly greater than that of any of these other males. For instance, it was nearly three-fold greater than for R4, the male he resided with in the original group just prior to fission, and who remained with Ss group where he competed with at least five other adult males.

Table 19.6. *Redtail birth rates (to June 1983)*

Group	Approx. size	Number of adult females	Number of infants born	Span of sample (mo)	Birth rate = infant/ female/ year
S (original)	35–50	15.5	30	59	0.39
Ss (large)	30–35	11	13	39	0.36
Sn (small)	15	5	12	39	0.74
BTP	25–35	9.5	21	63	0.42
TTK (Kan-yawara)	35	9.4	10	23	0.56

Female reproductive success

Birth rates nearly doubled in the small Sn group following fission, but remained essentially unchanged for the large Ss group (Table 19.6). An inverse correlation between birth rate and group size or with the number of adult females in the group is only apparent when comparing groups of extremes, e.g. S and Sn (Table 19.6). Although these variables may indirectly influence birth rate, there are others which may have a more important and direct impact.

The variable which correlates best with birth rates is the stability of male membership (Table 19.7; also see 'Aggression', below). The fewer the adult males that associated with the group or the longer only one male was present, the higher was the birth rate. Comparing the data in Table 19.7 reveals a weakly significant positive correlation between birth rate and the percentage of sample months in which only one male was present in the group ($r_s = 0.875, 0.10 > p > 0.05$). The inverse correlation between birth rate and the number of different males associated with the group per sample month is stronger ($r_s = -0.9, p = 0.05$).

Changes or differences in birth rates between groups were not obviously related to differences in survivorship of infants and young juveniles during the first two years of life. The groups compared here did not seem to differ in this respect, although the relative scarcity of complete counts of groups and determination of their age–sex composition prevents a detailed quantitative analysis of survivorship.

Intragroup aggression

The intragroup aggression considered here includes a wide range of interactions, such as supplantations, threat gestures, chases and, less frequently, physical contact with slapping or biting. The most

Table 19.7. *Birth rate and stability of male membership in redtail groups*

Group	Number of sample months	Number of adult males associated with group	% months with only one male in group	Number of males associated per sample month	Birth rate = infant/female/year
Sn (small)	25	1	100	0.04	0.74
TTK (Kanyawara)	16	1	100	0.063	0.56
BTP	40	8–12	77.5	0.20–0.30	0.42
Ss (large)	25	7–8	68	0.28–0.32	0.36
S (original)	41	16–21	63	0.39–0.51	0.39

Table 19.8. *Intragroup aggression in redtails: pre- vs. post-group fission*[a]

Group & year	Among females and juveniles			Adult male vs. adult females & juveniles			Male vs. male	Total of all 3 categories		
	No./ observ. hour	No./h/ group size	No./h/ no. of females	No./ observ. hour	No./h/ group size	No./h/ no. of females	No./ observ. hour	No./ observ. hour	No./h/ group size	No./h/ no. of females
S (original) 1975–80	135.5	3.2	8.7	51.4	1.2	3.3	68.4	255.3	6.0	16.5
Ss (large) 1980–3	128.7	4.0	11.7	80.6	2.5	7.3	116.8	326.1	10.0	29.6
Sn (small) 1980–3	16.0	0.9	3.2	10.7	0.6	2.1	5.3	32.0	1.9	6.4

[a]Weighted means of annual means × 1000.

aggressive acts involved adult males, particularly in two cases of infanticide.

Due to the problems of visibility in the rain forest, more than 50% of the aggressive encounters were only heard or very incompletely observed. Thus, although rates of aggressive encounters have been computed, these should perhaps be best interpreted as indices of frequency rather than highly accurate measures of rate. Any biases in these estimates are considered to be uniform between groups within Kibale.

Frequency of intragroup aggression underwent important changes following group fission (Table 19.8). No matter how the data are expressed, there was far less aggression within the small Sn group than in either the original or large Ss group. This decline was pronounced among all three categories of interacting participants (Table 19.8). The large Ss group, in contrast, experienced similar or higher rates of aggression among all classes of participants than the original group. The increases within Ss group, however, were most apparent in cases involving adult males against one another, particularly during the multi-male influx in 1980 and in cases of an adult male against adult females and/or juveniles. Thus, in terms of intragroup aggression, the small Sn group benefited from group fission, whereas the large Ss group experienced no improvement and in some ways was adversely affected by the fission.

It is suggested that the relatively high levels of aggression in the S and Ss groups compared with the low levels in the Sn group had a stressful impact on the females. This could have been an important, if not major, contributing factor to the low rates of reproduction among females in the S and Ss groups, in contrast to the higher birth rates among females of the Sn group.

Discussion

General

Group fission in species living in multi-male social groups such as *Macaca fuscata*, *M. mulatta*, and *Papio anubis* had several features in common. Those which might be applicable to one-male social groups are summarized as follows (Nash, 1976): (1) members who leave a group can be characterized in terms of social position; females are generally low in rank, peripheral, and may be closely related to one another; (2) fission occurs after a period of instability in male ranking relationships, but also depends on the establishment of a bond between males and females; (3) the group which leaves is subordinate and takes up a new range (see also Dunbar & Dunbar, 1974b, for one-male groups of *Colobus guereza*).

Fission in the redtail group did apparently occur after a relatively long period of instability in male relationships, characterized by multi-male influxes and short tenures. Although the actual fission was not observed, the fact that two males were associated with the large group and one with the small when sampling resumed suggests that fission may have occurred during a time of male instability. It could not be determined, however, whether division was based on rank, genealogy, or female choice of resident male.

Because Sn group was half the size of Ss group and with half the fighting power for defense of its territory, it was inevitably subordinate to Ss, as was shown in the result of intergroup conflicts. Nevertheless, it was the larger group which greatly expanded its range into new territory, and not Sn as might have been expected from studies of other species (Dunbar & Dunbar, 1974b; Nash, 1976; Melnick & Kidd, 1983). Sn's home range, in fact, covered much of the original range, including core areas.

A further point from these other studies is that group fission is usually a very gradual process. In the case of baboons it was more than 2.5 years after the initiation of fission before the small group was considered completely independent of the large group. The redtail fission was much faster, being complete within approximately 4–6 months.

The frequency of group fission may be rather higher in some of these other species than among redtails. However, the redtail case with one fission per 36.5 group-years is probably more typical of cercopithecids living in one-male social groups (e.g. *C. mitis* once per 44.3 group-years; T. M. Butynski, personal communication). For instance, all cases of fission in the two macaque species for which there are sufficient data come from artificially fed populations and cannot be considered representative of the natural state. The novel social system of gelada baboons, which have a very high concentration of harems foraging together in bands and herds, may contribute to their high incidence of group fission. Gelada units or harems were estimated to divide once every 6.4 to 8.3 years (0.12 to 0.156 per unit per year, Dunbar, 1984). In a 17-month study of five groups of *Colobus guereza*, fission was estimated to occur once every seven years per group, or 0.14 divisions per group per year (Dunbar & Dunbar, 1974b)). This relatively high rate is probably an artefact of the short duration of the study.

Comparison of group fission in redtails and blues

Redtails and blue monkeys (*C. mitis*) not only are closely related species, but both live in one-male social groups and in rain forests. These similarities warrant a more detailed comparison of the two case studies for which details are available, as well as with unpublished data for Kibale blues (T. M. Butynski and J. S. Lwanga, personal communication).

Fission in a group of blues living in the Kakamega forest of W Kenya (Cords & Rowell, 1986) was similar to that of the Kibale redtails in a number of ways. (1) Both groups were very large compared to others in their populations; for the Kibale blues, the group was large compared to others of the Ngogo sub-population. (2) The original group and the two new groups were similar in size in the two populations: 46 (33 and 13) for Kakamega blues; and 50 (35 and 15) for redtails. (3) There was instability in male membership and interrelations at and/or just prior to fission in all three populations. (4) The two new groups of Kakamega blues fought one another, with the larger group supplanting the smaller, as in redtails. Due to the disappearance of one splinter group in the Kibale blues, aggressive interactions were apparently absent. (5) The overlap in home ranges of the two new groups was considerable both for Kakamega blues (large group: 29.3% of its range; small: 44.3%), and for redtails (large: 31.5% and small: 63.6%). The only striking ecological difference between these two studies was that, in contrast to redtails, the new small group of Kakamega blues used a smaller proportion and less favorable part of the original range than did the large group, and it had no exclusive part of the previous core area. In contrast, one splinter group of blues at Kibale disappeared entirely from the study area, and the other expanded its range considerably (J. S. Lwanga, personal communication). (6) In Kakamega blues and Kibale redtails males had more exclusive access to the females following fission.

Perhaps the most significant similarity between the Kakamega blues and Kibale redtails was the increased reproductive success for females. Analysis of annual birth data for Kakamega blues from 1980 to 1986 (M. Cords, personal communication; Tsingalia & Rowell, 1984; Cords & Rowell, 1986) show that births per adult female were significantly higher (1) for the smaller splinter group in the two years (1985–6) following group fission than in the five years preceding it (1980–4, $U = 0$, $p = 0.047$, one-tailed); (2) for both splinter groups during the first year after division than in the preceding five years ($U = 4$, $p = 0.095$, one-tailed). Excluding females with suckling yearlings from the analysis (as recommended by M. Cords, personal communication), gave similar results. Birth rates were significantly higher for both groups in the two years post-fission than in the five years before division ($U = 2$, $p = 0.032$, one-tailed).

Competition, behavioural stress and birth rates

No obvious ecological advantages accrued to either of the two new redtail groups following fission. Daily travel distance apparently increased; there were no conspicuous improvements in range quality, and density remained similar, or increased slightly. The frequency of intergroup encounters increased dramatically for the large group and slightly so for the small group. The two new groups fought one another relatively frequently. If group fission did lead to ecological advantages, then they were subtle, such as improved foraging predictability and efficiency through more complete division of the home range, and a decrease in the number of individuals using a specific area at any one time.

Far more obvious than ecological changes were the behavioural shifts within the new groups, with most benefits attained by the small group. Although intragroup aggression remained high in the large group, it decreased greatly in the small group. Likewise, stability of male membership increased dramatically for the small group, while remaining very dynamic in the large group for at least seven months.

Correlated with these changes was an increased birth rate for the small group which was twice that of the original and large groups. This is in agreement with the observations on Kakamega blues described above. Similar results can be extracted from studies of group fission in *M. fuscata* and *P. anubis* which show increased birth rates in the smaller splinter groups (Table 19.9). Malik *et al.* (1985) also report greatly increased birth rates in the smaller splinter groups following three group fissions in *M. mulatta*. After one of the divisions, the birth rate also increased in the main group. A neighboring 'control' group showed no changes in its lower birth rate during the same period. Infant mortality in these groups was low. In gelada baboons, Dunbar (1984) found that larger harems, which are more prone to fission, have lower reproductive rates because of poor reproductive performance by low-ranking females. He speculated that this was due to physiological stress resulting from harassment by higher-ranking females. Furthermore, he estimated that a female's birth rate declines with declining status by about 0.048 births per year for each unit of decrease in dominance rank.

Table 19.9. *Group fission and birth rates in* Macaca fuscata *&* Papio anubis

Group	Date	Group size (approx.)		Ratio of infants to adult females[a]
M. fuscata				
Takasakiyama troop	May 1953		220	0.67
	Jan. 1956		370	0.46
	Feb. 1959		570	0.46
Fission				
	Jan. 1969	Main troop	550	0.49
		New branch troop	100	0.63
P. anubis				
Gombe B troop	1967–9 (18 mo)		49–58	0.43[b]
	Oct. 1969		53	0.47[c]
Fission				
	Oct. 1970	Main troop	34	0.55
		New H troop	20	0.80

[a] Data derived from Sugiyama (1960, Tables 1 & 3).
[b] Annual birth rate of 0.37 for B and C troops from Ransom (p. 51, 1981) seems to be in error, and was recalculated using his original data: 11 infants/17 females/1.5 yr = 0.43. Group counts are for May 1968 and May 1969.
[c] From Nash (Table 1, 1976). These are minimal birth rates for one year.

Further indirect evidence supporting the hypothesis that group fission leads to increased reproductive success is provided by a 17-month study of *Colobus guereza* in Ethiopia. During this period four non-fission groups underwent very little change in numbers, increasing by only one individual from a total of 25 to 26 monkeys (4% increase). In contrast, the two groups which resulted from fission increased by 33.3% from a total of 12 to 16 monkeys, and all as a result of births (Dunbar & Dunbar, 1974b).

Of possible relevance here is the high incidence of malformed infants at birth in a troop of *M. fuscata*. These malformations increased up to 40% per year as the troop increased in size, but then declined significantly beginning with the first year after the first troop fission (Furuya, 1969).

What seems to be a general phenomenon of decreasing birth rate or female reproductive success with increasing group size could be due to a variety of factors (see also van Schaik, 1983). Relatively larger groups and higher densities could lead to increased competition for food and/or decreased foraging efficiency. As a result, individuals, especially the lower ranking, are subjected to both social stress from aggression and harassment as well as to reduced physical fitness. These factors, in turn, could lead to lower birth rates. There is no evidence, however, that ecological stress may have influenced reduced birth rates, at least in the redtail troop.

Relatively larger group size could also affect social instability. For instance, in one-male groups, the greater the number of reproductive females present, the more attractive the group is likely to be to extragroup males seeking mates (e.g. Kibale redtails; Kakamega blues; geladas: Dunbar, 1979). With a high influx of extragroup males into redtail groups, aggression increased among males and also between males and females and juveniles. Stress from such aggression could lead to reduced birth rates. Likewise, Dunbar (1979, p. 72) suggests that the aggression and tension generated in a group of *Colobus guereza* by a number of additional peripheral adult males may have been so disruptive that no infants were born or successfully reared in the group over a period of four years. In yellow baboons, increased aggression by recent immigrant males against pregnant females can apparently induce reabsorption or abortion (Pereira, 1983). Studies of mice have shown reduced fertility in females copulating with more than one male in succession. It has been suggested that this pregnancy block is an adaptive response by females to periods of social instability (Dewsbury, 1982b).

Furthermore, with instability of male membership, the risks of male replacement and subsequent infanticide increase. High rates of infanticide also reduce female reproductive success and may lead to fission. For example, in the case of blue monkeys at the Ngogo site of Kibale, the original group experienced high birth rates before fission (Butynski, 1982b, personal communication). These high birth rates, however, were attributed in part to low infant survivorship, which in turn was related to the unusually high incidence of infanticide and infant disappearance following the frequent male replacements in this group. In this case, male instability (in terms of male tenure) and high levels of aggression among males and between males and adult females resulted in lower infant survivorship rather than lower birth rates (Butynski personal communication). Possibly as a result of the reduced female reproductive success this group divided.

In summary, the case of group fission in redtails might best be explained by the following working hypothesis. As the group increased in size, stress also increased, possibly as a result of increasing aggression among group members for food and among extragroup males competing with another for females. During multi-male influxes and after male replacement, there was also aggression between males and females and juveniles. The increased aggression, combined with instability in the adult male membership, led to stress and a decline in birth rate. As an adaptive response to conditions adverse to their reproductive success, females with their juveniles and infants divided into two new groups. The smaller group, with only five adult females, had no intrusions from extragroup males and a very pronounced reduction in intragroup aggression. Correspondingly, with greater social stability and decreased stress, the smaller group's birth rate doubled. The larger group, in contrast, continued to have a high rate of multi-male influx and instability of male membership during the first seven months post-fission, as well as high levels of intragroup aggression, and increased frequency of intergroup conflicts. Its birth rate remained low.

Regardless of differences between case studies, group fission must, of course, be ultimately viewed in terms of the costs and benefits to the females and their inclusive fitness, for it is they who divide into new groups. The most plausible working hypothesis at present is that group fission is likely in relatively large groups whenever female reproductive success is low, either through low birth rates or high infant mortality.

Genetic consequences

Group fission is the principal means of female dispersal in Kibale redtails and probably for most other cercopithecids living in matrilineal societies (e.g. Melnick & Kidd, 1983). Whether or not there is a matrilineal effect on the genetic consequences of group fission, as found for *M. mulatta* on Cayo Santiago (Chepko-Sade & Sade, 1979), such divisions of social groups are expected to accelerate subpopulational genetic differentiation beyond the rate expected by genetic drift alone. As a result, group fission may prove to be one of the most important single factors in understanding population genetics (Melnick & Kidd, 1983).

A case study of group fission in *M. mulatta* of the Himalayas showed that the smaller splinter group's genetic composition was representative neither of the population as a whole nor of the original parent group. It may have been sufficiently different to result in significant microgeographic genetic differentiation, e.g. as a founder group (Melnick & Kidd, 1983).

A factor not addressed in previous studies of the genetic effects of group fission is the importance of the frequency of group fission. For example, frequent division of social groups into new groups may lead, on average, to greater genetic similarity between groups. Conversely, infrequent group fission, as in Kibale redtails, may lead to greater genetic differentiation between groups by increasing the probability of genetic drift within groups. Thus, when a group eventually splits, it will have achieved a significant level of genetic distance from other social groups in the population. Much of this differentiation will depend on gene flow, largely mediated in redtails by turnover in male membership and their reproductive success.

Another genetic issue by this study is that of inbreeding. Long tenure by an adult male, such as male Curve in the new small group of redtails, raises the possibility of father–daughter matings. As the older and unrelated females die of 'old age', the resident male is left with progressively fewer mating opportunities other than with his own maturing daughters. In the small Sn group, the first of these probable daughters was reaching physical maturity in 1986 and had not yet reproduced. Thus, during the first three years after fission, the initial benefits to this small group in terms of increased birth rates may eventually be counterbalanced by inbreeding depression or decreased birth rates due to avoidance of father–daughter matings. Certainly, the individual females in the small group gained a reproductive advantage from group fission through increased birth rates. Presumably, their dispersing male offspring had an equally good chance of reproducing as did males dispersing from other groups. The future reproductive success of their maturing daughters, however, remains to be determined.

Summary of main results and conclusions
1. This study of group fission in redtails is among the first documentations of the phenomenon in a *Cercopithecus* and rainforest species.
2. Group fission was a gradual process spanning at least 4–6 months in which an unusually large group of nearly 50 individuals divided and formed two new permanent groups of 35 and 15 monkeys.
3. An estimate of the rate of group fission in Kibale redtails suggests that it is relatively uncommon: approximately once every 36.3 group-years.

4. Ecological consequences

Both new groups traveled further each day than did the original group one year prior to fission. The two new groups had similar daily travel distances and entered a similar number of 0.25 ha quadrats per day even though they differed in group size by two-fold.

The previous home range was divided into two new ranges. Contrary to what might be expected, the home range of the large group contained more surface area of the original range than that of the small group and moved into a larger area of new range.

The smaller group had a range and core area almost half the size of the large group, reflecting group size differences.

Population density increased after fission only as a consequence of increased intrusions by another group.

The rate of intergroup encounters and conflicts for both new groups increased. Despite their presumed close kinship, these two groups frequently engaged in territorial conflicts with one another.

Ecological advantages of group fission to the two new groups were not apparent. Foraging predictability and efficiency, however, may have improved following fission because division of the range meant that fewer individuals were using the same areas at any one time.

5. Intragroup aggression

Aggression was very much reduced in the new small group, but remained unchanged or increased slightly in the large group.

High incidence of aggression was often associated with multi-male influxes.

6. Reproductive consequences

The possible number of offspring sired per individual male increased in both new groups, but more so in the small than the large group.

Birth rate doubled in the small group, but remained unchanged in the large group. Similar results have been extracted from studies of fission in six other cercopithecids.

High birth rates correlate with high stability in male membership. Relatively long tenure by one male apparently enhances birth rates. High birth rates also correlate with low levels of intragroup aggression.

7.   It is hypothesized that:

increasing group size in redtails led to increased aggression over food and/or female mates;

with large numbers of adult females the group was particularly attractive to extragroup males, leading to multimale influxes and instability of male membership;

this social stress resulted in lowered birth rates;

adult females responded to this decline in their reproductive success by separating and forming two new groups;

although smaller groups may be less prone to instability of male membership and thus have higher birth rates than groups considered larger relative to the (sub)population, they may be less able to defend a territory, and more susceptible to inbreeding.

8.   Group fission is likely whenever reproductive success is low. Ultimately, group fission must be viewed in terms of the costs and benefits to the adult females and their inclusive fitness, for it is they who separate to form new groups.

Acknowledgments

Major fundings for the field work, data analysis, and writing was from the New York Zoological Society. Some of the data analysis was done while at the Center for Advanced Study in the Behavioral Sciences, funded in part by National Science Foundation BNS 76-22943 and the Alfred P. Sloan Foundation 82-2-10. Additional data were collected, analysed, and written up during the tenure of grants from the H. F. Guggenheim Foundation and the National Geographic Society (No. 2929-84) for studies of redtail monkey behavioural ecology. Permission to study in Kibale was granted by the President's Office of Uganda, Uganda National Research Council, and the Uganda Forest Department. The Department of Zoology, Makerere University provided our local affiliation in Uganda. Thanks are extended to Drs T. M. Butynski and M. Cords and Mr J. S. Lwanga for permission to refer to their unpublished results and to Drs T. M. Butynski, M. Cords, A. Gautier-Hion and T. Rowell for valuable comments on the manuscript.

III.20

Cercopithecus aethiops: a review of field studies

LAURENCE FEDIGAN and LINDA M. FEDIGAN

Introduction

Aims

To review, in any adequate way, the enormous and greatly varied literature on vervet monkeys, would require more time and talent than we have been allowed, as well as being beyond the intent and interests of the present chapter. The latter are to place the superspecies *C. aethiops* in some general relation to its arboreal congenerics by comparing aspects of its life-style in a variety of widely separated habitats, and drawing from these comparisons some characteristics or qualities that appear to explain the historic success of this most widespread of the *Cercopithecus* species.

The studies we have chosen to review are field studies from South Africa, East Africa, West Africa and the colonized Caribbean islands of St Kitts, Nevis and Barbados. These 50 plus studies range in time from the pioneering research of Hall, Brain, Gartlan and Struhsaker in the 1960s, to the most recent field studies which, building on those important early works, cover a wide spectrum of current theoretical interests. The studies we have of necessity omitted, those carried out with captive or caged groups, are even more numerous and cover an equally wide and interesting range of questions. Many of the field studies cited depend upon or refer to captive research, which has provided valuable precision and clarification of questions that require the control and detail of captive conditions. Thus, the decision to leave all captive studies out of the present discussion is arbitrary, resulting from the constraints of time and space, not from the feeling that such studies are less important or illuminating.

Table 20.1. *A summary of field studies of Cercopithecus aethiops*

Location, dates, references	Title and/or major aims	Major findings
South Africa		
Burman Bush Nature Reserve, Durban, 1972–3 (1) Basckin & Krige, 1973	Mother–infant interactions during the birth season in an urban group	Mothers with infants were more aggressive than others and sometimes formed coalitions with other mothers to chase males. Infants actively defended by mothers, not threatened until weaned. Mothers with infants often initiated group movement and attracted alloparents of all ages and both sexes
Burman Bush Nature Reserve, Durban, 1973–4 (2) Krige & Lucas, 1974	Aunting behaviour in urban troop of C. *aethiops*	Most allomothers were adult females, with infants of their own. Allomothers often removed infants forcibly from their mothers (kidnapped them). The infants tried to resist removal
Burman Bush Nature Reserve, Durban, 1974–8 (3) Henzi & Lucas, 1980	Movement between 3 groups of C. *aethiops*	12 males changed groups in 24 months. Migration peaked during the mating season along with aggression and wounding which was mostly directed at newcomers. Most males transferred to neighbouring groups, some with overlapping ranges and infrequent encounters
Private Reserve, N Transvaal, 1979 (4) Cambefort, 1981	Acquisition of new foods by P. *ursinus* and C. *aethiops*. Comparison of cultural transmission	*Papio* juveniles discovered 90% new foods, which quickly spread. Juveniles foraged separately on the periphery. C. *aethiops* discoveries at all ages but juveniles most (40%). Spread along social channels by 'pivotal' individuals; slower than *Papio*. C. *aethiops* mixed age–sex foraging explains differences
Burman Bush Nature Reserve, Durban, 1976–8 (5) Henzi, 1985	Significance of genital signalling in inter-male agonism in 2 C. *aethiops* groups	No seasonal variation or rank effect in scrotal color. Change at puberty. Penile erection usually provoked by external males. Males did not form alliances, interactions generally agonistic. May be a simpler system than a true multi-male one. As in other C. species, one 'alpha' male occupies a special central role
East Africa		
Lolui island. Lake Victoria (Uganda); Zimbabwe (S Rhodesia) (6) Hall & Gartlan, 1965 (7) Gartlan & Brain, 1968	(1) Ecology and demography of C. *aethiops* in rich and poor habitats (2) Comparison with C. *mitis*	Lolui – high resources, large groups, high density, small, strongly defended territories. Chobi – marginal, low in resources, smaller groups, lower density, larger ranges, more dispersed foraging with much inter-group vigilance. Less territorial display

Masai–Amboseli Game Reserve (Kenya), 1963–4 (8, 9, 10, 11) Struhsaker, 1967a,b,c,d	Ecology of *C. aethiops*; comparison with *Papio* and *C. mitis*; a complete catalogue of vervet monkey behaviour	Great differences in range size and patterns of neighbouring groups related to availability and distribution of resources. No relation between group size, territory or range. Some groups foraged coherently, others dispersed, territorial. Separated in niche from *C. mitis*. Competed with *Papio* for space and resources, yet shared sleeping sites, predator warning. *Papio* exploited open grassland better. *C. aethiops* losing habitat to elephants
Masai–Amboseli Game Reserve (Kenya), 1962–4 (12) Struhsaker, 1971	Relations between mother and infant *C. aethiops*	Compared to *Papio* sp. and *M. mulatta*, *C. aethiops* behaviourally precocious and weaned earlier. They mature at 3.5 to 4 years; 3.5 probable median age at first birth. Juvenile females allomother early but it is unclear if it increases survival rates
Zambesi River (Livingstone), Zambia, 1968–9 (13) Lancaster, 1972	Play mothering: the relations between juvenile females and young infants	1-yr-old females showed much allomothering: it declined in 2-year-olds, but increased at 3 years. Males showed very little interest in infants. Amount variable, and may increase efficiency of maternal care especially for primiparous mothers
Bole Valley (Ethiopia), 1972 (14) Dunbar & Dunbar, 1974a	Niche separation between *T. gelada*, *P. anubis* and *C. aethiops* in Ethiopia	*P. anubis* and *C. aethiops* prefer forest edge; *T. gelada* graminivorous, *P. anubis* also use more ground level resources than *C. aethiops*, but broad overlap. More apparent niche separation in these savannah species than in forest species
Diani Beach Forest (Kenya), 1972–3 (15) Moreno–Black & Maples, 1977	Niche separation of *P. cynocephalus*, *C. mitis*, *C. aethiops*, *Colobus angolensis*	Little competition. Polyspecific associations usually pacific or neutral, most aggression between 2 *C.* species. *Aethiops* best able to exploit disturbed and secondary growth, had the most varied diet including ornamental and cultivated plants
Amboseli National Park (Kenya), 1977–81 (16) Cheney et al., 1981	Competition among females, reproductive success and mortality	High ranking females did not raise more offspring than low ranks. Mortality was non-randomly distributed, low ranks suffered more from reduced access to resources, especially in the dry season, high ranks were subject to greater predation but unclear why
Amboseli National Park (Kenya), 1978–9 (17) Wrangham & Waterman, 1981	Relation of secondary compounds especially tannin, with feeding in *C. aethiops*	*Acacia* sp. main food resource. Selectivity often limited by availability. Tannin negatively related to food selection but condensed tannins in diets may increase water stress in droughts; may result in rank differences in mortality and reproduction

Continued

Table 20.1 (cont.)

Location, dates, references	Title and/or major aims	Major findings
Amboseli National Park (Kenya), 1978–9 [18] Wrangham, 1981	Effect of a severe water shortage on C. aethiops	*Continued* Males from a group lacking a water source 'invaded' the territory of neighbouring groups. Higher ranking females had access to preferred treeholes. Mortality was high in both sexes, greater in lower ranks
Samburu/Isiolo Reserve (Kenya), 1977–80 [19] Whitten, 1983	Diet and dominance in 2 groups of C. aethiops in seasonally dry habitats	*Acacia* sp. main resource for both groups. Distribution clumped. Dominance based on differences only where food items clumped. Where distribution of resources random, low ranks ate longer. Only defendable resources confer dietary benefits
Amboseli National Park (Kenya), 1977–8 [20] Cheney, 1981 [21] Cheney & Seyfarth, 1983 [22] Hauser et al., 1986	Intergroup encounters between 3 neighbouring groups of C. aethiops; territoriality, male emigration, female emigration, and group fission and fusion	All males plus some high ranking females displayed. Only low ranks interacted affinitively with other groups. Males herded their females but the latter sometimes formed coalitions to chase them. In 53% of encounters, only vocalizations exchanged. Differential aggression towards groups which exchanged males. Males habitually leave their natal groups by maturity. Females leaving cause group fissions and subsequent fusioning with larger groups if groups become too small to defend resources
Amboseli National Park (Kenya), 1978–80 [23] Lee, 1984a	Effects of maternal style, sex, rank and infant behaviour, on infant development	Infants independent at 12 weeks; females may mature sooner than males. Maternal styles inconsistent: rank, birth order and sex of the infant may produce differences. Availability of allomothers may mitigate the need for distinct or dichotomised maternal styles
Amboseli National Park (Kenya), 1985 [24] Brennan et al., 1985	Ecology and behaviour of C. aethiops in a tourist lodge habitat	Lodge C. aethiops had higher population densities than wild groups, showed more intragroup aggression, ate non-native foods provided by people, and behaved aggressively towards tourists; attacks reported
West Africa Parc National du Niokolo-Koba, Senegal, 1969 [25] Dunbar, 1974	Observations on the ecology and social organisation of the green monkey, C. sabaeus, in Senegal	Largely frugivorous during the study period (wet season), C. aethiops diet overlapped with P. papio. The greater home and day range of P. papio reduced competition. A birth peak occurred at the intense rainy season (Jan.–March). Diets and behaviour are similar to other populations in East Africa

Location, reference	Topic	Findings
Saloum Delta (Senegal), 1975 (26) Galat & Galat-Luong, 1976	Socio-ecology of C. aethiops colonizing a mangrove swamp	Formed groups for dispersed foraging. Territories defended mainly by vocal displays with loud calls. Fiddler crabs main staple, some 80% of time spent in mangroves
North Senegal (Sahel), 1975–6 (27) Galat & Galat-Luong, 1977	Demography and diet in a marginal habitat	Diet varied seasonally with availability of resources. After a low rainy season there was evidence of malnutrition and high mortality. Overall group numbers were stable during study
Parc National du W Niger (28) Poche, 1976	Distribution, ecology	C. aethiops tantalus appeared confined to gallery forest, was rarely seen far from water. Population density appears to be low
Kalamaloue National Park, Buffle Noir and S. Bakossi, Cameroun (29) Kavanagh, 1978a	Diet and feeding behaviour of C. aethiops tantalus in 3 habitats	Dietary range in the seasonally wet and dry sahelian savannah, guinea savannah and farmed forest was similar and similar to other areas, 1/3 time spent feeding on the ground. Less adept at eating flying insects in the trees than arboreal C. species and at eating insects on the ground than E. patas
Bakossi (Cameroun), 1974 (30, 31, 32) Kavanagh, 1978b, 1980a,b,	Adaptation of C. aethiops tantalus to cultivated 'farmed', rain forest	Flexibility in diet, group size and ranging, semi-terrestriality, cryptic coloration and variable response to predators, allow C. aethiops tantalus to colonize croplands, fallow and secondary growth. C. mona is restricted to intact forest
Cameroun (W Africa) 1974 (33) Kavanagh, 1981	Territoriality in 3 different locations and habitats	Expression varied from minimal to extreme. No correlation to group size or range but possibly to defendable resources or predation pressure. Only adult males took part in territorial defense
Mt Assirik, Parc National du Niokolo–Koba (Senegal), 1978–9 (34) Harrison, 1983c	Territorial behaviour and defence of food in seasonally wet/dry habitats	Encounters varied in kind and intensity with availability of preferred resources. Territoriality seemed associated with small, economically defendable ranges (3 sites) and with high population density (one site). Only adult males participated
Mt Assirik (Senegal) – see above (35, 36, 37) Harrison, 1983a, b, 1984a	Diet in a very seasonal habitat in relation to sex and optimal foraging strategies	Diet varied (see Table 20.2) according to availability and showed much flexibility. Increased selectivity when preferred foods available. Males fed longer than females, who reduced feeding time during lactation. All females fed similarly. Reducing energetic requirements may be an alternate strategy of optimization
Mt Assirik (Senegal) – see above (38) Harrison, 1985	Basic seasonal variation in activity budget	When not feeding, rested more than socialized. In dry season, activities changed to minimize extremes of temperature. Feeding synchrony increased with greater availability, especially in the wet season

Continued

Table 20.1 (cont.)

Location, dates, references	Title and/or major aims	Major findings
West Indies St Kitts, West Indies (39) Sade & Hildrech, 1965 (40) Poirier, 1972 (41) McGuire 1974 (42) Coppinger & Maguire, 1980 (43) Fedigan et al., 1984	The distribution, numbers and population dynamics of C. aethiops sabaeus on St Kitts	First established in the late 1600s, C. aethiops sabaeus have occupied all available habitats from intact forest on the volcanic slopes, through wooded ravines bordering cultivated fields, to the acacia scrubland of the semi-arid SW peninsula. Long considered a pest, populations remain large in spite of hunting, trapping and laboratory use
St Kitts, West Indies (44) McGuire 1974 (45) Fairbanks & Bird, 1978 (46) Chapman, 1985 (47) Chapman & Fedigan, 1984	Socio-ecology; diet, ranging patterns and adaptations of C. aethiops sabaeus to the island	Diets varied from sugar-cane, vegetables and fruit crops in the humid wooded areas to fibrous fruits and nuts in the semi-arid areas. Group size, range size, and foraging behaviour were variable. Territorial defence was not extreme and was associated with clumped, defendable resources in the areas of range overlap. Major predators were hunters, trappers and dogs and the vervets are shy, quiet and very vigilant
Barbados, West Indies (48) Denham, 1981 (49) Horrocks, 1986 (50) Horrocks & Hunte, 1986	Demography, distribution and life history characteristics	Group sizes and structure similar to those in Africa, but dry season birth peak. High fecundity: mean 1.2 per year, only the alpha female had 2 infants in the same year, mean interbirth interval was 11.8 months and mean age at weaning 8.6 months. Sexual maturity 36 months for females, 60 months for males
Vervet vocalisation studies Amboseli National Park (Kenya) (51) Struhsaker, 1967d (52) Cheney & Seyfarth, 1980 (53) Cheney & Seyfarth, 1981 (54) Seyfarth & Cheney, 1980 (55) Seyfarth & Cheney (1982) (56) Seyfarth et al., 1980a,b	Vocal communication C. aethiops; ontogeny, discrimination, classification, recognition, adaptive significance	Different calls used to designate classes of external danger, (e.g. terrestrial vs. aerial predators) alarm calls given by all age/sex classes indifferently, however, higher ranking individuals of both sexes alarm more frequently and longer. This appears to be unrelated to greater vigilance of arousal, but perhaps to greater vulnerability and more frequent 'spotting' of potential danger from the front of progressions. Kin and offspring likely benefit less from alarm calling than the caller. Mothers distinguish their infants' screams and respond quickly. Infants cue to their behaviour and quickly learn to discriminate.

The summaries and comparisons we have drawn fall into three very broad categories: ecology, or the question of where and how vervets get their living; demography, or how they grow, reproduce, and regulate group size; and social patterns, or how certain behaviours allow them to maintain and regulate cohesive groups under the great variety of conditions in which they live.

Format

To attempt to do justice to each of the studies cited in this review, and to avoid long and cumbersome point by point comparisons, we have first listed and summarized the major findings of each study in tabular form. Table 20.1 includes dates, locations, aims and the major findings.

We then summarize under the three headings, ecology, demography, and behaviour, what we feel are the generalizations that can be drawn from all the research listed. If these summaries at times seem to be too general, it is an unavoidable consequence of trying to find the common denominators of the varied studies.

Finally, we use these summaries, together with specific suggestions from particular studies, to explore the special behaviours and abilities of vervets, which may have enabled them to expand their range at times when their congenerics are on the retreat.

General description of vervets: morphology, distribution, social organization and lifeway

Vervets are small monkeys compared with other semi-terrestrial Cercopithecidae living in open habitats, and they exhibit only moderate sexual dimorphism in size, with the average adult female weighing 5.6 kg to the adult male's mean weight of 7 kg. Both sexes have similar greenish-gold fur, black faces fringed in white hair, large sharp canines and long expressive tails which are held aloft in various configurations during quadrupedal walking, and which can be braced or wrapped for support around trees or the mother's body by infants. Lacking the spectacular facial and body coloration of most of its forest dwelling congenerics, *C. aethiops* is none the less a rather elegant monkey with its delicate frame, rich fur, extravagant tail, and smooth, light-footed locomotor patterns both in the trees and on the ground. In East Africa, both sexes have abdominal skin tinted eggshell blue, which is clearly visible in the adult male scrotum. The colourful male genitalia, consisting of blue scrotum, red penis, red perianus and surrounding white hairs are prominently exhibited during dominance displays which have been referred to as flagging the 'red, white and blue'.

Vervets are the most widespread and abundant of all African monkeys (Struhsaker, 1967a). They can be found from Senegal to Somalia and from the southern border of the Sahara to the tip of South Africa. With the exception of the Congo Basin, the deserts of northern and southwest Africa and a few coastal areas, vervets inhabit savanna and woodland areas all over the African continent. What is more, *C. aethiops* is considered plentiful and abundant in many of the areas within its geographic range (Wolfheim, 1983), and is said by some to be the world's commonest species of monkey (e.g. Kavanagh, 1983b).

Any species with such an extensive distribution is likely to show local morphological variations, and indeed West, South, and East African vervets are sufficiently distinct that some authors (e.g. Dandelot & Prevost, 1972; Lernould, Chapter 4) have described *C. aethiops* as a superspecies consisting of four species: *C. sabaeus*, *C. aethiops*, *C. pygerythrus* and *C. tantalus*. However, other experts, while recognizing the major types as subspecies, have continued to treat *C. aethiops* as one highly variable, or regionally polymorphic species (e.g. Kingdon, 1974; Napier & Napier, 1985; Wolfheim, 1983). For the sake of consistency in the present compendium on the evolutionary biology of African guenons, we will follow Lernould's classification and refer to *Cercopithecus aethiops* as a superspecies. Common names, as well as taxonomic classification, are problematic for this monkey, since in various parts of Africa, *C. aethiops* are referred to colloquially as vervets, savannah monkeys, green monkeys (especially West Africa), tantalus monkeys (West Central Africa), grivets (East Africa) and so forth. For simplicity and consistency, whenever a *general* statement is made about these animals in the text or tables, they will be referred to either as *C. aethiops* or vervets. Kingdon (1971), has argued, on the basis of fur patterns, that vervets originated in southern and southwest Africa, moving north and east to exploit savannahs during dry periods, so that today we can recognize a pan-African cline with two extreme wings meeting in Uganda. Lernould (Chapter 4) classifies all of the vervets of southern and central Africa reaching up and east to Uganda, as *C. pygerythrus* (see his Figure 4.12). To the species *C. aethiops*, he allots the most limited and the most northerly distribution, in Ethiopia, the Sudan and northern Kenya. Lernould's distribution map for *C. tantalus* and *C. sabaeus* is more in agreement with other authors such as Kingdon (1971) in that *C. tantalus* is described as occurring in a belt across Africa from Uganda to Ghana, and *C. sabaeus* as being the

Table 20.2. *Group sizes, home ranges and diets of* C. aethiops

Study (see Table 20.1)	Group size	Range (hectares)	Diet % Fruit	% Flower	% Seed	% Foliage	% Inverts	% Other
13	$\bar{x} = 20$	$\bar{x} = 30$		17.6	50.6	18.7	7.4	5.7 (bark)
15	14	—	53.5	6.4	21.3	10.6	—	—
26	42	—	26.0	37.0	—	—	13.1	85.7 (all veg.)
27	33	138	—	—	—	—	mainly crabs	
29a	18	103	26.8	34.3	—	5.2	28.7	—
29b	76	56	61.7	11.3	—	16.8	7.3	
35	18–28	178	50.0	14.0	13.0	7.0	11.0	2.5 gum
3	$\bar{x} = 25$	$\bar{x} = 15$	—	—	—	—	—	—
4	36	—	—	—	—	—	—	—
6	$\bar{x} = 12$	25	—	—	—	—	—	—
7	6–21	—	—	—	—	—	—	—
9, 10	7–53	16	—	—	—	—	—	—
17	$\bar{x} = 20$	—	—	—	—	—	—	—
19	$\bar{x} = 39$	—	—	—	—	—	—	—
21	13–27	40	—	—	—	—	—	—
25	$\bar{x} = 12$	20	—	—	—	—	—	—
29c	11–18	13	—	—	—	—	—	—
			$n = 6$ $\bar{x} =$ 36.3	$n = 6$ $\bar{x} =$ 20.1	$n = 6$ $\bar{x} =$ 14.2	$n = 6$ $\bar{x} =$ 9.7	$n = 6$ $\bar{x} =$ 11.3	—

extreme western type, occurring west of the Volta River from Ghana to Senegal.

Cercopithecus aethiops live in multi-male, multi-female groups which range in size from approximately five individuals to 76 (mean group size is about 25, see Table 20.2) and with an adult sex ratio of 1.5 females to one male. Although several adult males live together in a social group, some studies report the presence of one adult male in each group who seems to occupy a specialized role and 'alpha' status, perhaps corresponding to the single breeding male position found in most other *Cercopithecus* species, or perhaps indicating a transitional form of 'age-graded' male society as postulated by Eisenberg, Muckenhirn & Rudran (1972) for vervets (see also Henzi, 1985). Males do, however, transfer between groups, whereas females remain in their natal groups. As in other primate species exhibiting male mobility and female philopatry, matrilineal kinship units are important structures in social life.

Vervets are eclectic in most aspects of their lifeway. Classified as semi-terrestrial, semi-arboreal, they seem equally adept on the ground as in the trees. Although they have been able to exploit open savannah areas for feeding in many parts of their range, vervets almost inevitably return to the safety of the trees to sleep at night. During their daily foraging trips, they eat a wide range of foods, including a variety of plant species and parts, as well as insects, birds' eggs, hatchlings, crustaceans and lizards. Described as opportunistic omnivores in diet (see Gautier-Hion, Chapter 15), they have been equally opportunistic in the range of habitats successfully exploited. The latter will be further described in the section on ecology.

Ecology
Studies of vervet ecology range in time from the early East African studies of the 1960s (Hall & Gartlan, 1965; Gartlan & Brain, 1968; Struhsaker, 1967a,b,c,d, 1971) which established the general patterns of ranging, feeding and territorial maintainance, to the most recent studies, particularly from West Africa, which have emphasized adaptations to new or marginal habitats and the testing of theories such as optimal foraging or resource competition (e.g. Harrison, 1983a,b,c, 1984a; Kavanagh, 1978a,b, 1980a,b). Thus, information ranges from general accounts of diet, food-getting behaviours, population size and density, and competition for resources, to detailed analyses of how one or two groups in a specific habitat make their living.

Throughout their range, in Africa and on the colonized Caribbean islands, vervets are found in a variety of habitats, from savannah with minimal tree cover, through open woodland to gallery and rain forest edge (Figure 20.1). Although preferring riverine woodland (Chism & Rowell, Chapter 21), they seem limited only by the need for water and sleeping trees. What sets them apart from other *Cercopithecus* species is their ability to live in so-called 'marginal' habitats (Struhsaker, 1967c; Poche, 1976; Galat & Galat-Luong, 1977). In particular, vervets are able to move into disturbed areas of secondary growth, including farming areas (Kavanagh, 1980b), special habitats such as mangrove swamps (Galat & Galat-Luong, 1976) and, like rhesus monkeys in India, to be able to adapt to and exploit human activities in tourist and urban parks (Basckin & Krige, 1973; Krige & Lucas, 1974; Henzi & Lucas, 1980; Brennan, Else & Altmann, 1985). In addition, *C. aethiops*, have successfully colonized several Caribbean islands onto which they were introduced, probably as pets, in the late 1600s. Large populations of these monkeys, probably of West African ancestry, are found on the islands

Figure 20.1. A group of green monkeys (*C. sabaeus*) in the lower Senegal Valley. Its home range included an *Acacia nilotica* forest flooded for some weeks during the rains, a gallery forest of *Ziziphus*, and an open tree savanna (from Galat & Galat-Luong, 1977; photo; A. Galat-Luong).

of St Kitts, Nevis and Barbados (see Table 20.1 for details). There too, as in Africa, they have adapted to human activity and become crop-raiders (Poirier, 1972; Horrocks & Hunte, 1986). In contrast, the arboreal mona monkey (*C. mona*) introduced on the island of Grenada has not been successful and may now have disappeared from the island.

In Africa, field studies of vervets have been concentrated in East and West Africa, with only a few long-term studies of 'urbanized' vervets in Burman Bush Reserve in Durban, coming from South Africa (Basckin & Krige, 1973; Krige & Lucas, 1974; Henzi & Lucas, 1980). Burman Bush is considered a 'rich' habitat where vervets can supplement a plentiful supply of natural and cultivated fruits with tourist handouts. Groups there are large, and the fecundity and survival rates are high, as is the resulting population density (Henzi & Lucas, 1980).

West Africa is notable for a number of studies of the ecology of *C. aethiops* in seasonally variable, marginal and new habitats (see Table 20.1). This body of research stresses variability in diet, the flexibility in what is eaten and when, the corresponding variability in how food is obtained, and the ability of *C. aethiops* to maintain themselves for considerable periods of time on relatively restricted and nutritionally poor, 'non-optimal' diets (Galat & Galat-Luong, 1976; Kavanagh,

Figure 20.2. Adult male *C. sabaeus* preying on a turtle-dove (*Strep-topelia senegalensis*) during the 1976 drought in the lower Senegal Valley (photo: A. Galat-Luong).

1978a; Harrison, 1984a). Variable ranging patterns, grouping patterns and intergroup relations (e.g. degree of territoriality), are seen as responses to these sometimes great fluctuations in resources (Galat & Galat-Luong, 1977; Kavanagh, 1981; Harrison, 1983c).

For example, in his study of optimal foraging of green monkeys on Mt Assirik in Senegal, Harrison (1984a, 1985) found that they adjusted major activities, such as foraging, to counter the effects of temperature extremes during the difficult dry season. Day range length and time spent foraging were correlated and both, in turn, varied with the density of available food. Thus, when the food was sparse, the vervets used an energetically 'cheap' strategy of reduced travel and activity, gradually switching to a more 'expensive' one in which the increased travel costs were balanced by higher returns on the energy expended, as food density increased. At the same time, feeding synchrony varied with the availability and type of resource and synchrony increased as more uncommon foods were eaten. Thus the Mt Assirik monkeys had not one but several strategies for optimizing foraging.

In very seasonally varied habitats, the 'low' season, almost always the dry season, is the one in which the animals are most at risk. They may, as Harrison found, reduce their activity levels and foraging travel time to palliate the effects of extreme heat and dryness and low

Figure 20.3. A group of *C. sabaeus* after the long 1976 drought in the Senegal Valley. The poor condition of the adult female is shown by its ragged coat and its prominent pelvis bones (photo: A. Galat-Luong).

resource levels (or low levels of preferred, high quality foods), and these 'cheap' strategies may be designed to get them through or to 'wait out' the dry season. The degree to which these 'waiting out' behaviours reflect stressful conditions perhaps can be seen in the results of the study by Galat & Galat-Luong (1976, 1977) of *C. aethiops* diet and demography in a marginal, seasonally dry, habitat in N. Senegal (see also Wrangham, 1981). In the second year of their study, during a prolonged dry season in which the monkeys found themselves in direct competition with Nile rats and which resulted in low levels of flowering and fruiting of many important species, the monkeys reduced their activity level to the point of lethargy, and in some cases, increased their consumption of gum and non-plant foods (including Nile rats and turtle-doves, Figure 20.2). They also showed clear signs of declining condition. Animals grew thin, their coats became ragged and one adult male became blind and disappeared (Figure 20.3). In all, seven individuals (six adults and one infant) out of a group of 47 disappeared during the end of the prolonged dry season, and the following year saw fecundity drop from 100% to approximately 25%. The authors suggest that the ability of *C. aethiops* to adjust fecundity, in particular to take advantage of periodic abundance by reproducing temporarily at high rates, is a contributor to their success.

In East Africa, vervets have been most commonly studied in the large parks, especially Amboseli National Park in Kenya. As in West Africa, vervets are found in a variety of habitats from quite open savannah scattered with fever trees providing shelter and food, to heavily treed riverine and gallery forest. Vervets in the parks have adapted well to local and tourist activities to the extent of becoming a pest around some tourist lodges (Brennan et al., 1985). The same themes of variability in foraging patterns, foods eaten, group size and range and intergroup relations are reported from East Africa as from West Africa. Flexibility in responding to fluctuating resources, the capability to exploit new resources in marginal or new habitats, tolerance for human settlement and activity, even hunting, all demonstrate the versatility of the vervet monkeys described in these studies. In particular, researchers note their ability to shift a whole range of behaviours in response to major changes in resource availability and distribution (see East African studies in Table 20.1).

While a group as a whole may alter activity or foraging strategies to account for seasonal fluctuations in conditions or resources as the Mt Assirik studies showed, social dominance as the regulator of individual access to resources has been the subject of several studies in Kenya (Cheney, Lee & Seyfarth, 1981; Wrangham & Waterman, 1981; Whitten, 1983). Wrangham (1981) found evidence of differential, rank-related mortality among females during a severe drought in Amboseli; high ranking females controlled access to the most available water-bearing trees; mortality was higher among lower ranking females. In this extreme situation, comparable perhaps to that described in the Galat & Galat-Luong study in Senegal (1977), the results of differential access to resources appear clear and decisive. Under more 'normal' conditions, rank-related, differential access is more difficult to demonstrate and its consequences correspondingly harder to show. For example, in a further study in Amboseli, Wrangham & Waterman (1981) found only two items out of nine compared, (the flowers of one species of acacia and the gum of another), which were eaten significantly more, in terms of feeding time, by high ranking animals than by lower ranking ones. One item, acacia seeds, which was plentiful, not particularly sought after and less clumped in its distribution, was eaten significantly more by low ranking females than high ranking ones.

Both preferred items constituted 'clumped' resources, which in a more extensive study by Whitten (1983) of competition for resources among female vervets in Samburu National Park, were found to be the major causes of overt feeding competition. Either it is only clumped, preferred resources which are readily defendable and worth the energy expended in supplanting other individuals, or competition for dispersed or secondary resources is not readily observable and measurable, even though present. As Whitten (see also Chapman, 1985) points out, most direct competition is avoided, inter-individual agonism is low and supplantation rates are not high, probably because subordinate females avoid clumped resources that are occupied by more dominant ones. They may move into a clump after dominant animals have left, or try to reach it before they arrive, only to leave when they approach, but such temporal 'sharing' probably results in their getting second bite at the choicest fruits and flowers, especially when these foods are readily depleted. Such competitive avoidance can lead to the dietary shifts described by Wrangham & Waterman (1981) and by Whitten (1983), and may result from different quality diets of the same items as well as diets composed of different items.

In the Samburu study, Whitten (1983) was able to show that seven out of 10 major food items (accounting for 80% of the total feeding time) had a clumped distribution and, further, that high ranking females had greater access to these foods and spent longer occupying the clumps than did lower ranks. By examining weights of captured animals she was able to establish that higher ranking females were, on average, heavier than lower ranks, and so could be assumed to be in better condition. Further, Whitten found that in one of the two study groups, higher ranking females conceived earlier, weaned sooner, and gave birth earlier the following birth season than most low ranking ones (see also Cheney *et al.*, 1981; Horrocks, 1986). From this she inferred that high ranking females have higher fecundity rates than low ranking females (and at least equivalent survival rates and reproductive lifespans) and thus forged a link between socially mediated dietary differences and differential lifetime reproductive success. However, as Whitten noted, such a link really requires lifetime reproductive data on known females, which is extremely difficult to obtain.

So, whilst there are some good data showing how social rank can regulate access to clumped, preferred, or essential resources (e.g. water) and how, in a time of great shortage or crisis, this differential access confers direct fitness benefits on high ranking animals, it is not yet clear how, or indeed if, under 'normal' conditions such benefits accrue, or do so sufficiently to be translated into measurable lifetime reproductive benefits.

In East Africa, as elsewhere (see Table 20.1), vervets are able to range sympatrically with other *Cercopithecus* species in areas of riverine and gallery forest and on the edges of the rain forest, to travel in mixed groups with *mona* monkeys in gallery forest (F. Bourlière, personal communication), to live alongside baboons and patas monkeys in open savannah, and to coexist with humans and their domesticated animals in disturbed and cultivated areas, even to the extent of becoming distinctively 'urbanized'. If versatility and opportunism seem to characterize the vervets' ability to adapt successfully to changing habitats and resources throughout subsaharan Africa, what makes them so versatile is not entirely clear. They are, or can be, catholic in their tastes, with diets varying from those with large amounts of tough, fibrous, relatively low quality foods, such as acacia pods, grasses and fibrous fruits, through the diets of soft fruits, flowers and seeds that characterize many other *Cercopithecus* species, to very specialized diets such as one comprised largely of fiddler crabs (see Table 20.2). In other words, although largely vegetarian, they are properly described as opportunistic omnivores. Given the opportunity, vervets will eat a fruit-based diet similar to that of other *Cercopithecus* species (see also Gautier-Hion, Chapter 15), but when forced to change, can and do so readily. These changes imply more than good digestion; the amount of time spent in searching for and processing food, the dispersion of the foraging group, the control of intergroup competition for preferred or limited resources, the often changeable relations with neighbouring groups expressed in variable overlap and territorial defence, are all necessary concomitants of a sometimes precarious lifeway. In addition, corresponding differences in group cohesion, predator response, and amount and type of vocal communication also have been associated with different food-getting situations. As Gartlan & Brain (1968) noted, 'habitat variability and atypical or impoverished environments are probably one of the main sources of social variability' (p. 257).

The proposed origin of vervets from a related type of swamp-dwelling guenon (Kingdon, 1971; Ruvolo, Chapter 7; Dutrillaux *et al.*, Chapter 9) which then evolved into a large number of specialized arboreal species, may leave vervets less able to compete under conditions favourable to arboreal specialists, but better able to survive and prosper under fluctuating conditions where the specialists' adaptations are less advantageous.

Since vervets are semi-terrestrial, they readily range and eat on the ground; being semi-arboreal, they also can exploit the foods available in the trees. A very minimal amount of trees will suffice; in the forest they have no advantage over other *Cercopithecus* species (e.g. *C. mitis* with which they are often sympatric), on the open savannah they are not as well adapted as the patas or baboons. However, being able to live in either situation, being able to switch between the two if habitats change, to manage on what is available and to modify group organization accordingly, makes vervets a very successful 'edge' or transition species. Historically and increasingly, edge areas or altered forest are expanding, so vervets are seen as a very successful 'colonizing' species. More specialized arboreal species are threatened by the loss of forests, as open country specialists are constrained by the conversion of savannah to cropland or grazing for domestic stock. Thus, while vervets may have lost some traditional habitats through changes in land patterns and use (Struhsaker, 1976), they have been able to exploit to their advantage the major changes in vegetation and land use that continue in Africa. Their spectacular success in rapidly colonizing and maintaining their populations in the Caribbean island of St Kitts, Nevis, and Barbados is an additional expression of these qualities of flexibility and versatility.

Demography

The ability of vervets to colonize new areas, whether islands in the Caribbean or newly created forest margins in Africa, is facilitated by several features of vervet demography.

Maturation is faster in vervets than in forest *Cercopithecus* species, being comparable to the rate of *E. patas*. Weaning may begin as early as 12 weeks and is usually completed by 8.5 months (Struhsaker, 1971; Lee, 1984a; Horrocks, 1986). Since vervet females frequently conceive in consecutive years, the conceptions take place while the mother is still suckling her previous infant (Horrocks, 1986). However, if the mothers do not conceive again in the same year, they may continue to suckle the current infant into the next year (Cheney *et al.*, 1981; Horrocks, 1986; Butynski, Chapter 16).

Age at first birth can be as low as 2.5 years ($\bar{x} = 3$), although females may not reach their full adult weight for at least another year. Males may be able to produce sperm by the age of 3 years, but they are not behaviorally mature until age 5 years, and may not reach full adult weight until the age of 6 plus. Thus, although males and females develop equally rapidly for the first 1.5 years, female development slows more rapidly than male development, with fully adult males reaching 1.25 times the weight of the adult females.

Although *C. aethiops* is considered to exhibit more of a birth peak than a true birth season, births are synchronized, with approximately 80% occurring during the same two or three month period. Births usually appear to be timed to occur just prior to or at the beginning of the season when resources are plentiful (Struhsaker, 1971; Lancaster, 1972; Wrangham, 1981; Galat & Galat-Luong, 1977; but see Gartlan & Brain, 1968; Dunbar, 1974; Horrocks, 1986; Butynski, Chapter 16). An interbirth interval of one year is most common; however, females who lose an infant soon after birth may conceive before the next mating season (Cheney *et al.*, 1981; Horrocks, 1986). Vervet females seem to be able to maintain a high rate of fecundity, at least over short periods of time. Long-term population studies of known free-ranging groups of vervets have yet to be published; however, it is clear that population growth rates can be rapid (e.g. Fedigan *et al.*, 1984). Even in marginal habitats, under stress from lack of resources, populations can be maintained or recover quickly when conditions improve (Galat & Galat-Luong, 1977; Wrangham, 1981).

The high survival rate of vervet offspring is another key to rapid population growth. As might be expected, mortality rates are variable, being low in unstressed populations with few predators, and high in groups under severe pressure from resource depletion, severe climatic stress (late rains, drought) or high rates of predation (Struhsaker, 1976; Wrangham, 1981; Horrocks, 1986). Earlier, Gartlan & Brain (1968) had suggested that under stressful conditions, subadult and adult males have a tendency to peripheralize. Presumably this would make them more vulnerable to whatever was causing the stress and reduce the stress on the central or core population of females, young and central males. However, in the best documented and more recent reports of vervet groups under stress, mortality was related to dominance rank and sex, and there was high infant mortality. There was no evidence of male peripheralization, although males in the drought situation reported by Wrangham (1981) did seek water in neighbouring ranges and two of four males who disappeared were found in neighbouring ranges. In their study of natality and mortality in three groups of vervets, Cheney *et al.* (1981; see also Whitten, 1983) found that dominant females were not more fecund or successful in raising offspring than subordinate females, but that the probable causes of mortality were different; high ranks being more vulnerable to predation, whereas low ranks were out-competed for resources. As they frequently exploit marginal habitats subject to extreme seasonal fluctuations in climate and resources, or to new predators such as hunters and domestic dogs, vervet populations often may be under severe stress. The flexibility in group size, territory, and foraging behaviours that allows them to exploit food resources beyond the reach of their congenerics, also aids rapid population growth or recovery, as well as maintenance of high population densities. On the other hand, where population densities and group size fall below the level at which resources can readily be defended, group fission and eventual group extinction may result (Hauser, Cheney & Seyfarth, 1986).

Social patterns

Several aspects of vervet monkey social behaviour, which have been well documented in field research, appear to contribute to the success of this species in terms of its abundance, distribution and persistence under conditions of environmental change and/or stress. For example, the communication patterns of vervets have been demonstrated to reflect learned cognitive abilities to categorize objects in the natural habitat and to be well adapted to the exchange of representational information about classes of environmental objects. Although such complex forms of communication may well exist in other species of non-human primates, it is only in the vervet monkey that they have been extensively documented (Cheney & Seyfarth, 1980; Seyfarth & Cheney, 1980, 1982; Seyfarth, Cheney & Marler, 1980a,b). For example, vervets give acoustically different alarms to different classes of external danger (large mammalian carnivores, raptors, snakes), and each alarm-call type is associated with a distinctive response, which represents appropriate escape behaviour for the particular type of predator involved (ground predator call: running up into the trees; aerial predator call: looking up; snake alarm call: looking down on the ground around them). Further, Cheney and Seyfarth's experimental field studies have demonstrated that vervets distinguish other individuals in their own group and in neighbouring groups on the basis of voice alone, and that they make hierarchical, non-egocentric classifications of others as belonging to particular kin groups or dominance positions. Such an ability to perceive, communicate, and act upon taxonomies of external objects and social others, is clearly an asset to a species which 'specializes' in flexible, opportunistic interactions with a changing environment, but which maintains cohesive, relatively standard social structures under a variety of conditions.

Since vervets are one of only a few species which continue to thrive when humans convert forest and savannah into agricultural land, it is also important to note their behaviour when foraging on the edges of cultivated fields or when actively crop-raiding. Under such conditions, several researchers have reported (Kavanagh, 1980b; Harrison, 1983a; Horrocks & Hunte, 1986) that vervets adopt totally unpredictable ranging patterns, cease to give loud alarm calls (which are counter-productive against human hunters) and use more quiet, non-locatable calls in their repertoire. As Kavanagh stated (1983b), it is as if these monkeys have learned to take up whispering. A few primatologists (Poirier, 1972; McGuire, 1974), also have suggested that vervets 'post sentinels' during crop-raids and while this may sound somewhat anthropomorphic, Horrocks & Hunte (1986) recently argued effectively that crop-raiding vervets in Barbados invariably monitor one, non-foraging, highly visible adult male who exposes the white of his chest to their view while positioning himself in the top of a nearby tree with a good view of the surrounding fields. At the first approach of a human or a dog, the male silently fades from view and the watching group quickly runs from the open field to the safety of the trees. Only when caught unawares by a nearby predator does the sentinel vocalize, thus warning the foragers and distracting the predator. Three hundred years of heavy human predation in the Caribbean may have selected for silent or 'whispering' vervets, but the monkeys also seem to be able facultatively to adjust the use of vocalizations to their own advantage.

Another example of apparently facultative behaviour in the vervet monkey is the expression and degree of territoriality. Some primates, such as hylobatids and callitrichids, are consistently territorial whatever the environmental context, but vervets are said to exhibit territorial behaviour which varies according to habitat type, season, predator pressure, distribution, abundance and defendability of resources and the history of relations between groups (e.g. Hall & Gartlan, 1965; Gartlan & Brain, 1968; Struhsaker, 1976; Cheney, 1981; Galat & Galat-Luong, 1976; Kavanagh, 1981; Harrison, 1983c; Chapman & Fedigan, 1984). To paraphrase Harrison, territorial behaviour in vervets is not a rigid maintenance of fixed boundaries, but a flexible response to variable conditions. Reports of encounters between groups range from descriptions of frequent and severe intergroup contact aggression, including wounding (Struhsaker, 1967b,d; Cheney, 1981) to the other extreme in which groups are reported sometimes to merge without agonism and groom and play together (Kavanagh, 1981; Harrison, 1983c). Although there is a trend for the East African vervets generally to be described as more territorial than the West African ones, even within one study site researchers report great variation (e.g. Cheney, 1981; Kavanagh, 1981; Harrison, 1983c). In West Africa, Kavanagh and Harrison targeted ecological variables, especially resource and predation patterns, as the causal factors in variable relations between groups. For example, Harrison found that territorial behaviour varied seasonally, according to the availability and distribution of favoured foods. Both Harrison and Kavanagh concluded that smaller groups living in ranges small enough to be economically defendable were more likely to be territorial than those living in larger ranges. Further, Kavanagh argued that groups in West Africa which are more subject to heavy human predation probably cannot afford extended vocal and visual territorial displays which catch the attention of hunters. At the East African study site of Amboseli, where known individuals and groups have been followed over several years, Cheney (1981) argued that the nature of intergroup encounters is strongly affected by the past history of male migration between groups. At Amboseli, group ranges are small ones in an open habitat, and each group is usually within sight of at least one other group in which individuals are recognized. Groups that frequently exchange males are less aggressive to each other, and may be thought of as forming a higher order 'community of groups'. Males usually transfer into a neighbouring group rather than a more distant one, and often into a group which already contains immigrants (brothers, peers) from their group of origin. Adult females have a major influence on the success of males in emigrating into their group and appear to be less aggressive to more familiar males from neighbouring groups.

On the island of St Kitts, we also found territorial behavior to be highly variable and concluded that knowledge of the history of intergroup relations was a major factor in understanding why some groups of C. *aethiops* vigorously defend their ranges against encroachment by conspecifics, whilst others ignore or interact amicably with encroaching neighbours. Whether one focuses upon variable ecological or historical correlates, the ability to adjust the defence of resources according to changing conditions is probably a selected behavioural capacity which plays an important role in the success of this species.

Further, it is probably not by coincidence that a species which exhibits relatively high fecundity and rapid maturation for a cercopithecine, also provides one of the most definitive examples among primates of an allomothering system in which the three parties involved, mothers, infants and allomothers, receive positive benefits which far outweigh potential costs. Hrdy (1976), in her review of alloparenting in primates, argued that in many species the infant is actually kidnapped and abused by the supposed 'caretaker', resulting in survival costs to the mother and infant. However, in the vervet monkey, researchers agree that allomothering is not abusive and mothers neither show reluctance to give up their infants to other caretakers, nor need to use aggression to regulate interactions and retrieve their offspring (Struhsaker, 1971; Lancaster, 1972; Lee, 1984a).

Many reports of vervet allomothering indicate that nulliparous females, especially older female siblings of the infant are the most frequent caretakers (e.g. Struhsaker, 1971; Lee, 1984a). However, other group members, including non-related juvenile males will also retrieve and care for infants. And in a unique report from South Africa, Krige & Lucas (1974) found that experienced *adult* females with offspring of their own will act as aunts and sometimes kidnap a resisting infant from a non-resisting mother. All other reports seem to agree with Lancaster's initial conclusion that all three parties cooperate and benefit from vervet allomothering: the mother receives relief from constant infant care, the infant is provided with additional caretakers, and the allomother gains highly important practice in the skills of infant care.

The birth weight of vervet infants is heavy in proportion to the mother's weight and the rapidly growing infant is a very real energetic burden to the mother. This burden can be alleviated when others take turns carrying the dependent infant during the first three months of life, the developmental time period when most allomothering occurs. The willingness of others to carry and care for the infant also gives the mother increased feeding opportunities and supports a system of early weaning and rapid independence. In turn, the capacity of vervet infants to mature rapidly and survive without constant maternal care in the first year of life, facilitates the adult female's ability to conceive and bear offspring on a yearly or biannual basis.

Conclusions

What is it, then, that makes *C. aethiops* such a successful primate? Dietary, demographic, and behavioural flexibility or opportunism are phrases that occur frequently in descriptions of vervets, as they have done so in this paper. But what exactly do the words opportunism and flexibility mean when we use them? Are they units of fitness? Or are they simply descriptions we use in a *post hoc* explanation of the observed variability in these animals? And, even if the latter is the case, as we believe it is, have we been able to describe the nature of the flexibility reported in so many studies?

One characteristic of *C. aethiops* that does single them out from other guenons is their ability to be completely at home both on the ground and in the trees. Being neither forest nor savannah specialists, they clearly have an advantage in a rapidly changing mosaic of habitats.

Along with another of their commonly reported traits, the lack of dietary specialization, this ability to exploit mixed and marginal habitats becomes doubly advantageous. If lack of specialization in habitat and dietary requirements gives *C. aethiops* greater ecological maneuverability than their congenerics, their demographic characteristics of rapid maturation, intense allomothering, and generally high fecundity, give them the capacity to adjust group and population size rapidly.

Thus we have a primate that can live in almost any habitat which provides a few trees, water, and a minimal diet, and that can increase its population size rapidly when conditions are favourable. What is needed to complete the picture of a successful colonizing primate are the behavioural capacities to make all the rest 'work'. Here we have suggested a few of the behavioural components that seem to facilitate the adaptive success of vervets. Social cohesion, that is, the ability to maintain the integrity of the core or reproductive center of the group, under conditions of stress, either environmental (e.g. resource depletion, extremes of climate) or demographic (e.g. high population density) is essential. Vervets do not seem to have individually unique behaviours to maintain group integrity, but the combination of group vigilance and defense, complex facultative communication skills, relatively low levels of severe intragroup agonism or competition even at high density, and the ability to survive and persist alongside conspecifics, congenerics and other potential competitors and predators, including humans, would appear to be implicated in their success.

In these respects, vervets are similar to the other abundant and widely distributed semi-terrestrial primates of the Old World, the macaques and the baboons. Especially, the rhesus macaque exhibits a similar ability to thrive in a variety of changing habitats, subsisting on almost any kind of food and colonizing new areas under human influence. Indeed, one would not be amiss to refer to monkeys such as vervets and macaques as the 'house sparrows' of the primate world. So, while most *Cercopithecus* monkeys were evolving as splendid specialists of the African rainforest, *C. aethiops*, seems to have specialized only in persistence under fluctuating conditions, and as a result, may prove to be one of the most successful 'survivors' on the changing African continent.

III.21

The natural history of patas monkeys

JANICE CHISM and THELMA E. ROWELL

Introduction

The Patas monkey (*Erythrocebus patas*) is frequently separated generically from the other guenons on morphological grounds because of its adaptations for cursorial locomotion. This separation now appears not to be justified by many of the taxonomic criteria discussed in this book (see Lernould, Chapter 4; Ruvolo, Chapter 7; Dutrillaux *et al.*, Chapter 9). The emphasis given to patas adaptations for terrestrial locomotion has often led to the idea that patas are behaviourally and ecologically very distinct from other guenons as well. A closer and longer look at patas natural history gives us reason to characterize this species as being more similar to other guenons than to the savannah-dwelling baboons.

Patas monkeys are found in a belt across Africa south of the Sahara and north of the equatorial forests. In East Africa their range used to extend as far south as the Serengeti National Park in northern Tanzania (Kingdon, 1971). In Kenya, in the last two decades, groups have been sighted as far south as the area of Amboseli (S. Altmann, personal communication), while at Gilgil, in the eastern Rift Valley, rare sightings of adult males have been reported.

The first major field study of patas was that conducted by Hall (1965a) in the Murchison Falls National Park in Uganda, on the north bank of the Victoria Nile. At the time of Hall's study the area was largely open grassland with occasional trees, although earlier it had been more wooded. The dominant grass, *Hyparrhenia* spp., grows to heights of more than 2 m and some areas were burned annually. Because of visibility problems, Hall concentrated his observations on groups which frequently moved and foraged on open erosion sheets.

Hall was conscious of the limitations of his observation methods and of the brevity of his study, which did not cover a full year, and was careful to separate his observations from his inferences about patas behaviour (a caution not always emulated in the frequent citations of his work).

Patas were also studied in the Waza National Park in northern Cameroon (Struhsaker, 1969; Struhsaker & Gartlan, 1970; Gartlan & Gartlan, 1973; Gartlan, 1975). These studies took advantage of the fact that, at the height of the dry season in that area, patas spent some time at waterholes daily so that social behaviour of groups at waterholes could be observed and filmed from a platform (Gartlan, 1974). During the wet season the patas moved out into the rest of the park, much of which is then inaccessible to vehicles, so that, at Waza also, observations were confined to part of the year.

Observations of the interactions of patas with their environment in any part of their range were thus fragmentary until the main study presented here. In Kenya, we observed a patas population throughout two successive annual cycles from 1979 to 1981. In addition, we observed a patas population in northern Ghana for about four weeks in 1978. This short study in West Africa has given us some feeling for the possible generality of the conclusions derived from the main study in Kenya.

The study areas

Our main study site was in the Laikipia District of Kenya, on the Equator (0° 9' N, 36° 40' E), at an altitude of 1900 m. This is close to the easternmost extent of the species range and may be close to the southern extreme of its range at present. The study area is a mosaic of *Acacia-Themeda* wooded grassland (Lind & Morrison 1974) ranging from treeless grassland areas to dense woodland. Most of the site was on the ADC Mutara Ranch, used by the Kenya government for the national stud herd of Boran cattle. The density of cattle at Mutara, about one head to four hectares, has always been low compared with ranches where cattle were raised primarily for meat. Tracks, fences and watertanks were the main human alterations to the landscape. Regular burning had taken place in the past, but had ceased except for small scale burns for firebreaks about 20 years prior to this study. Thus, the vegetation can be considered as relatively unaltered from the natural plant community for the area and the rainfall.

Wild ungulates on the ranch included Thompson's and Grant's gazelles, impala, waterbuck, eland, oryx, zebra, warthog and giraffe. Mammalian predators included lion, leopard, cheetah, wild dog, caracal, serval, two species of hyena and three species of jackal. Large raptors included the martial eagle and Verreaux' eagle owl. Other primates which had been observed on the ranch included vervets, baboons, Senegal bushbabies, Syke's monkeys and black and white colobus, although the last two species were not seen during the study. Baboons had been regularly shot as 'vermin' and there were no baboon groups within the home ranges of our two patas study groups. There were groups of baboons in adjacent areas.

The second study area was in northern Ghana, near the Red Volta River, at the village of Tenzugu (10° 41' N, 0° 49' W), at an altitude of 300 m. Tenzugu has long been an important center of ritual and traditional medicine (Fortes, 1945). The local Tallensi people consider many wild animals, including monkeys, to be 'vessels' which receive ancestral spirits. Thus, patas received some, but not complete, protection from hunting there.

The vegetation of the area is classified as guinea savannah woodland. The patas occupied a ring of hills surrounding the village, in a terrain distinguished by piles of huge granite boulders. The boulders made large areas inaccessible to grazing and so provided some protection to a variety of woody shrubs and trees. In less rocky areas the hills had been mostly deforested for firewood and regeneration was limited by burning and by grazing livestock. There, stunted specimens of trees were fairly common in the main, grassy cover. Tallensi homesteads were located in the flatter areas among the hills and these areas were densely planted with crops. Besides the patas monkeys, vervets were the only large mammal observed in the area during the study.

Rainfall

Rainfall records had been kept at Mutara for 53 years prior to our study (Figure 21.1). The average annual rainfall during that time was 630 mm (annual totals ranged from 330 mm to 1100 mm). Although, on average, the wettest months were April and November, and the driest month was January, variation in rainfall for each calendar month was striking. In addition, local variation in rainfall was such that one part of the study area could receive rain while an area a kilometre away remained dry.

The expectation for this part of Kenya is of two annual rainy periods: the 'long rains', which usually occur in April and May, and the 'short' rains, in November. Our study period included a year of slightly above average rain and a dry year; it was about average in terms of the distribution of rainfall over the year.

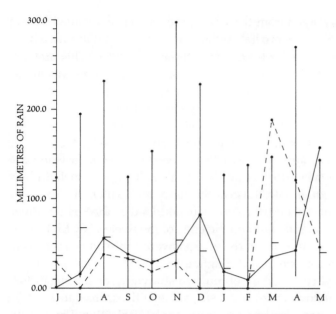

Figure 21.1. Monthly rainfall at A.D.C. Mutara Ranch, Kenya. Actual rainfall during the study is superimposed on monthly means and ranges calculated for the period 1927–79. Solid line: monthly rainfall, June 1979–May 1980; broken line: monthly rainfall, June 1980–May 1981.

Our study area, like most other areas where patas occur, could be described as having a seasonal rainfall pattern which is predictable in a general way, but highly variable, and therefore unreliable, in the short term. We might expect plants and animals living in such an environment to show both opportunism and some predictive seasonality.

The vegetation at Mutara responded with a green flush after a few rain showers. The quantity of plant growth in the area varied, not only with the amount of rainfall, but with how the rain was distributed on a day-to-day basis. Several days of slight rainfall adding up to 25 mm, for example, were less effective than the same 25 mm concentrated in one day's heavy rainfall.

The mean monthly minimum temperatures at Mutara varied from 8 to 13°C, the monthly maxima from 17 to 28°C. Daily temperature fluctuation was greatest in the dry months of January and February when cloud cover was least and night-time temperatures could drop to 0°C.

Annual rainfall at Bolgatanga, near Tenzugu, is about 1100 mm, with a single wet season from May to September. Monthly minimum temperatures ranged from 18 to 24°C, and the maxima ranged from 28 to 38°C. The higher temperatures at Tenzugu meant that evaporation was faster there than at Mutara (cf. Deshmukh, 1986). This probably contributed to the fact that although rainfall at Tenzugu was higher than at Mutara, the vegetation was no more luxuriant.

Animals

Two groups of patas were the main subjects of the study at Mutara. Other groups were occasionally encountered in the study area and groups to the north, east and southeast had ranges which overlapped to some extent with those of the study groups. On the west, the home ranges of our study groups were bounded by the Suguroi River; another group's range reached the edge of the river valley to the west of the river. Thus, the study groups were embedded in a patas population which extended throughout much of Laikipia District.

The size and composition of our two study groups during 1979-81 and during a follow-up study in 1983, along with census data for eight other Laikipia patas groups, are presented in Table 21.1. The population also included immature and adult males which lived apart from the heterosexual groups either as solitaries or in small, all-male associations. These males appeared to have large home ranges which overlapped those of several heterosexual groups. In addition, males constantly moved in and out of all-male associations and heterosexual groups (Chism & Rowell, 1986). We estimated the population density of animals in heterosexual groups in our study area as 1.4 per km². Of our two study groups, M2 was consistently more than twice as large as M1 during the main study. During the follow-up study in 1983 we found that M1 had been joined by females and juveniles of an adjacent, small group. This new group continued to use M1's old range.

The composition of the groups changed seasonally with the highly synchronized annual reproductive cycle (Chism, Rowell & Olson, 1984). Most infants were born in December and January and mortality of adult females and infants during our study was highest during the dry season. In some years the number of adult males in the groups increased briefly at the height of the conception period in July and August. Females and their immature offspring formed the stable core of the group, while the association of any particular male with a group was brief, ranging from a few days to about nine months.

At Tenzugu, two groups, with 40 and 27 members respectively, were followed for several days and a third group was followed briefly. Each group included a single adult male. The study was too short to determine accurately the age/sex composition of the groups. Grooming

Table 21.1. *Size and composition of patas groups*
Size and composition of main Mutara study groups

Date	Ad. ♂♂	Ad. ♀♀	Subad. ♂♂	Large juvs	Medium juvs	Small juvs	Infants	Total
7/80	1–3	9	0	2	0	3	6	21–23
7/81	1	4	0	2	0	6	3	16
7/83	1	14	0	1	5	8	12	41
7/80	1–7	19	0	4	4	6	14	48–54
7/81	1	16	0	5	1	12	12	47
7/83	4–10	22	0–5	2–5	7	14	16	68–74

Size and composition of non-study groups in Laikipia

Group/ date	Ad. ♂♂	Ad. ♀♀	Subad. ♂♂	Juveniles 1–3.5 yr	Infants 0–1 yr	Uncertain	Total
NR 1							
7/30/80	0	3	0	6	3	1	13
1/4/81	1	2	0	10	0		13
1/16/81	1	1	0	10	0	1	13
M 3							
5/ /80	1	11–12	0	7–8	7		27
3/18/81	1	10	0	10	9		30
4/1/81	1	10	0	10	9		30
M 4							
9/28/80	1	17–18	0	10–11	10		39
OP 2							
8/25/80	1	11	0	11	10		33
10/25/80	1	8–9	0	5	6–7	4	25
6/4/81	1	8	0	10	7		27
OP 3							
5/13/81	1	19	0	21–22	15–16		56
OP 4							
5/13/81	1	15	0	13	15		44
LH 1							
6/24/81	1	6	0	5	5		17
LH 2							
6/24/81	1	6	0	5	3	4	19

Size and composition of patas groups in Uganda and Cameroon

Site/source	No. groups counted	Ad. ♂♂	Ad. ♀♀	Subad. ♂♂	Juveniles	Infants	Total
Murchison Falls N.P.[a]	5	1–1	4–12	1–2	3–6	0–10	11–31
Waza N.P.[b]	5	1–2	3–17	0–0	0–7	3–15	7–34

[a]Data from Hall (1965a), Table III, the five complete group counts.
[b]Data from Struhsaker (1969) and Struhsaker & Gartlan (1970).

between males and females occurred frequently during the study, which suggests that a mating period may have been in progress (cf. 1979 observations at Mutara, Chism & Rowell, 1986). The youngest cohort of immatures was about six months old although there was one black infant (less than 2 months old) in the larger group.

Home range

Extent
Patas monkeys are difficult to habituate, as Hall (1965a) also noted. At Mutara, it took three months before we could regularly follow a group all day and six months before we could reliably observe the animals from 50 m. The home range sizes reported here are based on 116 day ranges mapped during the second year of the study, beginning in June 1980, when both groups were well habituated. Each group's day range was mapped for at least 5 days each month, a total of 61 days for M1 and 55 days for M2. Daily routes were drawn directly on maps, using compass coordinates and paced distances. The area which included all day ranges for both groups was divided into quadrats, each 318 m on a side. The home range of each group was calculated as the sum of the quadrats it was seen to enter. The home range of M1 was 2340 ha and that of M2 was 3200 ha (Figure 21.2). From additional day ranges mapped during the first year of the study and from the larger sample of days when groups were followed but not mapped, we know that these represent minimum estimates of the groups' actual home range area.

M1, which was less than one-half the size of M2 for much of the study, had a home range which was about three-fourths the size of that of M2. Given the variation in the size of our study groups over the period of the study, a more precise correlation between group and home range size would be unlikely.

The range of the two study groups overlapped extensively, so that 58% of M1's range and 44% of M2's range was shared with the other group. This degree of overlap appeared to be unusual since adjacent groups did not overlap as extensively or as frequently into the ranges of the study groups.

Day range
Patas usually started moving out of the sleeping area between 0730 and 0800 h. The time at which they stopped moving at night varied seasonally: during the dry season the groups were usually

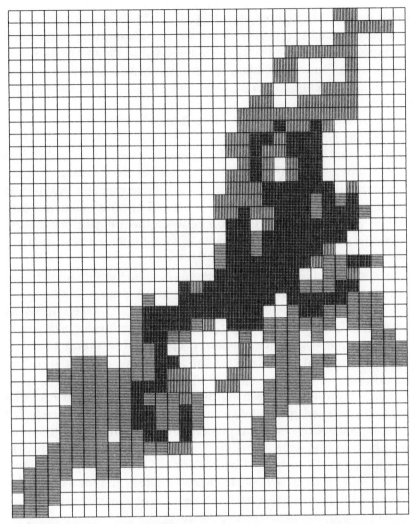

Figure 21.2. Home ranges of the study groups, M1 and M2, showing degree of overlap. Quadrats are 318 m × 318 m. Vertical shading: M1; horizontal shading: M2.

travelling and foraging until just before dark (about 1830 h), while during the wet season they often stopped moving an hour or more before sunset.

Day range length was calculated using a sample of 36 days for M1 and 30 days for M2 on which a group was followed and its route mapped for at least 9.5 h. The sample days were chosen from day ranges collected in the second year of the study to provide an even distribution throughout the year, including 3-6 days for each group for each month. The figures used were the actual distances the group travelled while being observed. For days when the group had moved from its sleeping area before being relocated in the morning, knowledge of its location the night before allowed us to calculate that no more than 0.5 km would have to be added to give a total average path length.

Group M1's day range varied from 1.38 km to 5.88 km with a mean of 3.83 km. Group M2's day range varied from 2.38 km to 7.5 km with a mean of 4.22 km (the longest day range observed, 10 km for M2, was not included here as it occurred on a day when the day range was not mapped). The difference in day range length between the groups was not significant (t-test, $p = 0.10$, df $= 64$) and does not seem commensurate with the difference in size between the groups if day ranges were determined by the availability of food for the number of group members.

In a sample of 571 scans of M1, evenly divided throughout the year, the most distant members were, on average, 96 m apart. In a comparable sample of 536 scans of M2, the most distant members were, on average, 179 m apart. The larger group thus had a greater path width, and the path width was roughly proportional to group size. The difference in path width between groups was greater than the difference in path length.

The daily route was usually determined by adult females, who typically initiated group movement, led progressions and selected sleeping sites. On a few occasions a resident male attempted to lead or shepherd a group, and sometimes, but not always, he would succeed in getting females to follow him. Males most often attempted to direct group movement when females were receptive. It appeared that on such occasions a male was trying to manoeuvre his group in relation to other patas groups or extragroup males.

It was sometimes clear that the daily route was determined by the monkeys' need to drink; this was the case when they moved long distances at the end of the day to reach water. On the other hand, during rainy periods, when the patas were drinking from puddles, they often did not visit water tanks at all for several days.

Home range use

Home range use was calculated from the number of times the daily path of a group passed through each quadrat. All parts of the home range were not used equally often: use varied from one to 15 visits to a quadrat (Figure 21.3).

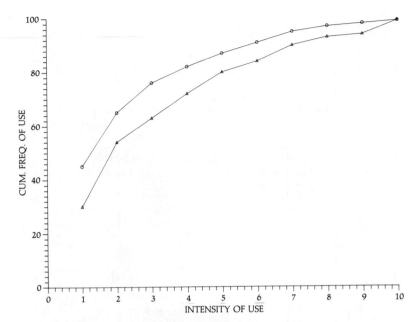

Figure 21.3. Comparison of cumulative home range use for each study group. △: Group M1, ○: Group M2.

For each group, the more frequently it visited a quadrat, the more likely that quadrat was also to be part of the other group's range (Figure 21.4(a), (b)). This is the reverse of the concept of a 'core area', which implies that each group should have most exclusive use of the most intensively used parts of its range.

With such range use patterns, it is not surprising that the two study groups encountered each other fairly often. In addition, the study groups had less frequent encounters with other, adjacent patas groups. Encounters were usually aggressive and sometimes involved one group chasing the other for a long distance (as much as 3 km). Females and juveniles were the chief antagonists; adult males usually took little part in encounters except during conception periods when resident males initiated encounters and actively participated in them.

Thus, the groups behaved as if they were territorial even though they were not able to sequester resources for exclusive use by their energetic defense of their home range. While it is difficult to imagine that an effective defense of such a large home range might be possible, one group was able to exclude another from an area briefly as, for example, when it was feeding on temporarily abundant food there.

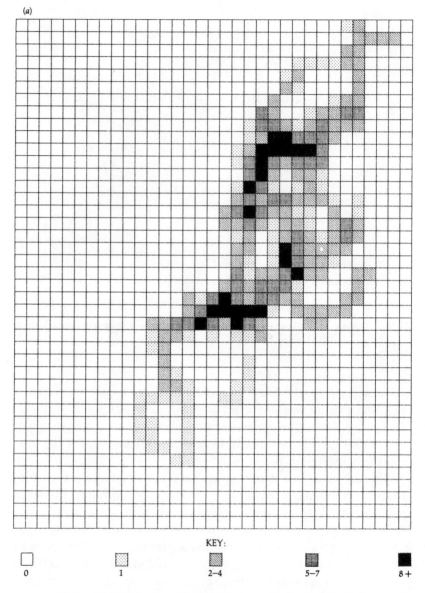

KEY:

☐ 0   ▧ 1   ▨ 2–4   ▦ 5–7   ■ 8+

Figure 21.4. Intensity of home range use. (*a*) Group M1; (*b*) Group M2. Density of stippling corresponds to the number of times each group was mapped in a particular quadrat.

(b)

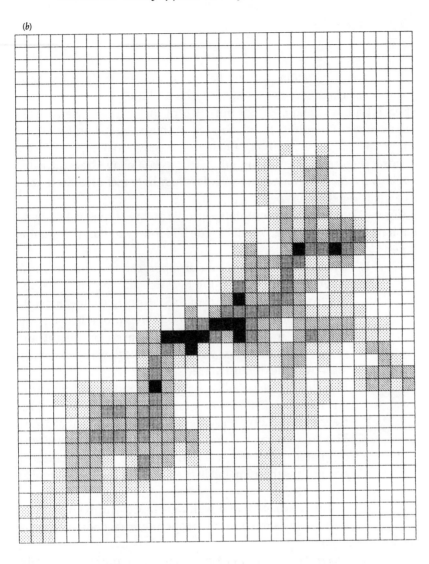

Habitat use

Vegetation

The vegetation within the home ranges of the two study groups was classified into six major types and the location and extent of each type was mapped. For purposes of this analysis, we counted the number of quadrats in which each of these vegetation types was either the predominant type, or one of the two major types in the quadrat (Figure 21.5). Although the 318 m × 318 m quadrats provided a relatively coarse grained analysis, few quadrats included more than two main vegetation types.

Figure 21.6 shows a representative profile of three types of woodland habitats in the study area. We distinguished four woodland vegetation types:

1. *Open acacia woodland* had an open canopy and no shrub layer (Figure 21.6). A one hectare sample plot contained 519 trees. A sample of 25 of these trees had a mean height of 4.2 m and the tree crown volume in one hectare was estimated to be about 13 100 m³, based on height and crown measurements. *Acacia drepanolobium* (or, rarely, *A. seyal*) was the dominant tree species.

2. *Dense acacia woodland* had a closed canopy and no shrub layer (Figure 21.6). A one hectare sample plot had 1121 trees in it and 25 of these trees had a mean height of 3.7 m. The tree crown volume for this type of woodland was estimated at 18 482 m³. We included dense stands of acacias less than 2 m tall in this category because of similar tree densities.

3. *Bushed woodland* had a well developed shrub layer and a tree density similar to that of open acacia woodland (Figure 21.6).

4. *Riverine woodland* was confined to watercourses and shrubs and climbers were well developed. The dominant tree species was *Acacia xanthophloea*.

Here we also distinguish two types of grassland:

5. *Grassland with scattered trees* had an average of 24 trees per hectare (usually *Acacia drepanolobium*).

6. *Grassland with few or no trees* had an average of three trees per hectare.

Boundaries between patches of different vegetation types, especially between woodland and grasslands, were often quite precise, so that

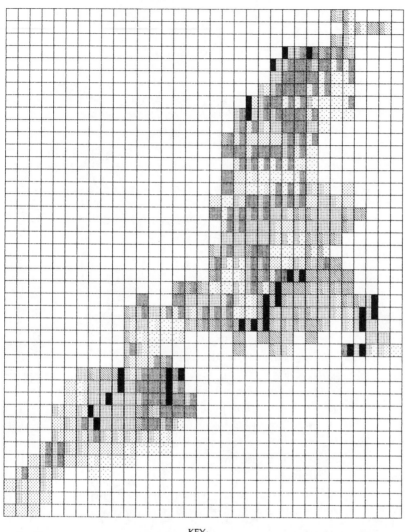

KEY:

GRASSLAND FEW OR NO TREES · GRASSLAND SOME TREES · OPEN WOODLAND · MIXED WOODLAND · DENSE WOODLAND · RIVERINE WOODLAND

Figure 21.5. Major vegetation types. Quadrats used by either study group are classified according to principal vegetation type.

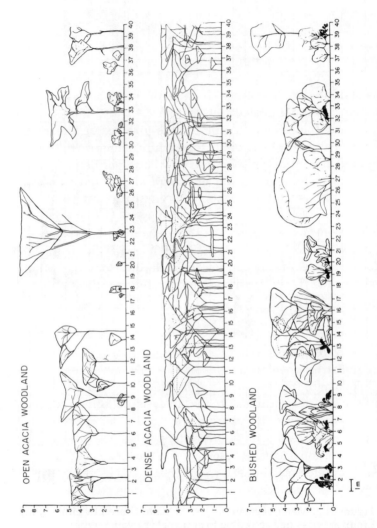

Figure 21.6. Vegetation profile. Profiles of three woodland habitat types, open acacia woodland, dense acacia woodland and bushed woodland, illustrating relative density and visibility within each.

Table 21.2. *Monthly distribution of half-hour samples during which habitat use was recorded for the study groups*

	Jan	Feb	Mar	Apr	May	Jun	Jul	Aug	Sep	Oct	Nov	Dec
M1	252	267	166	218	235	168	194	108	150	177	241	333
M2	211	204	133	180	155	141	100	178	197	110	91	297

Totals: M1 = 2509   M2 = 1997

margins and interiors of patches could be distinguished. The 100 m edge of a patch was designated as its margin.

On each day that the patas groups were followed at Mutara, we made an estimate every half hour of the amount of time (to the nearest five minutes) the group spent in each habitat type. Because of the extent of dispersion, a group was often spread over more than one habitat type at a time. If any member of the group was in a habitat type for at least five minutes during a half-hour period, the group was scored as having used that habitat in that period. Thus, scores of habitat use add up to more than 100% of the time available. During the study 4506 estimates of habitat use were made, 2509 for group M1 and 1997 for group M2. Table 21.2 shows the distribution of these estimates by month.

Preference for vegetation types

Patas were selective in their habitat use (Figure 21.7(a)). They strongly preferred open acacia woodland while they rarely entered bushed woodland, treeless areas and riverine woodland. In fact, the main rivers are not included in the analysis of home range use presented here because they were not entered at all on days when mapping occurred. Grassland with scattered trees was used at a rate that was proportional to its representation in the home ranges.

The strong preference shown for acacia woodland was a result of the patas' preference specifically for woodland margins (Figure 21.7(b)); the interiors were not used more than other, less favoured habitat types. Even at times of the year when the patas were feeding heavily on herbs, forbs or insects in grassland, they moved in the woodland margins, making brief forays into the grassland to feed.

Patas usually crossed open grassland only after long periods of scanning it from the edge of the woods. They then crossed in a 'leapfrog' fashion: one or two animals went ahead to a tree in the grassland, then the bulk of the group crossed to the tree and sat there,

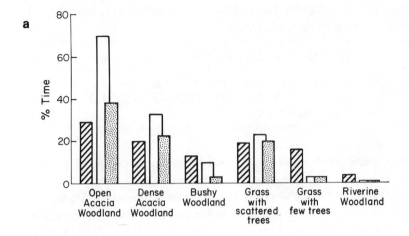

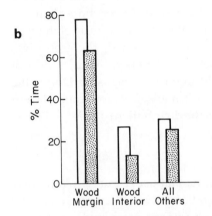

Figure 21.7. Habitat use by study groups at Mutara. (*a*) Use of vegetation types: hatched bars represent the proportion of each vegetation type in the quadrats used by the study groups; plain bars represent the proportion of time group M2 was in each vegetation type; stippled bars represent the proportion of time group M1 was in each vegetation type. (*b*) Relative proportion of time spent in woodland margin, woodland interior and all other vegetation types: plain bars represent M2; stippled bars represent group M1.

scanning again, while one or two animals moved ahead to another tree. This was repeated until the group again reached the margin of a wood. Often, one or two animals remained behind and scanned until the whole group had crossed, then raced across to catch up. The importance of the behaviour of the lead animals in these crossings was underscored by attempted crossings in which an initial movement into open grassland was aborted when the leaders suddenly ran back to the wood.

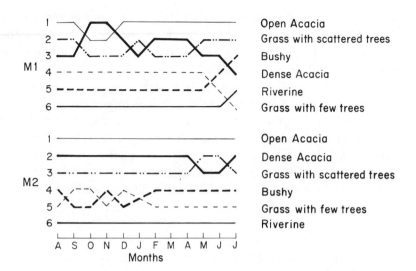

Figure 21.8. Habitat use during the year. Monthly rankings of the proportion of time spent in each habitat by the two study groups.

Seasonal habitat use

Choice of habitat did not differ dramatically over the year. Open acacia woodland, dense acacia woodland and grassland with scattered trees were almost always the most frequently used habitat types for both groups (Figure 21.8). Bushed woodland was seasonally important, when berries of shrubs such as *Carissa*, *Euclea* and *Lycium* were abundant. Riverine woodland and treeless grassland were little used at all times of the year.

Human artefacts

Patas were affected by human modifications to their range, all of which appeared to be improvements. There were 17 water tanks in the groups' home ranges. The groups drank from water tanks on most days, which may explain why they were able to avoid riverine vegetation so completely. The patas often sat in trees watching a water tank for long periods before approaching it to drink. Usually at least one animal remained in a nearby tree watching while the rest of the group drank. During rainy periods the patas used ephemeral water sources such as pools in gullies or rocky areas and did not visit water tanks. The patas spent 6% of observation time at water tanks.

Patas were recorded on or near fences, which often bordered tracks and roads, for 20% of observation time. Since water tanks were located at the junction of paddock fences, this may partially explain why the patas often travelled along fencelines. Where fencelines passed through woods, they provided an excellent 'margin': since trees and bushes were cleared for a few metres on each side of the fence, the straight, cleared fenceline provided good visibility for long distances, a feature which was taken advantage of by patas and observers alike.

A third result of human activity which was important to the patas was the presence of the prickly pear cactus, *Opuntia* sp. This had been introduced as cattle fodder and had colonized in three large patches, all near roads. Patas ate the young pads and fruit of the plant, and quadrats in which *Opuntia* had colonized were some of those with the highest use by both groups. Patas dealt with the resulting spines in their fingers by rubbing their hands in loose dirt.

Vertical habitat use

Patas occasionally jumped from tree crown to tree crown in dense acacia woodland, but the ends of the branches were generally too fragile to support the weight of an adult. For the most part, crowns of trees were not adjacent so travelling in the canopy was not possible. On the rare occasions when patas were observed in riverine forest moving in large *Acacia xanthophloea* trees, they jumped from tree to tree just as vervets do. Most of the time, however, patas moved from place to place on the ground.

The single tree species, *Acacia drepanolobium*, provided a major, year-round part of the patas diet, different parts of the trees being used at different seasons. Since most other items they consumed were available for only short times of the year, the patas were heavily dependent on acacias for food. They often fed in tree crowns but some of the most frequently eaten items could also be reached from the ground. Acacia gum was gnawed or scraped from tree trunks while standing bipedally on the ground or clinging to the trunk itself. As the monkeys walked under trees they would sometimes leap into the air to grab galls or insects from a branch overhead.

During the day, patas rested in clusters in a few trees. Females often rested on the ground when they had very young infants. Trees chosen for sleeping in at night were a different shape from those used for day resting. Day resting trees had wide, low, spreading crowns, while sleeping trees were not unusually tall but had narrow crowns with few branches near the ground. At night, each patas (except dependent infants) slept in a separate tree. Monkeys already settled in trees threatened others which tried to enter the same or even adjacent trees. The effect of this behaviour was that the groups were widely spread out at night, often over two hectares or more. A group never used the same sleeping trees on consecutive nights, so that it needed extensive woodland in its range just for sleeping.

Tenzugu

The Tenzugu patas spent much of their time on boulders and in the scrub and trees which grew among the boulders. Common tree species in the area included shea (*Butyrospermum parkii*), locust bean (*Parkia filicoidea*), baobab (*Adansonia digitata*), silk cotton (*Ceiba pentandra*), tamarind (*Tamarindus indica*), *Ficus* and *Acacia* spp. (Beque, cited in Fortes, 1945). The monkeys descended to cultivated fields as opportunity offered. The study at Tenzugu was not long enough to quantify their habitat use, but their movements were largely determined by human activities: the patas avoided people and their dogs and were attracted to crops.

Diet

The plants and animals which we saw patas eat are listed in Table 21.3. The most striking features of the Kenya patas' diet were, first, the high proportion of tree parts of the genus *Acacia*, and, second, the importance of insects in the diet. Since insects take longer to find and harvest than plant parts, the proportion of foraging effort devoted to insects was high. A third striking feature was the rarity of any parts of grass species in the diet at any time of year.

At Tenzugu, the patas were observed to eat the fruit of trees and bushes and to catch insects. Crop raiding appeared to provide a significant part of the diet. The patas raided crops frequently and were seen to eat guinea corn (*Sorghum vulgare*) seed heads and 'tiger nuts', the corms of an oil-rich species, *Cyperus esculentus*.

The area used by the Tenzugu patas groups during our study did not include any permanent surface water. We saw the monkeys drink from a rainwater pool in a rock, with one animal drinking at a time and threatening away others. We were told by villagers that there were long-lasting pools deep among the rocks, and that the patas used these pools all year around.

During the dry season the Tallensi obtain water daily from the Red Volta, about two kilometers away. If the home ranges of the patas groups in Tenzugu were of similar size to those recorded for patas in East Africa (Hall, 1965a and this study) they could also have used the river as a dry season water source.

Table 21.3. *Patas food items*

Plant foods
Key: New leaves (nl), flowers/flower buds (f/fb), fruit (fr), seeds (s), gum, galls,
stems (st), bulbs/corms (b/c)

Trees and shrubs
Acacia drepanolobium nl, f/fb, s, gum, galls
A. seyal nl, f/fb, s, gum, galls
A. xanthophloea f/fb, s, gum
Balanites aegyptiaca fr
Carissa edulis nl, fr
Euclea divinorum nl, fr
Lycium europaeum nl, f/fb, fr
Opuntia sp. f/fb, fr, st
Rhamnus staddo fr
Rhus natalensis fr
Schinus sp. f/fb, s
Scutia myrtina nl, f/fb, fr
Ximenia americanum fr

Herbs, forbs, etc.
Aloe kedongensis st
Asparagus aff. africanus s
A. buchananii b/c
A. falcatus b/c
Chenopodium opulifolium nl, f/fb, st
Commelina forskalaei s, st
C. reptans s, st
Cucumis aculeatus nl, fr, s
Datura stramonium part not identified
Hibiscus aponeurus f/fb, s
Lantana trifolium fr
Pavonia patens fr
Portulaca aff. foliosa b/c
Solanum incanum nl, f/fb
Trachyandra saltii nl, b/c

Grasses, sedges
Eleusine coracana nl
Themeda triandra nl, s, b/c
Cyperus blysmoides nl, b/c

Fungi
Agaricus campestris caps, st
Termitomyces microcarpus caps, st

Animal foods

Invertebrates
Ants
Caterpillars
Termites (winged phase)
Butterflies
Grasshoppers
Moths
Larvae of unidentified spp.

Vertebrates
Chameleons
Lizards
Nestling birds of unidentified spp.
Eggs (Crowned plover, *Stephanibyx coronatus*)

Predators

Patas interactions with predators have been described, else-where (Chism, Olson & Rowell, 1983) and will only be briefly sum-marized here. Not only adult males (Hall, 1965a) but also females, juveniles and even infants are highly vigilant at all times and may alert the rest of the group to the approach of a predator (or of other patas). Canids and humans were responded to with a low chutter audible over only about 50 m. This was usually followed by rapid flight into woodland rather than out across open areas. When a leopard or cheetah was detected, the group moved up into trees, gave loud alarm calls and looked in the direction of the predator. Males gave two-phase alarm barks to large felids, while females and juveniles gave high-pitched chirps. Females occasionally gave a high-pitched version of the male two-phase bark as an alarm call.

We analysed the habitat types in which predators were seen for 96 predator sightings recorded by the observers while with or near the patas groups. Most predator sightings (76%) were of jackals, mainly *Canis mesomelas*, the black-backed jackal. Jackals, which are probably mainly potential predators of young patas (Struhsaker & Gartlan, 1970; Chism *et al.*, 1983), were most often encountered in the margin of open acacia woodland (52% of all jackal sightings). While, overall, there was no evidence that patas were more likely to encounter predators in one habitat than another, cheetahs and wild dogs were most likely to be encountered in dense acacia woodland (67% and 100% of sightings respectively).

We never saw patas encounter predators in riverine woodland, although this was said to be prime leopard habitat. The patas entered these areas cautiously, however, after prolonged bouts of scanning from the edges of adjacent woodlands. Leopards were strongly sus-pected to have been responsible for the deaths of several adult females and their infants which disappeared overnight out of sleeping areas in woodland habitat (Chism *et al.*, 1983).

The only predator actually observed to kill patas during our study was the domestic dog. Frequencies with which dogs were seen in the various habitat types were similar to those of the patas groups' use of these areas. Our study groups made detours to avoid camps of herdsmen with their associated dogs and only used watertanks near such camps with great caution. By doing this the monkeys probably avoided some encounters with dogs but even after one group member was killed by dogs, the group returned to the area to feed for several days. This suggests food resources may be sufficiently restricted that in some seasons groups must risk encountering predators to use what is available.

At Tenzugu, humans and domestic dogs probably represented the only remaining potential predators of patas. The patas were chased almost daily, and often several times a day, by dogs, herdsboys and people working in the fields. The patas were alert to the behaviour of people and their dogs and sophisticated in the use of boulders to move out of reach of dogs and stones. After being chased, females and immatures used the 'moo' contact vocalization to re-establish contact with each other. Males were heard to use their two-phase alarm call at least once on every observation day.

Although the Tallensi tolerated the patas' crop raiding to a surprising degree, perhaps because of the monkeys' ritual significance, the adult male of the smaller study group was shot and killed just before the end of the study period. The shooting was said to have been retribution for a crop raid by the whole group a few days earlier. Killing of males may not have been a rare occurrence, and may partly explain why we saw no extragroup males in the area.

Vervets

Hall (1965b) noted that in Uganda, patas home ranges overlapped with those of vervets mainly in woodland habitats just beyond the river. In our Kenya study area, also, vervet home ranges were along water courses. Since the patas avoided these areas, encounters between the two species were rare. On a few occasions vervets were seen in acacia woodland areas within the home ranges of the patas groups. At water tanks patas were able to displace the vervets and keep them from drinking until the patas group left the area. On one such occasion, after the patas left, the vervets came to the tank and drank without any apparent concern for the presence of the observers, so that the ability of the patas to supplant the vervets was probably not due to the patas' greater habituation.

At Tenzugu, encounters between patas and vervets were frequent. We saw vervets chase and drive away patas from a patch of boulders, but on other occasions vervets seemed to withdraw as the patas approached. Once when patas supplanted vervets, the patas outnumbered the vervets. On another occasion patas and vervets alternately drove each other out of a fig tree in which each group fed for several hours. Thus, despite the patas' larger size, there was no clear dominance relationship between the two species at Tenzugu.

We did not see patas interact with primates other than vervets and humans at either of the study sites.

Discussion

Patas locomotor adaptations have been viewed as specializations to a mainly terrestrial way of life (e.g. Hall, 1965b). The results of the first long-term field study of patas presented here provide more detailed information on the ways in which patas have adapted to their savannah environment. Specifically, we have shown how locomotor adaptations are related to patterns of habitat use by this species, including feeding, ranging and predator-avoidance behaviours. Here we examine ways in which patas adaptations are similar to those of other savannah monkeys and how they are unique, and we compare them with the forest guenons, as well.

The term 'savannah' has been defined so broadly that it is now only useful to convey the antithesis of 'forest' (White, 1983). It is difficult to separate the term 'African savannah' from the idea of open rolling grasslands with a few trees on the horizon. Our observations of patas monkeys in East and West Africa demonstrate that patas are not a species particularly adapted to open grassland conditions, although under some circumstances they may be able to use this kind of habitat. Patas are more accurately viewed as primarily a woodland species, which, like vervets and olive baboons, two other classic 'savannah' primates, also spend much of their time in or near trees (Hall, 1965b; Hall & Gartlan, 1965; Rowell, 1966; Struhsaker, 1967b).

Patas are tall and lightly built. Like baboons (but unlike vervets) patas walk on their fingers rather than their flat hands, which increases the effective length of their forelimbs still further. These characters have been correctly interpreted as adaptations to moving on the ground. Patas are also said to be adapted for high-speed locomotion. While they are fast, for monkeys, patas could not possibly outrun a cheetah, for example. Rather than depending solely on high speed locomotion as a predator defense, we found that patas relied on a combination of vigilance, crypticity and rapid flight to avoid their predators.

Patas were vulnerable and were clearly nervous whenever they were far from trees (or, at Tenzugu, rocks). Their frequent visual scanning from high perches and their characteristic habit of standing bipedally to look around increases their chances of detecting a predator early enough to flee to the safety of the woodland.

Crypticity is also important as a predator defence. Females and immatures are inconspicuously coloured (as are many savannah antelopes). The more brightly coloured males can also be conspicuous in their behaviour, as when advertising their presence to other males. Adult males often appeared, however, to minimize their visibility by sitting low in tree crowns or by moving and sitting at the periphery of the group. Unlike vervet or baboon males, patas males do not give loud vocalizations when humans or canids approach their group. Most of the time the monkeys moved quietly and vocalized infrequently. In this regard, patas most resemble forest-floor guenons such as the de Brazza's monkey (Gautier-Hion & Gautier, 1978).

Patas night-resting behaviour was also distinct from that of vervets and baboons, which return night after night to one or a few sleeping sites and congregate in a few tall trees. Patas, in contrast, spread out widely at night and never spend two consecutive nights in the same area. This behaviour has been interpreted by Hall (1965a) and ourselves (Chism *et al.*, 1983) as an anti-predator, specifically an anti-leopard, adaptation.

The Kenya patas preferred to move along woodland margins which provided good visibility out across open grassland and ready access to the safety of trees. Given their dependence on trees or rocks as a refuge from predators, it is hard to imagine how patas could successfully occupy open grassland areas in which many of their predators were present.

Patas have very large home ranges, much larger than those of vervet groups of comparable size which live in the same areas (Hall, 1965b; Kavanagh, 1981). Patas home range size is similar to that of baboons, although baboons are much larger and typically live in larger groups. Patas also have long day ranges, again like those of baboons but unlike vervets. We believe that the cursorial adaptations of patas may be as important for covering the long distances necessary to the efficient exploitation of their large home ranges, as they are for high-speed escape from predators.

In some areas where patas occur, the ability to cover long distances efficiently may be related also to the location of water sources. Hall rarely saw the Murchison patas drink, but in Kenya the patas drank daily except during rainy periods. Although they show adaptations to rapid locomotion in a hot climate, there is no evidence that patas can go for long periods without water or that they have any special physiological adaptations for doing so (Mahoney, 1980; Kolka & Elizondo, 1983). At both of our study sites unusual sources of water were available, while at Waza patas were observed only at waterholes, so the question of the relation between patas ranging patterns and water sources remains unresolved. The observations of Gartlan and Struhsaker at Waza suggest, however, that during dry seasons in areas where water is scarce, patas will alter their day ranging patterns to stay near it as long as the food supply permits.

The patas' ability to exploit one or two species of *Acacia* for virtually all edible parts is similar to that reported for vervets (Wrangham & Waterman, 1981; Whitten, 1982). The importance of tree products in the diet is another way in which patas resemble other guenons. Baboons, in contrast, have developed an ability to exploit the monocotyledonous plants of the grasslands: grasses and bulbs form a significant part of their diet in such areas. Baboons have strong hands and fingers that allow them to dig up the underground storage parts of these plants as well as eating the aerial parts. Patas rarely ate any part of grass plants and dug only in soft earth.

Patas habitat use was influenced by several interacting factors. For example, our study groups' preference for open woodland over dense woodland, despite the greater crown volume and, therefore, potential greater food abundance of the latter, was probably due to the better visibility of open woodland along with the increased likelihood of encountering predators such as cheetahs and wild dogs in dense woodland.

Given the location of patas food resources and their anti-predator behaviours, it is not surprising that woodland margins should be so strongly preferred. Woodland margins provide a combination of good visibility, proximity of trees as refuges from predators and access to the food resources of both woodland and grassland, resulting in a preference that transcended seasonal food abundance in other habitat types.

Murchison Park, where Hall conducted his study of patas, was mainly grassland with a few scattered trees. How can we reconcile our view of patas as a woodland species with their presence in such a habitat? We think the answer may lie in the presence of geophytes, which are small but perennial plants which produce flowers and fruit although often completely hidden by tall grass.

Murchison Park was regularly burned and the combination of fire and elephants had greatly reduced the woodland. When experimental exclosures of fire and elephants were made in the park, within a year or two the plots had grown into small forests with many of the tree species characteristic of the nearby Budongo forest (Wheater, 1968; Spence & Angus, 1971). The Murchison patas thus appear to have had woodland food resources available to them. Similarly, at Tenzugu, patas ate products of trees much stunted by burning, cutting and browsing. In any case, Murchison appeared to be marginal habitat for patas and the species was more common further east, in *Acacia* woodland and thornscrub, notably near Soroti in the Teso District (T.R., personal observations). We do not know if any patas still survive in the park (now Kabalega Falls National Park).

In summary, the patas monkey can be regarded as having specialized in the use of patchy woodlands and adjacent open areas. Vervets, in contrast, have specialized in exploiting more continuous woodlands, such as gallery or riverine woodlands which extend throughout the African savannah region. Our observations suggest that, compared with vervets, patas are emancipated from the need for large, tall trees rather than from trees in general. In contrast to the baboons, which have developed the ability to use resources found only in grasslands, patas utilize scattered woodlands while retaining their guenon lifestyle. If guenons were originally woodland animals, as suggested by Pickford & Senut (Chapter 3), with respect to habitat, patas and vervets can be regarded as conservative. As guenons moved into closed canopy forest habitats as these became available, they found a rich habitat for species which rely on tree products as their main food.

We would like to end with a note on conservation. In East Africa, the range of patas seems to have been much reduced in the last twenty years, probably as a result of deforestation and the extension of agriculture. Our observations in Ghana suggest that patas are skilful crop raiders, and, though less visible and aggressive than baboons, their presence may not be tolerated by farmers. We expect that wherever woodland is removed and replaced with agriculture, patas populations will dwindle and be lost.

Acknowledgments

The 1979-81 research presented here was supported by NSF Grant No. BNS 7813037 to T. E. Rowell, and by grants from the American Express Foundation. Fieldwork in Ghana was supported by the American Express Foundation and the Department of Anthropology, University of California, Berkeley. 1979-81 field data were collected by J. Chism and D. K. Olson. Data and preliminary conclusions on habitat use patterns were previously presented by J. Chism and D. K. Olson at the American Association of Physical Anthropologists' 51st Annual Meeting, Eugene, Oregon, 1982.

The Department of Zoology, University of Nairobi, and the Institute of Primate Research, National Museums of Kenya, sponsored our fieldwork in Kenya. The Agricultural Development Corporation kindly gave us permission to work at ADC Mutara Ranch and provided us with housing. The Nairobi Herbarium staff, especially Ms Kabuye, assisted with identification of plant specimens.

The authors thank Thomas and Fleur Bower and Robin and Joan Slade for their hospitality and assistance during the field study in Kenya. In Ghana we were the guests of the Tengol of Tenzugu and his extensive family who assisted us in many ways. We thank Marina Cords for her comments on the manuscript. We are grateful for the technical assistance of Milton Rogers and William Rogers with the figures.

III.22

The social system of guenons, compared with baboons, macaques and mangabeys

THELMA E. ROWELL

Introduction

For the purposes of this discussion the Patas monkey (genus *Erythrocebus*) is included in the genus *Cercopithecus*, whose general common name is guenon. *Allenopithecus* and *Miopithecus* are also included among the guenons, with some reservation about the former because of its appearance, since little is known of its social system, and about the latter because it does seem to diverge rather widely from the rest of the guenons. I shall make generalisations from known species, with the full realisation that they may well turn out not to be justified as we learn more. One species, or superspecies, *Cercopithecus aethiops*, seems to differ from the rest of its genus on several points; it is perhaps ominous for the longevity of the following generalisations that it is also by far the most studied member of the genus. This does not pretend to be a complete review of all the very unevenly distributed information on the behaviour of cercopithecines; I shall emphasise comparisons that I can make from my own experience of several species in the wild and in captivity.

The first question is that of definition – what is social organization, how can it be described and characterised, and in what terms can one social system be compared with another? A social system is a set of rules of conduct for the members of a self-identifying group. Ultimately it should be understandable from observations of who does what to whom how often – but most species of guenons, and many papionine species, have not yet been studied in circumstances where individuals and relationships were known and behaviour quantified. In any case, 'who does what to whom how often' is in turn determined by who is available; by the demography of the population being considered.

Population structure will vary according to current predation pressure and food supply, and it may be grossly distorted by human intervention, in captivity or in groups provided with food, either intentionally or unintentionally. Thus it is unlikely that the structure, and hence the organisation of a single population will represent the capabilities of a whole species, and extrapolation from one population to the species as a whole is unlikely to be completely valid.

Where we have the necessary information, when we can describe who does what to whom how often in several populations, we immediately have too much. To compare quantitatively all sorts of interaction at once across many species, we need a multidimensional framework which is difficult to interpret. It is necessary to move up one integrative level, to be able to give a name to a perceived repeatable pattern of interindividual interactions, the set of rules which one might call a system (Hinde, 1983). I think there are, so far, no satisfactory shorthand characterisations of social systems. Those that are used frequently derive from human societies and are loaded with semantic baggage (as in 'harem'). They often refer to a single part of the whole system and so risk throwing the baby out with the bathwater (as in 'one-male group').

At lower integrative levels, we can look at the methods of communication which make up the interactions between individuals, and the description and relative frequencies of these can be reduced still further to the component motor patterns and other signals used. Experience has shown that these are likely to be stable, species-characteristic elements. They do not in themselves, however, tell us much about the higher integrative levels of social behaviour.

Those few species which have been studied in different environments seem to organize themselves somewhat differently in each. A good example is the variation in degree and form of territoriality of vervet monkeys (Fedigan & Fedigan, Chapter 20). This variation does not correspond with the taxonomic subdivisions proposed for the vervet group. Differences between populations of a species that are wild, provisioned, or caged provide another set of data about this point.

Two aspects of human behaviour must always be borne in mind when attempting comparisons between populations which have been described by different workers. The first is that the observer must be selective; people are not equally interested in all aspects of behaviour, and could not pay equal attention to all aspects if they were. Studies with a general, open-minded approach are no longer fashionable, and students now approach a new field site with specific, rather narrow questions derived from the current theoretical questions in their field. This can be a highly productive method, but it makes comparison between studies increasingly a task of comparing apples with oranges. Would observer A actually find the animals at observer B's site as different from his own as their published discussions of what they have seen would indicate? Or is it that, with different interests and expectations, they emphasised different aspects of essentially the same system? Comparisons by one person or team of different sites, as in the studies by Hall & Gartlan (1965), Gartlan & Brain (1968) and by Kavanagh (1981) of vervet monkeys, are as useful as they are rare. A second problem is the facility with which we can invent explanations of our observations: if we see a difference in reported behaviour and a difference in habitat at two sites, a causal relationship springs to mind and a new socio-ecological myth is born (e.g. Rowell, 1966). This approach has been criticised by Altmann (1974), Barash (1977) and Rowell (1979).

In summary, we have only partial answers to the most basic question: are there aspects of patterns of interactions between individuals which are species characteristics, in the sense that they will turn up in whatever environment members of that species find themselves, providing only that the appropriate social partners are available? Unless the answer is 'yes', any attempt to discuss social systems in a taxonomic framework is doomed to failure. Perhaps the best evidence for a 'yes' answer was provided by Kummer & Kurt (1963) when they commented on the behaviour of the hamadryas baboons at London Zoo described by Zuckerman (1932) with the hindsight provided by their own studies of wild hamadryas. It was possible to see that the London hamadryas were trying to establish normal bonds between individuals in the face of impossible demographic conditions. This was not just a wild vs. captive comparison, since Kummer (1956) had also studied a successfully organised captive group, whose behaviour generally paralleled that of the wild baboons in so far as it was possible to them. The hamadryas social system is unique in many ways, and I am not sure that the shreds of other social systems would be as recognisable under equivalent conditions.

Recent interest has centered on the determination of social organisation by the environment. A socio-ecologist would assume that any pervasive difference between species would be the result of an adaptive response to environmental selection pressure in the past. On the other hand, observations of a variety of monkeys in the forests of Cameroon led Struhsaker (1969) to suggest that there are basic differences in social organization between taxa, which might have arisen through genetic drift in ancestral stocks. More recently, he seems to have returned to a strictly adaptive position, and Struhsaker & Leland (1979) stressed differences in food resources in comparing guenons, mangabeys and colobines in the Kibale Forest of Uganda. There is some conflict between the view of social organization as being highly flexible and responsive to environmental fluctuation and to resulting demographic change (Rowell, 1967), and the idea that it can be modified by natural selection towards a precise adaptive match to a particular set of conditions (Clutton-Brock & Harvey, 1977). As our understanding of monkeys' habitats increases, so we become more aware of the instability of conditions in one place over time spans comparable to monkey life spans (Hubble & Foster, 1986), and precision adaptation seems increasingly unlikely.

It is my view that social organization serves very general purposes: animals which live in permanent groups gain protection from predation while being able to reduce somewhat their individual vigilance (Bertram, 1980). A permanent group also provides education for developing young, improving their responsiveness to the local environment (Rowell, 1976). A permanent group requires rules of conduct which will increase the predictability of the behaviour of its members and so perhaps reduce the amount of attention each individual needs to divert to group maintenance. What those rules are, precisely, may not matter within wide limits. A sociobiologist would expect the rules to be determined by selection pressure on reproductive behaviour (Alexander, 1974), but I am not at present convinced that such pressures can be simply enough defined to provide directional selection over long enough periods (Dunbar, 1979). Thus I would agree with Struhsaker's (1969) suggestion, and expect some differences between taxa in their social systems which have arisen by chance, which could be transmitted either genetically or culturally, and which could persist for long periods because their advantage lies in the consistency of the rules they provide rather than in the specifics of those rules. Members of a group in which the rules were changing would be at a disadvantage relative to members of other groups with stable social systems. If such a group did persist in spite of that disadvantage, a new set of arbitrary rules might stabilize and be perpetuated in its turn.

The comparison

At a most cursory level of inspection, the social behaviour of papionines and guenons seems different. Baboons, macaques, and even mangabeys are rowdy animals, seemingly continuously squabbling among themselves, and then making conspicuous gestures of appeasement or reconciliation. Thus baboons will lipsmack, present, embrace, or mount one another after a fight. In contrast, guenons seem discreet. Fighting is relatively rare, and affiliative behaviour is also subtle and infrequent. Reconciliation (in the sense of de Waal & Yoshihara, 1983) after a fight among patas monkeys might consist merely of approaching and sitting within a metre or so (York & Rowell, 1988). Both sorts of monkey live in highly cohesive groups. I suggest that the different first impressions they give reflect somewhat different mechanisms of social organization. There seem to be two parallel ways of coordinating social groups: the first is by the exchange of overt, specialized signals (gestures, noises); the second is by each individual monitoring others' movements and to what they are paying attention, and adjusting its own position accordingly (Rowell & Olson, 1983). All monkeys probably use both systems, but it seems that the baboons, macaques and mangabeys emphasize the first method: it was in macaques that elaborate signalling systems were first recognized (Altmann, 1962; Hinde & Rowell, 1962; Hall, 1962) and described, and the richness of their 'vocabulary' is immediately obvious after the briefest of observation. Guenons on the other hand put more emphasis on the second method, hence the impression they give of being discreet. Exchanges of overt specialized signals among them are relatively rare, and especially the elaborate gestures of appeasement and reconciliation such as presenting and embracing are rarely seen and may be absent in some species (Rowell, 1972). Vocal signals can be used in either system. Contact calls, which provide individual identification and reveal direction of movement are used in the 'monitor–adjust' system whenever visibility is poor, as when baboons or patas move through tall grass, and when most forest species move anywhere.

The most brightly coloured and elaborately patterned of mammals are guenons, while the baboons, macaques, and mangabeys have fur which is mainly a single colour, a drab grey, brown, or black. The papionines often have brightly coloured areas of naked skin which develop with maturity and can provide a varying signal: they may change their appearance seasonally or, as with perineal sexual swellings, with the hormonal state of the animal. Bright naked skin is rare among guenons. Some have small areas of blue face skin, and the scrotum of some species with white belly fur (vervets, patas, redtails, talapoins) becomes blue at maturity, but these structural colours are more or less invariable once they have developed, like coat colours. The bright colours of the guenons can be interpreted in terms of the monitor–adjust method of social organisation: they provide on the instant clear information about where the individual is, where it is looking, and which way it is going, all basic requirements of the 'monitor and adjust' system (see also Kingdon, 1980, Chapter 13). The generally duller, fuzzier coat patterns of juveniles correspond to the lesser importance attached by other group members to juvenile assessments of environmental changes, as noted by Seyfarth, Cheney & Marler (1980a) for predator alarm calls of vervets. The plain, quiet coat colours and sexual swellings of talapoins and swamp monkeys are more typical of papionines, and are anomalous among guenons. They lead one to expect a more papionine method of group coordination, but talapoins (there is no information on swamp monkeys) are not rowdy little monkeys, and subjectively I find them comparable to other guenons in their interaction patterns in captivity (see also Gautier-Hion, 1971a). They use contact calls extensively, every movement by an adult female in the wild being preceded and followed by a 'coo', often answered by another female; it may be that, in an unusually dense visual environment they rely on visual cues less and use auditory cues to monitor and adjust to each other's movements more than do other species. I observed that wild caught talapoins living in a cage with few visual barriers reduced their use of contact calls over several years, and contact calling before and after movements seemed to be lost in captive-reared animals. Increasing the number of branches in a talapoin cage immediately increased contact calls (A. Gautier-Hion, personal communication). Even this highly species-typical behaviour seems to be a response to a quality of the environment.

Whatever mechanism provides cohesion and coordination, the resulting groups of papionines or guenons have much in common. Females generally stay in the group into which they were born (though there are exceptions which Moore (1984) suggests may be more common than has been realized), so that groups are matrifocally organized; young males usually leave their natal group around adolescence. The size of groups varies, but within wide limits it is thought to be determined by the quantity and distribution of available food, both directly and as they affect demography. We know from artificially fed groups that macaques can maintain groups far larger than are ever seen naturally; similar opportunities have not been offered to guenons so it would be rash to assume that there is a characteristic group size which is smaller in guenons than in papionines. De Brazza's guenon (*Cercopithecus neglectus*) is the only cercopithecine which is seen in monogamous family groups (Gautier-Hion & Gautier, 1978), but in East Africa we have seen groups of twenty or so (Tsingalia, Rowell and Cords' observations in Kisere Forest, Kakamega National Reserve, Kenya) as was also reported by Kingdon (1971). De Brazza's will also live in larger groups in captivity. A mean group size typical of a species is a concept frequently used in theoretical discussions; it has little grounding in reality.

Female groups of blue monkeys and redtails (*Cercopithecus mitis* and *C. ascanius*) defend territories with well defined boundaries as do many other guenons (Struhsaker, 1967; Gautier & Gautier-Hion, 1983). Juveniles join in this defence and in fact often initiate group encounters. We are not sure whether male juveniles are involved: juveniles we have been able to identify have all been females. Patas have much larger home ranges which do overlap, but troop encounters follow a similar vigorous pattern when they occur, and one troop may chase another for some distance towards their boundary (Chism & Rowell, Chapter 21). De Brazza's were not observed to defend territories (Gautier-Hion & Gautier, 1978). Data on other species of guenon are lacking, but I predict that we shall find that defence of territory is usual among them, where populations are high enough to 'fill' the habitat. In contrast, the papionines are generally described as having overlapping home ranges, and in troop encounters the smaller troop of baboons or macaques typically withdraws before the larger. White-cheeked mangabeys have very large overlapping home ranges and use loud calls to maintain distances between groups of several hundred metres, in a habitat where visibility is about 50 m; both larger and smaller groups avoid one another (Waser, 1976). Defended boundaries have been described by Hamilton, Buskirk & Buskirk (1976) in a Namibian population of baboons with an extremely rigorous and limited habitat, but so far this case is exceptional. In this respect talapoins again provide a curious anomaly. Field studies describe well defined home ranges which may not even be contiguous with those of neighbouring troops, and no encounters between troops have been observed (Gautier-Hion, 1970, 1971b; Rowell, 1973; Rowell & Dixson, 1975). T. Butynski (personal communication) suggests that non-contiguous home ranges, which he found in a low-density population of blue monkeys, might be more common among guenons than has been suspected, since field studies are usually made in 'good' sites with high densities of monkeys.

Adult males (patas and blue monkeys) are not territorial and rarely join in boundary disputes of a troop in which they are resident. They may, during a mating period especially, take the opportunity provided by a boundary encounter to fight the male of the adjacent troop, but more usually blue monkey males, for example, clearly recognize the special status of the neighbouring resident males and distinguish them from non-residents by not threatening and chasing them on sight (Cords & Rowell, 1986; Chism & Rowell, 1986).

Adult male guenons are generally mutually intolerant in the presence of females. In captivity it is rarely possible to keep two fully adult males together with females, no matter how spacious the cage. In the wild most species are described as living in one-male groups; adult male patas and blue monkeys are vigilant for intruding males and vigorously chase off any they detect near the group in which they are resident. An intruder always defers to the resident and flees if chased. None the less, when females are receptive it is not uncommon to find a second, or more, males travelling with the group. This is perhaps attributable to the expedience of exhaustion for the resident male. He cannot overcome their persistence, and females do not seem to favour him over the newcomers as a mate (Tsingalia & Rowell, 1984; Cords *et al.*, 1986; Chism & Rowell, 1986). Vervets are exceptional guenons in that wild groups usually include more than one male, although one male is clearly deferred to by all the rest, and a male hierarchy is apparent. Talapoins also provide an exception, which will be considered later.

The mutual intolerance of adult male guenons leads to a one-male group composition, at least outside mating periods. This could in turn have the effect of limiting the size of female groups if the social system were determined by reproductive constraints. In contrast, papionines typically live in multi-male groups, in which hierarchical relationships between males are much complicated by the formation of alliances between males (e.g. Smuts, 1985). Male guenons have not been seen to form alliances, with the exception, once more, of the talapoin.

Since one male guenon lives with several females, it follows that most males spend most of their time away from female groups. The behaviour of resident males contrasts sharply with that of non-residents. Residents are conspicuous in their movements in trees, make frequent bouncing displays, and make species-characteristic loud rallying calls. Non-residents typically move quietly, hide in vegetation when approached, and rarely make the loud calls. Intruding males joining a group of females show individually varying behaviour between that typical of a non-resident and something approaching that of a resident. Clearly, resident males are much easier to observe, and little information is available on the behaviour of non-residents, even though most males must be non-resident most of the time. Single males may join and move with a troop of another species – redtails with blues in Kakamega, blues with redtails in Kibale (Struhsaker *et al.*, Chapter 24). A vervet male lived for long periods with baboons (Rowell, 1966). Adult male guenons are very often reported as being solitary, and whereas this must often be correct, we have found at Kakamega that an apparently solitary blue monkey male usually turns out, if observed for long enough, to be part of a rather widely spread association of males. They may be in visual contact with each other, at about 50 m apart, or they may be using auditory cues as their neighbour moves through leaves, so that they move together. The highly inconspicuous behaviour of non-resident males makes these groups very difficult to detect, and we could not do so until many members were habituated to our presence. Associations between particular male blue monkeys have persisted over several years, and from year to year the same patterns of avoidance between individuals have been seen, suggesting a continuity of their relationship. When one male became a resident in a female group, all others deferred to him whatever their previous relationship. Although these associations are very different from the more concentrated groups of females, they must be recognized as a significant social context for the adult males. They seem to represent an extreme in reliance on the monitor–adjust method of social organization rather than reliance on overt exchange of signals. Males touch each other extremely rarely, and aggressive contact is much more likely than friendly, although we have two or three times seen grooming between two male blue monkeys which were usually in association.

Sightings of apparently solitary patas males are more common than for forest species (Gartlan & Gartlan, 1973; Chism & Rowell, 1986), but male associations were also recognized, and given the generally more dispersed nature of all patas social groupings, the possibility that widely spread male associations are common in this species as well cannot be ruled out. Indeed I venture to predict they will be found to be common among guenon species when appropriate observations on habituated animals are carried out.

Solitary male papionines have occasionally been reported and even studied (Slatkin & Hausfater, 1976). Sugiyama (1976) describes non-resident adult male Japanese macaques, both solitary and in small groups, which are sufficiently generally known to have a name in Japanese (hanarezaru). Since several males can be resident in a female group of papionines, males are probably non-resident for a lesser part of their adult lives, however, and non-resident males must make up a smaller proportion of the adult population of papionines than of guenons. Talapoins, as usual, do not fit the general pattern. They live in large multi-male groups which at first sight seem like those typical of papionines. On closer analysis, however, some important differences emerged. Whereas in a baboon or macaque group interactions between males and females are common at all times of year, in a talapoin group they were not seen outside the brief mating period (Rowell & Dixson, 1975). Males moved together and interacted with males in other ways, including grooming together, while females interacted with females. Males and females were somewhat spatially separated, since males were consistently higher in the trees than females. The talapoin troop was actually two groups, male and female, moving through the forest together. This type of organisation is intermediate between that of the papionines, in which adult males and females form a single social unit in many respects, and that of other guenons, in which most males live separately from females, sometimes in associations with a very different style. Outside the mating period male and female talapoins are socially separate but spatially close, although not greatly intermingled. One could perhaps make an analogy between male and female talapoin subgroups and, for example, redtail and blue monkey troops moving in a mixed-species group.

The first interaction you are likely to see on approaching a baboon group is an adult female grooming an adult male. Interactions between males and females of all sorts are frequent, and individual males and females form long-lasting bonds (Strum, 1982; Smuts, 1985). This pattern reaches an extreme form among hamadryas baboons, where females of a harem group interact mainly with their male and little with each other (Kummer, 1968). In comparison, adult female guenons contact the resident male relatively infrequently unless they are receptive. Grooming is primarily an activity which occurs between females, whether adult or juvenile. Adult females are likely to be aggressive towards a male, especially if they suspect him of hurting an infant (Lancaster, 1973). While mangabey females became anxious and restrictive of their infants when the adult male was removed from their cage, patas females were indifferent (Chalmers, 1972). We have seen both blue and patas monkey female and juvenile groups live without an adult male for several days, without their behaviour being apparently disrupted. On the other hand, blue monkey females and juveniles interact vocally with their resident male quite frequently: a chorus of long croaks from females is followed by a male boom, and the group closes in towards the male following the loud 'pyow' call. Thus the resident male acts as a reference point for his group.

Older juvenile male patas in captivity, began to increase the time they spend alone at an age when females were most socially active (Rowell & Chism, 1986). In the wild they would leave the group soon after; there is no evidence that males are driven out of groups in the wild, in fact adult male patas seem quite inhibited about attacking juvenile males which harass them. It seems that behavioural changes associated with adolescence in male guenons may include reduction in contact behaviour and increase in personal space.

Conclusion

I find four main differences between papionine and cercopithecine social systems in so far as we know them at present. The first is in the method by which groups are organized and maintained, which in papionines emphasizes the frequent exchange of overt and often specialised communicative gestures, while among the cercopithecines the emphasis is on the monitor–adjust method. The most easily detectable evidence of this difference is the infrequency of specialized gestures of reconciliation among guenons. Both taxa have and use both methods of organization, but to different degrees.

The second difference is in the behaviour of adult males. Adult male guenons are mutually intolerant in the presence of females and so most seem to be unable to live in stable multi-male groups. Interactions between male papionines may be tense and rigid, but they can tolerate each other in the presence of females, and even form alliances and bonds with each other. In the absence of females, male guenons can live together, although not generally very close together.

The third difference is the defence of territories by female groups of guenons, which does not seem to be paralleled by any respect of the same boundaries by adult males. In contrast, groups of papionines have overlapping home ranges, and usually avoid each other wherever they meet.

The fourth difference is that male guenons interact much less with non-receptive females than do male papionines, and females do not seem to be consistently subordinate to them. In papionine groups males and females seem to be integrated into a single social organization to a much greater extent.

Some partial exceptions to these generalisations are already appearent, and may indicate fruitful areas for further research. Talapoins have provided partial exceptions several times in my argument, and often appear to be intermediate in their behaviour, as well as their appearance, between guenons and papionines. The swamp monkey also lives in large multi-male groups (Gautier, 1985), and I would predict that it will also prove to be intermediate, although perhaps in different ways. The other 'odd' species is the vervet, whose males will live together in female groups. The interesting thing about the exceptions is that they seem to vary independently: territoriality of females is not correlated with living in a one-male group; many male talapoins can live with females, but they do not form a single integrated social system. Some of the more obvious correlations with habitat can also be rejected: since the vervets often use the same habitat as baboons it is tempting to propose a habitat-induced convergence in social behaviour; but patas spend more time on the ground in more open country than do vervets, and yet their social organization seems to be more similar to that of forest species than does that of vervets.

Studies of papionines have for the most part concentrated on the rules governing social interactions within the group, and this has generally seemed a sufficient, complete area for investigation. In studies of guenons, on the other hand, the group has not seemed a sufficient object. Overt interactions within a guenon group are low key and infrequent most of the time, but on the other hand territorial defence draws attention to the interactions of adjacent groups and so onto the layout of the whole local population of females and juveniles. Adult males move between the female groups, and also have their own long-term social interactions apart from those of the females. At the same time, female groups of guenons associate with groups of other species in a way which has no parallel among the papionines (Gautier-Hion & Gautier, 1974; Gautier-Hion, Chapter 23; Cords, 1987b). This is not to suggest, of course, that papionine groups do not interact, nor that guenon groups have no internal organization. None the less, the different emphasis in studies probably reflect differences in style in the social behaviour of cercopithecines and papionines, and in attempting to compare social systems, style is in the end the matter of interest. We might characterize the papionine style of frequent interactions within a local social unit as intensive; in contrast, the cercopithecine style of relatively infrequent within-group communication, combined with rather frequent interactions between or across groups, can be characterized as extensive. This distinction between intensive and extensive social organization epitomises for me the difference between the behaviour of the two groups of monkeys as we presently understand it.

III.23

Polyspecific associations among forest guenons: ecological, behavioural and evolutionary aspects

ANNIE GAUTIER-HION

As soon as we started observing forest guenons in Gabon, we were impressed by the frequent occurrence and the stability of mixed-species troops of *Cercopithecus* monkeys (Gautier & Gautier-Hion, 1969). Shortly afterwards, Gartlan & Struhsaker's (1972) observations, carried out in Cameroon, established that the Makokou situation was not unique. Since then, polyspecific troops of guenons have been observed almost anywhere in rain forests, both in eastern (Marler, 1973; Struhsaker, 1975; Rudran, 1978a; Waser, 1980; Gautier, 1985; Cords, 1987b) and western Africa (Galat-Luong & Galat, 1978; Galat & Galat-Luong, 1985).

Despite the generality of the phenomenon, significant differences exist between and within communities, in the frequency of occurrence and the stability of mixed species troops. Some differences may reflect biased observations. Thus Struhsaker (1981b) compared census and longitudinal studies and found that the former exaggerate the percentage of polyspecific troops, while Waser (1982b) compared observed frequencies of association with theoretical predictions to determine whether or not polyspecific troops occurred by chance. Both authors have stressed the importance of community composition, demographic parameters and patterns of habitat use as factors influencing the probability and the nature of mixed-species troops. However, Struhsaker failed to find any strong ecological correlate and showed that observed differences between or within species for various ecological traits were not always related to expected changes in association rates, even when comparing two communities living 10 km from each other.

The variability observed in mixed troops also concerns the temporal variation in the frequency of their occurrence and in their persistence. On a diurnal scale, associations may peak in morning and evening, at midday, or they may show no rhythmicity at all; association may vary or not according to season and may or may not be stable from year to year (Gautier & Gautier-Hion, 1969; Gartlan & Struhsaker, 1972; Gautier-Hion & Gautier, 1974; Galat-Luong, 1979; Struhsaker, 1981b; Cords, 1987b).

Perhaps more important is the fact that the term 'time spent associated' covers two different meanings; one species troop can occur in association for 60% of its time with another species by being mixed either with two or three different troops every day (the common situation in East Africa) or with only one and the same troop of the other species (the case in Gabon). From a behavioural point of view, these situations are probably basically different.

Differences may also result from the fact that association patterns have not been studied explicitly but only indirectly during studies of one or the other primate species. Moreover attempts to find general trends seldom differentiated each species included in mixed troops. If expected benefits differ according to species and to the time of the day, every species should ideally be studied at the same period in and out of mixed troops, including one, two or more species.

Nevertheless, present data are consistent enough to assume that polyspecific associations are widespread among forest primates, very common in Africa and particularly well-developed within the genus *Cercopithecus* (see review in Struhsaker, 1981b). Within this genus, most authors agree that, at least for some species, associations are not the result of chance encounters (Gautier-Hion & Gautier, 1974; Struhsaker, 1981b; Waser, 1982b; Gautier-Hion, Quris & Gautier, 1983; Galat & Galat-Luong, 1985; Cords, 1987b). Consequently, I will not hark back to methodological problems which have been fairly discussed (Waser & Case, 1981; Waser, 1982b) and I will deliberately consider divergent results found at different sites as resulting from a wide range of possible situations and not as the consequence of anecdotal or biased observations.

A widespread but not generalized phenomenon

Census data on the frequency of occurrence of polyspecific troops come from four communities living in West Central Africa (Gabon, Gautier & Gautier-Hion, 1969; Cameroon, Gartlan & Struhsaker, 1972; Gautier, personal communication), in East Africa

Table 23.1. *Aggregated census data on the frequency of polyspecific troops'*
encounters

Genus	Super-species	Species	Number of encounters	% of poly-specific troops	References
Cercopithecus	—	*neglectus*	161	6	3, 5, 6, 7
Cercopithecus	*lhoesti*	*lhoesti*	15	27 ⎫	8
	lhoesti	*preussi*	27	7 ⎬ 12.5	2, 4
	lhoesti	*'solatus'*	14	7 ⎭	6
Cercopithecus	—	*talapoin*	69	48	5
Cercopithecus	—	*nigroviridis*	16	50	3
Cercopithecus	*nictitans*	*nictitans*	439	73 ⎫ 75.6	2, 5, 6, 7
	nictitans	*mitis*	103	87 ⎭	8
Cercopithecus	*cephus*	*cephus*	147	86 ⎫	5, 6, 7
	cephus	*erythrotis*	249	80 ⎪ 83.4	2
	cephus	*petaurista*	90	88 ⎬	1
	cephus	*ascanius*	170	84 ⎭	8
Cercopithecus	*mona*	*mona*	173	94 ⎫	8
	mona	*campbelli*	53	76 ⎬ 84.4	1
	mona	*pogonias*	191	77 ⎭	2, 5, 6, 7
Cercopithecus	—	*diana*	114	86	1

References: 1. Galat & Galat-Luong, 1985, Ivory Coast; 2. Gartlan & Struhsaker, 1972, Cameroun; 3. Gautier, 1985, Zaïre; 4. Gautier, Galat & Galat-Luong, unpublished data, Cameroun; 5. Gautier & Gautier-Hion, 1969, Gabon; 6. Gautier, Gautier-Hion & Loireau, unpublished data, Gabon; 7. Quris, 1976, Gabon; 8. Struhsaker, 1981b, Uganda.

(Uganda, Struhsaker, 1981b) and in West Africa (Côte d'Ivoire, Galat & Galat-Luong, 1985). Three categories can be recognized within the forest guenons studied to date (Table 23.1).

The first includes *C. neglectus*, the three forms of *C. lhoesti* and possibly *C. hamlyni*, which seem to avoid living in mixed troops; the second includes *C. (Miopithecus) talapoin* and probably *C. (Allenopithecus) nigroviridis* which are found associated in about half of encounters and appear to show no definite tendency either to form or to avoid association. A high rate of association characterizes all the other species studied: percentages of mixed troops vary from 73 to 94% for the members of superspecies *nictitans, cephus, mona* and *diana*, for a total of 1729 encounters. Despite differences in methods, the results obtained in different study sites are remarkably consistent. Yet we can ask whether ecological and behavioural traits allow us to explain differences between the three monkey categories defined.

Species which do not live in mixed-species troops

C. neglectus is the only species in this category whose ecology and behaviour has been studied (Gautier, 1975; Quris, 1976; Gautier-Hion & Gautier, 1978; Table 23.2). According to the milieu, animals spend up to 70% of their time at levels under five metres high and 20% on the ground. Small family groups use small home ranges and cover short distances each day without any consistent rhythm (Figure 23.1).

Resource exploitation is intensive: the same fruiting tree can be visited up to four times a day and every day for two weeks. Discretion characterizes fruit feeding and the wasting of fruit so characteristic of arboreal guenons does not occur.

The tactics used against predators is 'to see but not to be seen'. The monkeys are on the watch for danger and display a strategy of

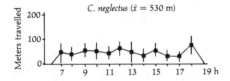

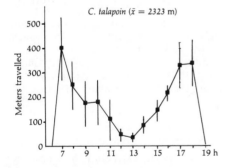

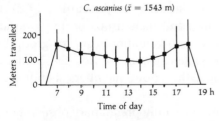

Figure 23.1. Mean daily travel and activity rhythm for two sympatric guenons of the riparian forest, *C. neglectus* and *C. talapoin* (Gautier-Hion & Gautier, 1978; Gautier-Hion, 1971b) and for *C. ascanius* (Cords, 1987b). Lines indicate standard deviations.

Table 23.2. *Behavioural and ecological features for some species of the three categories of guenons defined by their tendency to form mixed troops*

Species	Troop size	Strata use, %			Home range size (ha)	Daily travel (m)
		Lower	Middle	High		
C. neglectus	2–6	62 (30)	31	7	4–10	530
C. lhoesti°	2–7	++	?	?	?	?
*C. preussi°	2–8	85 (35)	15	5	?	?
*C. solatus°	5–15	85 (40)	20	–	?	?
*C. talapoin	60–100	75 (5)	25	–	110–140	2300
*C. nigroviridis°	>40	90 (?)	10		?	?
C. petaurista	4–24	37	59	4	41	?
C. campbelli	5–33	32	54	14	40	?
C. diana	5–30	10	57	33	93	?
C. cephus	5–22	30	61	9	30–90	1200
C. pogonias	11–18	11	67	22	55–100	1600
C. nictitans	11–28	3	69	28	55–80	1500
*C. ascanius	30–40	15	56	29	24	1445
*C. mitis	25	9	77	14	60.5	1300
¨C. ascanius	20–34	32	56.5	11.5	60	1545
¨C. mitis	21–45	35.5	48	16.5	38	1135

° = results based on sparse data. () = % time spent on the ground. Strata use: lower (<10 m); middle (10–20 m); high (>20 m) for * and ¨. For other species, lower = inferior levels; middle = canopy; high = emergent trees.
References: 1.–8, see Table 23.1; 9. Gautier-Hion & Gautier, 1978; 10. Gautier-Hion, 1971b; 11. Gautier-Hion, 1978; 12. Gautier-Hion & Gautier, 1974 and personal observations; 13. Gautier-Hion, 1980; 14. Struhsaker, 1978; 15. Cords, 1987b.

'adaptive silence'; cohesion calls are rarely given and the species does not include social alarm calls in its repertoire. When a predator is detected, monkeys flee silently onto the ground or alternatively freeze on a branch for up to five hours (Gautier-Hion & Gautier, 1978). If surprised at a short distance, the adult male drops on the ground, kicking his feet as a signal and diverts attention from his group by shaking branches and uttering violent barks towards the predator which he may eventually attack. Other group members silently stay motionless.

We lack data on C. *hamlyni* and have only preliminary observations on the three *lhoesti* species (Gartlan & Struhsaker, 1972; Struhsaker, 1981b; Gautier, unpublished data). These monkeys, like C. *neglectus*, spend a great deal of time on the ground and are mainly observed less than five metres from the ground (Table 23.2; Figure 23.2). As in C.

Diet, %			Antipredatory behaviours			
Plant material		Prey				References
Fruit	Others		Calls	Flight	Freezing	
74.5	20.5	5	♂ only	ground	yes	7, 9
?	?	?	rare	ground	?	8
?	?	?	rare	ground	?	2, 4
?	?	?	rare	ground	?	6
61	3	36	yes	trees	no	10
81	2	17	yes	ground	no	3
77	15.5	7.5	yes	trees	sometimes	1
78	6.5	15.2	yes	trees	no	1
76	19.5	4.5	yes	trees	no	1
78	10	12	yes	trees	sometimes	11, 12, 13
82	2	16	yes	trees	no	11, 12, 13
70.5	20	9.5	yes	trees	no	11, 12, 13
43.5	32	24.5	yes	trees	no	14
45	34.5	20.5	yes	trees	no	14
62	13	25	yes	trees	no	15
55.5	27.5	17	yes	trees	no	15

Figure 23.2. *C. lhoesti lhoesti* is a semi-terrestrial species, mainly observed in lower forest strata; it is rarely found in mixed troops (photo: F. Bourlière, Epulu, Zaïre).

neglectus, cohesion calls are hardly heard during troop movements (see Gautier, Chapter 12). Although these species possess alarm calls, they are not frequently given. In most situations, only the adult male utters the barks. Similar unobtrusive tactics also probably characterize the new species *C. salongo* from Zaïre which is described as an 'elusive species, semi-terrestrial, hard to observe' (Kuroda, Kano & Muhindo, 1985).

Species occasionally found in mixed-species troops

Both *C. talapoin* and *C. nigroviridis* live in large multi-male troops (Gautier-Hion, 1971b; Gautier, 1985) whose cohesion is ensured by fairly elaborate vocal repertoires (Gautier, 1974 and unpublished data). They both inhabit lower strata of swamp forest; *C. nigroviridis* is also found on the ground (Table 23.2). *C. talapoin* is the lightest arboreal guenon, while the swamp monkey is the lightest terrestrial one.

Talapoin troops are characterized by a large home range and extensive daily movements which may be related to their very insectivorous diet, a tendency possibly shared by the swamp guenon (Figure 23.1). Confronted with predators, the strategy of these monkeys rests upon detecting it by uttering calls which spread within the whole troop. Mobbing and flight will follow depending on the situation. Talapoin monkeys flee through the trees and swamp monkeys over the ground. Both species have stable sleeping sites located near rivers which animals dive into as soon as a predator is detected.

Species regularly found in mixed-species troops

These species are all arboreal and the forest strata they use overlap considerably (Table 23.2); they live in middle-sized troops most often including only one fully adult male; their daily movements and the size of their home range do not differ much, while their daily activity rhythm is most often bimodal (Figure 23.1).

Their mainly frugivorous diet is closely similar among species of the same community and most plant species are eaten by every participating monkey species. Interspecific differences relate mainly to folivorous tendencies and/or to the quality and quantity of insects eaten (see Gautier-Hion, Chapter 15).

The vocal repertoires of these species were found to be homologous both in the number of signals given and in their functional significance (Gautier, 1969, 1975, 1978; Struhsaker, 1970; Marler, 1973; Gautier & Gautier, 1977). All species have developed similar tactics in order to avoid predators, i.e. alarm calls uttered by adult females and juveniles reinforced by the loud barks of the adult male and followed by flight through the trees (Table 23.2).

To what extent do differences pointed out among the three monkey categories preclude or enhance the formation of mixed-species troops?

As far as we know, guenons which avoid living in mixed troops have in common their semi-terrestrial habits, a well-developed sexual dimorphism in body weight (Gautier-Hion & Gautier, 1985), their habit of living in relatively small-sized troops, their intensive use of habitat, and possibly also their anti-predatory tactics based upon silence and concealment. Most are described as elusive monkeys and it is not mere chance that two of them have escaped the attention of scientists until recently. To avoid drawing attention applies as much to movements, as to vocal communication or anti-predatory behaviour. An 'adaptive silence' is only compatible with small-sized social units. Any increase in troop size would obviously reduce the efficiency of such a strategy and thus preclude life in mixed troops. In fact, numerous observations have shown that de Brazza monkeys actively avoid joining troops of other species (Gautier-Hion & Gautier, 1978).

As suggested by Struhsaker (1981b) for *C. lhoesti*, the monkey eagle (*Stephanoetus coronatus*) does not appear to be a major predator for de Brazza monkeys and other semi-terrestrial guenons which are probably mainly taken by scansorial predators. It is likely that the latter have influenced the evolution of their overall strategies of discretion.

Two constraints linked with their small body size seem to condition the strategies of talapoin monkeys: their insectivorous diet and their vulnerability to a potentially large range of predators. It is therefore advantageous for these tiny monkeys to live in large troops because increased numbers ensure better vigilance and predator detection, while the use of a more extensive foraging area offers greater opportunity for catching mobile prey and allows for a less rapid depletion of resources. It is easy to understand why talapoins are never found together with de Brazza monkeys although they both inhabit riparian forests: talapoin troops may move up to 550 m/h, i.e. the average distance travelled each day by a *C. neglectus* group (Figure 23.1) and they display quite opposite foraging tactics. In fact, talapoins are only found mixed with those arboreal guenons which occasionally join them for a few hours, separating again as the talapoins move on. The latter's pattern of movement is furthermore constrained by their customary return to the same traditional sleeping sites whose location

Figure 23.3. Adult female *C. talapoin* with her clinging infant and her juvenile (only the twining tail of the latter can be seen), at their sleeping place above a river. Large home range and the use of traditional stable sleeping sites preclude this species from forming long-lasting polyspecific associations (photo: A. R. Devez, Makokou, Gabon).

above water is quite likely a response to predation pressure (Gautier-Hion, 1971b; Figure 23.3).

All known species of arboreal *Cercopithecus* are most often found in polyspecific troops and a persistently monospecific troop has rarely been seen (Figure 23.4). These species differ little with respect to body weight (range 3 to 8 kg), dietary characteristics, troop size, strategies of habitat use or anti-predatory defence against the crowned-hawk eagle.

Consequently, a long-lasting polyspecific association tends to take place only among those species that are closely related ecologically and behaviourally. Dietary overlap in plant food is considered as the most

Figure Figure 23.4. *C. diana* mainly lives in forest canopy and emergent trees. It is found in mixed troops in 86% of encounters (here with a *Colobus polykomos*; photo: A. Galat-Luong, Taï Forest, Ivory Coast).

obvious and necessary condition for the formation and persistence of mixed troops (Struhsaker, 1981b). However the case of the de Brazza monkeys which share 75% of fruit species with sympatric guenons shows that it is not a sufficient condition. An association requires that not only dietary needs, but also foraging strategies should not differ fundamentally: the more similar these are, the greater is the likelihood that an association might develop. Thus the proverb 'birds of a feather flock together' seems to apply better to guenons' polyspecific associations than the principle of competitive exclusion (Gause, 1934). It is none the less necessary to ask whether or not the life in mixed troops leads to a shift of species-specific ecological niches and whether such shifts increase or decrease niche overlap.

Niche overlap among associated species

Only a comparative study of a species living either in monospecific or in mixed-species troops within the same community can allow us to assess an eventual shift of ecological niche. Two such studies have been so far carried out, one in Gabon in a community of three arboreal guenons: *C. nictitans*, *C. pogonias*, and *C. cephus* (Gautier-Hion *et al.*, 1983), the other in Kenya, where only *C. mitis* and *C.*

ascanius occur (Cords, 1987b). Although the time spent in association between species at either site did not fundamentally differ, the quality and pattern of association were not the same. In Kenya, each of the two focal troops, of *C. mitis* and *C. ascanius*, associated mostly with each other but it could also be found with other troops during the same day; associations were most frequent during the middle of the day and might disband several times each day; the modal duration of these associations was less than one hour.

In Gabon, great differences were found between troops living on an area of about 1 km²; one *C. cephus* troop (called CC1, Table 23.3) lived mainly on its own for five years; during the same time, another *C. cephus* troop (CC2) increased the time spent in mixed troops from 18% to 45.5%, while a *C. nictitans* troop and a *C. pogonias* troop were hardly found apart from each other. In mixed troops preferential mixings occurred; the bispecific troop *C. nictitans–C. pogonias* and the trispecific troop including the latter and *C. cephus* (CC2) being the most frequent association (Table 23.3). In the trispecific troop, *C. pogonias* and *C. nictitans* never disbanded, while the *C. cephus* troop associated with these two species for several consecutive days and nights before leaving on its own for several days.

Patterns of habitat use by the *C. cephus* monospecific troop, the *C. pogonias–C. nictitans* bispecific troop and the trispecific troop differed significantly (Table 23.4). In the trispecific troop, the *C. cephus* monkeys ranged over a larger area, increased the distance travelled daily, were found in upper strata, visited more hectares each day and retraced their steps less. *C. pogonias* and *C. nictitans*, when associated with *C. cephus*, used a smaller area and tended to be found in lower strata; like *C. cephus*, they increased their daily range, visited more hectares and retraced their steps less. However, differences were less significant than for *C. cephus*.

Significant shifts were also found in the ratios of use of the three structurally different forest types. While *C. cephus* alone tended to avoid the highest forest with open understorey (F1, Table 23.4) and searched for the densest understorey habitat (F3), the reverse was found for *C. nictitans* and *C. pogonias*. Within the trispecific troop, two trends were observed: a tendency for the three species to colonize the tallest forest more than expected, especially during the main daily periods of fruit feeding (in the morning and in the evening); and an enhanced tendency for *C. pogonias* and *C. nictitans* to colonize dense undergrowth during the midday resting period, which was also the main period for insect foraging, together with a decreased tendency for *C. cephus* to do so.

Table 23.3. *Percentage occurrence of mixed troops and their composition. CC1: C. cephus troop 1; CC2: C. cephus troop 2; CP: C. pogonias troop; CN: C. nictitans troop*

Troops	C. cephus 1	C. cephus 2	C. pogonias	C. nictitans
% time associated				
1972	5	18	97	97
1976	10	40	100	100
1977	6.5	45.5	100	100
Mixed troops' composition	1977 + CP + CN (4%)	+ CP + CN (42.5%)	+ CN (51%)	+ CP (51%)
	+ CP + CN + CC2 (2.5%)	+ CP + CN + CC1 (2.5%)	+ CN + CC2 (42.5%)	+ CP + CC2 (2.5%)
	—	—	+ CN + CC1 (4%)	+ CP + CC1 (4%)
	—	—	+ CN + CC1 + CC2 (2.5%)	+ CP + CC1 + CC2 (2.5%)

Table 23.4. *Patterns of habitat use according to troop composition. F1: highest forest with clear undergrowth; F3: lowest forest with very dense undergrowth; F2: intermediate type between F1 and F3*

Troops		C. cephus	C. cep. + C. pog. + C. nict.	C. pog. + C. nict.
Home range size, ha		60	119	148
Daily travel, m		1295	1980	1825
Area visited daily, ha		11	24	20
Renewal index		0.42	0.87	0.83
Mean height, m		15 21.5	24.3 27.6	26.3 26.6
Ratios of use of forest types				
All day	F1	0.51	0.98	1.11
	F2	1.11	1.13	0.99
	F3	1.27	0.74	0.75
Fruit feeding period	F1	0.57	1.34	1.00
	F2	1.36	1.07	1.12
	F3	1.03	0.25	0.76
Resting period	F1	0.43	0.74	0.98
	F2	0.82	1.23	1.12
	F3	1.57	0.93	0.39

The blue and redtail monkeys observed by Cords in Kenya are the two ecological counterparts of *C. nictitans* and *C. cephus*. At this site, contrary to what happened in Gabon, redtails had a larger home range, travelled more and covered a greater surface area than blues. When mixed with redtails, blues tended to increase their daily travel and covered a greater surface area each day, while redtails were found to increase their visits to more open forests and to cover more area not previously visited by blues on the same day.

Thus, contrary to what might be expected, polyspecific association tends to reduce interspecific differences in patterns of habitat use, as already found for fruit diets (Struhsaker, 1981b). This results from interspecific adjustments which differ according to the period of the day and the predominant activity of the moment.

Are mixed troops a mere aggregation of individuals or is there any kind of interspecific social organization?

Interspecific social interactions are rare within mixed troops and aggressive interactions predominate (Struhsaker, 1981b; Cords, 1987b). However, the number of aggressive bouts reported by Struhsaker in mixed troops does not fundamentally differ from those quoted for intraspecific aggression within each species (Struhsaker & Leland, 1979). In fact, differences related mainly to the relative importance of aggressive and friendly behaviours. In Kakamega forest, the latter constituted about 20% of all interactions between blues and redtails (Cords, 1987b). Although this rate is lower than that which was observed within each species, it is far from being insignificant in terms of behavioural evolution.

Interspecific vocal exchanges are more developed than physical interactions: alarm calls are spread by phonoresponses throughout the mixed-species troop; whichever species calls first, intermingled cohesion calls by females and juveniles of all species punctuate troop activity. Still more remarkable is countercalling between adult males (Gautier, 1969, 1975; Struhsaker, 1970; Marler, 1973; Gautier-Hion & Gautier, 1974; Cords, 1987b) in which there is a synchronized vocal output of loud calls by associated males (Figure 23.5). Whether these loud calls are a mechanism which imposes the own specific-mate recognition system (SMRS, Paterson, 1985) and lead to syngamy or have evolved for the purpose of preventing hybridization (isolating mechanism, *sensu* Mayr, 1963) remains to be experimentally investigated. In either case, the mixing of calls by associated males may only be a question of challenge between males of different species to which

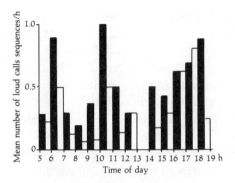

Figure 23.5. Diurnal variation in countercalling between the *C. pogonias* adult male (black bars), and the *C. nictitans* adult male (open bars), of the same bispecific troop. Total number of loud calls sequences: *C. pogonias* = 89, *C. nictitans* = 53 (from Gautier-Hion & Gautier, 1974).

troop members of every species respond selectively. In fact, strong differences in patterns of calling according to species have been observed. Gautier (1969) has shown that, in bispecific troops of *C. nictitans* and *C. cephus* in Gabon, the *C. nictitans* male was always the first to call while the *C. cephus* male joined it only in two out of three cases. The same was found in Cameroun for *C. nictitans* and *C. erythrotis* (Struhsaker, 1970) and in East Africa for *C. mitis* and *C. ascanius* (Marler, 1973; Cords, 1987b; Struhsaker, personal communication).

Striking interspecific differences were also found in the trispecific troop already mentioned. In this troop, the *C. pogonias* male maintained at all times the highest level of vocal activity (Table 23.5), participating in 58% of the sequences heard while the *C. cephus* male only participated in 18%. Signals of leaving at dawn (given in 91% of cases) were given first by the *C. pogonias* male. The *C. cephus* male called in only 6% of cases and never initiated the sequences. When the *C. cephus* troop joined the *C. pogonias–C. nictitans* one, loud calls were generally heard: when this happened the *C. pogonias* male participated in 90% of cases and was the first caller in all cases. He was also the first to react when a monkey eagle was in view; the *C. cephus* male never vocalized first in this context. By contrast, when the danger came from the ground, female and juvenile *C. cephus* gave the first alarm calls and this induced the calling of adult male *C. cephus* and *C. pogonias* in the same proportion. In all situations, the *C. nictitans* male showed an intermediate position.

Table 23.5. *Vocal activity of the adult males of the trispecific troop (number of sequences)*

Contexts	C. pogonias		C. nictitans		C. cephus	
	% calling	% first caller	% calling	% first caller	% calling	% first caller
All sequences (366)	55	77	43	18	18	5
Dawn sequences (16)	69	64	50	36	6	0
Troops junction (13)	100	100	38	0	15	0
Troops disjunction (11)	0	—	0	—	0	—
Intertroop spacing (11)	100	80	47	20	6	0
Avian predators (9)	100	89	100	11	100	0
Scansorial predators (36)	100	44.5	100	8.5	100	47

No detailed study has been published to date on this subject for other study sites. However Cords (1987b) pointed out that interspecific vocal exchanges between males punctuate the formation of mixed troops in only 27% of cases and that both blue and redtail males participated equally.

Interactions between polyspecific troops

Gautier (1969) has suggested that intertroop spacing among *C. cephus* troops mixed with *C. nictitans* was assumed by the latter species and that there exists a true acoustic parasitism by *C. cephus* which should be partly explained by the fact that its loud calls possess a carrying distance much lower than those of *C. nictitans*. It was also argued that *C. cephus* and *C. ascanius* were not territorial species (Gautier, 1969; Struhsaker, 1970; Marler, 1973). However, both *C. cephus* and *C. ascanius* displayed territorial behaviours (Struhsaker & Leland, 1979; Gautier & Gautier-Hion, 1983; Cords, 1987b).

At Kakamega, Cords does not mention countercalling between mixed troops. A different situation occurred in Gabon, where it is not rare that loud calls given at dawn by a mixed troop release loud calls of two or more neighbouring mixed troops. Differential rates of calling according to species characterize these vocal exchanges. During close intertroop encounters between the focal trispecific troop and a neighboring one, the *C. cephus* males of both mixed troops rarely took part in territorial vocal battles (Table 23.5); by contrast, in the focal troop, the *C. pogonias* male was always the first caller and participated in all countercalling, while the *C. nictitans* male only joined in about half the cases. In the neighbouring troop, the *C. nictitans* male assumed the prime position, and vocalized in 82% of cases, while the *C. pogonias* male called only in 39%.

All these facts suggest a kind of interspecific social organization which includes both a sharing of roles and mutualistic behaviours, while a strong component of local tradition appears to characterize the organization of such troops. In the focal mixed troop, the *C. pogonias* male that called the first, initiated the troop movements and punctuated with its calls the mixed troop formation, appeared to have a leading role; this latter being taken by the *C. nictitans* male in the neighbouring troop. In both troops, as in those previously observed (Gautier, 1969), the *C. cephus* males behaved as 'followers'. The sharing of roles was most obvious in predatory situations, showing that the *C. pogonias* male was the most watchful for birds of prey while both *C. cephus* females and the *cephus* male were on the look out for terrestrial predators. Encounters between neighbouring troops evidenced interspecific territorialism; even in this case, *C. pogonias* and/or *C. nictitans* took the prime position in intertroop spacing while, in most cases the spacing of *C. cephus* troops depended upon the vocal behaviours of the two other species.

Evidence for interspecific behavioural attraction

Waser (1987) has stressed the fact that interspecific behavioural attraction was '*prima facie* evidence that selection has favoured association'. Three gross categories of behavioural interspecific relations could be envisaged, concerning not only the formation of associations but also their maintenance: (1) active avoidance of encounters: this has been observed in guenons for *C. neglectus* only and reminds what Rodman (1973) has described in Bornean primates, where the presence of one species in one place decreased the probability to find another one; (2) neutral encounters: when crossing paths or when attracted by the same stimulus, monkeys of different species may aggregate temporarily; this probably characterizes most chance encounters (Waser, 1982); (3) active searching for encounters, which implies that monkeys are able to modify their behaviour in order to join troops of other species and to maintain interspecific cohesion.

The examination of responses given by members of mixed troops to loud calls of associated males, and of their spatial patternings, emphasizes the active role played by every species to associate. In the monkey community of Gabon, it was observed that dawn calls of the first calling male (either *C. pogonias* or *C. nictitans*) were sufficient to initiate movements of all mixed troop members. Two alternative

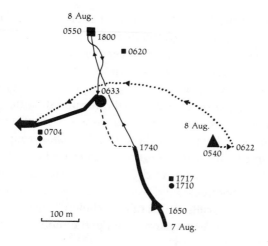

Figure 23.6. Movements of *C. nictitans* (broken line), *C. pogonias* (solid line), *C. cephus* (dotted line) and mixed troops (large solid line). On 7 August, the *C. nictitans* and *C. pogonias* monkeys were mixed all the day; they separated at about 1740 h and slept less than 200 m apart from each other (sleeping sites of *C. pogonias*: large black square, of *C. nictitans*: large black circle). The *C. cephus* troop was alone and slept 400–500 m from the others (sleeping site: large black triangle). At 0550 h (8 August), the *C.pogonias* monkeys resumed their activity and moved towards the *C. nictitans* which waited at their place. The *C. cephus* moved in an opposite direction. At 0620 h, the adult male *C. pogonias* gave a first sequence of loud calls (small dark square). No countercalling was given by the other males. Two minutes later, the *C. cephus* changed their course and quickly directed their steps towards the *C. pogonias*. The *C. nictitans* were still staying silently. The *C. pogonias* met with them about 11 min after calling and both troops began to move together. At the same time, the *C. cephus* travelled 600 m and arrived close to the bispecific troop. The three troops mixed and the three adult males uttered loud calls at 0704 h, initiated by the *C. pogonias* male (two observers were present, one following the *C. nictitans* and *C. pogonias* troops, the other the *C. cephus* one) (from Gautier-Hion & Gautier 1974).

patterns were observed (Gautier-Hion & Gautier, 1974): members of the species calling could wait before setting off for the arrival of 'their' associated troop which might have been sleeping up to 200 m away, or they could move in its direction as 'searching for it' (Figure 23.3). In both cases, members of the first species calling were in the front of movement. Spectacular changes in movements of a troop which was not mixed before were also observed. Thus, a quick joining could be induced in *C. cephus* members by the loud calls of a *C. pogonias* male, even at a distance of 500 m (Figure 23.6).

Most often, after the midday resting period or after an alarm episode, a mixed troop only resumed its movement after loud calls of the prime calling species. *C. cephus* members were never observed to initiate such resumption of daily movements nor leading the way. Still more impressive was the behaviour observed when one species belonging to a mixed troop was engaged in an intraspecific territorial battle. Three such intraspecific conflicts were observed between members of a monospecific troop of *C. cephus* and those of a *C. cephus* troop mixed with *C. pogonias* and *C. nictitans* (Gautier & Gautier-Hion, 1983). In these situations, both *C. pogonias* and *C. nictitans* waited up to one hour, sitting nearby in the trees, for the conflict to end. Afterwards the *C. pogonias* male gave loud calls which rallied 'their' *C. cephus*, and the activity of the mixed troop was resumed while the monospecific one went off on its own.

Such level of responsiveness and attraction was not described in other monkey communities. According to Cords (personal communication), they are not shown among *C. mitis* and *C. ascanius* at Kakamega. Conversely, in Taï forest, Galat (personal communication) observed that in mixed troops *C. diana*, whose adult male was the first caller, was also the first species to move after the midday resting period, while *C. campbelli* and *C. petaurista* always followed it. To be a 'follower' does not imply to have a passive role: in the few documented cases available, *(cephus)* members, that is, the smallest *Cercopithecus* in the communities involved (and probably the more recently evolved, see Kingdon, Chapter 11; Dutrillaux *et al.*, Chapter 9) appeared to strive for association more actively than all other species, and could be the one that benefited the most. In Gabon, the tendency for *C. cephus* to form mixed troops is so great that two troops of this species were able to live together within a single mixed troop, and were suspected to compete between them to associate (Gautier & Gautier-Hion, 1983).

Additional information concerns the composition of 'relict troops' found where hunting pressure is high (Gautier & Gautier-Hion, 1969, and unpublished data). Such troops may include no more than four animals, each of them belonging to a different species. However, no *C. neglectus* was ever observed in such groupings. Finally, it is not rare to observe a juvenile arboreal guenon frequenting for several consecutive days the juvenile 'clan' of a troop of a different species before rejoining its own. Another remarkable example is given by Struhsaker, Butynski & Lwanga (Chapter 24), who observed a *C. mitis* male living in a redtail troop for six years.

Struhsaker (1981b) has pointed out the possible role of interspecific play in the subsequent formation and maintenance of interspecific affinities. It is likely that a young *C. pogonias*, which has spent three years of its life playing with *C. nictitans* monkeys of its own age, developed a pattern of conspecific recognition different from that of a *C. neglectus* who has only been living with its parents and one sibling and avoided other guenons. The evidence for strong interspecific behavioural attraction among arboreal guenons allows us to assume that selection pressure for associations has been strong; this raises two questions: what is the functional significance of this life style and did it play any role in guenon evolution?

What is the functional significance of mixed troops of forest guenons?

Two alternative explanations have been repeatedly brought forward to explain the adaptive functions of polyspecific associations: a reduction of predation pressure and an improvement in foraging efficiency (see a review in Cords, 1987b), in favour of mixed troops vs. single species troops. Let us examine in turn these two possibilities, keeping in mind that they are not mutually exclusive.

Although most field workers have suspected that potential predators are more efficiently detected by mixed troops than by single species troops' members, no definite evidence has yet been presented, for three main reasons. First, only large diurnal birds of prey have usually been considered because it was assumed that the crowned-hawk eagle was the only significant predator of forest guenons. This fact, has however, never been properly established: indeed, a recent study of the leopard's diet in the Taï Forest, carried out by Hopp-Dominik (1984), showed that monkeys rank second among the favoured prey categories of this large cat, just following forest antelopes: 19.5% of the leopard's scats included arboreal guenons' remains. Second, far too often, the predatory behaviour of a given species was inferred from the 'alarm' behaviour of its potential prey. This is a questionable procedure; predator–prey interactions can only be studied reliably by studying predators, as pointed by Waser (1987). Finally, anti-predatory benefits gained from living in mixed troops have sometimes been estimated by relating the occurrence of mixed troops with the number of times an eagle was seen flying over the monkeys; or the number of unsuccessful eagle attacks (Freeland, 1977; Waser, 1980; Cords, 1987b). This method does not take into account the fact that every participating species in a mixed troop is neither equally exposed to an attack by a given predator, nor to the same extent. Furthermore, it does not take into consideration the possibility of a coevolution of the behaviour of both predator and prey. This might have led eagles, for instance, to attack monospecific troops more often than mixed ones (Struhsaker, 1981b). Consequently, only a comparison of the rate of successful captures of a given species of monkey in monospecific and in mixed troops could provide convincing evidence of anti-predatory benefits. It takes a lot of time to get such figures: during 1500 hours of observation of the same community, we only observed four successful captures by eagles, all of *C. cephus*; at least three of them involved individuals living in monospecific troops (Gautier-Hion *et al.*, 1983). This is no more than a mere indication.

Another method for collecting indirect evidence is to compare what happens in different forest types, not equally suitable for predators' approach. For instance, monkeys living in a forest with a dense understorey, the canopy of which is shrouded with hanging lianas, have less chance of being taken by eagles than others living in a high forest with a more open canopy. Indeed, *C. cephus* has been found to be highly dependent on the physical structure of the forest, selectively searching for the densest understorey, though this habitat was the poorest in fruit resources. When associated with *C. pogonias* and *C. nictitans*, *C. cephus* was seen to enter more open and 'richer' forest (Gautier-Hion *et al.*, 1981, 1983). This shift in habitat could be explained by the increased protection against predators afforded by the vigilant behaviour of associated species (especially *C. pogonias*, see above) and/or by an actual defence. Following an attack by the eagle upon one *C. cephus* monkey, I saw on one occasion the *C. nictitans* adult male of a large pentaspecific troop, actively driving the eagle away (followed by a *Lophocebus albigena* adult male).

It is worth noticing at this stage that, in all study areas, the smallest forms of guenons, belonging to the (*cephus*) superspecies that are potentially the most vulnerable to predators, are also those which display the strongest tendency to live in mixed troops. This might be indirect evidence of the role of predation (Struhsaker, 1981b).

The diet of species most commonly seen associated has been investigated by a number of authors, both in monospecific and in polyspecific troops. The most consistent result is that there is a significant positive correlation among frugivorous monkeys between the percentage of time species spent together and the percentage of dietary overlap in plant food (Struhsaker, 1978, 1981b; Waser, 1980), while the same is not found for animal food (Struhsaker, 1981b; Gautier-Hion *et al.*, 1983; Cords, 1987b). On the other hand, no consistent relationship has been found between overall fruit availability and the frequency of occurrence of mixed troops or with the time spent feeding on fruit (Freeland, 1977; Waser, 1980; Struhsaker, 1981b) although Struhsaker observed that associations are more frequent when monkeys eat rare and 'preferred' foods. However, periods of fruit scarcity do not necessarily lead to a disbanding of mixed troops and, generally speaking, the timing of association does not correlate with food (Gautier-Hion *et al.*, 1983).

The possibility that living in mixed troops might increase the foraging efficiency of the species associated, and lead to better nutritional conditions, has been investigated in the trispecific troop already mentioned, by analysing ranging patterns together with fruit availability. It has been shown that association results in a change of habitat use including a less intensive exploitation and a better selection of the areas with the largest supply of fruiting species, at the expense of a very low extra cost of locomotion. The new areas exploited every day were larger and the richest forest types were significantly selected. For instance, there was a better exploitation of the high forest with open undergrowth that harboured 1.4 times more fruiting trees and twice as many fruiting species as the forest with dense understorey. Cords (1987b) also found some evidence that participation to mixed troops increased foraging efficiency and/or the quality of diet, at least for *C. ascanius*.

Thus polyspecific life-style appears to enhance both food resources exploitation and predator avoidance. Consequently, it may be viewed as a mechanism which has been fashioned by natural selection, either for increasing the quality and quantity of food available – in this case, predator avoidance would then become only an effect (*sensu* Williams, 1966), or for reducing the predation pressure – in this case, better exploitation of food resources could be an incidental consequence. It is not yet possible to settle the question. But how can interspecific foraging have evolved and how can it enhance foraging efficiency?

In Gabon, diets of *C. nictitans*, *C. pogonias*, and *C. cephus* have been noted to diverge most when there is less fruit available (Gautier-Hion, 1980 and Chapter 15). This suggests that competition for fruit does occur. However, no correlation has been found between the time the three species spend associated and the availability of fruit; this implies that mixed troops are able to withstand the decrease of a food supply without disbanding, and/or that the cost of competition is smaller than that of predation.

The evolution of interspecific foraging may be seen as a response to the extreme diversity of plant resources and to the very diverse and more or less unpredictable phenological patterns which characterize a rain forest. This leads to a variety of species distribution patterns, and thus of food patch sizes; consequently there exists a variety of ways by which animals can acquire food items. It can therefore be assumed that, within such a complex environment, a troop of 60 monkeys, belonging to three different species neither too dissimilar nor similar, but representing a subtle gradient of morphological attributes, would be better able to exploit successfully the available resources than three mono-specific troops of 20 individuals each, or a single monospecific troop of 60 members (Gautier-Hion & Gautier, 1979; Waser, 1987).

Has polyspecific social organization any adaptive value for the participating species? It is not possible to answer this question, as the gain in fitness eventually provided by this optional life-style has not yet been measured. However, it seems worth mentioning in this respect that the monospecific troop of *C. cephus* we observed at Makokou for nine years increased its membership less than the neighbouring con-specific one, which lived part of the time in association with other species and which enlarged its home range, overlapping more and more the surface area covered by the monospecific troop (Gautier-Hion & Gautier, 1974; Sourd, 1983).

Did polyspecific associations play a role in the speciation of the genus *Cercopithecus*?

Although a number of forms still remained unstudied, two general remarks can be made about guenons. First, the rate of speciation and subspeciation has been less important in semi-terrestrial species than in arboreal ones; second, whereas most (if not all) arboreal species can live in mixed troops, the semi-terrestrial guenons apparently rarely (if ever) do so. Are these relationships between, on the one hand, 'conservative' semi-terrestrial habits and inability to live in mixed troops, and on the other hand, polyspecific life style and active speciation, just a matter of chance?

As seen in Chapter 9, Dutrillaux *et al.*'s interpretation of the kary-ological evolution of guenons implies the hybridization of individuals with different karyotypes. Hybridizations between present wild populations seem the rule near the boundary zone between species (see Lernould, Chapter 4). Recently, hybrids between fully sympatric species have been observed in polyspecific troops, in peculiar demographic situations characterized by differences in the relative abundance of parent species' troops. These hybrids are fertile (at least females) and they apparently benefit from some advantages over both parent species (Struhsaker *et al.*, Chapter 24).

One can therefore wonder whether there is a relationship between polyspecific life, hybridization and speciation, whether the polyspecific associations are a recent evolutionary event (on a geological time scale), or whether it is a much older way of life which could have had incidental effects on speciation, at least in certain situations, such as changes of climatic conditions and their subsequent effect on habitat production (see also chapter 24).

In other words, are polyspecific associations just an elaborate strategy for 'optimally' exploiting the environment (resources, defence against predators) or has this peculiar life style had any effect on guenon speciation through hybridization? Further field and laboratory work will have to elucidate these points.

Conclusions

A great deal of work remains to be done before the meaning of polyspecific associations of forest guenons can be further clarified. What emerges now is the variability of patterns relating not only to the frequency of occurrence of mixed troops, but also, and perhaps more meaningfully, to their nature.

Their structure indeed varies in a number of respects, especially:

the degree of 'faithfulness' within mixed troops, i.e. the tendency for one troop of one species to mix or not with a single troop of another species, thus inducing more or less complex interspecific relationships, learning processes and protocultural behaviours;

the asymmetry of behaviours between mixed species, leading to different roles at the species level;

the asymmetry of costs and benefits both in nature and in extent, with the most active species – generally the smallest one – benefiting the most from the association.

In the foregoing pages, I emphasized the more stable polyspecific associations, disregarding those which occur by mere chance. I have also excluded from my analysis associations between guenons and other forest monkeys, such as mangabeys (Waser, 1980) and colobines (e.g. Struhsaker, 1975; Galat & Galat-Luong, 1985). Yet these associations are most interesting and also contribute to shedding light on the possible determinants of mixed groups of monkeys.

The study of a Peruvian community (Terborgh, 1983) definitely established that the African situation, and the Gabon case in particular, can no longer be considered as unique. It is noteworthy that Terborgh also considered that compatibility of life styles 'is a major factor inducing the formation and maintenance of mixed troops in Platyrrhine monkeys', and that there exists an asymmetry of behaviours and roles among mixed species, as well as unbalanced costs and benefits.

It must not be forgotten, however, that the monospecific and polyspecific ways of life of forest monkeys always represent alternative tactics which are adopted in turn, probably to 'get the better' of a very complex environment. But this flexibility, which provides another example of the unusual plasticity of primate behaviour, might have also helped to lessen the reproductive barriers between species belonging to an actively speciating genus, and in so doing it might also have contributed to influence the pace of speciation.

Acknowledgments

My thanks go to Profs F. Bourlière and G. Richard, Drs M. Cords, D. Lachaise, S. McEvey, and T. Struhsaker for their helpful comments.

III.24

Hybridization between redtail
(*Cercopithecus ascanius schmidti*)
and blue (*C. mitis stuhlmanni*)
monkeys in the Kibale Forest,
Uganda

THOMAS T. STRUHSAKER, THOMAS M. BUTYNSKI
and JEREMIAH S. LWANGA

Introduction

Hybridization among African monkeys has been recorded both in captivity (Gray, 1972) and in the wild. Most of the cases described from the field appear to occur in areas or zones where secondary contact has occurred between two normally allopatric species or subspecies due to the breakdown of ecological barriers (Lernould, Chapter 4). Probable examples of this kind of hybridization are those between *Papio hamadryas* and *P. anubis* (Kummer, Goetz & Angst, 1970; Nagel, 1971; Shotake, 1981); *P. anubis* and *P. cynocephalus* (Maples, 1972; Samuels & Altmann, 1986); *P. anubis* and *Theropithecus gelada* (Dunbar & Dunbar, 1974c); *Cercopithecus cephus* and *C. erythrotis* (same superspecies, if not conspecific, Struhsaker, 1970); *C. albogularis* and *C. mitis stuhlmanni* (same superspecies, if not conspecific, Booth, 1968).

More perplexing are the cases of hybridization between two distinct species which occur sympatrically in several localities without hybridizing: crosses between *C. mona* and *C. p. pogonias* (Struhsaker, 1970) and between redtails, *C. ascanius schmidti* and blues, *C. mitis stuhlmanni*. Among the African guenons, hybridization is best known for the latter. The first description (Aldrich-Blake, 1968) was a fertile adult female hybrid living in a group of blue monkeys in Budongo forest, Uganda.

She was accompanied by an infant who was very blue-like in appearance and probably the result of a backcross with a male blue. Hybrids between these two species have also been recorded from the Itwara Forest of W Uganda (J. F. Oates, personal communication); and with a different subspecies, *C. mitis doggetti*, in the Gombe Stream National Park in Tanzania (Clutton-Brock, 1972).

This chapter describes hybridization between redtails and blues of Kibale where both parental species and the hybrids have been studied in depth over many years. This study provides insight into the factors favouring hybridization between two distinct species that typically live sympatrically over large areas without interbreeding. An understanding of why and how the normal mechanisms of reproductive isolation may break down helps to clarify problems of species recognition, individual identity, reproductive strategies, and the possible role of hybridization in speciation.

Methods

Studies of primates have been conducted in the Kibale Forest, W Uganda since 1970. The majority of observations have been made at two study sites: Kanyawara (0°34'N, 30°21'E, about 4 km²) and Ngogo (0°30'N, 30°25'E; about 7 km², Figure 24.1). Although the hybrids were first discovered at Ngogo on 13 May 1975 by Struhsaker, detailed and systematic studies on them were not begun until 1978. Methods used were similar to those employed in earlier and contemporary studies of red colobus, redtail and blue monkeys (Struhsaker, 1975, 1980; Butynski, 1982b).

Detailed observations were made on three hybrids and their offspring living in four different social groups of redtails (Figures 24.1, 24.2). These social groups were systematically observed for a total of 1505 h (Struhsaker; 828 h from January 1978 to April 1986; Butynski: 391 h from November 1979 to September 1984; Lwanga: 286 h from October 1984 to December 1985). During these periods, all social and feeding behaviour was recorded *ad libitum*. Attention was focused on the hybrids, but actual observation time on them was far less than the total time spent with their redtail social groups.

Results

Hybridization apparently occurred only in the Ngogo study area. The three known hybrids, one male and two females, each lived in different redtail social groups (Figures 24.1, 24.2), and appeared to be of similar age, i.e. within 1–3 years. One of the females produced

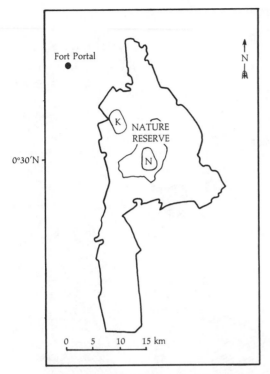

Figure 24.1. Location of the two major study areas in Kibale Forest Reserve: K (Kanyawara) and N (Ngogo).

three infants sired by male redtails, while the other female's three offspring were sired by a male blue. At the Kanyawara study area, a fourth hybrid (adult male) was seen three times by Butynski in August and November 1983. This individual must have been transient as he was seen neither before nor afterwards despite thousands of observation hours at that site by many qualified observers.

The parental species, redtails and blues, are similar in diet, vocalizations and social systems, with generally 15–35 members living in matrilineal, one-male groups which defend territories (Struhsaker, 1977, 1980, Chapter 18; Butynski, 1982b, personal communication; Cords, 1987a, Chapter 17). These two species, however, are strikingly different (1) in physical appearance: blues are silvery dark grey while redtails are tinged rufous with white underparts and distinct facial pattern (Figures 24.3, 24.4); (2) in size: redtails weigh 3–4 kg while blues weigh 4–7.5 kg; (3) in chromosome number: redtail 66 and blue 72; (4) in the number of acrocentric chromosomes: redtails 12 and blues 18 (Hill, 1966). Dutrillaux *et al.* (Chapter 9), however, has found that the

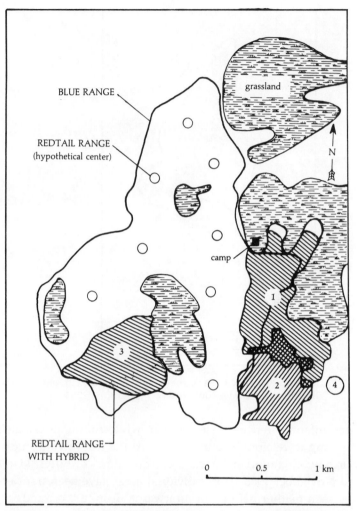

Figure 24.2. Ngogo study area and ranges of the three groups of
redtails containing hybrids: 1 (BTP), 2 (RAT), 3 (SW), and 4 (East).
Male H of the RAT group immigrated into the adjacent East group.

chromosome number of blues is 70 and the number of acrocentric
chromosomes is 18 in redtails and 20 in blues. He also found that the
two species differ by four chromosomal rearrangements.

The Ngogo hybrids were distinct in appearance and clearly intermediate between blues and redtails (Figures 24.5, 24.6). They tended to
look more like blues, however, in their larger size, generally darker and
less distinct color pattern, and general body configuration. Characters

Figure 24.3. Adult male redtail from Kanyawara (photo: L. Leland).

Figure 24.4. Adult male blue from Kanyawara (photo: L. Leland).

Figure 24.5. Adult female hybrid MBT on right and adult female redtail on left (photo: L. Leland).

Figure 24.6. Adult female hybrid MBT on left and juvenile redtail on right (photo: L. Leland).

Figure 24.7. Young juvenile backcross between hybrid female MBT and adult male blue RD (photo: L. Leland).

derived from redtails were the faint nose spot (prominent in redtails) and brown tail (orange in redtails).

The six backcrosses, four of which survived long enough to be clearly observed, were difficult to distinguish from the species with whom the hybrids bred. The three backcrosses born to hybrid MBT and whose genotype was assumed to be 75% blue greatly resembled blue monkeys (Figure 24.7). The one who survived infancy (and believed to be female) only vaguely resembled a redtail in having a very faint nose patch and a tail whose proximal half was dark brown. The only distinguishing character of the three redtail backcrosses born to hybrid PT (one male and two of undetermined sex) was that, instead of a clear and sharply demarcated dark line through the white cheek whiskers as in redtails, the lower half of the cheek whiskers was medium grey without the distinct black line. The tails of these redtail backcrosses appeared to fall within the range of colour variation for redtails, although they may have been slightly darker.

The density of blue monkeys at the Ngogo site where the hybrids were found differed significantly from that at the Kanyawara site (Butynski, unpublished data). Blue density at Ngogo was only 5–7 individuals per km^2, or about one-tenth that of Kanyawara, and lower than that described for any other population. By comparison, the redtail population at Ngogo was about 75–100 per km^2. Only one blue social group lived within the 7 km^2 Ngogo study area (although two others were occasionally seen on the periphery); whereas at Kanyawara, group density was about 3 per km^2. Furthermore, many more solitary male blues were seen at Ngogo than at Kanyawara. From Nov. 1978 to July 1980, they were contacted 16 times at Ngogo but not at all at Kanyawara; but from 1981 to 1984 at least five solitary (but probably transient) male blues were seen at Kanyawara (Butynski, unpublished data). The density of solitary male blues at Kanyawara is estimated to be about one-seventh that of Ngogo (Butynski, unpublished data).

Such a high density of solitary male blues at Ngogo, combined with the extremely low density of social groups there, suggest that competition for mates was particularly intense. Other supporting data are found in the social dynamics of blue groups in the two study areas. Male tenure in the social group of blues at Ngogo ($\bar{x} = 5.8$ months) was considerably shorter than for seven groups at Kanyawara ($\bar{x} \geq 41.1$ months, Butynski, unpublished data). Furthermore, several cases of infanticide by new resident males were observed and strongly suspected in the Ngogo group, but not in any of the Kanyawara groups (Butynski, 1982b). By shortening interbirth intervals, infanticide can increase the infanticidal male's opportunities for reproduction during relatively short tenures (Hrdy, 1979). Finally, we suggest that hybridization is a further reflection of competition among male blues for mates at Ngogo. Mating with a female redtail is considered to be an alternative reproductive strategy of male blues in situations where there is a relative shortage of conspecific mates and competition between males for females is particularly intense.

The actual process of hybridization has not been observed. Possible attempts, however, were made by an adult male blue monkey (called RD) who in June 1980 joined the BTP redtail group containing a nulliparous old subadult female hybrid called MBT. He and MBT disappeared from the group between 1 June and 12 August 1986. Early in his stay, RD chased adult female redtails in what may have been unsuccessful attempts at copulation. During the six years he was with the group, however, he was seen to copulate only with the hybrid female (see section on sexual behavior and reproduction).

If, within six years the male blue was still unable to copulate successfully with any of the redtail females in the BTP group, then such a strategy to mate with redtails is unlikely to be prevalent among solitary male blues. No blue monkey associated permanently with either of the other redtail groups containing hybrids, or any other redtail groups observed. On two different days (20 June 1981 and 24 April 1986) a solitary male blue temporarily associated with the SW group of redtails of which adult female hybrid PT was a member. It may be that hybridization takes place during such brief, opportunistic encounters.

Social behaviour of hybrids

The hybrids were socially integrated into their natal redtail social groups. The adult male hybrid (H) was apparently born in the RAT group where he was observed as a subadult from January 1978 to July 1979. When this group was next observed in November 1979, male H was missing. He was next located in the adjacent East group in December 1981, where he remained at least until January 1987. Hybrid females MBT and PT remained in what were believed to be their natal groups, BTP and SW groups respectively.

Hybrids both groomed and were groomed by redtails. The two female hybrids showed apparent interest in infants of redtail females by approaching and sitting next to them, and also by grooming their mothers. Similarly, redtail adult females and juveniles showed strong interest in the infants of hybrid females. Such interest in infants is typical behaviour of both redtails and blues. The young backcrosses, resulting from female hybrid matings with either redtails or blues, frequently played with redtail peers in their respective social groups.

Intergroup aggression

Hybrids, particularly the two females, actively defended territories with their fellow redtail group members against neighbouring groups of redtails. They behaved in a manner indistinguishable from redtails . At no time did they or the redtails fight blue monkeys in any way that might be interpreted as territorial.

The larger size of hybrids compared with redtails appeared to give them an advantage in aggressive encounters, since redtail females from opposing groups usually retreated when confronted by a female hybrid. Whether their size influenced the outcome of territorial conflicts, however, is not clear. In a subsample of 32 encounters (T. Struhsaker, unpublished data) for the BTP group against one neighbouring group 'S' (which eventually split into two groups, see Chapter 19), the BTP group clearly won 25%, lost 44% and had inconclusive outcomes or draws in 31% of the cases. These data are somewhat misleading, however, as the BTP group was often avoided by these other groups when it intruded into their ranges and core areas. The

BTP group, in fact, expanded its range to encompass much of the S group's range following fission. Although the larger size of the female hybrid may have influenced this range expansion, it also coincided with the union of the much larger male blue monkey RD with the BTP group, and the fission of the S group, both factors which were likely to have influenced intergroup relations.

Intragroup aggression

Aggression between hybrids and redtails within the group was infrequent. Only 14 encounters were observed by Struhsaker and Butynski, and hybrids won or were dominant each time. Four of these involved the hybrid male when he was either an old subadult or adult: twice he slapped towards a juvenile as it approached him; and one time each he chased an adult male and subadult redtail. Of the ten encounters recorded for the two adult female hybrids, five were supplantations, two were chases, and three were attacks with a slap. Adult female, subadults, and juveniles were the recipients nine times and a young juvenile once. The context was clearly over food on four occasions; food or space on another; proximity to the hybrid's infant twice; and undetermined three times.

Three of the redtail and blue backcrosses had four aggressive or dominance interactions with redtails. Two of the encounters appeared to be related to gaining the attention of the backcross's mother, such as when jabbing at or supplanting the redtail being groomed by the hybrid mother. Once the juvenile blue backcross (SOB) kept two juvenile redtails away from a *Monodora myristica* fruit on which it fed, in a clear example of priority of access to food.

Priority of access to food was the most frequent context of aggression or dominance interactions between the hybrids, backcrosses, and redtails, and best demonstrated from Struhsaker's notes of 17 December 1981. In this example, an adult female redtail prevented a juvenile from gaining access to a large, ripe *Monodora* fruit which had a wide gap in its epicarp, exposing the edible mesocarp. (Because this fruit has a very tough, thick epicarp which redtails seem unable to chew through, they must usually rely on the larger mangabeys, *Cercocebus albigena*, to open them. These monkeys eat only part of the mesocarp before leaving the fruit still attached to the tree, and thus available to redtails.) Both the adult female and juvenile redtail ran away from the fruit, however, when the adult female hybrid PT approached and began feeding on it.

She was soon joined by her two redtail backcross offspring: an old infant and young juvenile. They fed beside her, all three alternating by sticking their heads into the large fruit to eat the mesocarp. Two other young juvenile redtails tried to approach this fruit, but were prevented access by the hybrid and her two offspring. Nine minutes after beginning their feeding bout on this fruit, female PT and her two offspring left. Upon their departure, two young juveniles and one old infant redtail descended upon the same individual fruit to feed.

Not surprisingly, the adult male blue monkey RD, who resided in the BTP redtail group, was dominant over all its members. Only two agonistic encounters were observed between him and the hybrid female MBT and their backcross offspring. He had priority of access over a cluster of ripe figs on one occasion; and the young juvenile backcross quickly ran past him in clear avoidance in another.

Sexual behaviour and reproduction

The three hybrids at Ngogo were seldom seen to engage in sexual behaviour. However, their eight bouts of sexual mounting (each bout may consist of more than one mount) comprised 23.5% of all 34 bouts recorded for the three redtail groups in which they resided. This disproportionately high figure can probably be attributed to our biased sampling which concentrated on the hybrids.

The hybrid male H performed two copulations: one complete mount (with pause and presumed ejaculation) and one which may have been complete either on the same or a different adult female redtail in the RAT group (his presumed natal group) in June and September 1978. Hybrid female MBT of the BTP group had two complete copulations with the male blue monkey RD; two bouts of incomplete mounts with a redtail called SD; and a complete mount with a redtail male of undetermined identity (L. Leland, personal communication). In the SW group, female hybrid PT had one incomplete mount with a male redtail. Thus, of the total of eight bouts, six were with redtails and two with blues. All copulations employed behavioural patterns typical of redtails and blues. (Copulation is qualitatively identical in the two species). Female MBT approached and presented sexually to both male redtails and blues.

MBT's daughter SOB, a blue backcross, was the only backcross old enough to engage in sexual behaviour at the time of writing. She had an incomplete copulation with a juvenile redtail in August 1986 when she was 4 years, 9 months old.

The two female hybrids are known to have produced at least six offspring. Female PT gave birth to three infants who were clearly sired by male redtails. Her first infant, a male called DCB, was estimated to have been born between January and March 1979, or perhaps earlier. DCB was last seen in the SW group as an old juvenile male in July 1985. He may have emigrated from the group between then and when the group was next observed in late January 1986. PT's next infant, OPT, which was born between March and mid-May 1981, disappeared and presumably died between 17 July and 14 November 1982 at approximately 14–19 months of age. PT's next known infant, OPT II, was born betwen 5 and 16 June 1984 and was still alive (as was PT) in January 1987.

All three infants of female hybrid MBT were sired by a blue male, presumably male RD who was the only resident blue male in the BTP group throughout this study. Her first infant, SOB, was born between 1 and 15 November 1981 and was still alive as a juvenile in January 1987[1]. Her second known infant was born during the first ten days of May 1985, but was dead on 12 or 13 May. She carried the corpse for three days. MBT gave birth again between 1 and 22 April 1986, but this infant had disappeared when MBT was next observed on 29 May 1986. If MBT's disappearance in 1986 was due to death, which seems likely, then she produced only one surviving offspring[1].

It could not be determined if male hybrid H was able to breed successfully. Careful checks of the RAT group (from 1978 to July 1985) and the East group (from 1981 to May 1986), however, revealed no indication of backcrosses even though redtail reproduction was good in both groups in which male H resided[2].

The interbirth intervals for female hybrids appear to be considerably longer than for female redtails, and perhaps slightly longer than for female blues. The most accurate interbirth interval for hybrid female PT was 37–39 months, between her second and third infants. The interval between her first and second, although less accurate, is estimated to be between 24 and 25 months. PT was still without a new infant on 12 January 1987, 30.5 months after the birth of her third infant[3]. For female MBT there are two fairly accurate estimates of interbirth intervals: 42 months and 11 months. The latter, of course, was due to the early death (10 days old) of her second infant. It is well established for many primate species that neonatal deaths appreciably shorten interbirth intervals.

The interbirth interval for redtails in Kibale is 17.8 months (SD ± 6.2, range 11.5–28 months, Struhsaker in Cords, 1987a). This is half that of hybrids (35 months). Intervals for blues are less certain, but they are probably longer than 24 months (Cords, 1987a). Although the few data suggest that hybrid females have longer interbirth intervals than redtails, one cannot rule out the possibility of unobserved neonatal deaths (one can, of course, argue the same for our determination of these intervals for redtails and blues).

Reproductive rates (number of births/adult female/year) suggest similar trends. At Ngogo these rates were approximately 0.33 for the hybrids, 0.42 for blues, and 0.48 for redtails. At Kanyawara, however, reproductive rates of female blues were considerably lower (0.20).

Feeding ecology

Redtails and blue monkeys ate many of the same species-specific plant foods. For example, during 10 monthly samples in which a group of redtails and a group of blues with overlapping home ranges were observed simultaneously at the Kanyawara site, dietary overlap of plant foods was 33.8%. This similarity in diet was greater than for any of the other nine monkey-species pair combinations compared there (Struhsaker, 1978). It is not surprising, therefore, that the hybrids and backcrosses shared many plant foods in common with both redtails and blues.

Of particular interest are those foods which were shared with only one or the other parental species, but not both. Table 24.1 demonstrates that the hybrids ate some foods in common only with redtails, and others only with blues. Other hybrids' foods were rarely eaten by either of the parental species.

The results in Table 24.1 clearly suggest that there is a strong genetic component to dietary preferences. For example, foods eaten primarily by the hybrids and blues, but not at all or rarely by redtails, were unlikely to have been selected through imitation or social learning, since the hybrids had very infrequent and short-duration contact with blues. A vivid example is provided by Struhsaker's notes of 18 August 1980. Hybrid female MBT fed extensively on the floral buds and flowers of *Premna angolensis* (25 bites in 2.5 min) during which time eight different redtails of her group moved past her through the same tree without pausing even to inspect the floral parts.

Similarly, arguments for a strong genetic component are supported by examples of foods eaten by hybrids and backcrosses, but rarely or not at all by either parental species, e.g. *Cordia millenii* fruit, *Zanha golungensis* young leaves, and floral buds of *Voacanga thouarsii*. The consumption of such foods may represent cases of genetic codominance analogous to the ABO blood types (Farnsworth, 1978) in which the heterozygote – or hybrid in this case – is qualitatively different from

Table 24.1. *A partial and qualitative comparison of selected food items for redtails (n ≥ 3000 feeding scores), blues (n = 5757), hybrids and backcrosses (BC, n = 275). Rare scores for redtails and blues are given in (); the first figure corresponds to Ngogo site, the second to Kanyawara site; —: no feeding score*

Species	Items	Redtails	Blues	Hybrids
Conopharyngia holstii (K. Schum.) Stapf	Flowers	yes	—	yes
Randia urcelliformia Hiern.	Fruit	yes	(1.6–0)	yes
Piper guineense Schumach. & Thonn.	Fruit	yes	—	yes + BC
Chrysophyllum albidum G. Don	Fruit	yes	—	rare
Urera cameroonensis Wedd.	Young leaves	—	yes	yes
Funtumia latifolia (Stapf) Schlechter	Floral galls	—	yes	yes
Premna angolensis Guerke	Floral buds, flowers	(2–2)	yes	yes
Balanites wilsoniana Dawe & Sprague	Mature leaves	(0–7)	yes	yes
Spathodea nilotica/ campanulata P. Beauv.	Young leaves	(1–0)	yes	yes + BC
Cordia millenii Bak.	Fruit	(2–1)	(2–0)	yes + BC
Zanha golungensis Hiern.	Young leaves	(2–0)	rare	yes + BC
Voacanga thouarsii Roem. & Schult.	Floral buds	—	—	yes

either parental species, rather than a blend of the two. Diet is probably developed by the genetic propensities being shaped through trial and error learning, social learning (imitation) and the dietary options available.

Discussion

Within the limitations of reproductive compatibility, the frequency of occurrence of hybridization is likely to depend on the advantages and disadvantages to the parental species. Because female redtails make a greater physiological investment than do male blues, emphasis is given to an analysis of the female cost/benefit ratio.

Advantages

Male blues are thought to hybridize with female redtails as an alternative mating strategy when intrasexual competition for conspecific mates is particularly intense: hybridization is better than low or no reproduction with conspecifics. In terms of reproductive physiology, male blues invest little. It is still unclear, however, why redtail females mate with blue males when redtail males are available.

The ultimate advantages to either sex which hybridizes lies in their reproductive success and inclusive fitness. How fertile the matings are between blues and redtails is not known. But the fact that hybrid females (and possibly males) are fertile, and that female hybrids will mate with either parental species allow potential reproductive advantages.

Hybrids have several other advantages. One obvious benefit is that they and at least the blue backcrosses are substantially larger than redtails and dominant over them. This dominance is expressed primarily in terms of priority of access to food. Their larger size also means that they and their accompanying offspring are less likely to be supplanted from limited food resources by other primate species. On one occasion mangabeys were seen supplanting redtails, but not the hybrid and her offspring, from a fruiting tree (L. Leland, personal communication). Although there are no data, a larger body size may also mean that the infants of female hybrids are less susceptible to attack by infanticidal male redtails following male replacement (Struhsaker, 1977). Although hybrids often participate very actively and aggressively in territorial defense, it is not clear whether their larger size influences the outcome of these encounters.

The hybrids also have an advantage over redtails in feeding on a wider array of plant species and parts. This broader feeding niche may reduce food competition between hybrids and other group members.

Disadvantages

The cost of hybridization to a male blue is low in terms of gamete production. However, the investment of time necessary for him to achieve successful matings with female redtails may be very substantial. We have not yet seen a copulation between a male blue and female redtail, even though the blue male RD was resident in the BTP group of redtails for six years. It may require many years for female redtails to accept a male blue. Alternatively, hybridization may occur under peculiar circumstances or with particular individual male blues. If so, this may account for the similarity in age of the three known hybrids.

Another disadvantage to both male and female alike is that matings between blues and redtails may be less fertile than between conspecifics. Furthermore, the hybrids appeared to have lower reproductive success. In mammals, male hybrids are generally infertile more often than females (Dobzhansky, 1970). The one hybrid male appears not to have successfully reproduced at all. Whether this was due to relatively greater rejection by female redtails or to partial or complete infertility is not clear. The female hybrids, although fertile, had longer interbirth intervals and probably lower reproductive success than redtail females in Kibale. The 50% mortality of infant backcrosses appears high compared with redtails, although this difference is less certain because of incomplete data on redtail infant survivorship after six months of age.

Under most conditions the disadvantages of hybridization apparently outweigh the advantages, thus accounting for the scarcity of hybrids. Although the hybrids may have certain advantages over redtails due to their larger body size and over both parental species because of a broader food niche, these seem to be outweighed by the lower reproductive success of the hybrids, particularly the male, in comparison with redtails. There may be little difference in reproductive success between female hybrids and the Ngogo blues, however. Nevertheless, when male–male competition for mates is extremely high, the only opportunity that some males have to reproduce may be with hybrids or with females of a more abundant and closely related species. In the present case it would seem to be a far better strategy for a male blue to mate successfully with a hybrid or redtail female than never to mate during his lifetime. This should be the case even if the fitness of the resultant offspring is relatively low. In contrast, there appear to be few, if any, reproductive advantages for a female redtail to mate with a male blue. Such matings are more likely to lower her reproductive success.

While sightings of only four different hybrids have been made in Kibale, there could have been many more backcrosses in the population which were overlooked because of phenotypic masking.

Although the F1 backcrosses are difficult to distinguish from the male parent, and F2 backcrosses presumably even more so, the presence of predominantly blue backcrosses living in redtail groups should be conspicuous. Predominantly redtail backcrosses living in redtail groups would, however, be readily overlooked. In any event, the potential for introgression, particularly among redtails, remains high.

Behavioural plasticity, individual identity, and species recognition

Hybridization and the social integration of hybrids into their natal redtail groups demonstrate the potential behavioural plasticity of redtails and blues toward one another. All forms of social behaviour occur between these two species and their hybrids (also see Struhsaker, 1981b). Color patterns, vocalizations, postures and other characters that may enhance species recognition do not preclude intimate social interactions between redtails and blues.

The usual barriers to hybridization by these closely related species apparently break down when there is a shortage of females in relation to males for one of the species. This shortage of females may also be reflected in a lower population density of the same species. A situation similar to that of Ngogo was found in the Gombe National Park, Tanzania where *C. mitis* were found in low densities compared with redtails and solitary male *mitis* were common. The majority of redtail groups there contained one or more hybrids (Clutton-Brock, 1972). In contrast, blues outnumbered redtails in the Budongo Forest of W Uganda and here the hybrid female and her backcross infant lived with a group of blues (Aldrich-Blake, 1968). Mayr (1963) mentions similar cases for other vertebrate groups.

The breakdown of reproductive barriers, combined with the total social integration of hybrids within their natal groups of redtails, raises important questions concerning individual and species identity. In some interactions there appears to be an important acquired or learned component of social affiliations, while in others, genetic propensities may prevail. In support of the latter is the case of female hybrid MBT who copulated with both redtail and blue males. When the blue male RD joined her redtail group, she groomed, consorted with, and produced three infants with him. However, none of the female redtails in her group were seen to have any physical contact with him in six years. In contrast MBT's and PT's close social ties with the female and juvenile redtails of their groups support the idea of social learning. Although all three hybrids seemed well integrated in their natal groups, the male hybrid H may have had difficulty in finding mates once outside his natal group. While social learning presumably had strong influence in shaping his individual identity as a redtail, his more blue-like appearance may have inhibited unfamiliar redtail females from mating with him.

None of the hybrids showed any indication of being attracted to members of the one blue group with whom they shared their range. Nor did we see any aggression or indication of territorial defense between blues and hybrids. Although the hybrid female MBT and hybrid male H lived in adjacent groups of redtails, no interactions were seen between them. Clearly the hybrids behaved socially more as if they were redtails than blues, even though their general appearance was more like a blue. These observations argue for the great importance of learning in the development of social affinities and species identity. In contrast, genetics may play a greater role in shaping food preferences, and in some cases, may influence mate choice.

A final example of behavioural plasticity and individual species identity concerns the blue male RD. He resided in the BTP group of redtails for six years along with a series of resident male redtails. During this time he moved with them throughout their daily ranging patterns and even assisted in their defense of territorial boundaries against neighbouring groups of redtails. During his first year in the group, RD occasionally chased female redtails in what might have been attempts at copulation. One of the most prominent indications of his plasticity in behaviour concerned his loud calls. He gave several bouts of the pyow call each day, as is typical for resident males in blue groups, although he gave far more pyows per bout than most resident male blues. Solitary males do not usually give this call. Even more striking was the fact that RD gave the boom call, typical too for blue males, but in response to the grunts of female and juvenile redtails rather than to those of blues. Solitary male blues have never been heard to boom. Thus, male RD responded to a similar set of cues given by a different species. In other words, he appeared to have adopted a redtail group, even though his affiliative interactions were restricted to a female hybrid and their offspring.

Speciation by hybridization

Although speciation through hybridization occurs rather widely among plants, it has not been verified for higher animals (Mayr, 1963; Dobzhansky, 1970; Wright, 1978). Introgression could cause appreciable genetic modification, but it is considered unlikely to result in speciation (Wright, 1978).

In the case of redtail × blue hybrids, the possibility of hybridization *per se* leading to speciation is contingent upon whether or not the male hybrids are fertile. Even if the male hybrids are infertile, however, one could imagine that if backcrosses are fertile, speciation could occur through a combination of introgression, polyploidy, the Founder Effect, and genetic drift. A scarcity of mates in one of two closely related species occurring in small, isolated populations, such as is likely to have occurred in forest patches formed during the Pleistocene interpluvials, or as may be occurring now as forests are being fragmented by humans, could lead to a high incidence of hybridization and extensive introgression. Differences in food habits, as shown for the Kibale hybrids, might result in ecological differentiation from the parental species. Certainly, hybridization under these hypothetical conditions offers possibilities for subspeciation, if not speciation (see Dutrillaux *et al.*, Chapter 9, regarding karyotypic hybridization).

Summary

1. Data are summarized from an eight-year study of three hybrid crosses between redtail and blue monkeys and their six backcrosses in the Kibale Forest of W Uganda.
2. Hybrids occurred primarily in a restricted part of the forest where densities of blue monkeys were low, the number of males per female was high, and competition for mates among male blues was intense. Hybridization with redtail females is thought to be an alternative mating strategy of male blues.
3. Hybrids were socially integrated into their natal redtail groups. They engaged in all forms of social behaviour and assisted in territorial defense against neighbouring groups.
4. The two hybrid females were fertile and each produced three young. One hybrid mated only with redtail males and the other copulated with both redtails and blues but only produced offspring with a male blue who resided in her redtail group for six years.
5. *Advantages of hybridization*
   (a) When competition for conspecific females is intense, male blues gain a reproductive advantage by hybridizing.
   (b) Hybrids are larger and dominant to redtails. This gave the hybrids and their offspring priority of access to food.
   (c) Hybrids were less likely to be supplanted from food by other primate species than were redtails.
   (d) Infants of hybrids may be less prone to infanticidal attacks because of their mothers' larger size and blue-like appearance.
   (e) Hybrids fed on a wider array of plant species and parts than did redtails giving them more choices and possibly less competition.

6. *Disadvantages of hybridization*
   (a) Probable rejection of male blues as mates by most female redtails.
   (b) The hybrid male may have been infertile as he had very low, if any, reproductive success; he might also have been rejected by female redtails.
   (c) Hybrid females had lower reproductive success than female redtails, but perhaps not lower than blues. This was due to longer interbirth intervals and possibly lower infant survivorship.
7. Introgression may be more common than indicated by the frequency of hybrids and backcrosses because F2 backcrosses may often be indistinguishable from the parental species.
8. Behavioural plasticity, individual identity and species recognition are discussed. Although all of these are influenced both by genetics and learning, social affinities and individual and species identity seem to be more strongly affected by social experience. Food habits and mate choice, on the other hand, may be more influenced by genetic factors.
9. Speculation is offered on the possible role of hybridization in speciation/subspeciation, with an emphasis on the combined interaction of introgression, the Founder Effect, and genetic drift.

Acknowledgements

Funding for this study was from the New York Zoological Society. Some of the data were collected, analyzed and written up during the tenure of grants from the H. F. Guggenheim Foundation and the National Geographic Society (No. 2929–84) to T. T. Struhsaker for studies of redtail monkey behavioural ecology. Permission to study in Kibale was granted by the President's Office of Uganda, Uganda National Research Council, and the Uganda Forest Department. The Department of Zoology, Makerere University, provided our local affiliation in Uganda. The assistance of all these parties is gratefully acknowledged. Special thanks go to Lysa Leland for valuable comments and discussion.

Notes added in proof—December 1987

1. SOB was still alive (female, 7 yrs old). Her hybrid mother MBT and blue father RD had not reappeared.

2. Hybrid adult male H was seen in the redtail East group: no redtail backcrosses were seen.

3. Hybrid female PT and her offspring OPTII (3 yrs old) were seen. PT has given birth to another infant in late August (i.e. 38 months after the previous infant OPTII).

Concluding comments: problems, old and new

THE EDITORS

To retrace the evolutionary biology of any group of living organisms needs a fairly good knowledge of its fossil record and past living conditions, and an understanding of both the internal and the external factors that 'shape' its evolution, within the constraints of its own history.

The evolutionary biology of primates in general, and of guenons in particular, makes no exception to this rule, and this justifies our attempt to bring together in this book the viewpoints of so many disciplines of the life sciences. We are not in a position to say whether this initiative was premature or not, but we are confident that the cross-fertilization of ideas, which took place at Paimpont and has since continued, has, to some extent at least, clarified the situation, and will contribute to stimulating new lines of research.

Before commenting any further on some of the most salient points made in this book, two general remarks must be made. First of all, when trying to interpret any trait of the life history of living animals, it is mandatory to adopt a truly integrated approach. African guenons' biology will always be better understood if past is not divorced from present, form from function, and behaviour from ecology. This principle is well illustrated by most chapters of this book.

Secondly, we must always remain aware of the dangers of premature generalizations. Of course, any student of natural phenomena is, quite naturally, 'looking for patterns' when facing a complex situation, and any devotee of theoretical biology dreams of 'models' with the largest possible heuristic value. However, nature seldom obeys simple rules; most of the time, there are alternative solutions to what we consider ourselves to be simple problems – if only because all taxa have neither the same genetic potential, nor the same history.

These provisos being kept in mind, what kind of conclusions can be drawn from this provisional 'state of the art' review, and what kind for future research do the foregoing chapters suggest?

The guenons' fossil record still remains sparse, and most fossil specimens are fragmentary. The convergence of viewpoints between paleontologists is therefore all the more remarkable. M. Leakey (Chapter 1), M. Pickford and B. Senut (Chapter 3) all consider that the evidence at our disposal suggests that early cercopithecoids were probably semi-terrestrial and inhabited woodland and open country, their modern diversity in forests representing a geologically recent phenomenon. In all probability, the large number of cycles of major forest spread and retreat during the last 2–3 Myr, as discussed by A. Hamilton (Chapter 2), favoured the colonization of rain forests by guenons, and their recent radiation. But the ecological determinants of forest colonization by early cercopithecoids still remain obscure. Did they find at the forest margins, and later in the forest galleries and refuges, the kind of food they preferred, while their original woodlands and savannas were turning rapidly into drier steppes, unable to sustain their populations during a large part of the year? This is likely. In this perspective, both the use of palaeoclimatological indicators, and a better knowledge of the ecology of marginal *Cercopithecus aethiops* populations, and of the semi-terrestrial forest guenons, should help to clarify the problem.

Strange as it may appear, the very knowledge of the recent, and even present-day distribution of many forms of guenons remains far from satisfactory, as shown by M. Colyn, J. M. Lernould and J. Oates' chapters. There is still room for a good deal of field work. Western and central Africa having for a long time been considered as the 'White Man's Grave', zoological investigations here lagged behind those carried out in the neotropics and southeast Asia. In most parts of Africa, detailed studies of primate distribution did not in fact begin before the middle of the present century. Unfortunately, the original distribution of the various forest categories had, already at that time, been greatly altered by human activities, and was but a poor reflection of what it was a century or two before. The situation has again deteriorated since that time, and it is often no longer possible to retrace the original limits of distribution and habitat of some forms (see J. Chism and T. Rowell, Chapter 21).

The interpretation of the present geographical distribution of the various forms raises further problems. The role of the various kinds of barriers to dispersal, and the location of the former Pleistocene refuge areas, are not always easy to establish. Let us consider, for instance, the case of large rivers. Such watercourses might influence the dispersal of forest monkeys in two different ways. In a very large forested area, they do act as true barriers to dispersal for primate species unable to swim. But when river flow was reduced during dry climatic episodes, they were more easily crossed; moreover, the gallery forests along their banks acted as refuges for many forest species, which later expanded their ranges when climatic conditions improved. This is what probably happened in the Zaïre Major Fluvial Refuge. Similarly, a mountain range whose 'alpine' zone now represents an absolute barrier to dispersal for forest species, may well have been the core area of a forest refuge during a wet episode when it was completely covered by forests.

Besides the natural barriers, those due to human activities are also very effective in preventing guenons' dispersal and changing the structure of guenon communities. Some areas have been so repeatedly cleared for cultivation and burned by traditional man that they have been transformed into more or less open forest–savanna mosaics. We do know that in such situations some high forest species like the diana monkey very quickly disappear, whereas others thrive in the second growth vegetation. It is therefore fitting to remember that all rain forest refugia cannot be attributed to the climatic effects of Pleistocene glaciations; some of them, at least, might be due to more recent causes (Connor, 1986, A. Hamilton, Chapter 2).

The three chapters of this book dealing with the classification and distribution of guenons not only emphasize how difficult it is to reach a taxonomic consensus in this highly variable group, but also raises a number of fundamental questions on speciation processes. The status of many forms, of subspecific and even specific rank, still remains ill defined, to say the least; it is therefore hard to avoid the conclusion that guenons, the forest species in particular, are still in an active stage of speciation.

This gives an added value to the contributions of the geneticists (M. Ruvolo, T. Turner *et al.*, B. Dutrillaux *et al.*, Chapters 7–9). Although no more diverse serologically than other genera of Cercopithecidae, the supergenus *Cercopithecus* is exceptional among primates in its extreme karyotypic heterogeneity, especially its arboreal forms. The number of chromosomes steadily increases from the most 'primitive' (which include the 'satellite' subgenera *Allenopithecus*, *Miopithecus*, and *Erythrocebus*) to the most variable forms of arboreal guenons. Among the latter, the evolution of three 'species clusters' (the *mona*, *cephus* and *nictitans* superspecies) have probably taken place very recently in the same forest environments. The likelihood of such a 'sympatric' speciation episode has led Dutrillaux and his co-workers to propose the concept of 'populational evolution', according to which a common pool of chromosomal modifications might have been divided unevenly into a number of sympatric populations, with hybridization occurring between different karyotypes. This is made all the more plausible as representatives of these 'incipient species' often live in occasional or even semi-stable multi-specific groups whose member species sometimes interbreed in natural conditions. These hybrids can, in turn, backcross with one or both parent species, and produce fertile offspring. This particular kind of introgressive hybridization could increase the spread of genes from one form into the gene pool of another and modify the rate of evolution within the species cluster. The observations of T. Struhsaker *et al.* (Chapter 24) suggest that female hybrids of redtails and blue monkeys enjoy distinct advantages. If this is confirmed, and substantiated by observations on other species, this would add weight to Dutrillaux's model of speciation.

In any case, the little we know at present of the possible role of hybridization in primate speciation processes should stimulate field workers to look more systematically for wild hybrid individuals, and encourage studies on captive hybrids. Moreover a study of the 'strength' of barriers both genetic and behavioural, should be initiated. It is indeed likely that complete isolation between members of an actively speciating species cluster does not take place suddenly but by steps. As long as a population of 'incipient species' has some possibility left to interbreed with related forms, the door is left open for more variation to occur. Rich and heterogeneous environments such as tropical rain forests, where a number of closely related forms can coexist in adjacent very similar niches should favour such a process – all the more so when the behavioural barriers to interspecific mating are far from absolute.

One of the unpredicted results of the various studies here presented by geneticists, morphologists (J. Kingdon; R. Martin & A. MacLarnon, Chapters 10, 11), and behavioural scientists (J.-P. Gautier, Chapter 12), is their agreement as regards the probable phylogeny of the group. Minor discrepancies excepted, all of them agree to consider the savanna guenons and the semi-terrestrial forest species as ancestral, and the true forest guenons as recently derived. This view entirely agrees with that of paleontologists. This is all the more remarkable since these authors reached their conclusions quite independently.

The structure and dynamics of natural populations of guenons remain the least known aspects of their natural history, with the partial exception of the vervet (L. Fedigan & L. Fedigan, Chapter 20). In most cases we do not even know whether the populations studied are increasing, decreasing or are more or less stable. This is very unfortunate, as a fair knowledge of the demography of specific populations living in undisturbed environments is necessary to follow each cohort of individuals to see how it reacts to short-term environmental changes. An understanding of how population structure and dynamics might influence the evolution of individual life history traits and of emergent patterns at the population level is also dependent upon demographic knowledge.

Unfortunately, the prospects of rapidly improving this situation remain rather dim. The demography of long-lived mammals, particularly arboreal ones living in 'closed' rain forest habitats, is notoriously difficult to study. Unless one works with individuals of known age, regularly followed by reliable teams of observers, there is little hope of accurately assessing the sex-ratio of troop members at different ages, their age–specific birth and death rates, and the emigration and immigration rates.

Long-term studies of protected guenon populations in natural habitats will also be necessary to understand the factors regulating population and troop sizes. However, repeated census will provide significant results only if they are carried out at the same time as studies of resource availability. This means that a closer cooperation with plant and animal ecologists working at the same study sites is to be encouraged. The most recent investigations carried out in the humid tropics all definitely suggest that the quality of food is as important as its amount and availability for predominantly frugivorous animals such as guenons (A. Gautier-Hion, Chapter 15). This is particularly obvious during the 'lean seasons' corresponding to the drier months of the year. At such times, a few 'keystone plant resources' are able to sustain the frugivorous mammal and bird communities through the period of seasonal scarcity, and in so doing they set the carrying capacity of the forest. In Peruvian Amazonia, these 'emergency rations' are basically represented by palm nuts, figs and nectar (Terborgh, 1986). However, the situation is probably different for forest guenons as there are only about 50 palm species in the whole of Africa, against 1140 and 1150 respectively for the neotropics and the indo-australian region (Corner, 1966) and figs are seldom eaten by *Cercopithecus*, at least in West Africa. It is worth noticing, for example, that among the 57 kinds of fruit seen eaten for an annual cycle, by *C. cephus*, not a single species of *Ficus* or of palm is mentioned (Sourd & Gautier-Hion, 1986). There are also important differences in species richness of plants between different African sites: for instance, over 100 tree species per hectare were enumerated near Makokou (Gabon), and only 33 in the Kibale Reserve (Uganda). Obviously the 'keystone plant resources' must greatly vary from one site to the other. As a matter of fact, the whole question of guenon nutrition needs to be given increased consideration. The ontogeny of feeding habits, the determinants of food selection by adults at different seasons of the year, and in various physiological conditions (pregnancy and lactation) remain to be studied and, here again, field observations must be combined with experiments in controlled captive conditions.

T. Butynski's contribution on seasonality of reproduction represents a definite advance in our knowledge of guenons' reproductive biology. The fact that all species so far studied have birth seasons and peaks, although some populations exhibit year-round mating and birth in areas of high rainfall, is particularly interesting. It provides a further argument in favour of the importance of the harshness and duration of the 'lean season', even in the apparently most favourable environments. Much more field and laboratory work needs to be carried out, however, on guenon reproduction. We still need to disentangle better the respective roles of climatic, nutritional and social factors in determining the onset of the breeding seasons. Many other important questions also need to be answered. How can synchronous female receptivity be explained? Do kin relationships play a role in mate selection? To what extent, and in which ways, is excessive inbreeding prevented? What is the cost of reproduction to females and males? What are the lifetime reproductive successes of both sexes?

The section on social behaviour is one of the lengthiest of the present book, and it emphasizes the attention paid by field workers to this aspect of the natural history of guenons during the past two decades. Many classic viewpoints on the social organization and social structure have now been called into question. More than ever, diversity, and to some extent flexibility, must remain key words for those studying the socio-ecology of this group of monkeys. The former distinction made between small 'closed' uni-male troops in forest guenons, and larger, more 'open', often multi-male troops in the *aethiops* superspecies, needs to be seriously qualified. The transfer of adult males, and sometimes females, from one group to the other seems to be the rule rather than the exception, ensuring a regular gene flow between reproductive units. Although there is usually only one fully adult male in each group of 'harem polygynous' guenons, as many as six can be present at any one time on a temporary basis in the Kibale redtails (T. Struhsaker, Chapter 18). Male tenure is extremely variable in duration too, ranging from less than one month to at least five years. Among the redtails and blue monkeys of the Kakamega Forest (Kenya), individual males may mate in at least two different troops within a single breeding season, and over a period of two to three years (M. Cords, Chapter 17). It is therefore next to impossible to know which male actually sired offspring in such social groups. The option for non-resident adult males to live temporarily in rather loose, widely spread associations (T. Rowell, Chapter 22) is worth more attention.

Another important problem, far from being settled, is what determines the splitting of troops in forest and savanna guenons. The observations made on Kibale redtails (T. Struhsaker & L. Leland, Chapter 19) suggest that once a troop's nutritional needs exceed the carrying capacity of its home range, it tends to split in two 'daughter troops'. The very fact that the smaller daughter troop, whose size and home range were approximately half those of the larger one, doubled its reproductive rate over the two years following splitting, while the reproductive rate of the larger daughter troop remained the same, suggest that the original troop had overexploited the resources of its former range. However, as long as actual measurements of both food consumption and food availability are not carried out, such a conclusion cannot be considered as more than plausible speculation. Besides the ecological determinants of group splitting, the social determinants must not be overlooked; the nature and quality of relationships between troop members are continuously changing, and these changes may also lead to troop fission.

The frequency with which forest guenons live in polyspecific groups is one of their most remarkable characteristics (A. Gautier-Hion, Chapter 23). These mixed troops have now been observed everywhere in the African rain forests, from western to eastern Africa, through Gabon and Zaïre. Many of them have little in common with more or less random temporary mixing of sympatric species. They are semi-stable associations of particular species, some of which are even more often in mixed parties than in monospecific troops. Such long-lasting associations only occur between species closely related ecologically and behaviourally, and a kind of supraspecific social organization apparently may exist. As shown by the observations carried out during the last few years, the polyspecific lifestyle has some definite advantages for associated species; the polyspecific groups make more efficient use of the resources of their common home range, and more easily detect the approach of terrestrial and aerial predators. An important side-effect of polyspecific life is that, at least in certain demographic and distribution contexts, it facilitates hybridization.

Obviously, we are still far from hearing the last word on the evolutionary biology of African guenons, but progress has been made in some promising directions. We are now reasonably certain that this group of monkeys is still in an active stage of evolution, and some of the highly variable forest forms belong to an incipient stage of speciation. This has a correspondence with an unusual karyotypic heterogeneity, a definite tendency to live in mixed troops, and a propensity to hybridize. Are all these traits causally related, and to what extent? This needs now to be explored more closely. The study of 'evolution in action' in a group of living primates no longer remains an inaccessible dream . . . as long as monkeys and their habitats are and continue to be protected.

References

Adamson, D. A., Gasse, F., Street, F. A. & Williams, M. A. J. (1980). Late Quaternary history of the Nile. *Nature, London*, **288**, 50–5.

Agwu, C. D. C. & Beug, H. J. (1984). Palynologische Untersuchungen an marinen Sedimenten vor der Westafrikanischen Küste. *Palaeoecology of Africa*, **16**, 37–52.

Aldrich-Blake, F. P. G. (1968). A fertile hybrid between two *Cercopithecus* species in the Budongo Forest, Uganda. *Folia primatologica*, **9**, 15–21.

Aldrich-Blake, F. P. G. (1970). The ecology and behaviour of the blue monkey *Cercopithecus mitis stuhlmanni*, PhD dissertation, University of Bristol.

Alexander, R. D. (1974). The evolution of social behavior. *Annual Review of Ecology and Systematics*, **5**, 324–82.

Allen, J. A. (1925). Primates collected by the American Museum Congo Expedition. *Bulletin of the American Museum of Natural History*, **47**, 283–499.

Allen, R. C. & Buettner-Janusch, J. (1973). Red cell and serum proteins of patas monkeys, *Cercopithecus* (= *Erythrocebus*) *patas*. *Folia primatologica*, **20**, 321–30.

Altmann, S. A. (1962). A field study of the sociobiology of rhesus monkeys, *Macaca mulatta*. *Annals of the New York Academy of Science*, **102**, 338–435.

Altmann, S. A. (1974). Baboons, space, time, and energy. *American Zoologist*, **14**, 221–48.

Amtmann, E. (1966). Zur Systematik Afrikanischer Streifenhörnchen der Gattung *Funisciurus*. Ein Beitrag zur Problematik klimaparalleler Variation und Phänetik. *Bonner zoologische Beiträge*, **17**, 1–44.

Andrew, R. J. (1963). The origin and evolution of the calls and facial expressions of the Primates. *Behaviour*, **20**, 1–109.

Andrew, R. J. (1965). The origins of facial expressions. *Scientific American*, **213**, 88.

Andrews, P. J. (1981a). The origin of the Cercopithecoidea. In *Catalogue of Primates in the British Museum (Natural History) and elsewhere in the British Isles, II: Family Cercopithecidae, Subfamily Cercopithecinae*, ed. P. H. Napier, pp. 202. London: British Museum (Natural History).

Andrews, P. J. (1981b). Species diversity and diet in monkeys and apes during the Miocene. In *Aspects of human evolution*, ed. C. B. Stringer, pp. 25–62. London: Taylor & Francis.

Andrews, P. J. & Aiello, L. (1984). An evolutionary model for feeding and positional behaviour. In *Food acquisition and processing in Primates*, ed. D. J. Chivers, B. A. Wood & A. Bilsborough, pp. 429–66. New York: Plenum Press.

Andrews, P. J., Groves, C. P. & Horne, J. F. M. (1975). Ecology of the Lower Tana River Floodplain (Kenya). *Journal of the East African Natural History Society*, **151**, 1–31.

Andrews, P. J. & Van Couvering, J. A. H. (1975). Palaeoenvironments in the East African Miocene. In *Approaches to Primate Paleobiology, Contributions to Primatology*, Vol. 5, ed. F. S. Szalay, pp. 62–103. Basel: Karger.

Arambourg, C. (1959). Vertébrés continentaux du Miocène supérieur de l'Afrique du Nord. *Mémoire du Service de la Carte Géologique, Algérie*, nouvelles séries, Paléontologie, **4**, 1–161.

Arnold, M. L., Honeycutt, R. L., Baker, R. J., Sarich, V. M. & Jones, J. K. Jr (1982). Resolving a phylogeny with multiple data sets: a systematic study of phyllostomoid bats. *Occasional Papers of the Museum, Texas Technical University*, **77**, 1–15.

Assémien, P., Filleron, J. C., Martin, L. & Tastet, J. P. (1970). Le Quaternaire de la zone littorale de Côte d'Ivoire. *Bulletin Asequa*, **25**, 65–78.

Atlas de la République unie du Cameroun (1979). Paris: l'Institut Géographique National.

Atlas National du Sénégal (1977). Paris: l'Institut Géographique National.

Atlas of Uganda (1967). 2nd edition. Entebbe: Lands and Surveys Department.

Aubréville, A. (1949). *Climats, forêts et désertification de l'Afrique tropicale*. Paris: Société d'Editions Géographiques, Maritimes et Coloniales.

Avise, J. C. (1976). Genetic differentiation during speciation. In *Molecular evolution*, ed. F. Ayala, Sunderland, MA: Sinauer Press.

Barash, D. P. (1977). *Sociobiology and behavior*. New York: Elsevier.

Barnicot, N. A. & Hewett-Emmett, D. (1971). Red cell and serum proteins of talapoin, patas, and vervet monkeys. *Folia primatologica*, **15**, 65–76.

Basckin, D. R. & Krige, P. D. (1973). Some preliminary discussions on the behaviour of an urban group of vervet monkeys (*Cercopithecus aethiops*) during the birth season. *Journal of Behavioural Science*, **1**, 287–96.

Bauchot, R. (1982). Brain organization and taxonomic relationships in Insectivora and Primates. In *Primate brain evolution: Methods and concepts*, ed. E. Armstrong & D. Falk, pp. 163–75. New York: Plenum Press.

Baverstock, P. R., Cole, S. R., Richardson, B. J. & Watts, C. H. S. (1979). Electrophoresis and cladistics. *Systematic Zoology*, **28**, 214–20.

Bearder, S. K. & Martin, R. D. (1980). *Acacia* gum and its use by bushbabies, *Galago senegalensis* (Primates: Lorisidae). *International Journal of Primatology*, **1**, 103–28.

Becker, P. H. (1982). Species specificity in bird sounds. In *Acoustic communication in birds*, ed. D. E. Kroodsma & E. H. Miller, pp. 214–44. New York: Academic Press.

Benefit, B. R. & Pickford, M. (1986). Miocene fossil cercopithecoids from Kenya. *American Journal of Physical Anthropology*, **69**, 441–64.

Benoist, J. (1981). Morphological distances between Israeli groups of various origin. *Journal of Human Evolution*, **10**, 599–603.

Bernstein, I. S. (1970). Some behavioural elements of Cercopithecoidea. In *Old World monkeys: evolution, systematics and behaviour*, ed. J. R. Napier and P. H. Napier, pp. 263–95. New York: Academic Press.

Bertram, B. C. R. (1980). Vigilance and group size in ostriches. *Animal Behaviour*, **28**, 278–86.

Bertrand, M. (1969). The behavioural repertoire of the stumptail Macaque. *Bibliotheca Primatologica*, **11**. Basel: Karger.

Bielicki, T. & Charzewksi, J. (1977). Sex differences in the magnitude of statural gains of offsprings over parents. *Human Biology*, **49**, 265–77.

Birchette, M. G. (1981). Postcranial remains of *Cercopithecoids*. *American Journal of Physical Anthropology*, **54**, 201.

Bishop, W. W., Miller, J. A. & Fitch, F. J. (1969). New Potassium-Argon age determinations relevant to the Miocene fossil mammal sequence in East Africa. *American Journal of Science*, **267**, 669–99.

Blancou, L. (1958). Notes biogéographiques sur les Mammifères de l'A.E.F. *Bulletin de l'Institut d'Etudes Centrafricaines*, Brazzaville, No. 15–16, 7–42.

Bonnefille, R. (1976). Palynological evidence for an important change in the vegetation of the Omo basin between 2.5 and 2 million years ago. In *Earliest man and environments in the Lake Rudolf basin*, ed. Y. Coppens, F.C. Howell, G. L. Isaac & R. E. F. Leakey, pp. 421–31. Chicago: University of Chicago Press.

Bonnefille, R. (1979). Méthode palynologique et reconstitutions paléoclimatiques au Cénozoique dans le Rift Est Africain. *Bulletin de la Société Géologique de France*, **21**, 331–42.

Bonnefille, R. (1983). Evidence for a cooler and drier climate in the Ethiopian uplands towards 2.5 Myr ago. *Nature, London*, **303**, 487–91.

Bonnefille, R. (1984). The evolution of the East African environment. In *The evolution of the East Asian environment, Vol. II*, ed. R. O. Whyte, pp. 579–612. University of Hong Kong: Centre for Asian Studies.

Bonnefille, R. & Letouzey, R. (1976). Fruits fossiles d'*Antrocaryon* dans la vallée de l'Omo (Ethiopie). *Adansonia*, **16**, 65–82.

Bookstein, F. L. (1978). *The measurement of biological shape and shape change*. Berlin: Springer-Verlag.

Booth, A. H. (1954a). A note on the colobus monkeys of the Gold and Ivory Coasts. *Annals and Magazine of Natural History*, 12th series, **7**, 857–60.

Booth, A. H. (1954b). The Dahomey Gap and the mammalian fauna of the West African forests. *Revue de Zoologie et Botanique Africaines*, **50**, 305–14.

Booth, A. H. (1955). Speciation in the mona monkeys. *Journal of Mammalogy*, **36**, 434–49.

Booth, A. H. (1956a). The distribution of primates in the Gold Coast. *Journal of the West African Science Association*, **2**, 122–33.

Booth, A. H. (1956b). The Cercopithecidae of the Gold and Ivory Coasts: geographic and systematic observations. *Annals and Magazine of Natural History*, 12th series, **9**, 476–80.

Booth, A. H. (1957). Observations on the natural history of the olive colobus monkey, *Procolobus verus* (van Beneden). *Proceedings of the Zoological Society of London*, **129**, 421–30.

Booth, A. H. (1958a). The Niger, the Volta and the Dahomey Gap as geographical barriers. *Evolution*, **12**, 48–62.

Booth, A. H. (1958b). The zoogeography of West African primates: a review. *Bulletin de l'Institut français d'Afrique noire*, série A, **20**, 587–622.

Booth, C. P. (1968). Taxonomic studies of *Cercopithecus mitis* Wolf (East Africa). *National Geographic Society Research Reports*, **37**, 51.

Bouchain, C. (1985). Etude comparée des vocalisations de cohésion-contact chez *Cercopithecus pogonias* et chez *Cercopithecus ascanius* et leurs deux hybrides. Diplôme d'études, Université de Rennes.

Bourlière, F. (1979). Significant parameters of environmental quality for nonhuman primates. In *Primate ecology and human origins*, ed. I. S. Bernstein & E. O. Smith, pp. 23–46. New York: Garland Press.

Bourlière, F. (1985). Primate communities: their structure and role in tropical ecosystems. *International Journal of Primatology*, **6**, 1–26.

Bourlière, F. & Hadley, M. (1970). The ecology of tropical savannas. *Annual Review of Ecology and Systematics*, **1**, 125–52.

Bourlière, F., Hunkeler, C. & Bertrand, M. (1970). Ecology and behaviour of Lowe's guenon (*Cercopithecus campbelli lowei*) in the Ivory Coast. In *Old World monkeys: evolution, systematics and behaviour*, ed. J. R. Napier & P. H. Napier, pp. 297–350. New York: Academic Press.

Bourlière, F., Morel, G. & Galat, G. (1976). Les grands mammifères de la Basse Vallée du Sénégal et leurs saisons de reproduction. *Mammalia*, **40**, 401–12.

Bramblett, C. A., DeLuca Pejaver, L. & Drickman, D. J. (1975). Reproduction in captive vervet and Sykes' monkeys. *Journal of Mammalogy*, **56**, 940–6.

Brennan, E. J., Else, J. G. & Altmann, J. (1985). Ecology and behaviour of a pest primate: vervet monkeys in a tourist lodge habitat. *African Journal of Ecology*, **23**, 35–44.

Bristowe, W. S. (1929). Mating habits of spiders with special reference to problems surrounding sexual dimorphism. *Proceedings of the Zoological Society*, London **188**, 389.

Brockelman, W. Y. & Gittins, S. P. (1984). Natural hybridization in the *Hylobates lar* species group: implications for speciation in gibbons. In *The lesser apes: Evolutionary and behavioural biology*, ed. H. Preuschoft, D. J. Chivers, W. J. Brockelman & N. Creel, pp. 498–532. Edinburgh: Edinburgh University Press.

Brown, F. H. & Feibel, C. S. (1986). Revision of the lithostratigraphic nomenclature in the Koobi Fora region, Kenya. *Journal of the Geological Society, London*, **143**, 297–310.

Brown, W. M., Prager, E. M., Wang, A. & Wilson, A. C. (1982). Mitochondrial DNA sequences of primates: tempo and mode of evolution. *Journal of Molecular Evolution*, **18**, 225–39.

Burton, F. D. (1972). Anatomy of the hand in Catarrhine monkeys as related to manipulative behaviour. *Dissertation Abstracts International*, **30B**, 3957.

Butynski, T. M. (1982a). Vertebrate predation by primates: a review of hunting patterns and prey. *Journal of Human Evolution*, **11**, 421–30.

Butynski, T. M. (1982b). Harem-male replacement and infanticide in the blue monkey (*Cercopithecus mitis stuhlmanni*) in the Kibale Forest, Uganda. *American Journal of Primatology*, **3**, 1–22.

Butzer, K. W., Isaac, G. L., Richardson, J. L. & Washbourn-Kamau, C. K. (1972). Radiocarbon dating of East African lake levels. *Science, NY*, **175**, 1069–76.

Byrne, R. W. (1981). Distance vocalizations of Guinea baboons (*Papio papio*) in Senegal: an analysis of functions. *Behaviour*, **78**, 283–312.

Calder, W. A. (1984). *Size, function and life history*. Cambridge, Mass: Harvard University Press.

Cambefort, J. P. (1981). A comparative study of culturally transmitted patterns of feeding habits in the chacma baboon, *Papio ursinus*, and the vervet monkey, *Cercopithecus aethiops*. *Folia primatologica*, **36**, 243–63.

Carroll, R. W. (1986). Status of the Lowland Gorilla and other Wildlife in the Dzanga-Sangha Region of Southwestern Central African Republic. *Primate Conservation*, **7**, 38–41.

Chalmers, N. R. (1968). The visual and vocal communication of free-living mangabeys in Uganda. *Folia primatologica*, **9**, 258–80.

Chalmers, N. R. (1972). Cooperative aspects of early infant development in some captive Cercopithecines. In *Primate socialisation*, ed. F. Poirier, pp. 63–82. New York: Random House.

Chance, M. R. A. (1956). Social structure of a colony of *Macaca mulatta*. *British Journal of Animal Behaviour*, **4**, 1.

Chance, M. R. A. (1962). An interpretation of some agonistic postures. The role of 'cut-off' acts and postures. *Symposium of the Zoological Society of London*, **8**, 71–99.

Chapin, J. P. (1923). Ecological aspects of bird distribution in tropical Africa. *American Naturalist*, **58**, 106–25.

Chapman, C. A. (1985). The influence of habitat on behaviour in a group of St. Kitts green monkeys. *Journal of Zoology, London*, **206**, 311–20.

Chapman, C. A. & Fedigan, L. M. (1984). Territoriality in the St. Kitts vervet, *Cercopithecus aethiops*. *Journal of Human Evolution*, **13**, 677–86.

Charles-Dominique, P. (1977). *Ecology and behaviour of nocturnal prosimians*. London: Duckworth.

Charney, J. G., Stone, P. H. & Quirk, W. J. (1975). Drought in the Sahara: a biogeophysical feedback mechanism. *Science, NY*, **187**, 434–5.

Cheney, D. L. (1981). Intergroup encounters among free-ranging vervet monkeys. *Folia primatologica*, **35**, 124–46.

Cheney, D. L., Lee, P. C. & Seyfarth, R. M. (1981). Behavioral correlates of non-random mortality among free-ranging female vervet monkeys. *Behavioral Ecology & Sociobiology*, **9**, 153–61.

Cheney, D. L. & Seyfarth, R. M. (1980). Vocal recognition in free-ranging vervet monkeys. *Animal Behaviour*, **28**, 362–7.

Cheney, D. L. & Seyfarth, R. M. (1981). Selective forces affecting the predator alarm calls of vervet monkeys. *Behaviour*, **76**, 25–61.

Cheney, D. L. & Seyfarth, R. M. (1983). Nonrandom dispersal in free-ranging vervet monkeys: social and genetic consequences. *The American Naturalist*, **122**, 392–412.

Chepko-Sade, B. D. & Sade, D. S. (1979). Patterns of group splitting within matrilineal kinship groups, a study of social group structure in *Macaca mulatta* (Cercopithecidae: Primates). *Behavioural Ecology and Sociobiology*, **5**, 67–86.

Cherry, L. M., Case, S. M., Kunkel, J. G. & Wilson, A. C. (1979). Comparison of frogs, humans and chimpanzees. *Science, NY*, **204**, 435.

Cherry, L. M., Case, S. M., Kunkel, J. G., Wyles, J. S. & Wilson, A. C. (1982). Body shape metrics and organismal evolution. *Evolution*, **38**, 914–33.

Cherry, L. M., Case, S. M. & Wilson, A. C. (1978). Frog perspective on the morphological difference between humans and chimpanzees. *Science, NY*, **200**, 209–11.

Chism, J., Olson, D. K. & Rowell, T. E. (1983). Diurnal births and perinatal behavior among wild patas monkeys: Evidence of an adaptive pattern. *International Journal of Primatology*, **4**, 167–84.

Chism, J. & Rowell, T. E. (1986). Mating and residence patterns of male patas monkeys. *Ethology*, **72**, 31–9.

Chism, J., Rowell, T. E. & Olson, D. K. (1984). Life history patterns of female patas monkeys. In *Female Primates: Studies by women primatologists*, ed. M. Small, pp. 175–92. New York: Alan R. Liss.

Chivers, D. J. & Hladik, C. M. (1980). Morphology of the gastro-intestinal tract in Primates: comparisons with other mammals in relation to diet. *Journal of Morphology*, **166**, 337–86.

Chivers, D. J., Wood, B. A. & Bilsborough, A. (1984). *Food acquisition and processing in Primates*. New York: Plenum Press.

Church, R. J. Harrison (1980). *West Africa*, 8th edition. London: Longman.

Clausen, H. S. (1964). Notes on the distribution of West African cyprinodonts, with special reference to a hitherto unnoticed ichthyogeographical boundary. *Videnskabelig Meddelelser fra Dansk Naturhistorisk Forening i Kjobenhavn*, **126**, 323–6.

CLIMAP project members (1976). The surface of the Ice-Age Earth. *Science, NY*, **191**, 1131–6.

CLIMAP project members (1984). The last interglacial ocean. *Quaternary Research*, **21**, 123–224.

Clutton-Brock, T. H. (1972). Feeding and ranging behaviour in the red colobus. PhD dissertation, University of Cambridge.

Clutton-Brock, T. H. (ed.) (1977a). *Primate ecology: studies of feeding and ranging behaviour in lemurs, monkeys and apes*. New York: Academic Press.

Clutton-Brock, T. H. (1977b). Some aspects of intraspecific variations in feeding and ranging behaviour in Primates. In *Primate ecology: studies of feeding and ranging behaviour in lemurs, monkeys and apes*, ed. T. H. Clutton-Brock, pp. 539–66. New York: Academic Press.

Clutton-Brock, T. H., Guinness, F. E. & Albon, S. D. (1982). *Red deer, behavior and ecology of two sexes*. Chicago: University of Chicago Press.

Clutton-Brock, T. H. & Harvey, P. H. (1977). Primate ecology and social organisation. *Journal of Zoology, London*, **183**, 1–39.

Coe, M. J., Cumming, D. H. & Phillipson, J. (1976). Biomass and production of large African herbivores in relation to rainfall and primary production. *Oecologia*, **22**, 341–54.

Coetzee, J. A. (1967). Pollen analytical studies in East and Southern Africa. *Palaeoecology of Africa*, **3**, 1–146.

Colyn, M. M. (1986). Les Mammifères de forêt ombrophile entre les rivières Tshopo et Maïko (Région du Haut-Zaïre). *Bulletin de l'Institut royal des Sciences naturelles de Belgique: Biologie*, **56**, 21–6.

Colyn, M. M. (1987). Les Primates des forêts ombrophiles de la Cuvette du Zaïre: Interprétations zoogéographiques des modèles de distribution. *Revue de Zoologie africaine*, **101**, 183–96.

Colyn, M. M. & Verheyen, W. N. (1987a). Distributions géographiques relatives de *Cercopithecus wolfi elegans* et de *Cercopithecus wolfi wolfi* dans le bassin du Zaïre – remarques sur les limites respectives de ces deux sous-espéces avec celles d'autres formes de la superespèce *mona*. *Revue de Zoologie africaine*, **101**, in press.

Colyn, M. M. & Verheyen, W. N. (1987b). Considérations sur la provenance de l'holotype de *Cercopithecus mitis maesi* Lönnberg, 1919 (Primates, Cercopithecidae) et description d'une nouvelle sous-espèce: *Cercopithecus mitis heymansi*. *Mammalia*, **51**, 271–81.

Colyn, M. M. & Verheyen, W. N. (1987c). *Colobus rufomitratus parmentieri*, une nouvelle sous-espèce du Zaïre (Primates, Cercopithecidae). *Revue de Zoologie africaine*, **101**, 126–32.

Connor, E. F. (1986). The role of Pleistocene forest refugia in the evolution and biogeography of tropical biotas. *Trends in Ecology and Evolution*, **1**, 165–8.

Coppinger, R. P. & Maguire, J. P. (1980). *Cercopithecus aethiops* of St. Kitts. A population estimate based on human predation. *Caribbean Journal of Science*, **15**, 1–7.

Corbet, G. B. & Hanks, J. (1968). A revision of the elephant shrews, family Macroscelididae. *Bulletin of the British Museum: Natural History, Zoology*, **16**, 47–111.

Cords, M. (1984). Mating patterns and social structure in redtail monkeys (*Cercopithecus ascanius*). *Zeitschrift für Tierpsychologie*, **64**, 313–29.

Cords, M. (1987a). Forest guenons and patas monkeys: male–male competition in one-male groups. In *Primate Societies*, ed. B. B. Smuts, D. L. Cheney, R. M. Seyfarth, R. W. Wrangham & T. T. Struhsaker, pp. 98–111. Chicago: University of Chicago Press.

Cords, M. (1987b). Mixed-species association of *Cercopithecus* monkeys in the Kakamega forest. *University of California: Publications in Zoology*, **117**, pp. 1–109.

Cords, M., Mitchell, B. J., Tsingalia, H. M. & Rowell, T. E. (1986). Promiscuous mating among blue monkeys in the Kakamega Forest, Kenya. *Ethology*, **72**, 214–26.

Cords, M. & Rowell, T. E. (1986). Group fission in blue monkeys of the Kakamega Forest, Kenya. *Folia primatologica*, **46**, 70–82.

Cords, M. & Rowell, T. E. (1987). Birth intervals of *Cercopithecus* monkeys of the Kakamega Forest, Kenya. *Primates*, **28**, 277–81.

Corner, E. F. H. (1966). *The natural history of palms*. London: Weidenfeld & Nicholson.

Corruccini, R. S. (1975). Multivariate analysis in biological anthropology: some considerations. *Journal of Human Evolution*, **4**, 1–19.

Cott, H. B. (1940). *Adaptive coloration in animals*. London: Methuen.

Crandall, L. S. (1964). *The management of wild mammals in captivity*. Chicago: The University of Chicago Press.

Crane, J. (1967). Combat and its ritualization in fiddler crabs (Ocypodidae) with species reference to *Uca rapax*. *Zoologica*, **52**, 5876.

Creel, N. & Preuschoft, H. (1984). Systematics of the lesser apes: a quantitative taxonomic analysis of craniometry and other variables. In *The lesser Apes. Evolutionary and behavioural biology*, ed. H. Preuschoft, D. J. Chivers, W. J. Brockelman & N. Creel, pp. 562–613. Edinburgh: Edinburgh University Press.

Creel, N. & Preuschoft, H., Brockelman, W. Y. & Chivers, D. J. (1984). Pathways of speciation: some conclusions. In *The lesser Apes. Evolutionary and behavioural biology*, ed. H. Preuschoft, D. J. Chivers, W. Y. Brockelman & N. Creel, pp. 633–85. Edinburgh: Edinburgh University Press.

Cronin, J. E., Vincent, M. S. (1976). Molecular evidence for dual origin of mangabeys among Old World monkeys. *Nature, London*, **260**, 700–2.

Crook, J. H. (1966). Gelada baboon herd structure and movement. In *Play exploration and territory in mammals*, ed. P. A. Jewell & C. Loizos. London: Symposium of the Zoological Society, 18, 237–58.

Curry-Lindahl, K. (1956). Ecological studies on mammals, birds, reptiles and amphibians in the eastern Belgian Congo. *Annales du Musée royal du Congo belge, Sciences Zoologiques*, Part I, **42**, 1–78.

Dandelot, P. (1959). Note sur la classification des Cercopithèques du groupe *aethiops*. *Mammalia*, **23**, 357–68.

Dandelot, P. (1962). Confrontation d'un Primate africain, *Cercopithecus mitis* Wolf, avec son habitat. *Compte rendu des séances de la Société de Biogéographie*, **343**, 23–35.

Dandelot, P. (1965). Distribution de quelques sous-espèces de Cercopithecidae en relation avec les zones de végétation de l'Afrique. *Zoologica Africana*, **1**, 167–76.

Dandelot, P. (1968). *Preliminary identification manual for Mammals, Primates: Anthropoidea*. Washington DC: Smithsonian Institution Press.

Dandelot, P. (1971). Suborder Anthropoidea. In *The Mammals of Africa: an identification manual*, ed. J. Meester & H. W. Setzer, pp. 5–45. Washington DC: Smithsonian Institution Press.

Dandelot, P. & Prevost, J. (1972). Contribution à l'étude des primates d'Ethiopie (Simiens). *Mammalia*, **36**, 607–33.

Dechamps, R. & Maes, F. (1985). Essai de reconstitution des climats et des végétations de la basse vallée de l'Omo au Plio-Pleistocène à l'aide de bois fossiles. In *L'environnement des hominidés au Plio–Pléistocène*, ed. M. M. Beden *et al.*, pp. 175–222. Paris: Masson.

de Heinzelin, J., Haesaerts, P. & Howell, F. C. (1976). Plio–Pleistocene formations of the lower Omo basin with particular reference to the Shungura Formation. In *Earliest man and environments in the Lake Rudolf Basin*, ed. Y. Coppens, F. C. Howell, G. L. Isaac & R. E. F. Leakey, pp. 24–49. Chicago: University of Chicago Press.

Delany, M. J. & Happold, D. C. D. (1979). *Ecology of African Mammals*. London: Longman.

Delson, E. (1974). The oldest known fossil Cercopithecidae. *American Journal of Physical Anthropology*, **41**, 474–5.

Delson, E. (1975a). Evolutionary history of the Cercopithecidae. In *Approaches to primate palaeobiology. Contributions to Primatology*, Vol. 5, ed. F. Szalay, pp. 167–217. Basel: Karger.

Delson, E. (1975b). Palaeoecology and zoogeography of the Old World monkeys. In *Primate Functional Morphology and Evolution*, ed. R. H. Tuttle, pp. 37–64. The Hague: Mouton.

Delson, E. (1977). A new species of ?*Prohylobates* from the early Miocene of Libya. *American Journal of Physical Anthropology*, (Abstract), **47**, 126.

Delson, E. (1979). *Prohylobates* (Primates) from the early Miocene of Libya: a new species and its implications for cercopithecid origins. *Geobios*, **12**, 725–33.

Denham, W. W. (1981). History of green monkeys in the West Indies. Part I. Migration from Africa. *Journal of the Barbados Museum and Historical Society*, **36**, 211–371.

de Ploey, J. (1968). Quaternary phenomena in the Western Congo. In *Means of correlation of Quaternary successions, 8*, Proceedings of the VIIth International Congress of the Association for Quaternary Research, pp. 501–17. University of Utah Press.

Deputte, B. L. (1986). Ontogénèse du Cercocèbe à joues blanches en captivité (*Lophocebus albigena*). Thèse de Doctorat, Université de Rennes.

Deputte, B. L. & Goustard, M. (1980). Copulatory vocalizations of female macaque (*Macaca fascicularis*): variability factors analysis. *Primates*, **21**, 83–99.

Deshmukh, I. (1986). *Ecology and tropical biology*. Oxford: Blackwell.

DeVore, I. & Washburn, S. L. (1963). Baboon ecology and human evolution. In *African ecology and human evolution*, ed. F. C. Howell & F. Bourlière, pp. 335–67. New York: Wenner Gren Foundation.

Devred, R. (1958). La végétation forestière du Congo belge et du Ruanda-Urundi. *Bulletin de la Société Royale de Botanique de Belgique*, 409–68.

de Waal, F. B. M. & Yoshihara, D. (1983). Reconciliation and redirected affection in rhesus monkeys. *Behaviour*, **85**, 224–41.

Dewsbury, D. A. (1982a). Ejaculate cost and male choice. *American Naturalist*, **119**, 601–10.

Dewsbury, D. A. (1982b). Pregnancy blockage following multiple-male copulation or exposure at the time of mating in deer mice, *Peromyscus maniculatus*. *Behavioural Ecology and Sociobiology*, **11**, 37–42.

Diamond, A. W. & Hamilton, A. C. (1980). The distribution of forest passerine birds and Quaternary climatic change in tropical Africa. *Journal of Zoology, London*, **191**, 379–402.

Dingle, H. & Khamala, C. P. M. (1972). Seasonal changes in insect abundance and biomass in an East African grassland with reference to breeding and migration in birds. *Ardea*, **60**, 216–21.

Dittus, W. P. J. (1984). Toque macaque food calls: a semantic communication concerning food distribution in the environment. *Animal Behaviour*, **32**, 470–7.

Dobzhansky, T. (1970). *Genetics of the evolutionary process.* New York: Columbia University Press.

Dorst, J. & Dandelot, P. (1970). *A field guide to the larger mammals of Africa.* London: Collins.

Dorward, D. C. & Payne, A. I. (1975). Deforestation, the decline of the horse, and the spread of the tsetse fly and trypanosomiasis (*Nagana*) in nineteenth century Sierra Leone. *Journal of African History*, **16**, 239–56.

Dracopoli, N. C. (1981). Population Genetics of Kenyan Vervet Monkeys (*Cercopithecus aethiops*). PhD thesis, University of London.

Drake, R., van Couvering, J. A., Pickford, M. & Curtis, G. (1987). K–Ar geochronology of Early Miocene volcanic strata and associated vertebrate and early hominoid fossil localities: Rusinga and Mfwangano Islands, Uyoma Peninsula and Karungu, Western Kenya. *Quarterly Journal of the Geological Society of London*, in press.

Dunbar, R. I. M. (1974). Observations on the ecology and social organization of the green monkey, *Cercopithecus sabaeus*, in Senegal. *Primates*, **15**, 341–50.

Dunbar, R. I. M. (1977). Feeding ecology of Gelada baboon: a preliminary report. In *Primate ecology: Studies of feeding and ranging behaviour in lemurs, monkeys and apes*, ed. T. H. Clutton-Brock, pp. 252–73. London: Academic Press.

Dunbar, R. I. M. (1979). Population demography, social organization, and mating strategies. In *Primate ecology and human origins*, ed. I. S. Bernstein & E. O. Smith, pp. 65–88. New York: Garland.

Dunbar, R. I. M. (1980). Demographic and life history variables of a population of gelada baboons (*Theropithecus gelada*). *Journal of Animal Ecology*, **49**, 485–506.

Dunbar, R. I. M. (1984). *Reproductive decisions, an economic analysis of gelada baboon social strategies.* Princeton: Princeton University Press.

Dunbar, R. I. M. (1987). Demography and reproduction. In *Primate societies*, ed. B. Smuts, D. L. Cheney, R. M. Seyfarth, R. W. Wrangham & T. T. Struhsaker, pp. 240–9. Chicago: University of Chicago Press.

Dunbar, R. I. M. & Dunbar, E. P. (1974a). Ecological relations and niche separation between sympatric terrestrial primates in Ethiopia. *Folia primatologica*, **21**, 36–60.

Dunbar, R. I. M. & Dunbar, E. P. (1974b). Ecology and population dynamics of *Colobus guereza* in Ethiopia. *Folia primatologica*, **21**, 188–208.

Dunbar, R. I. M. & Dunbar, E. P. (1974c). On hybridization between *Theropithecus gelada* and *Papio anubis* in the wild. *Journal of Human Evolution*, **3**, 187–92.

Dunbar, R. I. M. & Dunbar, E. P. (1975). Social dynamics of Gelada baboons. *Contributions to primatology*, **6**, 1–157. Basel: Karger.

Duncan, T. & Stuessey, T. F., eds (1984). *Cladistics: perspectives on the reconstruction of evolutionary history.* New York: Columbia University Press.

Dupuy, A. R. (1971). Statut actuel des Primates au Sénégal. *Bulletin de l'Institut français d'Afrique noire*, série A, **33**, 467–78.

Dupuy, A. R. (1972). Une nouvelle espèce de Primate pour le Sénégal: le cercopithèque hocheur, *Cercopithecus nictitans* (L.). *Mammalia*, **36**, 306–7.

Dutrillaux, B. (1986). Le rôle des chromosomes dans l'évolution: une nouvelle interprétation. *Annales de Génétique*, **29**, 69–75.

Dutrillaux, B., Biemont, M. C., Viegas-Péquignot, E. & Laurent, C. (1979). Comparison of the karyotypes of four Cercopithecoidae: *Papio papio, P. anubis, Macaca mulatta* and *M. fascicularis. Cytogenetics Cell Genetics,* **23,** 77–83.

Dutrillaux, B. & Couturier, J. (1986). Principes de l'analyse chromosomique appliquée à la phylogénie: l'exemple des Pongidae et des Hominidae. *Mammalia,* **50,** 22–37.

Dutrillaux, B., Couturier, J. & Chauvier, G. (1980). Chromosomal evolution of 19 species or sub-species of Cercopithecinae. *Annales de Génétique,* **23,** 133–43.

Dutrillaux, B., Couturier, J., Muleris, M., Lombard, M. & Chauvier, G. (1982). Chromosomal phylogeny of forty-two species or sub-species of cercopithecoids (Primates, Catarrhini), *Annales de Génétique,* **25,** 96–109.

Dutrillaux, B., Couturier, J., Muleris, M., Rumpler, Y. & Viegas-Péquignot, E. (1986b). Relations chromosomiques entre sous-ordres et infra-ordres, et schéma évolutif général des Primates. *Mammalia,* **50,** 108–23.

Dutrillaux, B., Couturier, J. & Viegas-Péquignot, E. (1986a). Evolution chromosomique des Platyrhiniens. *Mammalia,* **50,** 56–81.

Dutrillaux, B., Viegas-Péquignot, E., Couturier, J. & Chauvier, G. (1978). Identity of euchromatic bands from man to Cercopithecidae. *Human Genetics,* **45,** 283–96.

Dutrillaux, B., Webb, G., Muleris, M., Couturier, J. & Butler, R. (1984). Chromosome study of *Presbytis cristatus*: presence of a complex Y-auto-some rearrangement in the male. *Annales de Génétique,* **27,** 148–53.

Eck, G. (1976). Cercopithecoidae from the Omo group deposits. In *Earliest man and environments in the Lake Rudolf Basin,* ed. Y. Coppens, F. C. Howell, G. L. Isaac & R. E. F. Leakey, pp. 332–44. Chicago: University of Chicago Press.

Eck, G. (1987). Further fossil specimens of the genus *Cercopithecus* from the Shungura Formation. In *Les faunes Plio–Pleistocenes de la Basse Vallée de L'Omo (Ethiopie),* ed. Y. Coppens & F. C. Howell, Tome 2. Paris: CNRS.

Eck, G. & Howell, F. C. (1972). New fossil *Cercopithecus* material from the lower Omo Basin, Ethiopia. *Folia primatologica,* **18,** 325–55.

Eck, G. & Jablonski, N. G. (1984). A reassessment of the taxonomic status and phyletic relationships of *Papio baringensis* and *Papio quadratirostris* (Primates: Cercopithecidae). *American Journal of Physical Anthropology,* **65,** 109–34.

Eisenberg, J. F., Muckenhirn, N. A. & Rudran, R. (1972). The relation between ecology and social structure in primates. *Science, NY,* **176,** 863–74.

Eisentraut, M. (1973). *Die Wirbeltierfauna von Fernando Poo und Westkamerun. Bonner Zoologische Monographien,* **3.** Bonn: Zoologisches Forschungsinstitut und Museum Alexander Koenig.

Eldredge, N. & Cracraft, J. (1980). *Phylogenetic patterns and the evolutionary process: Method and theory in comparative biology.* New York: Columbia University Press.

Elftman, H. & Manter, J. (1935). The evolution of the human foot with especial reference to the joints. *Journal of Anatomy,* **70,** 56–67.

Else, J. G. (1985). Captive breeding of vervet monkeys (*Cercopithecus aethiops*) in harems. *Laboratory Animal Science,* **35,** 373–5.

Else, J. G., Eley, R. M., Suleman, M. A. & Lequin, R. M. (1985). Reproductive biology of Sykes and blue monkeys (*Cercopithecus mitis*). *American Journal of Primatology,* **9,** 189–196.

Else, J. G., Eley, R. M., Wangula, C., Worthman, C. & Lequin, R. M. (1986).

Reproduction in the vervet monkey (*Cercopithecus aethiops*): II. Annual menstrual patterns and seasonality. *American Journal of Primatology*, **11**, 333–42.

Emmons, L., Gautier-Hion, A. & Dubost, G. (1983). Community structure of the frugivorous-folivorous forest mammals of Gabon. *Journal of Zoology, London*, **199**, 209–22.

Emory, G. R. (1976). Attention, orientation and status hierarchy in Mandrills and Gelada baboons. *Behaviour*, **59**, 70–89.

Epple, G., Golob, N. F., Cebul, M. S. & Smith, A. B. (1981). Communication by scent in some Callitrichidae (Primates); an interdisciplinary approach. *Chemical Senses*, **6**, 377–90.

Fairbanks, L. A. & Bird, J. (1978). Ecological correlates of interindividual distance in the St. Kitts vervet (*Cercopithecus aethiops sabaeus*). *Primates*, **19**, 605–14.

Farnsworth, M. W. (1978). *Genetics*. New York: Harper & Row.

Farris, J. S. (1972). Estimating phylogenetic trees from distance matrices. *American Naturalist*, **106**, 645–68.

Farris, J. S. (1981). Distance data in phylogenetic analysis. In *Advances in cladistics*, ed. V. A. Funk & D. R. Brooks, pp. 3–23. New York: The New York Botanical Garden, Bronx.

Fay, J. M. (1985). Range extensions for four *Cercopithecus* species in the Central African Republic. *Primate Conservation*, **6**, 63–8.

Fedigan, L. M., Fedigan, L., Chapman, C. A. & McGuire, M. T. (1984). A demographic model of colonization by a population of St Kitts vervets. *Folia primatologica*, **42**, 194–302.

Felsenstein, J. (1982). Numerical methods for inferring evolutionary trees. *The Quarterly Review of Biology*, **57**, 379–404.

Felsenstein, J. (1985a). Phylogenies and the comparative method. *American Naturalist*, **125**, 1–15.

Felsenstein, J. (1985b). Phylogenetic inference package. *Phylip Newsletter*, **7**.

Ferris, S. D., Sage, R. D., Huang, C.-M., Nielsen, J. T., Ritte, V. & Wilson, A. C. (1983). Flow of mitochondrial DNA across a species boundary. *Proceedings of the National Academy of Sciences, USA*, **80**, 2290–4.

Ferris, S. D., Wilson, A. C. & Brown, W. H. (1981). Evolutionary tree for apes and humans based on cleavage maps of mitochondrial DNA. *Proceedings of the National Academy of Sciences, USA*, **78**, 2432–6.

Fleagle, J. G. (1983). Locomotor adaptations of Oligocene and Miocene hominoids and their phyletic implications. In *New Interpretations of Ape and Human Ancestry*, ed. R. L. Ciochon & R. S. Corruccini, pp. 301–24. New York: Plenum Press.

Fleagle, J. G. (1985). Size and adaptation in Primates. In *Size and scaling in Primate biology*, ed. W. L. Jungers, pp. 2–20. New York: Plenum Press.

Fleagle, J. G. & Mittermeier, R. A. (1980). Locomotor behavior, body size and comparative ecology of seven Surinam monkeys. *American Journal of Anthropology*, **52**, 301–2.

Flohn, H. (1973). Antarctica and the global Cenozoic evolution: a geophysical model. *Palaeoecology of Africa*, **8**, 37–53.

Flohn, H. & Nicholson, S. (1980). Climatic fluctuations in the arid belt of the 'Old World' since the last glacial maximum: possible causes and future implications. *Palaeoecology of Africa*, **12**, 3–21.

Fortes, M. (1945). *The dynamics of clanship among the Tallensi*. London: Oxford University Press.

Fourtau, R. (1920). *Contribution à l'étude des vertébrés miocènes de l'Egypte.* Cairo: Government Press.

Frade, F. (1949). Algunas novidades para a faune da Guine Portuguesa (aves e mamiferos). *Annais Junta des Missoes Geograficas e de Investigacoes Coloniais Lisboa*, **4**, 165–86.

Frechkop, S. (1938). Mammifères. (Exploration du Parc National Albert, mission G. F. de Witte, 10, Institut des Parcs Nationaux du Congo.)

Frechkop, S. (1943). *Mammifères.* (Exploration du Parc National Albert, mission S. Frechkop, 1, Institut des Parcs Nationaux du Congo).

Fredoux, A. & Tastet, J. P. (1976). Apport de la palynologie à la connaissance paléogéographique du littoral ivoirien entre 8000 et 12 000 ans BP. Seventh African Micropalaeontological Colloquium, Ile-Ife, Nigeria.

Freeland, W. J. (1977). Blood sucking flies and primates polyspecific associations. *Nature, London*, **269**, 801–2.

Freeland, W. J. (1979). Mangabey (*Cercocebus albigena*) social organization and population density in relation to food use and availability. *Folia primatologica*, **32**, 108–24.

Funaioli, V. & Simonetta, A. M. (1966). The mammalian fauna of the Somali Republic. Status and conservation problems. *Monitore Zoologico Italiano, Firenze, Supplemento*, 285–347.

Furuya, Y. (1969). On the fission of troops of Japanese monkeys. II: General view of troop fission of Japanese monkeys. *Primates*, **10**, 47–70.

Galat, G. (1975). Eco-éthologie de *Cercopithecus sabaeus. Rapport, Centre ORSTOM*, Dakar.

Galat, G. (1977). Enquête sur les mammifères de Lobaye. Recensements et densité des primates et observations sur l'écologie de *Colobus pennanti oustaleti. Rapport de Mission en Lobaye, I, Empire Centrafricain.* Adiopodoumé, Côte d'Ivoire: ORSTOM.

Galat, G. (1978). Données écologiques sur les singes de la région de Bozo. *Rapport de Mission à Bozo, Empire Centrafricain.* Adiopodoumé, Côte d'Ivoire: ORSTOM.

Galat, G. (1983). Socioécologie du singe vert (*Cercopithecus aethiops sabaeus*) en référence de quatre Cercopithecinés sympatriques (*Cercocebus atys, Cercopithecus campbelli, C. diana, C. petaurista*) d'Afrique de l'Ouest. Thèse de Doctorat, Université de Paris 6.

Galat, G. & Galat-Luong, A. (1976). La colonisation de la mangrove par *Cercopithecus aethiops sabaeus* au Sénégal. *La Terre et Vie*, **30**, 3–30.

Galat, G. & Galat-Luong, A. (1977). Démographie et régime alimentaire d'une troupe de *Cercopithecus aethiops sabaeus* en habitat marginal au nord Sénégal. *La Terre et la Vie*, **31**, 557–77.

Galat, G. & Galat-Luong, A. (1985). La communauté de primates diurnes de la forêt de Taï, Côte d'Ivoire. *Revue d'Ecologie (Terre Vie)*, **40**, 3–32.

Galat-Luong, A. (1975). Notes préliminaires sur l'écologie de *Cercopithecus ascanius schmidti* dans les environs de Bangui (R.C.A.). *La Terre et la Vie*, **29**, 288–97.

Galat-Luong, A. (1979). Interactions interspécifiques chez les primates diurnes du Parc National de Taï, Côte d'Ivoire. *Rapport IUET*: Abidjan.

Galat-Luong, A. (1983). *Socio-écologie de trois colobes sympatriques*, Colobus badius, C. polykomos *et* C. verus *du Parc National de Tai.* Thèse de Doctorat, Université de Paris.

Galat-Luong, A. & Galat, G. (1978). *Abondance relative et associations plurispécifiques des primates diurnes du Parc National de Taï, Côte d'Ivoire.* Abidjan: Centre d'Adiopodoumé ORSTOM.

Galat-Luong, A. & Galat, G. (1979). Conséquences comportementales de perturbations sociales répétées sur une troupe de mones de Lowe, *Cercopithecus campbelli lowei*, de Côte-d'Ivoire. *La Terre et la Vie*, **33**, 49–58.

Gartlan, J. S. (1969). Sexual and maternal behaviour of the vervet monkey, *Cercopithecus aethiops*. *Journal of Reproduction and Fertility, Supplement*, **6**, 137–50.

Gartlan, J. S. (1974). *Ecology and behavior of the patas monkey*. New York: Rockefeller University Film Service (16 mm film).

Gartlan, J. S. (1975). Adaptive aspects of social structure in *Erythrocebus patas*. *Proceedings of the 5th International Congress of Primatology*, Tokyo.

Gartlan, J. S. & Brain, C. K. (1968). Ecology and social variability in *Cercopithecus aethiops* and *C. mitis*. In *Primates; studies in adaptation and variability*, ed. P. C. Jay, pp. 253–92. New York: Holt, Rinehart & Winston.

Gartlan, J. S. & Gartlan, S. (1973). Quelques observations sur les groupes exclusivement mâles chez *Erythrocebus patas*. *Annales de la Faculté des Sciences du Cameroun*, **12**, 121–44.

Gartlan, J. S., McKey, D. B., Waterman, P. G., Mbi, L. N. & Struhsaker, T. T. (1980). A comparative study of the phytochemistry of two African rain forests. *Biochemical Systematics and Ecology*, **8**, 401–22.

Gartlan, J. S. & Struhsaker, T. T. (1972). Polyspecific associations and niche separation of rain-forest anthropoids in Cameroon, West Africa. *Journal of Zoology, London*, **168**, 221–66.

Gasse, F., Rognon, R. & Street, F. A. (1980). Quaternary history of the Afar and Ethiopian Rift lakes. In *The Sahara and the Nile*, ed. M. A. J. Williams & H. Faure, pp. 361–400. Rotterdam: A. A. Balkema.

Gause, G. F. (1934). *The struggle for existence*. Baltimore: Williams & Williams.

Gautier, J.-P. (1969). Emissions sonores d'espacement et de ralliement par deux cercopithèques arboricoles. *Biologica gabonica*, **5**, 117–45.

Gautier, J.-P. (1971). Etude morphologique et fonctionelle des annexes extralaryngées des Cercopithecinae: liaison avec les cris d'espacement. *Biologia gabonica*, **7**, 229–67.

Gautier, J.-P. (1974). Field and laboratory studies of the vocalizations of talapoin monkeys (*Miopithecus talapoin*). *Behaviour*, **49**, 1–64.

Gautier, J.-P. (1975). Etude comparée des systèmes d'intercommunication sonore des Cercopithecinés forestiers africains. Thèse de Doctorat, Université de Rennes.

Gautier, J.-P. (1978). Répertoire sonore de *Cercopithecus cephus*. *Zeitschrift für Tierpsychologie*, **46**, 113–69.

Gautier, J.-P. (1985). Quelques caractéristiques écologiques du singe des marais, *Allenopithecus nigroviridis* Lang 1923. *Revue d'Ecologie (Terre Vie)*, **40**, 331–42.

Gautier, J.-P. & Gautier-Hion, A. (1969). Les associations polyspécifiques chez les Cercopithecidae du Gabon. *La Terre et la Vie*, **23**, 164–201.

Gautier, J.-P. & Gautier, A. (1977). Communication in Old World monkeys. In *How Animals Communicate*, ed. T. E. Sebeok, pp. 890–964. Bloomington: Indiana University Press.

Gautier, J.-P. & Gautier-Hion, A. (1982). Vocal communication within a monkey group: an exhaustive analysis by biotelemetry. In *Primate communication*, ed. C. T. Snowdon, C. H. Brown & M. R. Petersen, pp. 5–29. London: Cambridge University Press.

Gautier, J.-P. & Gautier-Hion, A. (1983). Comportement vocal des mâles adultes et organisation supraspécifique dans les troupes polyspécifiques de Cercopithèques. *Folia primatologica*, **40**, 161–74.

Gautier, J.-P. & Gautier-Hion, A. (1988). Quavering parameters of calls may constitute a perceptual basis for specific and individual recognition among forest guenons. In *New approaches towards an understanding of primate vocal communication*, ed. D. Todt, P. Goeldeking & D. Symmes, in press.

Gautier, J.-P., Loireau, J. N. & Moysan, F. (1986). Distribution, ecology, behaviour and phylogenic affinities of a new species of guenon, *Cercopithecus (lhoesti) 'solatus'*, Harrison 1984. XIth Congress of the International Primatological Society, Göttingen.

Gautier-Hion, A. (1968). Étude du cycle annuel de reproduction du talapoin (*Miopithecus talapoin*), vivant dans son milieu naturel. *Biologia Gabonica*, **4**, 163–73.

Gautier-Hion, A. (1970). L'organisation sociale d'une bande de talapoins (*Miopithecus talapoin*) dans le nord-est du Gabon. *Folia primatologica*, **12**, 116–41.

Gautier-Hion, A. (1971a). Répertoire comportemental du talapoin (*Miopithecus talapoin*). *Biologica Gabonica*, **7**, 295–391.

Gautier-Hion, A. (1971b). L'écologie du talapoin du Gabon (*Miopithecus talapoin*). *La Terre et la Vie*, **25**, 427–90.

Gautier-Hion, A. (1973). Social and ecological features of talapoin monkeys – Comparisons with sympatric Cercopithecines. In *Comparative ecology and behaviour of primates*, ed. R. P. Michael & J. H. Crook, pp. 147–70. London: Academic Press.

Gautier-Hion, A. (1975). Dimorphisme sexuel et organisation sociale chez les Cercopithecinés forestiers africains. *Mammalia*, **39**, 365–74.

Gautier-Hion, A. (1978). Food niches and coexistence in sympatric primates in Gabon. In *Recent advances in primatology, Vol. II*, ed. D. J. Chivers & J. Herbert, pp. 270–86. New York: Academic Press.

Gautier-Hion, A. (1980). Seasonal variations of diet related to species and sex in a community of *Cercopithecus* monkeys. *Journal of Animal Ecology*, **49**, 237–69.

Gautier-Hion, A. (1983). Leaf consumption by monkeys in western and eastern Africa: a comparison. *African Journal of Ecology*, **21**, 107–13.

Gautier-Hion, A. (1984). La dissémination des graines par les cercopithecidés forestiers africains. *Revue d'Ecologie (Terre Vie)*, **39**, 159–65.

Gautier-Hion, A., Duplantier, J.-M., Emmons, L., Feer, F., Heckestweiler, P., Moungazi, A., Quris, R. & Sourd, C. (1985b). Coadaptation entre rythmes de fructification et frugivorie en forêt tropicale humide du Gabon: mythe ou réalité. *Revue d'Ecologie (Terre Vie)*, **40**, 405–34.

Gautier-Hion, A., Duplantier, J.-M., Quris, R., Feer, F., Sourd, C., Decoux, J.-P., Dubost, G., Emmons, L., Erard, C., Heckestweiler, P., Moungazi, A., Roussilhon, C. & Thiollay, J.-M. (1985a). Fruit characters as a basis of fruit choice and seed dispersal in a tropical forest vertebrate community. *Oecologia*, **65**, 324–37.

Gautier-Hion, A. & Gautier, J.-P. (1974). Les associations polyspécifiques des Cercopithèques du plateau de M'passa, Gabon. *Folia primatologica*, **22**, 134–77.

Gautier-Hion, A. & Gautier, J.-P. (1976). Croissance, maturité sexuelle et sociale, reproduction chez les cercopithécinés forestiers africains. *Folia primatologica*, **26**, 165–84.

Gautier-Hion, A. & Gautier, J.-P. (1978). Le singe de Brazza: une stratégie originale. *Zeitschrift für Tierpsychologie*, **46**, 84–104.

Gautier-Hion, A. & Gautier, J.-P. (1979). Niche écologique et diversité des espèces sympatriques dans le genre *Cercopithecus*. *Revue d'Ecologie (Terre Vie)*, **33**, 493–507.

Gautier-Hion, A. & Gautier, J.-P. (1985). Sexual dimorphism, social units and ecology among sympatric forest guenons. In *Human sexual dimorphism*, ed. J. Guesquière, R. D. Martin & F. Newcombe, pp. 61–77. London: Taylor & Francis.

Gautier-Hion, A., Gautier, J.-P. & Quris, R. (1981). Forest structure and fruit availability as complementary factors influencing the habitat use by a troop of monkeys (*Cercopithecus cephus*). *Revue d'Ecologie (Terre Vie)*, **35**, 511–36.

Gautier-Hion, A., Quris, R. & Gautier, J.-P. (1983). Monospecific *vs* poly-specific life: a comparative study of foraging and antipredatory tactics in a community of *Cercopithecus* monkeys. *Behavioural Ecology and Sociobiology*, **12**, 325–35.

Geissmann, T. A. (1987). A sternal gland in the siamang gibbon. *International Journal of Primatology*, **8**, 1–15.

Ghiglieri, M. P. (1984). *The Chimpanzees of Kibale Forest*. New York: Columbia University Press.

Giresse, P. & Lanfranchi, R. (1984). Les climats et les océans de la région congolaise pendant l'Holocène. Bilans selon les échelles et les méthodes de l'observation. *Palaeoecology of Africa*, **16**, 77–88.

Girolami, L. (1985). Sexual behavior of Samango monkeys (*Cercopithecus albogularis*). Paper presented at the symposium 'Biologie, phylogénie et spéciation chez les cercopithèques forestiers', Paimpont, France, 20–22 August 1985.

Glander, K. (1981). Feeding patterns in mantled howling monkeys. In *Foraging behaviour: ecological, ethological and psychological approaches*, ed. A. C. Kamil & T. D. Sargent, pp. 231–59. New York: Garland Press.

Goodman, M., Braunitzer, G., Staugl, A. & Schrank, B. (1983). Evidence of human origins from haemoglobins of African apes. *Nature, London*, **303**, 546–8.

Goodman, M., Koop, B. F., Czelusnizk, J., Weiss, M. L. & Slightom, J. L. (1984). The η-globin gene: its long evolutionary history in the η-globin gene family of mammals. *Journal of Molecular Biology*, **180**, 803–23.

Gould, S. J. (1966). Allometry and size in ontogeny and phylogeny. *Biological Reviews*, **41**, 587–640.

Gray, A. P. (1972). *Mammalian hybrids. A check-list with bibliography*, 2nd edition. Farnham Royal: Commonwealth Agricultural Bureaux.

Gray, P. & Wolfe, L. (1980). Height and sexual dimorphism of stature among human societies. *American Journal of Physical Anthropology*, **53**, 441–56.

Green, S. (1975). Variation of vocal pattern with social situation in the Japanese monkey (*Macaca fuscata*): a field study. In *Primate Behavior*, Vol. 4, ed. L. A. Rosenblum, pp. 1–102. New York: Academic Press.

Grimm, R. J. (1967). Catalogue of sounds of the pigtailed macaque (*Macaca nemestrina*). *Journal of Zoology, London*, **152**, 361–73.

Grove, A. T. & Warren, A. (1968). Quaternary landforms and climate on the south side of the Sahara. *Geographical Journal*, **134**, 194–208.

Grubb, P. (1973). Distribution, divergence and speciation of the drill and mandrill. *Folia primatologica*, **20**, 161–77.

Grubb, P. (1978a). The potto (*Perodicticus potto*: Primates, Lorisidae) in Nigeria and adjacent territories. *Bulletin de l'Institut français d'Afrique noire*, série A, **40**, 909–13.

Grubb, P. (1978b). Patterns of speciation in African mammals. *Bulletin of the Carnegie Museum of Natural History*, **6**, 152–67.

Grubb, P. (1982). Refuges and dispersal in the speciation of African forest mammals. In *Biological Diversification in the Tropics*, ed. G. T. Prance, pp. 537–53. New York: Columbia University Press.

Guthrie, R. D. (1971). A new theory of mammalian rump patch evolution. *Behaviour*, **38**, 132–45.

Gyldenstolpe, N. (1928). Zoological results of the Swedish Expedition to Central Africa, 1921. *Arkiv för zoologi*, **20**, 1–76.

Haddow, A. J. (1952). Field and laboratory studies on an African monkey (*Cercopithecus ascanius schmidti*). *Proceedings of the Zoological Society of London*, **122**, 34–52.

Haffer, J. (1977). Pleistocene speciation in Amazonian birds. *Amazonia*, **6**, 161–91.

Hagaman, R. M. & Morbeck, M. E. (1984). Data transformations in multivariate morphometric analyses. *Journal of Human Evolution*, **13**, 225–45.

Haimoff, E. H., Gittins, S. P., Whitten, A. J. & Chivers, D. J. (1984). A phylogeny and classification of gibbons based on morphology and ethology. In *The lesser apes. Evolutionary and behavioural biology*, ed. H. Preuschoft, D. J. Chivers, W. Y. Brockelman & N. Creel, pp. 614–32. Edinburgh: Edinburgh University Press.

Haines, R. W. (1958). Arboreal or terrestrial ancestry of placental mammals. *Quarterly Review of Biology*, **33**, 1–23.

Hall, J. B. & Swaine, M. D. (1981). Classification and ecology of vascular plants in a tropical rain forest: forest vegetation in Ghana. *Geobotany, Vol. 1*, ed. M. J. A. Werger. The Hague: Junk.

Hall, K. R. L. (1962). The sexual, agonistic, and derived social behaviour patterns of the wild chacma baboon (*Papio ursinus*). *Proceedings of the Zoological Society, London*, **139**, 283–327.

Hall, K. R. L. (1965a). Behaviour and ecology of the wild patas monkey, *Erythrocebus patas*, in Uganda. *Journal of Zoology, London*, **148**, 15–87.

Hall, K. R. L. (1965b). Ecology and behaviour of baboons, patas and vervet monkeys in Uganda. In *The Baboon in Medical Research*, ed. H. Vagtborg, pp. 43–61. Austin: University of Texas Press.

Hall, K. R. L., Boelkins, R. C. & Goswell, M. S. (1965). Behaviour of patas monkeys, *Erythrocebus patas*, in captivity, with notes on the natural habitat. *Folia primatologica*, **3**, 22–49.

Hall, K. R. L. & DeVore, I. (1965). Baboon social behaviour. In *Primate behaviour: field studies of monkeys and apes*, ed. I. DeVore, pp. 53–110. New York: Holt, Rinehart & Winston.

Hall, K. R. L. & Gartlan, J. S. (1965). Ecology and behaviour of the vervet monkey, *Cercopithecus aethiops*, Lolui Island, Lake Victoria. *Proceedings of the Zoological Society of London*, **145**, 37–57.

Hall, R. L. (1978). Sexual dimorphism for size in seven Nineteenth Century Northwest Coast populations. *Human Biology*, **50**, 159–71.

Haltenorth, T. & Diller, H. (1985). *Mammifères d'Afrique et de Madagascar*. Neuchâtel: Delachaux & Niestlé.

Hamilton, A. C. (1972). The interpretation of pollen diagrams from highland Uganda. *Palaeoecology of Africa*, **7**, 45–149.

Hamilton, A. C. (1976). The significance of patterns of distribution shown by forest plants and animals in tropical Africa for the reconstruction of upper Pleistocene palaeoenvironments: a review. *Palaeoecology of Africa*, **9**, 63–97.

Hamilton, A. C. (1981). The Quaternary history of African forests: its relevance to conservation. *African Journal of Ecology*, **19**, 1–6.

Hamilton, A. C. (1982). *Environmental History of East Africa*. London: Academic Press.

Hamilton, A. C. & Perrott, R. A. (1979). Aspects of the glaciation of Mount Elgon. *Palaeoecology of Africa*, **11**, 153–61.

Hamilton, A. C. & Perrott, R. A. (1980). Modern pollen deposition on a tropical African mountain. *Pollen et Spores*, **22**, 437–68.

Hamilton, A. C., Taylor, D. & Vogel, J. C. (1986). Early forest clearance and environmental degradation in south-west Uganda. *Nature, London*, **320**, 164–7.

Hamilton, W. J. III & Arrowood, P. C. (1976). Copulation vocalizations of chacma baboons (*Papio ursinus*), gibbons (*Hylobates hoolock*), and humans. *Science, NY*, **200**, 1405–9.

Hamilton, W. J. III, Buskirk, R. E. & Buskirk, W. H. (1976). Defence of space and resources by chacma (*Papio ursinus*) baboon troops in an African desert and swamp. *Ecology*, **57**, 1264–72.

Happold, D. C. D. (1985). Geographical ecology of Nigerian mammals. *Annales du Musée royal de l'Afrique centrale, Sciences Zoologiques*, **246**, 5–49.

Harding, R. S. O. (1984). Primates of the Kilimi area, northwest Sierra Leone. *Folia primatologica*, **42**, 96–114.

Harrington, J. E. (1974). Olfactory communication in *Lemur fulvus*. In *Prosimian Biology*, ed. R. D. Martin, G. A. Doyle & A. C. Walker, pp. 331–46. London: Duckworth.

Harrison, M. J. S. (1982). The behavioural ecology of green monkeys (*Cercopithecus sabaeus*), at Mt Assirik, Senegal. PhD dissertation, University of Stirling.

Harrison, M. J. S. (1983a). Patterns of range use by the green monkey, *Cercopithecus sabaeus*, at Mt Assirik, Senegal. *Folia primatologica*, **41**, 157–79.

Harrison, M. J. S. (1983b). Age and sex differences in the diet and feeding strategies of the green monkey, *Cercopithecus sabaeus*. *Animal Behaviour*, **31**, 969–77.

Harrison, M. J. S. (1983c). Territorial behaviours in the green monkey, *Cercopithecus sabaeus*, seasonal defense of local food supplies. *Behavioral Ecology and Sociobiology*, **12**, 85–94.

Harrison, M. J. S. (1984a). Optimal foraging strategies in the diet of the green monkey, *Cercopithecus sabaeus*, at Mt Assirik, Senegal. *International Journal of Primatology*, **5**, 435–72.

Harrison, M. J. S. (1984b). Poster presentation. Xth Congress of the International Primatological Society, Nairobi, Kenya.

Harrison, M. J. S. (1985). Time budget of the green monkey, *Cercopithecus aethiops*: some optimal strategies. *International Journal of Primatology*, **6**, 351–76.

Harrison, M. J. S. (1988). A new species of guenon (genus *Cercopithecus*) from Gabon. *Journal of Zoology, London*, in press.

Harrison, M. J. S. & Hladik, C. M. (1986). Un primate granivore, le Colobe noir dans la forêt du Gabon. Potentialité d'évolution du comportement alimentaire. *Revue d'Ecologie (Terre Vie)*, **41**, 281–98.

Harrison, T. (1982). *Small bodied apes from the Miocene of East Africa*. PhD dissertation, University of London.

Harrison, T. (1986). New fossil anthropoids from the Middle Miocene of East Africa and their bearing on the origin of the Oreopithecidae. *American Journal of Physical Anthropology*, **71**, 265–84.

Harvey, P. H. & Mace, G. M. (1982). Comparisons between taxa and adaptive trends: Problems of methodology. In *Current Problems in Sociobiology*, ed. King's College Research Group, pp. 343–61. Cambridge: Cambridge University Press.

Hastenrath, S. & Kutzbach, J. E. (1983). Paleoclimatic estimates from water and energy budgets of East African lakes. *Quaternary Research*, **19**, 141–53.

Hauser, M. D., Cheney, D. L. & Seyfarth, R. M. (1986). Group extinction and fusion in free-ranging vervet monkeys. *American Journal of Primatology*, **11**, 63–77.

Hays, J. D., Imbrie, J. & Shackleton, N. J. (1976). Variations in the Earth's orbit: pacemaker of the ice ages. *Science, NY*, **194**, 1121–32.

Hecky, R. E. & Degens, E. T. (1973). *Late Pleistocene–Holocene chemical stratigraphy and paleolimnology of the Rift Valley lakes of Central Africa.* Technical Reports of the Woods Hole Oceanographic Institute.

Heinroth-Berger, K. (1959). Beobachtungen an handaufgezogene Mantelpavianenen *(Papio hamadryas)*. *Zeitschrift für Tierpsychologie*, **16**, 706–732.

Hennig, W. (1950). *Grundzüge einer Theorie der Phylogenetischen Systematik.* Berlin: Deutscher Zentralverlag.

Hennig, W. (1965). Phylogenetic systematics. *Annual Review of Entomology*, **10**, 97–116.

Hennig, W. (1966). *Phylogenetic systematics.* Urbana: University of Illinois Press.

Henzi, S. P. (1985). Genital signalling and the co-existence of male vervet monkeys *(Cercopithecus aethiops pygerythrus)*. *Folia primatologica*, **45**, 129–47.

Henzi, S. P. & Lucas, J. W. (1980). Observations on the inter-troup movement of adult vervet monkeys *(Cercopithecus aethiops)*. *Folia primatologica*, **33**, 220–35.

Henzi, S. P. & Lawes, M. (1987). Breeding season influxes and the behaviour of adult male samango monkeys *(Cercopithecus mitis albogularis)*. *Folia primatologica*, in press.

Herzog, M. D. & Hohmann, G. M. (1984). Male loud calls in *Macaca silenus* and *Presbytis johnii*: a comparison. *Folia primatologica*, **43**, 189–97.

Heymans, J. C. (1975). Contribution à la détermination des Primates de la région du Haut-Zaïre (République du Zaïre): Tableaux synoptiques. *Les Naturalistes Belges*, **56**, 73–82.

Hill, W. C. O. (1966). *Primates. Comparative anatomy and taxonomy, Vol. VI. Catarrhini, Cercopithecoidea, Cercopithecinae.* Edinburgh: Edinburgh University Press.

Hill, W. C. O. & Booth, A. H. (1957). Voice and larynx in African and Asiatic colobinae. *Journal of Bombay Natural Society*, **54**, 309–21.

Hinde, R. A. (ed.) (1983). *Primate social relationships.* Oxford: Blackwell Scientific publications.

Hinde, R. A. & Rowell, T. E. (1962). Communication by posture and facial expression in the rhesus monkey, *(Macaca mulatta)*. *Proceedings of the Zoological Society, London*, **138**, 1–21.

Hladik, C. M. (1977a). A comparative study of the feeding strategies of two sympatric species of leaf monkeys: *Presbytis senex* and *P. entellus*. In *Primate Ecology: studies of feeding and ranging behaviour in lemurs, monkeys and apes*, ed. T. H. Clutton-Brock, pp. 481–501. New York: Academic Press.

Hladik, C. M. (1977b). Chimpanzees of Gabon and chimpanzees of Gombe: some comparative data on the diet. In *Primate Ecology: studies of feeding and ranging behaviour in lemurs, monkeys and apes*, ed. T. H. Clutton-Brock, pp. 324–53. New York: Academic Press.

Hladik, A. (1978). Phenology of leaf production in rain forest of Gabon: Distribution and composition of food for folivores. In *The ecology of arboreal folivores*, ed. G. G. Montgomery, pp. 51–71. Washington, DC: Smithsonian Institution Press.

Hladik, C. M. (1981). Diet and evolution of feeding strategies among forest Primates. In *Omnivorous Primates: gathering and hunting in human evolution,* ed. R. S. O. Harding & G. Teleki, pp. 215–54. New York: Columbia University Press.

Hodun, A., Snowdon, C. T. & Soini, P. (1981). Subspecific variation in the long calls of the Tamarin, *Saguinus fuscicollis*. *Zeitschrift für Tierpsychologie*, **57**, 97–110.

Hohmann, G. M. & Herzog, M. P. (1985). Vocal communication in lion tailed macaques (*Macaca silenus*). *Folia primatologica*, **45**, 148–78.

Homewood, K. M. (1976). Ecology and behaviour of the Tana river Mangabey (*Cercocebus galeritus galeritus*). PhD dissertation, University of London.

Homewood, K. M. (1978). Feeding strategy of the Tana mangabey (*Cercocebus galeritus galeritus*) (Mammalia: Primates). *Journal of Zoology, London*, **186**, 375–91.

Hooijer, D. (1963). Miocene Mammalia from Congo. *Annales du Musée royal de l'Afrique centrale, Sciences Géologiques*, **46**, 1–77.

Hopkins, B. (1962). Vegetation of the Olokemeji Forest Reserve, Nigeria, I. General features of the reserve and the research sites. *Journal of Ecology*, **50**, 559–98.

Hopp-Dominik, B. (1984). Etude du spectre de proies de la panthère, *Panthera pardus* dans le Parc National de Taï, en Côte d'Ivoire. *Mammalia*, **48**, 476–87.

Hopwood, A. T. (1933). Miocene primates from British East Africa. *Annals and Magazine of Natural History*, Series 10, **11**, 96–8.

Horrocks, J. A. (1986). Life-history characteristics of a wild population of vervets (*Cercopithecus aethiops sabaeus*) in Barbados, West Indies. *International Journal of Primatology*, **7**, 31–47.

Horrocks, J. A. & Hunte, W. (1986). Sentinel behavior in vervet monkeys: who sees whom first? *Animal Behaviour*, **34**, 1566–7.

Horwich, R. H. (1976). The whooping display in Nilgiri langur: an example of daily fluctuations superimposed on a general trend. *Primates*, **17**, 419–31.

Hoshino, J. (1985). Feeding ecology of mandrills (*Mandrillus sphinx*) in Campo Animal Reserve, Cameroon. *Primates*, **26**, 248–73.

Hoshino, J., Mori, A., Kudo, H. & Kawai, M. (1984). Preliminary report on the grouping of mandrills (*Mandrillus sphinx*) in Cameroon. *Primates*, **25**, 295–307.

Howells, W. W. (1966). Population distances: biological, linguistic, geographical, and environmental. *Current Anthropology*, **7**, 531–5.

Howells, W. W. (1984). Introduction. In *Multivariate statistical methods in physical anthropology*, ed. G. N. Van Vark & W. W. Howells, pp. 1–11. Dordrecht, Holland: Reidel Publishing Co.

Hrdy, S. B. (1974). Male–male competition and infanticide among the langurs (*Presbytis entellus*) of Abu, Rajasthan. *Folia primatologica*, **22**, 19–58.

Hrdy, S. B. (1976). Care and exploitation of nonhuman primate infants by conspecifics other than the mother. In *Advances in the study of behaviour*, volume 6, ed. D. S. Lehrman, R. A. Hinde & E. Shaw, pp. 101–58. New York: Academic Press.

Hrdy, S. B. (1979). Infanticide among animals: A review, classification and examination of the implications for the reproductive strategies of females. *Ethology and Sociobiology*, **1**, 13–40.

Hrdy, S. B. (1981). *The woman that never evolved*. Cambridge, Mass.: Harvard University Press.

Hsu, K. J., Montadert, L., Bernoulli, D., Cita, M. B., Erickson, A., Garrison, R. E., Kidd, R. B., Melières, F., Müller, C. & Wright, R. (1977). History of the Mediterranean salinity crisis. *Nature, London*, **267**, 399–403.

Hubbell, S. P. & Foster, R. B. (1986). Biology, chance and history and the structure of tropical rain forest tree communities. In *Community ecology*, ed. J. M. Diamond & T. J. Case. New York: Harper & Row.

Hunkeler, C., Bourlière, F., and Bertrand, M. (1972). Le comportement de la Mone de Lowe (*Cercopithecus campbelli lowei*). *Folia primatologica*, **17**, 218–36.

Hurni, H. (1981). Simien Mountains-Ethiopia: palaeoclimate of the last cold period (Late Würm). *Palaeoecology of Africa*, **13**, 127–37.

Hylander, W. L. (1975). Incisor size and diet in Anthropoids with special references to Cercopithecidae. *Science, NY*, **189**, 1095–8.

Imbrie, J. & Imbrie, J. Z. (1980). Modeling the climatic response to orbital variations. *Science, NY*, **207**, 943–53.

International Zoo Yearbook (1961, 1963, 1965, 1967, 1968, 1969, 1970, 1971, 1972, 1973, 1974, 1975, 1977, 1978, 1979, 1980, 1981, 1982, 1983). Mammals bred in captivity and multiple generation births. London: The Zoological Society of London.

ISCN (1978). *An international system for human cytogenetic nomenclature. Birth defects: original article series*, Vol. 14, No. 8 (The National Foundation, New York 1978); also in *Cytogenetics Cell Genetics*, **21**, 309–404.

Jackson, G. & Gartlan, J. S. (1965). The flora and fauna of Lolui Island, Lake Victoria. *Journal of Ecology*, **53**, 573–97.

Janson, C. (1983). Adaptation of fruit morphology to dispersal agents in a Neotropical forest. *Science, NY*, **219**, 187–9.

Jeannin, A. (1936). *Les Mammifères Sauvages du Cameroun*. Paris: Paul Lechevalier.

Jolly, C. (1966). Introduction to the Cercopithecoidea with notes on their use as laboratory animals. *Symposia of the Zoological Society of London*, **17**, 427–57.

Jones, C. (1966). Stomach contents and gastro-intestinal relationships of monkeys collected in Rio Muni, West Africa. *Mammalia*, **34**, 107–17.

Jones, C. & Sabater Pi, J. (1968). Comparative ecology of *Cercocebus albigena* (Gray) and *Cercocebus torquatus* (Kerr) in Rio Muni, West Africa. *Folia primatologica*, **9**, 99–113.

Jones, T. S. (1950). Notes on the monkeys of Sierra Leone. *Sierra Leone Agricultural Notes*, No. 22.

Jouventin, P. (1975). Les rôles des colorations du Mandrill. *Zeitschrift für Tierpsychologie*, **39**, 455–62.

Joysey, K. A. & Friday, A. E., eds (1982). *Problems of phylogenetic reconstruction*. London: Academic Press.

Jungers, W. L. (ed.) (1985). *Size and scaling in primate biology*. New York: Plenum Press.

Jurgens, U. (1979). Vocalizations as an emotional indication. A neuroethological study on the squirrel monkeys. *Behaviour*, **69**, 88–117.

Kadomura, H. (1982). Summary and conclusions. In *Geomorphology and environmental changes in the forest and savanna of Cameroon*, ed. H. Kadomura, pp. 99–106. Japan: Hokkaido University.

Kavanagh, M. (1978a). The diet and feeding behavior of *Cercopithecus aethiops tantalus*. *Folia primatologica*, **30**, 30–63.

Kavanagh, M. (1978b). Monkeys' new life in the forest. *New Scientist*, **77**, 515–17.

Kavanagh, M. (1980a). Invasion of the forest by an African savannah monkey: behavioural adaptations. *Behaviour*, **73**, 238–60.

Kavanagh, M. (1980b). Selective pressures and the socio-ecology of a West African cercopithecine monkey. *Tropical Ecology and Development*, **5**, 377–82.

Kavanagh, M. (1981). Variable territoriality among Tantalus monkeys in Cameroon. *Folia primatologica*, **36**, 76–98.

Kavanagh, M. (1983a). Birth seasonality in *Cercopithecus aethiops*: A social advantage from synchrony? In *Perspectives in Primate Biology*, ed. P. K. Seth, pp. 89–98. New Delhi: Today and Tomorrow's Printers.

Kavanagh, M. (1983b). *A complete guide to monkeys, apes and other primates.* London: Jonathan Cape.

Kavanagh, M. (1986). Invasion of the forest by an African savannah monkey: behavioural adaptations. *Behaviour*, **73**, 238–60.

Kawamoto, Y., Shotake, T. & Nozawa, K. (1982). Genetic differentiation among three genera of Cercopithecidae. *Primates*, **23**, 272–86.

Kay, R. F. (1975). The functional adaptations of primate molar teeth. *American Journal of Physical Anthropology*, **43**, 195–216.

Kay, R. F. (1978). Molar structure and diet in extant Cercopithecidae. In *Development, function and evolution of teeth*, ed. P. M. Butler & K. A. Joysey, pp. 309–39. London: Academic Press.

Kay, R. F. & Covert, H. H. (1984). Anatomy and behaviour of extinct Primates. In *Food acquisition and processing in primates*, ed. D. J. Chivers, B. A . Wood & A. Bilsborough, pp. 467–508. New York: Plenum Press.

Keay, R. W. J. (1953). An outline of Nigerian vegetation, 2nd edn. Lagos: Government Printer.

Keay, R. W. J. (1959). Derived savanna – derived from what? *Bulletin de l'Institut Français d'Afrique Noire*, **21**, 427–38.

Kemp, A. C. (1973). A study of the ecology, behaviour and systematics of *Tockus* hornbills (Aves: Bucerotidae). *Transvaal Museum Memoir* No. 20.

Kendall, R. L. (1969). An ecological history of the Lake Victoria basin. *Ecological Monographs*, **39**, 121–76.

Keverne, E. B. (1980). Olfaction in the behaviour of non-human Primates. *Symposium of the Zoological Society of London*, **45**, 313–27.

Kingdon, J. S. (1971). *East African Mammals: an Atlas of Evolution in Africa. Vol. 1.* London: Academic Press.

Kingdon, J. S. (1980). The role of visual signals and face patterns in African forest monkeys (guenons) of the genus *Cercopithecus*. *Transactions of the Zoological Society, London*, **35**, 425–75.

Kingdon, J. S. (1982). *East African Mammals; an Atlas of Evolution in Africa, Vol. III, Part C, Bovids.* London: Academic Press.

Kitchen, F. D. & Bearn, A. B. (1965). The serum group specific component in non-human primates. *American Journal of Human Genetics*, **17**, 42–50.

Kleiber, H. (1961). *The fire of life. An introduction to animal energetics.* New York: Wiley.

Klein, D. F. (1978). The diet and reproductive cycle of a population of vervet monkeys (*Cercopithecus aethiops*). PhD dissertation, New York University.

Kleiman, D.G. & Eisenberg, J. F. (1973). Comparisons of canid and felid social systems from an evolutionary perspective. *Animal Behaviour*, **21**, 637–59.

Knight, R. S. & Siegfried, W. R. (1983). Inter-relationships between type, size and colour of fruits and dispersal in Southern African trees. *Oecologia*, **56**, 405–12.

Kock, D. (1969). Die Verbreitung der Primaten in Sudan. *Zeitschrift für Säugetierkunde*, **34**, 193–216.

Kolka, M. A. & Elizondo, R. S. (1983). Thermoregulation in *Erythrocebus patas*: A thermal balance study. *Journal of Applied Physiology*, **55**, 1603–8.

Konishi, M. (1978). Evolution of design features in the coding of species specificity. *American Zoologist*, **18**, 57–72.

Koop, B. F., Goodman, M., Xu, P., Chan, K. & Slightom, J. L. (1986). Primate η-globin DNA sequences and man's place among the great apes. *Nature, London*, **319**, 234–8.

Kowamoto, Y., Shotake, T. & Nozawa, K. (1982). Genetic differentiation among three genera of Cercopithecidae. *Primates*, **23**, 272–86.

Krige, P. D. & Lucas, J. W. (1974). Aunting behaviour in an urban troop of *Cercopithecus aethiops*. *Journal of the Behavioral Sciences*, **2**, 55–61.

Krige, P. D. & Lucas, J. W. (1975). Behavioural development of the infant vervet monkey in a free-ranging troop. *Journal of the Behavioral Sciences*, **2**, 151–60.

Kruskal, J. B. (1964a). Multidimensional scaling by optimising goodness of fit to a nonmetric hypothesis. *Psychometrika*, **29**, 1–27.

Kruskal, J. B. (1964b). Nonmetric multidimensional scaling: A numerical method. *Psychometrika*, **29**, 115–29.

Kuhn, H.-J. (1964). Zur Kenntniss von Bau und Funktion des Magens der Schlankaffen (Colobinae). *Folia primatologica* **2**, 193–221.

Kuhn, H.-J. (1965). A provisional checklist of the mammals of Liberia. *Senckenbergiana Biologica*, **46**, 321–40.

Kummer, H. (1956). Sozialverhalten bei eine Mantelpaviangruppe. *Revue Suisse de zoologie*, **63**, 288–97.

Kummer, H. (1968). *Social organisation of hamadryas baboons*. Basel, Karger.

Kummer, H. (1970). Behavioural characters in Primate taxonomy. In *Old World monkeys: evolution, systematics and behaviour*, ed. J. R. Napier & P. H. Napier, pp. 25–36. New York: Academic Press.

Kummer, H., Goetz, W. & Angst, W. (1970). Cross-species modifications of social behavior in baboons. In *Old World monkeys: evolution, systematics and behaviour*, ed. J. R. Napier & P. H. Napier, pp. 351–64. New York: Academic Press.

Kummer, H. & Kurt, F. (1963). Social units of a free-living population of hamadryas baboons. *Folia primatologica*, **1**, 4–19.

Kuroda, S., Kano, T. & Muhindo, K. (1985). Further information on the new monkey species, *Cercopithecus salongo* Thys van den Audenaerde 1977. *Primates*, **26**, 325–33.

Lack, P. C. (1986). Diurnal and seasonal variation in biomass of arthropods in Tsavo East National Park, Kenya. *African Journal of Ecology*, **24**, 47–51.

Lancaster, J. B. (1971). Play-mothering: The relations between juvenile females and young infants among free-ranging vervet monkeys (*C. aethiops*). *Folia primatologica*, **15**, 161–82.

Lancaster, J. B. (1972). Play-mothering: the relations between juvenile females and young infants among free-ranging vervet monkeys. In *Primate socialization*, ed. F. E. Poirier, pp. 83–104. New York: Random House.

Lancaster, J. B. (1973). Coalitions among adult females in a group of free-ranging vervet monkeys. *Proceedings of the IVth international congress of primatology*. Basel: Karger.

Lancaster, J. B. & Lee, R. B. (1965). The annual reproductive cycle in monkeys and apes. In *Primate behavior: Field studies of monkeys and apes*, ed. I. DeVore, pp. 486–513. New York: Holt, Rinehart & Winston.

Langdale-Brown, I., Osmaton, H. A. & Wilson, J. G. (1964). *The vegetation of Uganda*. Entebbe: Government Printer.

Laurent, R. F. (1973). A parallel survey of equatorial amphibians and reptiles in Africa and South America. In *Tropical Forest Ecosystems in Africa and South America*, ed. B. J. Meggers, E. S. Ayensu & W. D. Duckworth, pp. 259–66. Washington, DC: Smithsonian Institution Press.

Lawick-Goodall, J. van (1968). The behaviour of free-living chimpanzees in the Gombe Stream Reserve. *Animal Behaviour Monographs*, **1**, 161–311.

Leakey, L. (1958). Problems relating to fossil man. *The Leech*, **28**, 116–19.

Leakey, M. G. (1982). Extinct large colobines from the Plio-Pleistocene of Africa. *American Journal of Physical Anthropology*, **58**, 157–72.

Leakey, M. G. (1985). Early Miocene cercopithecids from Buluk, Northern Kenya. *Folia primatologica*, **44**, 1–14.

Leakey, M. G. & Delson, E. (1987). Fossil Cercopithecidae from the Laetoli Beds, Tanzania. In *The Pliocene site of Laetoli, northern Tanzania*, ed. M. D. Leakey & J. M. Harris, pp. 91–107. Oxford: Oxford University Press.

Leakey, R. E. F. (1969). New Cercopithecidae from the Chemeron Beds of Lake Baringo, Kenya. *Fossil Vertebrates of Africa*, **1**, 53–70.

Le Boeuf, B. J. (1974). Male–male competition and reproductive success in elephant seals. *American Zoologist*, **14**, 163–76.

Lebrun, J. & Gilbert, G. (1954). Une classification écologique des forêts du Congo. *Publications de l'Institut National pour l'Etude Agronomique du Congo belge*, série scientifique, **63**, 89 pp.

Ledbetter, D. H. (1981). Chromosomal evolution and speciation in the genus *Cercopithecus* (Primates, Cercopithecinae). PhD dissertation, The University of Texas, Austin. University Microfilms, Ann Arbor, Michigan.

Lee, E. T. (1974). The shape oriented dissimilarity of polygons and its application to the classification of chromosome images. *Pattern Recognition*, **6**, 47–60.

Lee, E. T. (1976). An application of fuzzy sets to the classification of geometric figures and chromosome images. *Information Sciences*, 10, 95–114.

Lee, P. C. (1984a). Early infant development and maternal care in free-ranging vervet monkeys. *Primates*, **25**, 36–47.

Lee, P. C. (1984b). Ecological constraints on the social development of vervet monkeys. *Behaviour*, **91**, 245–62.

Lee, P. C. (1988). Nutrition, fertility and maternal investment in primates. *Journal of Zoology, London*, in press.

Lee, P. C., Brennan, E. J., Else, J. G. & Altmann, J. (1986). Ecology and behaviour of vervet monkeys in a tourist lodge habitat. In *Primate ecology and conservation*, ed. J. G. Else & P. C. Lee, pp. 229–235. Cambridge: Cambridge University Press.

Leland, L., Struhsaker, T. T. & Butynski, T. M. (1984). Infanticide by adult males in three primate species of the Kibale Forest, Uganda: a test of hypotheses. In *Infanticide, comparative and evolutionary perspectives*, ed. G. Hausfater & S. B. Hrdy, pp. 151–72. New York: Aldine Publishing Co.

Lenglet, G. L. (1984). Sexual dimorphism of the weight of the skull in African Cercopithecidae (Primates, Catarrhini). *Folia primatologica*, **42**, 134–46.

Lestrel, P. E., Kimbel, W. H., Prior, G. W. & Fleischmann, M. L. (1976). Size and shape of the Hominoid distal femur: Fourier analysis. *American Journal of Physical Anthropology*, **46**, 281–90.

Leutenegger, W. (1978). Scaling of sexual dimorphism in body size and breeding systems in primates. *Nature, London*, **272**, 610–11.

Leutenegger, W. (1982). Scaling of sexual dimorphism in body weight and canine size in primates. *Folia primatologica*, **37**, 163–76.

Leutenegger, W. & Kelly, J. T. (1977). Relationship of sexual dimorphism in canine size and body size to social, behavioral and ecological correlates in anthropoid primates. *Primates*, **18**, 117–36.

Lewontin, R.C. (1974). *The genetic basis of evolutionary change*. New York: Columbia University Press.

Lind, E. M. & Morrison, M. E. S. (1974). *East African vegetation*. London: Longman.

Lindburg, D. G. (1987). Seasonality of reproduction in primates. In *Comparative Primate Biology*, 2-B, ed. G. Mitchell & J. Erwin. New York: Alan R. Liss.

Livingstone, D. A. (1967). Postglacial vegetation of the Ruwenzori Mountains in Equatorial Africa. *Ecological Monographs*, **37**, 25–52.

Livingstone, D. A. (1971). A 22,000-year pollen record from the plateau of Zambia. *Limnology & Oceanography*, **16**, 349–56.

Livingstone, D. A. (1975). Late Quaternary climatic change in Africa. *Annual Review of Ecology and Systematics*, **6**, 249–80.

Livingstone, D. A. (1980). Environmental changes in the Nile headwaters. In *The Sahara and the Nile*, ed. M. A. J. Williams & H. Faure, pp. 339–59. Rotterdam: Balkema.

Livingstone, D. A. (1981). Deep drilling in African lakes. *Palaeoecology of Africa*, **13**, 121.

Livingstone, D. A., Rowland, M. & Bailey, P. E. (1982). On the size of African riverine fish faunas. *American Zoologist*, **22**, 361–9.

Lönnberg, E. (1919). Contributions to the knowledge about the monkeys of Belgian Congo. *Revue de Zoologie africaine*, **7**, 107–54.

McDermid, E. M. & Ananthakrishnan, R. (1972). Red cell enzymes and serum proteins of *Cercopithecus aethiops* (South African Green monkey). *Folia primatologica*, **17**, 122–31.

McDermid, E. M., Vos, G. H. & Downing, H. J. (1973). Blood groups, red cell enzymes and serum proteins of baboons and vervets. *Folia primatologica*, **19**, 312–26.

MacDougall, I. & Watkins, R. (1985). Age of the hominoid-bearing sequence at Buluk, northern Kenya. *Nature, London*, **318**, 175–8.

McGrew, W. C., Baldwin, P. J. & Tutin, C. E. G. (1981). Chimpanzees in a hot, dry and open habitat: Mt Assirik, Senegal, West Africa. *Journal of Human Evolution*, **10**, 227–44.

McGuire, M. T. (1974). The St Kitts Vervet. *Contributions to Primatology*, **1**, 1–199.

Machado, A. de Barros (1965). Acerca do achado en Angola de una forma racial inédita do *Cercopithecus ascanius* (Audebert, 1799). *Boletim da Academia das Ciéncias de Lisboa*, **37**, 182–98.

Machado, A. de Barros (1969). *Mamiferos de Angola ainda nâo citados ou pouco conhecidos.* Publicaçoes culturais, Companhia de Diamantes de Angola, Lisboa, **46**, 93–232.

McHenry, H. M. & Corruccini, R. S. (1975). Distal humerus in hominoid evolution. *Folia primatologica*, **23**, 227–44.

MacInnes, D. G. (1943). Notes on East African Miocene primates. *Journal of the East African Natural History Society*, **17**, 141–81.

McKey, D. B. (1978). Soils, vegetation and seed eating by black *Colobus* monkeys. In *The ecology of arboreal folivores*, ed. G. G. Montgomery, pp. 423–37. Washington, DC: Smithsonian Institution Press.

MacLarnon, A. M., Chivers, D. J. & Martin, R. D. (1986a). Gastro-intestinal allometry in primates and other mammals including new species. In *Primate Ecology and Conservation*, ed. J. G. Else & P. C. Lee, pp. 75–85. Cambridge: Cambridge University Press.

MacLarnon, A. M., Martin, R. D., Chivers, D. J. & Hladik, C. M. (1986b). Some aspects of gastro-intestinal allometry in primates and other mammals. In *Définition et origines de l'Homme*, ed. M. Sakka, pp. 293–302. Paris: Editions du CNRS.

McMahon, R. P. (1977). Aspects of the behaviour of samango monkeys (*Cercopithecus (mitis) albogularis labiatus*). MSc thesis, University of Natal.

McMahon, T. A. & Bonner, J. T. (1983). *On size and life.* New York: Scientific American Books.

Mahalanobis, P. C. (1936). On the generalized distance in statistics. *Proceedings of the National Institute of Science, India,* **2**, 49–55.

Mahoney, S. A. (1980). Cost of locomotion and heat balance during rest and running from 0° to 55° C in a patas monkey. *Journal of Applied Physiology,* **49**, 789–800.

Mainguet, M., Canon, L. & Chemin, M. C. (1980). Le Sahara: géomorphologie et paléogéomorphologie éoliennes. In *The Sahara and the Nile,* ed. M. A. J. Williams & H. Faure, pp. 17–35. Rotterdam: Balkema.

Malbrant, R. (1952). *Faune du Centre Africain Français.* Paris: Paul Lechevalier.

Malbrant, R. & Maclatchy, A. (1949). *Faune de l'Equateur Africain Français.* Mammifères, 2. Encyclopédie Biologique. Paris: Paul Lechevalier.

Maley, J. & Livingstone, D. A. (1983). Extension d'un élément montagnard dans le sud du Ghana (Afrique de l'Ouest) au Pléistocène Supérieur et à l'Holocène inférieur: premières données polliniques. *Comptes Rendus des Séances de l'Académie des Sciences, Paris,* Série II, 1287–92.

Malik, I., Seth, P. K. & Southwick, C. H. (1985). Group fission in free-ranging rhesus monkeys of Tughlaqabad, Northern India. *International Journal of Primatology,* **6**, 411–22.

Maples, W. R. (1972). Systematic reconsideration and a revision of the nomenclature of Kenya baboons. *American Journal of Physical Anthropology,* **36**, 9–20.

Marler, P. (1957). Specific distinctiveness in the communication signals of birds. *Behaviour,* **11**, 13–39.

Marler, P. (1961). The logical analysis of animal communication. *Journal of Theoretical Biology,* **1**, 295–317.

Marler, P. (1965). Communication in monkeys and apes. In *Primate behavior: field studies of monkeys and apes,* ed. I. DeVore, pp. 544–85. New York: Holt, Rinehart & Winston.

Marler, P. (1968). Aggregation and dispersal: two functions in primate communication. In *Primates: studies in adaptation and variability,* ed. P. Jay, pp. 428–38. New York: Holt, Rinehart & Winston.

Marler, P. (1970). Vocalizations of East African monkeys. I. Red Colobus. *Folia primatologica,* **13**, 81–91.

Marler, P. (1972). Vocalizations of East African monkeys. II. Black and White Colobus. *Behaviour,* **42**, 175–97.

Marler, P. (1973). A comparison of vocalizations of red tailed monkeys and blue monkeys, *Cercopithecus ascanius* and *C. mitis,* in Uganda. *Zeitschrift für Tierpsychologie,* **33**, 223–47.

Marsh, C. W. (1978). Tree phenology in a gallery forest on the Tana River, Kenya. *East African Agricultural and Forestry Journal,* **43**, 305–16.

Marshall, J. T. & Marshall, E. R. (1976). Gibbons and their territorial songs. *Science,* **193**, 235–7.

Marshall, J. T., Sugardjito, J. & Markaya, M. (1984). Gibbons of the lar group: relationships based on voice. In *The Lesser Apes. Evolutionary and behavioural biology,* ed. H. Preushoft, D. J. Chivers, W. Y. Brockelman & N. Creel, pp. 533–41. Edinburgh: Edinburgh University Press.

Martin, R. D. (1978). Towards a new definition of Primates. *Man,* **3**, 377–401.

Martin, R. D. (1980). Adaptation and body size in primates. *Zeitschrift für Morphologie und Anthropologie,* **71**, 115–24.

Martin, R. D. (1984). Body size, brain size and feeding strategies. In *Food acquisition and processing in Primates*, ed. D. J. Chivers, B. A. Wood & A. Bilsborough, pp. 73–104. New York: Plenum Press.

Martin, R. D., Chivers, D. J., MacLarnon, A. M. & Hladik, C. M. (1985). Gastrointestinal allometry in primates and other mammals. In *Size and scaling in Primate biology*, ed. W. L. Jungers, pp. 61–89. New York: Plenum Press.

Maslow, A. H. (1936). The role of dominance in the social and sexual behaviour of infra-human primates. *Journal of Genetics and Psychology*, **49**, 161–338.

Maslow, A. H. (1948). Dominance quality and social behaviour in infra-human primates. *Journal of the Society of Psychology*, **11**, 313–24.

Matschie, P. (1912). Beschreibungen einiger neuer Meerkatzen des 'Musée du Congo Belge'. *Revue de Zoologie africaine*, **1**, 433–42.

Matschie, P. (1913). Neue Affen aus Afrika-Nebst einigen Bemerkungen über bekannte Formen. *Annales de la Société royale zoologique et malacologique de Belgique*, **47**, 45–81.

Mayr, E. (1963). *Animal species and evolution*. Cambridge: Harvard University Press.

Mayr, E. (1969). *Principles of systematic zoology*. New York: McGraw Hill.

Mayr, E. (1974). *Populations espèces et évolution*. Paris: Hermann.

Melnick, D. J. & Kidd, K. K. (1983). The genetic consequences of social group fission in a wild population of rhesus monkeys (*Macaca mulatta*). *Behavioural Ecology and Sociobiology*, **12**, 229–36.

Miller, R. E., Murphy, J. V. & Mirsky, I. A. (1959). Relevance of facial expression and posture as cues in communication of affect between monkeys. *Archivs of Genetic Psychiatry*, **1**, 48–488.

Milton, K. (1979). Factors influencing leaf choice in howler monkeys: a test for some hypotheses of food selection by generalists herbivores. *American Naturalist*, **114**, 362–78.

Milton, K. (1982). Dietary quality and demographic regulation in a howler monkey population. In *The Ecology of a Tropical Forest*, ed. E. G. Leigh, Jr, A. S. Rand & D. M. Windsor, pp. 273–89. Washington, DC: Smithsonian Institution Press.

Milton, K. (1983). Morphometric features as tribal predictors in North Western Amazonia. *Annals of Human Biology*, **10**, 435–40.

Misonne, X. (1963). *Les rongeurs du Ruwenzori et des régions voisines*. (Exploration du Parc National Albert, 14, Institut des Parcs Nationaux du Congo.)

Monard, A. (1938). Résultats de la mission scientifique du Dr Monard en Guinée Portugaise 1937–38. *Arquivos do Museu Bocage*, **9**, 121–50.

Montgomery, G. G. & Sunquist, M. E. (1978). Habitat selection and use by two-toed and three-toed sloths. In *The Ecology of Arboreal Folivores*, ed. G. G. Montgomery, pp. 329–59. Washington, DC: Smithsonian Institution Press.

Moore, J. (1984). Female transfer in primates. *International Journal of Primatology*, **5**, 537–90.

Moreau, R. E. (1966). *The bird faunas of Africa and its islands*. London: Academic Press.

Moreau, R. E. (1969). Climatic changes and the distribution of forest vertebrates in West Africa. *Journal of Zoology, London*, **158**, 39–61.

Moreno-Black, G. & Maples, W. R. (1977). Differential habitat utilization of four *Cercopithecidae* in a Kenyan forest. *Folia primatologica*, **27**, 87–107.

Mörike, D. (1973). Verhalten einer Gruppe von Dianameerkatzen im Frankfurter Zoo. *Primates*, **14**, 263–300.

Morrison, M. E. S. (1968). Vegetation and climate in the uplands of south-western Uganda during the Later Pleistocene Period. 1. Muchoya Swamp, Kigezi District. *Journal of Ecology*, **56**, 363–84.

Morrison, M. E. S. & Hamilton, A. C. (1974). Vegetation and climate in the uplands of south-western Uganda during the Later Pleistocene Period, II. Forest clearance and other vegetational changes in the Rukiga Highlands during the last 8000 years. *Journal of Ecology*, **63**, 1–31.

Mott, C. S., Turner, T. R. & Else, J. G. (1984). Genetic differentiation in three populations of Kenya *Cercopithecus mitis*. *American Journal of Physical Anthropology*, **63**, 197.

Moynihan, M. (1967). Comparative aspects of communication in New-World Primates. In *Primate Ethology*, ed. D. Morris, pp. 236–65. London: Weidenfeld & Nicholson.

Muleris, M., Couturier, J. & Dutrillaux, B. (1981). Le caryotype de *Cercopithecus (mona) campbelli campbelli*. Comparaison avec les autres cercopithèques et l'homme. *Annales de Génétique*, **24**, 137–40.

Muleris, M., Gautier, J.-P., Lombard, M. & Dutrillaux, B. (1985). Etude cytogénétique de *Cercopithecus wolfi*, *Cercopithecus erythrotis*, et d'un hybride *Cercopithecus ascanius* × *Cercopithecus pogonias grayi*. *Annales de Génétique*, **28**, 75–80.

Muleris, M., Couturier, J. & Dutrillaux, B. (1986). Phylogénie chromosomique des Cercopithecoidea. *Mammalia*, **50**, No. spécial, 38–52.

Mykytowycz, R. (1972). The behavioural role of the mammalian skin glands. *Naturwissenschaften*, **59**, 133–9.

Nagel, U. (1971). Social organization in a baboon hybrid zone. *Proceedings of the third Congress of Primatology*, Vol. 3, 48–57.

Napier, J. R. (1960). Studies of the hands of living primates. *Proceedings of the Zoological Society, London*, **134**, 647–57.

Napier, J. R. (1970a). Palaeoecology and Catarrhine evolution. In *Old World Monkeys*, ed. J. R. Napier & P. H. Napier, pp. 55–95. New York: Academic Press.

Napier, J. R. (1970b). *The roots of mankind*. London: George Allen & Unwin.

Napier, J. R. & Napier, P. H. (1967). *A handbook of living primates*. London: Academic Press.

Napier, J. R. & Napier, P. H. (1985). *Natural history of the primates*. London: British Museum (Natural History).

Napier, P. H. (1981). *Catalogue of Primates in the British Museum (Natural History); Part II: Family Cercopithecidae, Subfamily Cercopithecinae*. London: British Museum (Natural History).

Napier, P. H. (1985). *Catalogue of Primates in the British Museum (Natural History); Part III: Family Cercopithecidae, Subfamily Colobinae*. London: British Museum (Natural History).

Nash, L. T. (1976). Troop fission in free-ranging baboons in the Gombe Stream National Park, Tanzania. *American Journal of Physical Anthropology*, **44**, 63–77.

Nash, L. T. (1983). Reproductive patterns in galagos (*Galago zanzibaricus* and *Galago garnettii*) in relation to climatic variability. *American Journal of Primatology*, **5**, 181–96.

National Atlas of Kenya (1970). 3rd edition. Nairobi: Survey of Kenya.

Nei, M. (1975). *Molecular population genetics and evolution*. New York: Elsevier.

Nei, M. (1985). Stochastic aspects of nucleotide evolution. Paper presented at

'*Evolutionary perspectives and the new genetics*', University of Michigan Medical School, Ann Arbor, Michigan, 17–18 June, 1985.

Newman, J. D. (1985). Squirrel monkey communication. In *Handbook of squirrel monkey research*, ed. L. A. Rosenblum & C. L. Coe, pp. 99–126. New York: Plenum Press.

Newman, J. D. & Symmes, D. (1982). Inheritance and experience in the acquisition of primate acoustic behaviour. In *Primate communication*, ed. C. T. Snowdon, C. H. Brown & M. R. Petersen, pp. 259–78. New York: Cambridge University Press.

Nummelin, M. (1986). The seasonal fluctuations of forest floor insect densities on the areas of different forestry practices in Kibale Forest, western Uganda. *Abstract*. International Conference on Tropical Entomology – Nairobi.

Oates, J. F. (1981). Mapping the distribution of West African rain-forest monkeys: issues, methods and preliminary results. *Annals of the New York Academy of Sciences*, **376**, 53–64.

Oates, J. F. (1982). In search of rare forest primates in Nigeria. *Oryx*, **16**, 431–6.

Oates, J. F. (1985). The Nigerian guenon, *Cercopithecus erythrogaster*: ecological, behavioral, systematic and historical observations. *Folia primatologica*, **45**, 25–43.

Oates, J. F. (1986). *Action plan for African primate conservation: 1986–1990*. New York: IUCN/SSC Primate Specialist Group.

Oates, J. F. & Jewell, P. A. (1967). Westerly extent of the range of three African lorisoid primates. *Nature, London*, **215**, 778–9.

Oates, J. F. & Trocco, T. F. (1983). Taxonomy and phylogeny of black-and-white colobus monkeys. *Folia primatologica*, **40**, 83–113.

Omar, A. and DeVos, A. (1971). The annual reproductive cycle of an African monkey (*Cercopithecus mitis kolbii* Neuman). *Folia primatologica*, **16**, 206–15.

Oppenheimer, J. R. (1977). Communication in New World monkeys. In *How animals communicate*, ed. T. E. Sebeok, pp. 851–89. Bloomington: Indiana University Press.

Otis, J. S., Froehlich, J. W. & Thorington, R. W. (1981). Seasonal and age-related differential mortality by sex in the mantled howler monkey, *Alouatta palliata*. *International Journal of Primatology*, **2**, 197–205.

Otte, D. (1974). Effects and functions in the evolution of signalling systems. *Annual Review of Ecology and Systematics*, **5**, 385–417.

Oxnard, C. (1973). *Form and pattern in human evolution: Some mathematical, physical and engineering approaches*. Chicago: University of Chicago Press.

Oxnard, C. (1975). *Uniqueness and diversity in human evolution: Morphometric studies of Australopithecines*. Chicago: University of Chicago Press.

Paterson, H. E. H. (1985). The recognition concept of species. In *Species and speciation*, ed. E. S. Vrba, pp. 21–9. Pretoria: Transvaal Museum.

Pearson, K. (1926). On the coefficient of racial likeness. *Biometrika*, **18**, 105–17.

Penrose, L. S. (1954). Distance, size and shape. *Annals of Eugenics*, **18**, 337–43.

Pereira, M. E. (1983). Abortion following the immigration of an adult male baboon (*Papio cynocephalus*). *American Journal of Primatology*, **4**, 93–8.

Perret, D. I., Smith, P. A., Potter, D. D., Mistlin, A. S., Head, A. S., Milner, A. D. & Jeeves, M. A. (1984). Behavioural and natural responses to faces in the macaque. *Primate Eye*, **25**, 13–14.

Perrott, R. A. & Street-Perrott, F. A. (1982). New evidence for a late Pleistocene wet phase in northern intertropical Africa. *Palaeoecology of Africa*, **14**, 57–83.

Peters, R. H. (1983). *The ecological implications of body size*. Cambridge: Cambridge University Press.

Pickford, M. (1975). *Stratigraphy and palaeoecology of five late Cainozoic formations in the Kenya Rift Valley*. PhD dissertation, University of London.

Pickford, M. (1981). Preliminary Miocene mammalian biostratigraphy for Western Kenya. *Journal of Human Evolution*, **10**, 73–97.

Pickford, M. (1982). New higher primate fossils from the Middle Miocene deposits at Majiwa and Kaloma, Western Kenya. *American Journal of Physical Anthropology*, **58**, 1–19.

Pickford, M. (1983). Sequence and environments of the lower and middle Miocene hominoids of Western Kenya. In *New interpretations of ape and human ancestry*, ed. R. L. Ciochon & R. S. Corruccini, pp. 421–39. New York: Plenum Press.

Pickford, M. (1984). Kenya palaeontology gazetteer – Vol. 1, Western Kenya. *Special Publications of the National Museums of Kenya*, 1–282.

Pickford, M. (1985). A new look at *Kenyapithecus* based on recent collections from Western Kenya. *Journal of Human Evolution*, **14**, 113–43.

Pickford, M. (1986). The geochronology of Miocene higher primate faunas of East Africa. In *Primate evolution*, ed. J. Else & P. Lee, pp. 19–33. Cambridge: Cambridge University Press.

Pickford, M., Senut, B., Hadoto, D., Musisi, J. & Kariira, C. (1986). Nouvelles découvertes dans le Miocène inférieur de Napak, Ouganda Oriental. *Comptes Rendus des Séances de l'Académie des Sciences, Paris*, Série II, **302**, 47–52.

Pilgrim, G. E. (1915). New Siwalik primates and their bearing on the question of the evolution of man and the anthropoids. *Recent Geological Survey, India*, **45**, 1–74.

Poche, R. M. (1976). Notes on primates in Parc National du W. du Niger, West Africa. *Mammalia*, **40**, 187–98.

Poirier, F. E. (1968). Nilgiri langur (*Presbytis johnii*) territorial behaviour. *Primates*, **9**, 351–64.

Poirier, F. E. (1972). The St Kitts green monkey (*Cercopithecus aethiops sabaeus*): ecology, population dynamics, and selected behavioral traits. *Folia primatologica*, **17**, 20–55.

Pokras, E. M. & Mix, A. C. (1985). Eolian evidence for spatial variability of late Quaternary climates in tropical Africa. *Quaternary Research*, **24**, 137–49.

Post, D. G. (1982). Feeding behavior of yellow baboons (*Papio cynocephalus*) in the Amboseli National Park, Kenya. *International Journal of Primatology*, **3**, 403–30.

Powell, J. (1983). Interspecific cytoplasmic gene flow in the absence of nuclear gene flow: evidence from *Drosophila*. *Proceedings of the National Academy of Science, USA*, **80**, 492–5.

Prell, W. L., Hutson, W. H., Williams, D. F., Bé, A. W. H., Geitzenauer, K. & Molfino, B. (1980). Surface circulation of the Indian Ocean during the last glacial maximum, approximately 18,000 yr B.P. *Quaternary Research*, **14**, 309–36.

Prost, J. H. (1965). Methodology and gait analysis of gaits of monkeys. *American Journal of Physical Anthropology*, **23**, 215–40.

Quris, R. (1973). Emissions sonores servant au maintien du groupe social chez *Cercocebus galeritus agilis*. *La Terre et la Vie*, **27**, 232–67.

Quris, R. (1975). Ecologie et organisation sociale de *Cercocebus galeritus agilis* dans le nord-est du Gabon. *La Terre et la Vie*, **29**, 337–98.

Quris, R. (1976). Données comparatives sur la socioécologie de huit espèces de Cercopithecidae vivant dans une même zone de forêt primitive périodiquement inondée (N-E Gabon). *La Terre et la Vie*, **30**, 193–209.

Quris, R. (1980). Emissions vocales de forte intensité chez *Cercocebus galeritus*: structure, caractéristiques spécifiques et individuelles, modes d'émission. *Mammalia*, **44**, 35–50.

Quris, R., Gautier, J.-P., Gautier, J.-Y. & Gautier-Hion, A. (1981). Organisation spatio-temporelle des activités individuelles et sociales dans une troupe de *Cercopithecus cephus*. *Revue d'Ecologie (Terre Vie)*, **35**, 37–53.

Rahm, U. (1966). Les Mammifères de la forêt équatoriale de l'Est du Congo. *Annales du Musée royal de l'Afrique centrale, Sciences Zoologiques*, **149**, 39–121.

Rahm, U. (1970). Ecology, zoogeography and systematics of some African forest monkeys. In *Old World monkeys: evolution, systematics and behaviour*, ed. J. R. Napier & P. H. Napier, pp. 589–626. London: Academic Press.

Ransom, T. W. (1976). Ecology and social behaviour of baboons (*Papio anubis*) at the Gombe National Park. PhD dissertation, University of California, Berkeley.

Rea, D. K. & Schrader, H. (1985). Late Pliocene onset of glaciation: ice-rafting and diatom stratigraphy of North Pacific DSDP cores. *Palaeogeography, Palaeoclimatology, Palaeoecology*, **49**, 313–25.

Redican, W. K. (1975). Facial expressions in non-human primates. *Primate Behaviour*, Vol. 4, ed. L. A. Rosenblum, pp. 184–94. New York: Academic Press.

Relethford, J. H. & Hodges, D. C. (1985). A statistical test for differences in sexual dimorphism between populations. *American Journal of Physical Anthropology*, **66**, 55–61.

Rhodesia: Its natural resources and economic development (1965). Salisbury: Collins.

Richard, A. (1985). *Primates in nature*. New York: Freeman & Co.

Richman, B. (1976). Some vocal distinctive features used by gelada monkeys. *Journal of the Acoustic Society of America*, **60**, 718–24.

Ripley, S. (1967). Intertroop encounters among Ceylan grey langurs. In *Social communication among primates*, ed. S. A. Altmann, pp. 237–53. Chicago: University of Chicago Press.

Ritchie, J. C., Eyles, C. H. & Haynes, C. V. (1985). Sediment and pollen evidence for an early to mid-Holocene humid period in the eastern Sahara. *Nature, London*, **314**, 352–5.

Robbins, C. B. (1978). The Dahomey Gap – a reevaluation of its significance as a faunal barrier to West African high forest mammals. *Bulletin of the Carnegie Museum of Natural History*, **6**, 168–74.

Roche, J. (1971). Recherches mammalogiques en Guinée forestière. *Bulletin du Museum National d'Histoire Naturelle*, série 3, **16**, 737–81.

Rodgers, W. A. & Homewood, K. M. (1982). Species richness and endemism in the Usambara mountain forests, Tanzania. *Biological Journal of the Linnean Society*, **18**, 197–242.

Rodgers, W. A., Owen, C. F. & Homewood, K. M. (1982). Biogeography of East African forest mammals. *Journal of Biogeography*, **9**, 41–54.

Rodman, R. S. (1973). Synecology of Bornean Primates: I – a test for interspecific interaction in spatial distribution of five species. *American Journal of Physical Anthropology*, **38**, 655–60.

Rodman, R. S. & Cant, J. G. H. (1984). *Adaptation for foraging in nonhuman primates*. New York: Columbia University Press.

Rollinson, J. M. M. (1975). Interspecific comparisons of locomotor behaviour and prehension in eight species of African forest monkeys. PhD dissertation, University of London.

Rose, M. D. (1973). Quadrupedalism in Primates. *Primates*, **14**, 337–58.

Rose, M. D. (1983). Miocene hominoid post-cranial morphology: monkey-like, ape-like, neither or both? In *New interpretations of ape and human ancestry*, ed. R. L. Ciochon & R. S. Corruccini, pp. 405–17. New York: Plenum Press.

Rosenzweig, M. L. (1968). Net primary productivity of terrestrial communities: Prediction from climatological data. *The American Naturalist*, **102**, 67–74.

Rosevear, D. R. (1953). *Checklist and atlas of Nigerian mammals.* Lagos: Government Printer.

Rossignol-Strick, M. (1983). African monsoons, an immediate climate response to orbital insolation. *Nature, London*, **304**, 46–9.

Rossignol-Strick, M. & Duzer, D. (1979). West African vegetation and climate since 22,500 BP from deep-sea cores. *Pollen et Spores*, **21**, 105–34.

Rossignol-Strick, M., Nesteroff, W., Olive, P. & Vergnaud-Grazzini, C. (1982). After the deluge: Mediterranean stagnation and sapropel formation. *Nature, London*, **295**, 105–10.

Rowell, T. E. (1966). Forest living baboons in Uganda. *Journal of Zoology, London*, **149**, 344–64.

Rowell, T. E. (1967). Variability in the social organisation of primates. In *Primate Ethology*, ed. D. Morris, pp. 219–35. London: Weidenfeld & Nicholson.

Rowell, T. E. (1970). Reproductive cycles of two *Cercopithecus* monkeys. *Journal of Reproduction and Fertility*, **22**, 321–38.

Rowell, T. E. (1972). Organisation of caged groups of *Cercopithecus* monkeys. *Animal Behaviour*, **19**, 625–45.

Rowell, T. E. (1973). Social organization of wild talapoin monkeys. *American Journal of Physical Anthropology*, **38**, 593–8.

Rowell, T. E. (1976). Growing up in a monkey group. *Ethos*, **3**, 22–36.

Rowell, T. E. (1977). Reproductive cycles of the talapoin monkey (*Miopithecus talapoin*). *Folia primatologica*, **28**, 188–202.

Rowell, T. E. (1979). How would we know if social organization were not adaptive? In *Primate ecology and human origins*, ed. I. S. Bernstein and E. O. Smith, pp. 1–22. New York: Garland.

Rowell, T. E. & Chism, J. (1986). The ontogeny of sex differences in the behaviour of patas monkeys. *International Journal of Primatology*, **7**, 83–107.

Rowell, T. E. & Dixson, A. F. (1975). Changes in the social organisation during the breeding season of wild talapoin monkeys. *Journal of Reproduction and Fertility*, **43**, 419–34.

Rowell, T. E. & Hinde, R. A. (1962). Vocal communication by the rhesus monkeys (*Macaca mulatta*). *Proceedings of the Zoological Society of London*, **138**, 279–94.

Rowell, T. E. & Olson, D. K. (1983). Alternative mechanisms of social organisation in monkeys. *Behaviour*, **86**, 31–54.

Rowell, T. E. & Richards, S. M. (1979). Reproductive strategies of some African monkeys. *Journal of Mammalogy*, **60**, 58–69.

Rubenstein, D. I. (1980). On the evolution of alternative mating strategies. In *Limits to action: the allocation of individual strategies*, ed. J. E. R. Straton, pp. 65–100. New York: Academic Press.

Ruddiman, W. F. & Duplessy, J. C. (1985). Conference on the least deglaciation: timing and mechanism. *Quaternary Research*, **23**, 1–17.

Rudran, R. (1978a). Socioecology of the blue monkeys (*Cercopithecus mitis stuhlmanni*) of the Kibale forest, Uganda. *Smithsonian Contributions to Zoology*, **249**, 1–88.

Rudran, R. (1978b). Intergroup dietary comparisons and folivorous tendencies of two groups of blue monkeys, *Cercopithecus mitis*. In *The ecology of arboreal folivores*, ed. G. G. Montgomery, pp. 483–504. Washington, DC: Smithsonian Institution Press.

Rumpler, Y. & Dutrillaux, B. (1986). Evolution chromosomique des Prosimiens. *Mammalia*, **50**, 82–107.

Ruvolo, M. (1983). Genetic evolution in the African guenon monkeys Primates, Cercopithecinae). PhD dissertation, Harvard University. University Microfilms, Ann Arbor, Michigan.

Sabater Pi, J. (1972). Contribution to the ecology of the *Mandrillus sphinx* Linnaeus 1758 of Rio Muni (Republic of Equatorial Guinea). *Folia primatologica*, **17**, 304–19.

Sade, D. S. & Hildrech, R. W. (1965). Notes on the green monkey (*Cercopithecus aethiops sabaeus*), on St Kitts, West Indies. *Caribbean Journal of Science*, **5**, 67–79.

Sadleir, R. M. F. S. (1969). *The ecology of reproduction in wild and domestic mammals*. London: Methuen.

Salzano, F. M., Callegari-Jacques, S. M., Franco, M. H. L. P., Hutz, M. H., Weimer, T. A., Silva, R. S., Da Rocha, F. J. (1980). The Caingang revisited: blood genetics and anthropometry. *American Journal of Physical Anthropology*, **53**, 513–24.

Samuels, A. & Altmann, J. (1986). Immigration of a *Papio anubis* male into a group of *Papio cynocephalus* baboons and evidence for an *anubis–cynocephalus* hybrid zone in Amboseli, Kenya. *International Journal of Primatology*, **7**, 131–8.

Sanderson, I. T. (1940). The mammals of the North Cameroons forest area. *Transactions of the Zoological Society of London*, **24**, 623–725.

Sanghvi, L. D. (1953). Comparison of genetical and morphological methods for a study of biological differences. *American Journal of Physical Anthropology*, **11**, 385–404.

Sarich, V. M. & Wilson, A. C. (1967). Rates of albumin evolution in primates. *Proceedings of the National Academy of Science, USA*, **58**, 142–8.

Sarnthein, M. (1978). Sand deserts during glacial maximum and climatic optimum. *Nature, London*, **272**, 43–6.

Sarnthein, M., Telzlaff, G., Koopmann, B., Wolter, K. & Pflaumann, U. (1981). Glacial and interglacial wind regimes over the eastern subtropical Atlantic and North-West Africa. *Nature, London*, **293**, 193–6.

Sayer, J. A. & Green, A. A. (1984). The distribution and status of large mammals in Benin. *Mammal Review*, **14**, 37–50.

Schaik, C. P. van (1983). Why are diurnal Primates living in groups? *Behaviour*, **87**, 120–43.

Schaik, C. P. van & Noordwijk, M. A. van (1985). Interannual variability in fruit abundance and the reproductive seasonality in Sumatran long-tailed macaques (*Macaca fascicularis*). *Journal of Zoology, London*, **206**, 533–49.

Schapiro, S. J. (1985). Reproductive seasonality: birth synchrony as female–female competition and cooperation in captive *Cercopithecus aethiops* and *C. mitis*. PhD dissertation, University of California, Davis.

Schilling, A. (1980). Bases morphologiques et comportementales de la communication olfactive chez les Prosimiens. Thèse de Doctorat, Université de Paris VI.

Shiøtz, A. (1967). The treefrogs (Rhacophoridae) of West Africa. *Spolia Zoologica Musei Hauniensis*, **25**, 1–346.

Schlichte, H.-J. (1978). The ecology of two groups of blue monkeys, *Cercopithecus mitis stuhlmanni*, in an isolated habitat of poor vegetation. In *The ecology of arboreal folivores*, ed. G. G. Montgomery, pp. 505–17. Washington, DC: Smithsonian Institution Press.

Schlosberg, H. (1952). The description of facial expression in terms of two dimensions. *Journal of Experimental Psychology*, **44**, 229–37.

Schouteden, H. (1947). De Zoogdieren van Belgish-Congo en van Ruanda-Urundi. *Annales du Musée royal du Congo belge*, **2** (3), 1–576.

Schmidt-Nielsen, K. (1984). *Why is animal size so important?* Cambridge: Cambridge University Press.

Schwagmeyer, P. L. (1985). Multiple mating and intersexual selection in thirteen-lined ground squirrels. In *The biology of ground dwelling squirrels*, ed. J. O. Murie & G. R. Michener, pp. 275–93. Lincoln, Nebraska: University of Nebraska Press.

Schwarz, E. (1926). Die Meerkatzen der *Cercopithecus aethiops* Gruppe. *Zeitschrift für Säugetierkunde*, **1**, 28–47.

Schwarz, E. (1928a). Stadien der Artbildung. Die geographischen und biologischen Formen der Mona-Meerkatze (*Cercopithecus mona* Schreber). *Zeitschrift für induktive Abstammungs- und Vererbungslehre*, **2**, 1299–319.

Schwarz, E. (1928b). Notes on the Classification of the African Monkeys in the Genus *Cercopithecus*, Erxleben. *Annals and Magazine of Natural History*, 10th series, **1**, 649–63.

Schwarz, E. (1928c). Die Sammlung Afrikanischer Affen im Congo-Museum. *Revue de Zoologie et de Botanique africaine*, **16**, 105–52.

Schwarz, E. (1932). Der Vertreter der Diana-Meerkatze in Zentral-Afrika. *Revue de Zoologie et de Botanique africaine*, **21**, 251–4.

Scorer, J. (1980). Some factors affecting the feeding ecology and socio-biology of the samango monkey, *Cercopithecus albogularis schwarzi* Roberts, 1931. MSc thesis, University of Pretoria.

Senut, B. (1986a). Upperlimb postcranial elements of Miocene cercopithecoids from East Africa: implications for function and taxonomy. *Primate Report*, **14**, 87.

Senut, B. (1986b). Long bones of the primate upper limb: monomorphic or dimorphic? *Human Evolution*, **1**, 7–22.

Servant, M. & Servant-Valdary, S. (1980). L'environnement quaternaire du bassin du Tchad. In *The Sahara and the Nile*, ed. M. A. J. Williams & H. Faure, pp. 133–62. Rotterdam: Balkema.

Seyfarth, R. M. & Cheney, D. L. (1980). The ontogeny of vervet monkey alarm calling behaviour: a preliminary report. *Zeitschrift für Tierpsychologie*, **54**, 37–56.

Seyfarth, R. M. & Cheney, D. L. (1982). How monkeys see the world: a review of recent research on East African vervet monkeys. In *Primate communication*, ed. C. T. Snowdon, C. H. Brown & M. R. Petersen, pp. 239–52. Cambridge: Cambridge University Press.

Seyfarth, R. M. & Cheney, D. L. (1984). The acoustic features of vervet monkey grunts. *Journal of Acoustical Society of America*, **75**, 1623–8.

Seyfarth, R. M., Cheney, D. L. & Marler, P. (1980a). Vervet monkey alarm calls: semantic communication in a free-ranging primate. *Animal Behaviour*, **28**, 1070–94.

Seyfarth, R. M., Cheney, D. L. & Marler, P. (1980b). Monkey response to three different alarm calls: evidence of predator classification and semantic communication. *Science, NY*, **210**, 801–3.

Shackleton, N. J. & Opdyke, N. D. (1973). Oxygen isotope and paleomagnetic stratigraphy of Equatorial Pacific Core V2-238: oxygen isotope temperatures and ice volumes on a 10^5 and 10^6 year scale. *Quaternary Research*, **3**, 39–55.

Shackleton, N. J. *et al.* (1984). Oxygen isotope calibration of the onset of ice-rafting and history of glaciation in the North Atlantic region. *Nature, London*, **307**, 620–3.

Shotake, T. (1981). Population genetical study of natural hybridization between *Papio anubis* and *P. hamadryas*. *Primates*, **22**, 285–308.

Sibley, C. G. & Ahlquist, J. E. (1984). The phylogeny of the hominoid primates, as indicated by the DNA–DNA hybridization. *Journal of Molecular Evolution*, **20**, 2–15.

Simons, E. L. (1967). Review of the phyletic interrelations of Oligocene and Miocene Old World Anthropoidea. In *Evolution des Vertébrés. Problèmes actuels de paléontologie. Colloques Internationaux du Centre National de la Recherche Scientifique*, Paris, **163**, 597–602.

Simons, E. L. (1969). A Miocene monkey (*Prohylobates*) from northern Egypt. *Nature, London*, **223**, 687–9.

Simons, E. L. (1970). The deployment and history of Old World monkeys (Cercopithecidae, Primates). In *Old World monkeys: evolution, systematics and behaviour*, ed. J. R. Napier & P. H. Napier, pp. 97–137. New York: Academic Press.

Simons, E. L. (1971). A current review of the interrelationships of Oligocene and Miocene Catarrhini. In *Dental morphology and evolution*, ed. A. Dahlberg, pp. 193–208. Chicago: University of Chicago Press.

Simons, E. L. (1972). *Primate evolution: an introduction to man's place in nature*. New York: MacMillan.

Simons, E. L. (1974). *Parapithecus grangeri* (Parapithecidae; Old World Higher Primates): new species from the Oligocene of Egypt and initial differentiation of Cercopithecoidea. *Postilla*, **166**, 1–12.

Simons, E. L. (1985). Origins and characteristics of the first hominoids. In *Ancestors: the hard evidence*, ed. E. Delson, pp. 37–41. New York: Alan R. Liss.

Sinclair, A. R. E. (1978). Factors affecting the food supply and breeding season of resident birds and movements of Palaearctic migrants in a tropical African savannah. *Ibis*, **120**, 480–97.

Slatkin, M. & Hausfater, G. (1976). A note on the activities of a solitary male baboon. *Primates*, **17**, 311–22.

Sly, D. L., Harbaugh, S. W., London, W. T. & Rice, J. M. (1983). Reproductive performance of a laboratory breeding colony of patas monkeys (*Erythrocebus patas*). *American Journal of Primatology*, **4**, 23–32.

Smith, D. G. (1980). Paternity exclusion in six captive groups of rhesus monkeys (*Macaca mulatta*). *American Journal of Physical Anthropology*, **53**, 243–9.

Smuts, B. B. (1985). *Sex and friendship in baboons*. New York: Aldine.

Smythe, N. (1986). Competition and resource partitioning in the guild of neotropical terrestrial frugivorous mammals. *Annual Review of Ecology and Systematics*, **17**, 169–88.

Smythe, N., Glanz, W. E. & Leigh, E. G., Jr (1982). Population regulation in some terrestrial frugivores. In *The Ecology of a Tropical Forest*, ed. E. G. Leigh, Jr, A. S. Rand & D. M. Windsor, pp. 227–38. Washington, DC: Smithsonian Institution Press.

Sneath, P. H. A. (1967). Trend surface analysis of transformation grids. *Journal of Zoology, London*, **151**, 65–122.

Sneath, P. H. A. & Sokal, R. R. (1973). *Numerical taxonomy. The principles and practice of numerical classification.* San Francisco: W. H. Freeman & Co.

Snowdon, C. T., Hodun, A., Rosenberger, A. L. & Coimbra-Filho, A. F. (1986). Long calls structure and its relation to taxonomy in Lion Tamarins. *American Journal of Primatology*, **11**, 253–61.

Sourd, C. (1983). Etude des modes d'exploitation des ressources fruitières par *Cercopithecus cephus* au cours d'un cycle annuel. Thèse de Doctorat, Université de Rennes.

Sourd, C. & Gautier-Hion, A. (1986). Fruit selection by a forest guenon. *Journal of Animal Ecology*, **55**, 235–44.

Southwick, C. H., Beg, M. A. & Siddiqui, M. R. (1965). Rhesus monkeys in North India. In *Primate behavior*, ed. I. DeVore, pp. 111–59. New York: Holt, Rinehart & Winston.

Sowunmi, M. A. (1981a). Nigerian vegetational history from the late Quaternary to the present day. *Palaeoecology of Africa*, **13**, 217–34.

Sowunmi, M. A. (1981b). Aspects of late Quaternary vegetational changes in West Africa. *Journal of Biogeography*, **8**, 457–74.

Spence, D. H. N. & Angus, A. (1971). African grassland management: Burning and grazing in Murchison Falls National Park. In *The scientific management of plant and animal communities for conservation*, ed. E. Duffey. Oxford: Blackwell.

Spolsky, C. & Uzzell, T. (1984). Natural interspecies transfer of mitochondrial DNA in amphibians. *Proceedings of the National Academy of Sciences, USA*, **81**, 5802–5.

Stein, R. & Sarnthein, M. (1984). Late Neogene events of atmospheric and oceanic circulation offshore Northwest Africa: high resolution record from deep-sea sediments. *Palaeoecology of Africa*, **16**, 9–36.

Stern, B. R. & Smith, D. G. (1984). Sexual behavior and paternity in three captive groups of rhesus monkeys (*Macaca mulatta*). *Animal Behaviour*, **32**, 23–32.

Stoltz, L. P. (1972). The size, composition and fissioning in baboon troops (*Papio ursinus* Kerr 1792). *Zoologica Africana*, **7**, 367–78.

Struhsaker, T. T. (1967a). Social structure among vervet monkeys (*Cercopithecus aethiops*). *Behaviour*, **29**, 83–121.

Struhsaker, T. T. (1967b). Behavior of vervet monkeys. *University of California Publications in Zoology*, **82**, 1–64.

Struhsaker, T. T. (1967c). Ecology of vervet monkeys (*Cercopithecus aethiops*) in the Masai–Amboseli Game Reserve, Kenya. *Ecology*, **48**, 891–904.

Struhsaker, T. T. (1967d). Behavior of vervet monkeys and other cercopithecines. *Science, NY*, **156**, 1197–203.

Struhsaker, T. T. (1967e). Auditory communication among vervet monkeys (*Cercopithecus aethiops*). In *Social communication among primates*, ed. S. A. Altmann, pp. 238–324. Chicago: University of Chicago Press.

Struhsaker, T. T. (1969). Correlates of ecology and social organization among African cercopithecines. *Folia primatologica*, **11**, 80–118.

Struhsaker, T. T. (1970). Phylogenetic implications of some vocalizations of *Cercopithecus* monkeys. In *Old World Monkeys: evolution, systematics and behaviour*, ed. J. R. Napier & P. H. Napier, pp. 365–444. New York: Academic Press.

Struhsaker, T. T. (1971). Social behaviour of mother and infant vervet monkeys (*Cercopithecus aethiops*). *Animal Behaviour*, **19**, 233–50.

Struhsaker, T. T. (1973). A recensus of vervet monkeys in the Masai–Amboseli Game Reserve, Kenya. *Ecology*, **54**, 930–32.

Struhsaker, T. T. (1975). *The red colobus monkey.* Chicago: University of Chicago Press.

Struhsaker, T. T. (1976). A further decline in numbers of Amboseli vervet monkeys. *Biotropica,* **8**, 211–14.

Struhsaker, T. T. (1977). Infanticide and social organization in the redtail monkey (*Cercopithecus ascanius schmidti*) in the Kibale Forest, Uganda. *Zeitschrift für Tierpsychologie,* **45**, 75–84.

Struhsaker, T. T. (1978). Food habits of five monkey species in the Kibale Forest, Uganda. In *Recent Advances in Primatology,* Vol. II, ed. D. J. Chivers & J. Herbert, pp. 225–48. New York: Academic Press.

Struhsaker, T. T. (1980). Comparison of the behaviour and ecology of red colobus and redtail monkeys in the Kibale Forest, Uganda. *African Journal of Ecology,* **18**, 33–51.

Struhsaker, T. T. (1981a). Vocalizations, phylogeny and paleogeography of red Colobus monkeys (*Colobus badius*). *African Journal of Ecology,* **1**, 265–83.

Struhsaker, T. T. (1981b). Polyspecific associations among tropical rain forest primates. *Zeitschrift für Tierpsychologie,* **57**, 268–304.

Struhsaker, T. T. & Gartlan, J. S. (1970). Observations on the behavior and ecology of the patas monkey (*Erythrocebus patas*) in the Waza Reserve, Cameroon. *Journal of Zoology, London,* **161**, 49–63.

Struhsaker, T. T. & Leland, L. (1979). Socioecology of five sympatric monkey species in the Kibale forest, Uganda. *Advances in the Study of Behaviour,* **9**, 159–228.

Struhsaker, T. T. & Leland, L. (1977). Colobines: Infanticide by adult males. In *Primate Societies,* ed. B. B. Smuts, D. L. Cheney, R. M. Seyfarth, R. W. Wrangham & T. T. Struhsaker, pp. 93–7. Chicago: The University of Chicago Press.

Struhsaker, T. T. & Leland, L. (1985). Infanticide in a patrilineal society of red colobus monkeys. *Zeitschrift für Tierpsychologie,* **69**, 89–132.

Strum, S. C. (1982). Agonistic dominance in male baboons: an alternative view. *International Journal of Primatology,* **3**, 175–202.

Sugiyama, Y. (1960). On the division of a natural troop of Japanese monkeys Takaskiyama. *Primates,* **2**, 109–48.

Sugiyama, Y. (1976). Life history of male Japanese monkeys. In *Advances in the study of behavior, Vol. 7,* ed. J. S. Rosenblatt, R. A. Hinde, E. Shaw & C. Beer, pp. 255–284. New York: Academic Press.

Sugiyama, Y., Yoshiba, K. & Parthasarathy, M. D. (1965). Home range, mating season, male group and intertroop relationship in Hanuman langur (*Presbytis entellus*). *Primates,* **6**, 73–106.

Szalay, F. S. & Decker, R. L. (1974). Origins, evolution and function of the tarsus in Late Cretaceous Eutheria and Paleocene Primates. In *Primate locomotion,* ed. F. A. Jenkins, pp. 223–259. New York: Academic Press.

Szalay, F. S. & Delson, E. (1979). *Evolutionary history of the primates.* New York: Academic Press.

Talbot, M. R. (1980). Environmental responses to climatic change in the West African sahel over the past 20 000 years. In *The Sahara and the Nile,* ed. M. A. J. Williams & H. Faure, pp. 37–62. Rotterdam: Balkema.

Talbot, M. R. (1983). Lake Bosumtwi, Ghana. *Nyame Akuma,* **23**, 11.

Talbot, M. R. & Delibrias, G. (1980). A new late Pleistocene-Holocene water-level curve for Lake Bosumtwi, Ghana. *Earth and Planetary Science Letters,* **47**, 336–44.

Talbot, M. R. & Hall, J. B. (1981). Further late Quaternary leaf fossils from Lake Bosumtwi, Ghana. *Palaeoecology of Africa,* **13**, 83–92.

Talbot, M. R., Livingstone, D. A., Palmer, P. G., Maley, J., Melack, J. M., Delibrias, G. & Gulliksen, S. (1984). Preliminary results from sediment cores from Lake Bosumtwi, Ghana. *Palaeoecology of Africa*, **16**, 173–92.

Tassy, P. & Pickford, M. (1983). Un nouveau mastodonte zygolophodonte (Proboscidea, Mammalia) dans le Miocène inférieur d'Afrique Orientale: systématique et paléontologie. *Géobios*, **16**, 53–77.

Templeton, A. R. (1980). The theory of speciation via the founder principle. *Genetics*, **94**, 1011–38.

Templeton, A. R. (1981). Mechanisms of speciation – A population genetics approach. *Annual Review of Ecology and Systematics*, **12**, 23–48.

Terborgh, J. (1983). *Five New World primates: a study in comparative ecology.* Princeton: Princeton University Press.

Terborgh, J. (1986). Community aspects of frugivory in tropical forests. In *Frugivores and seed dispersal*, ed. A. Estrada & T. H. Fleming, pp. 371–84. Dordrecht: W. Jung.

Thomas, H. & Petter, G. (1986). Révision de la faune de Mammifères du Miocène supérieur de Menacer (ex-Marceau), Algérie: discussion sur l'âge du gisement. *Géobios*, **19**, 357–73.

Thompson, D'Arcy, W. (1917). *On growth and form* (2nd Edition, 1952). London: Cambridge University Press.

Thorington, R. W. & Groves, C. P. (1970). An annotated classification of the Cercopithecoidea. In *Old World Monkeys: evolution, systematics and behaviour*, ed. J. R. Napier & P. H. Napier, pp. 629–47. New York: Academic Press.

Thys van den Audenaerde, D. F. E. (1977). Description of a monkey skin from east-central Zaïre as a probable new monkey species (Mammalia, Cercopithecidae). *Revue de Zoologie africaine*, **91**, 1000–10.

Tiercelin, J. J., Renaut, R. W., Delibrias, G., Le Fournier, J. & Bieda, S. (1981). Late Pleistocene and Holocene lake level fluctuations in the Lake Bogoria basin, northern Kenya rift valley. *Palaeoecology of Africa*, **13**, 105–20.

Tinbergen, N. (1951). *The study of instinct.* Oxford: Oxford University Press.

Tobien, H. (1986). An early upper Miocene (Vallesian) *Mesopithecus* premolar from Rheinhessen, FRG. *Primate Report*, **14**, 139.

Tsingalia, H. M. & Rowell, T. E. (1984). The behaviour of adult male blue monkeys. *Zeitschrift für Tierpsychologie*, **64**, 253–68.

Turner, T. R. (1981). Blood protein variation in a population of Ethiopian vervet monkeys (*Cercopithecus aethiops aethiops*). *American Journal of Physical Anthropology*, **55**, 225–32.

Turner, T. R., Mott, C. S. & Maiers, J. (1986). Genetic and morphological studies on two species of Kenyan Monkeys, *C. aethiops* and *C. mitis.* *Proceedings of the Xth International Congress of Primatology.* London: Cambridge University Press.

Tutin, C. E. G. (1980). Reproductive behaviour of wild chimpanzees in the Gombe National Park, Tanzania. *Journal of Reproduction and Fertility*, **28**, 43–57.

Tutin, C. E. G. & Fernandez, M. (1985). Foods consumed by sympatric populations of Gorillas and Chimpanzees in Gabon. *International Journal of Primatology*, **6**, 27–43.

Tuttle, R. H. (1969). Terrestrial trends in the hands of the Anthropoidea. *IInd International Congress of Primatology, Atlanta*, Vol. 2, 192–200.

Van Campo, E., Duplessy, J. C. & Rossignol-Strick, M. (1982). Climatic conditions deduced from 150-kyr oxygen isotope-pollen record from the Arabian Sea. *Nature, London*, **296**, 56–9.

Vandenbergh, J. G. & Drickamer, L. C. (1974). Reproductive coordination among free-ranging rhesus monkeys. *Physiology and Behavior*, **13**, 373–6.

Van Donk, J. (1976). An ^{18}O record of the Atlantic Ocean for the entire Pleistocene. *Memoirs of the Geological Society of America*, **145**, 147–64.

Van Zinderen Bakker, E. M. (1982). African palaeoclimates 18 000 years BP. *Palaeoecology of Africa*, **15**, 77–99.

Van Zinderen Bakker, E. M. & Coetzee, J. A. (1972). A re-appraisal of late-Quaternary climatic evidence from tropical Africa. *Palaeoecology of Africa*, **7**, 151–91.

Verdcourt, B. (1963). The Miocene non-marine Mollusca from Rusinga Island, Lake Victoria and other localities in Kenya. *Palaeontographica*, **A 121**, 1–37.

Verdcourt, B. (1984). Discontinuities in the distribution of some East African land snails. In *World Wide Snails: Biogeographical studies on non-marine Mollusca*, eds. A. Solem & A. C. van Bruggen, pp. 134–55. Leiden: E. J. Brill & Dr W. Backhuys.

Verheyen, W. N. (1962). Contribution à la craniologie comparée des primates. *Annales du Musée royal de l'Afrique centrale, Sciences Zoologiques*, série 8, **105**, 1–256.

Verheyen, W. N. (1963). New data on the geographical distribution of *Cercopithecus (Allenopithecus) nigroviridis* Pocock, 1907. *Revue de Zoologie et de Botanique africaine*, **68**, 393–6.

Verschuren, J. (1972). *Contribution à l'écologie des Primates, Pholidota, Carnivora, Tubulidentata et Hyracoidea (Mammifères)*. (Exploration du Parc National des Virunga, 3). Fondation pour favoriser les Recherches scientifiques en Afrique.

Vine, I. (1970). Communication by facial visual signals. In *Social behaviour in birds and mammals*, ed. J. H. Crook, pp. 279–354. New York: Academic Press.

Von Koenigswald, G. H. R. (1969). Miocene Cercopithecoidea and Oreopithecoidea from the Miocene of East Africa. *Fossil Vertebrates of Africa*, **1**, 39–51.

Wallis, S. J. (1978). The sociology of *Cercocebus albigena johnstoni* (Lyddeker): An arboreal, rain forest monkey. PhD dissertation, University of London.

Ward, J. H. (1963). Hierarchical grouping to optimize an objective function. *Journal of the American Statistical Association*, **58**, 236–44.

Waser, P. M. (1974). Intergroup interactions in a forest monkey: The mangabey *Cercocebus albigena*. PhD dissertation, The Rockefeller University.

Waser, P. M. (1975). Experimental playbacks show vocal mediating of intergroup avoidance in a forest monkey. *Nature, London*, **255**, 56–8.

Waser, P. M. (1976). *Cercocebus albigena*: site attachment, avoidance and intergroup spacing. *American Naturalist*, **110**, 911–35.

Waser, P. M. (1977). Feeding, ranging, and group size in the mangabey *Cercocebus albigena*. In *Primate ecology: studies of feeding and ranging behaviour in lemurs, monkeys and apes*, ed. T. H. Clutton-Brock, pp. 183–222. London: Academic Press.

Waser, P. M. (1980). Polyspecific associations of *Cercocebus albigena*: geographic variation and ecological correlates. *Folia primatologica*, **33**, 57–76.

Waser, P. M. (1982a). The evolution of male loud calls among mangabeys and baboons. In *Primate communication*, ed. C. T. Snowdon, C. H. Brown & M. R. Petersen, pp. 117–43. New York: Cambridge University Press.

Waser, P. M. (1982b). Primate polyspecific associations: do they occur by chance? *Animal Behaviour*, **30**, 1–8.

Waser, P. M. (1987). Interactions among Primate societies. In *Primate societies*, ed. S. Smuts, D. L. Cheney, R. M. Seyfarth, R. W. Wrangham & T. T. Struhsaker, pp. 210–226. Chicago: Universityof Chicago Press.

Waser, P. M. & Case, T. J. (1981). Monkeys and matrices: on the coexistence of omnivorous primates. *Oecologia*, **49**, 102–8.

Waterman, P. G. (1984). Food acquisition and processing as a function of plant chemistry. In *Food acquisition and processing in primates*, ed. D. J. Chivers, B. A. Wood & A. Bilsborough, pp. 177–212. New York: Plenum Press.

Waterman, P. G. & Choo, G. M. (1981). The effects of digestibility reducing compounds in leaves on food selection by some colobines. *Malaysian Applied Biology*, **10**, 147–62.

Webb, P. W. (1984). Form and function in fish swimming. *Scientific American*, **251** (1), 72–82.

Werger, M. J. A. & Coetzee, B. J. (1978). The Sudano-Zambezian Region. In *Biogeography and ecology of southern Africa*, ed. M. J. A. Werger, pp. 301–422. The Hague: Junk.

Western, D. & Lindsey, W. K. (1984). Seasonal herd dynamics of a savanna elephant population. *African Journal of Ecology*, **22**, 229–44.

Wheater, R. J., n.d. The second five year plan, Murchison Falls National Park, Uganda National Parks (unpublished, quoted in R. M. Laws, I. S. C. Parker & R. C. B. Johnstone (ed.), *Elephants and their habitats*. Oxford: Clarendon Press, 1968.)

White, F. (1981). The history of the Afromontane archipelago and the scientific need for its conservation. *African Journal of Ecology*, **19**, 33–54.

White, F. (1983). *The vegetation of Africa*. Paris: Unesco.

White, F. & Werger, M. J. A. (1978). The Guineo-Congolian transition to southern Africa. In *Biogeography and ecology of southern Africa*, ed. M. J. A. Werger, pp. 599–620. The Hague: Junk.

Whitten, P. L. (1982). Female reproductive strategies among vervet monkeys. PhD dissertation, Harvard University.

Whitten, P. L. (1983). Diet and dominance among female vervet monkeys (*Cercopithecus aethiops*). *American Journal of Primatology*, **5**, 139–59.

Wickler, W. (1967). *The sexual code*. New York: Doubleday.

Wiley, E. O. (1981). *Phylogenetics: The theory and practice of phylogenetic systematics*. New York: John Wiley & Sons.

Williams, G. C. (1966). *Adaptation and natural selection*. Princeton: Princeton University Press.

Williamson, P. G. (1985). Evidence for an early Plio–Pleistocene rainforest expansion in East Africa. *Nature, London*, **315**, 487–9.

Wilson, A. C. (1976). Gene regulation in evolution. In *Molecular evolution*, ed. F. Ayala. Sunderland, MA: Sinauer Press.

Wilson, W. L. & Wilson, C. C. (1975). Species specific vocalizations and the determination of phylogenetic affinities of the *Presbytis aygula melalophos* group in Sumatra. In *Contemporary primatology*, ed. S. Kondo, M. Kawai & A. Ehara, pp. 439–63. Basel: S. Karger.

Windrow, B. (1973). The 'Rubber-Mask' technique – I. Pattern measurement and analysis. *Pattern Recognition*, **5**, 175–97.

Wolfheim, J. H. (1983). *Primates of the world. Distribution, abundance and conservation*. Seattle: University of Washington Press.

Woodruff, F., Savin, S. M. & Douglas, R. G. (1981). Miocene stable isotope record: a detailed Pacific Ocean study and its paleoclimatic implications. *Science, NY*, **212**, 665–8.

Wrangham, R. W. (1981). Drinking competition in vervet monkeys. *Animal Behaviour*, **29**, 904–10.

Wrangham, R. W. & Waterman, P. G. (1981). Feeding behaviour of vervet monkeys on *Acacia tortilis* and *Acacia xanthophloea*: with special reference to reproductive strategies and tannin production. *Journal of Animal Ecology*, **50**, 715–31.

Wright, S. (1965). The interpretation of population structure by F-statistics with special regard to systems of mating. *Evolution*, **19**, 395–420.

Wright, S. (1978). *Evolution and the genetics of populations*, Vol. 4, *Variability within and among natural populations*. Chicago: University of Chicago Press.

Yemane, K., Bonnefille, R. & Faure, H. (1985). Palaeoclimatic and tectonic implications of Neogene microflora from the Northwestern Ethiopian highlands. *Nature, London*, **318**, 653–6.

York, A. D. & Rowell, T. E. (1988). Reconciliation following aggression in patas monkeys (*Erythrocebus patas*). *Animal Behaviour*, in press.

Zadeh, L. (1973). Outline of a new approach to the analysis of complex systems and decision processing. In *Multiple Criteria Decision Making*, ed. M. Zeleny & J. Cochrane, pp. 686–725. Columbia, SC: University of South Carolina Press.

Zeeve, S. R. (1985). Swamp monkeys of the Lomako Forest, Central Zaïre. IUCN/SSC. *Primate Specialist Group Newsletter*, **5**, 32–3.

Zuckerman, S. (1932). *Social life of monkeys and apes*. London: Paul Trench & Trubne.

AUTHOR INDEX

TAXONOMIC INDEX

SUBJECT INDEX

(Except where other species are named, all entries refer to guenons.)